ANALYTIC
FUNCTION THEORY

ANALYTIC
FUNCTION THEORY

BY

EINAR HILLE

VOLUME II

CHELSEA PUBLISHING COMPANY
NEW YORK, N.Y.

SECOND EDITION

The present, Second Edition of Volume II is a fourth
(corrected) printing of the First Edition, originally
published in 1962. It is published at New York, N.Y. in
1987 and is printed on 'long-life' acid-free paper.

CIP

Library of Congress Cataloging in Publication Data

Hille, Einar, 1894-1980
      Analytic function theory.

      Includes bibliography.
      1. Analytic functions.   I. Title
QA331.H54   1987   515   73-647
ISBN 0-8284-0270-1   (v. 2)

Printed in the United States of America

Foreword

Volume II of *Analytic Function Theory*, now laid before the mathematical public, is a direct continuation of Volume I. This applies to the numbering of chapters, sections, theorems, and formulas as well as to subject matter. The latter is on the whole more advanced than in Volume I. For the reader's convenience, a table of contents for Volume I is included in Volume II.

The subject matter of an introductory text in function theory is almost uniquely determined. In a book of intermediate character, such as the present one, the situation is different. Less of the included material need be canonical and more can be chosen according to the author's interest and preference. Such choice was freely exercised while the book was in the making, and in retrospect it is not always easy to justify this inclusion or that omission.

Of the ten chapters in Volume II, five will probably be accepted as canonical, namely, Chapters 10, Analytic Continuation; 13, Elliptic Functions; 14, Entire and Meromorphic Functions; 15, Normal Families; and 17, Conformal Mapping. Considerations of space naturally make it necessary to restrict even canonical material, and grave omissions are unavoidable.

The theory of Riemann surfaces is not developed for its own sake in this treatise. The brief mention in Chapters 10 and 12 (Algebraic Functions) is sufficient for our purposes but it may not suffice for all readers. There are, however, monographs where the interested student will find what he wants. The discussion in Chapter 12 aims to give the student a working knowledge of algebraic singularities through the use of Newton's diagram and the implicit function theorem. The treatment of algebraic integrals is mainly preparatory to the elliptic case.

The instructor who looks for material to supplement Volume I will find enough in the six chapters already mentioned and he will perhaps make less use of the other four chapters. These are: 11, Singularities and Representation of Analytic Functions; 16, Lemniscates; 18, Majorization; and 19, Functions Holomorphic in a Half-Plane. These chapters, either by themselves or combined with Chapters 14 and 17, form material for an intermediate or advanced course in function theory and, naturally, these topics are nearer to the active front of the subject than is the rest of the book.

Suggestions for collateral reading are appended to the various chapters. More literature is quoted than in Volume I, especially more research papers. In order to keep the Bibliography at the end of the volume within reasonable bounds, it has been restricted to books.

I am indebted to colleagues, too many to mention individually, for advice, bibliographical references, and sundry information. My gratitude goes in the first place to Ernest C. Schlesinger, who read the whole manuscript in detail and suggested many corrections and improvements. Warm thanks are due to my friends Shizuo Kakutani and C. T. Ionescu Tulcea for active help in every form, including numerous discussions of theorems and non-theorems, and also to G. Kurepa, who made it possible for me to write Chapter 18 on the shores of the blue Adriatic. Last but not least, I wish to thank Ginn and Company for excellent cooperation, friendly interest, and most efficient handling of the book.

It remains for me to wish this child Godspeed! It will now have to fend for itself and stand or fall on its own merits.

EINAR HILLE

New Haven, Connecticut

Contents

Volume I

1 · NUMBER SYSTEMS

2 · THE COMPLEX PLANE

3 · FRACTIONS, POWERS, AND ROOTS

4 · HOLOMORPHIC FUNCTIONS

5 · POWER SERIES

6 · SOME ELEMENTARY FUNCTIONS

7 · COMPLEX INTEGRATION

8 · REPRESENTATION THEOREMS

9 · THE CALCULUS OF RESIDUES

Volume II

10 · ANALYTIC CONTINUATION

11 · SINGULARITIES AND REPRESENTATION OF ANALYTIC FUNCTIONS

12 · ALGEBRAIC FUNCTIONS

13 · ELLIPTIC FUNCTIONS

14 · ENTIRE AND MEROMORPHIC FUNCTIONS

15 · NORMAL FAMILIES

16 · LEMNISCATES

17 · CONFORMAL MAPPING

18 · MAJORIZATION

19 · FUNCTIONS HOLOMORPHIC IN A HALF-PLANE

Symbols

1. Set theory:

$a \in A$	a is an element of the set A; a belongs to A
$B \subset A$	B is a subset of A
$C \cap D$	Intersection of sets C and D
$C \cup D$	Union of sets C and D
$\bar{S}$	Closure of the set S
S'	The derived set of a given set S
$\mathbf{C}(S)$	The complement of S
∂S	The boundary of S
Int (S)	The interior of S

2. Complex variables:

$a = \Re(a + bi)$	a is the real part of $a + bi$
$b = \Im(a + bi)$	b is the imaginary part of $a + bi$
$z = x + iy$	$x = \Re(z)$; $y = \Im(z)$
$\bar{z} = x - iy$	The conjugate of z
$|z| = (x^2 + y^2)^{\frac{1}{2}}$	The absolute value of z
arg z	The argument of z (differs from arc $\tan \frac{y}{x}$ by a multiple of π)
C	The field of complex numbers

3. Curves, domains, regions:

C is a "scroc"	C is a simple closed rectifiable oriented curve
C_i	Interior of C
C_e	Exterior of C
$C*$	$C \cup C_i$
D	Domain: open, arcwise connected set
R	Region: a domain plus a subset of its boundary
$[z_1, z_2]$	Closed interval or closed line segment with endpoints z_1 and z_2
$[z_1, z_2, \cdots, z_n]$	Polygonal line joining $z_1, z_2, \cdots, z_n$ in this order
$\Pi: [z_1, z_2, \cdots, z_n, z_1]$	Oriented simple closed polygon with vertices at z_1, $z_2, \cdots, z_n$
$d(z_1, z_2)$	Euclidean distance of z_1 and z_2
$\chi(z_1, z_2)$	Chordal distance of z_1 and z_2
$d(S_1, S_2)$	Distance between the sets S_1 and S_2

4. Function spaces:

$C[a, b]$	Set of functions continuous in the interval $[a, b]$
$BV[a, b]$	Set of functions of bounded variation in $[a, b]$
$C[D]$	Set of functions continuous in the domain D
$CB[D]$	The subset of bounded functions of $C[D]$
$H[D]$	Set of functions holomorphic in D
$HB[D]$	The subset of bounded functions of $H[D]$
$\|f\|$	Norm of f

5. Additional function classes:

A, **B**	Entire functions of genus zero or one with real negative zeros
$B(0)$	Functions bounded and holomorphic in $\Re(z) > 0$
$\mathbf{E}(\omega_1, \omega_3)$	Elliptic functions of period $2\omega_1$ and $2\omega_3$
$H_2 = H_2(D)$	Functions holomorphic in the unit disk (D or K) with boundary values in $L_2(0, 2\pi)$
$H_p(D)$	Ditto in $L_p(0, 2\pi)$, $1 \leqq p$
$L_p(a, b)$	Measurable functions $f(t)$ such that $|f(t)|^p$ is integrable over (a, b) in the sense of Lebesgue
$L_2 H[D]$	Functions holomorphic in a domain D such that $|f(x + iy)|^2$ is integrable over D
M, **M**$_0$	Laplace-Stieltjes integrals absolutely convergent in $\Re(z) > 0$, $\geqq 0$
$\mathscr{S}, \mathscr{U}$	Univalent functions holomorphic inside, outside the unit disk

Analytic Function Theory

Volume II

10

ANALYTIC CONTINUATION

10.1. Introduction. In Volume I we have encountered large classes of differentiable functions. At this juncture it is desirable to re-examine briefly how the functions are introduced, and what further steps should be taken to complete the study.

There is an underlying pattern of successive extensions of the class of functions considered. We started with the complex field C, to which we added the function z. The *least algebra* containing C and z is the set $\mathfrak{P}$ of *polynomials* in z. These functions are holomorphic in the finite plane and have a pole at infinity. The *least field* containing C and z is the set $\mathfrak{R}$ of *rational functions* of z. Such functions are holomorphic save for poles in the extended plane, and $C \subset \mathfrak{P} \subset \mathfrak{R}$.

This is as far as we can get using only arithmetical operations. Any further extension requires limiting processes. We get the largest class of functions holomorphic in the finite plane by adjoining to $\mathfrak{P}$ all limit functions of convergent sequences of polynomials, the convergence being uniform on every compact set. This is the algebra $\mathfrak{E}$ of *entire functions*, and clearly every entire function is obtainable in this manner. If $f(z) \in \mathfrak{E} \ominus \mathfrak{P}$, then $z = \infty$ is an isolated essential singular point of $f(z)$ and does not belong to the domain of existence of $f(z)$. Thus, in passing from $\mathfrak{P}$ to $\mathfrak{E}$ the domain of existence of the elements is reduced from the extended to the finite plane.

Similarly, the class $\mathfrak{R}$ is not maximal with respect to the property that its elements are holomorphic save for poles in the finite plane. The maximal class is the field $\mathfrak{M}$ of *meromorphic functions*. Every element of $\mathfrak{M}$ is the limit of a suitably chosen sequence of rational functions. If $f(z) \in \mathfrak{M} \ominus \mathfrak{R}$, then either $f(z)$ has infinitely many poles having a single limit point at infinity, or else it has a finite number of poles and $z = \infty$ is an isolated essential singular point. The point at infinity does not belong to the domain of existence of such a function. Again, in passing from $\mathfrak{R}$ to $\mathfrak{M}$ the domain of existence of the elements is reduced from the extended plane to the finite plane.

All the functions of $\mathfrak{M}$ are single-valued, but we cannot study such a function in detail without considering the inverse function, which is, normally, infinitely many-valued. We know that such a study leads to new types of singularities, such as algebraic and logarithmic branch points.

If $f(z) \in \mathfrak{R}$, the inverse function has only a finite number of singularities, which are algebraic branch points or poles. In Chapter 12 we shall make a study of this case and, more generally, of the properties of *algebraic functions*, that is, the roots of equations of the form

$$P(z, w) = 0,$$

where $P(z, w)$ is a polynomial in the two variables z and w. The reader will find some information concerning inverse functions of meromorphic functions in Section 14.7.

The extension problem does not end with the introduction of the class $\mathfrak{M}$. If we apply limiting processes to subsets of elements of $\mathfrak{P}$ or of $\mathfrak{R}$, we can very well get functions which do not belong to $\mathfrak{M}$. Such functions will have singularities other than poles in the finite plane. The domain of existence may then omit part of the plane, and the functions need not be single-valued. An important problem considered in Section 10.4 is that of determining what point sets can serve as domains of existence of single-valued analytic functions and of constructing analytic functions with preassigned domains of existence. A somewhat simpler problem is that of constructing holomorphic functions which are holomorphic in a given domain D. Such functions will then have a domain of existence containing D.

For a couple of special cases we are in possession of the solution of the last-named construction problem. If D is the open disk $|z - a| < R$, then the set of functions holomorphic in D coincides with the set of power series in the variable $(z - a)$ having the radius of convergence $\geq R$. This class $H[D]$ evidently coincides with the limit functions of sequences $\{P_n(z)\} \subset \mathfrak{P}$, where the convergence is uniform on interior concentric disks. More generally, as long as D is a bounded simply-connected domain, the functions of $H[D]$ are limits of sequences of polynomials. See Section 16.6.

The assumption that D is simply-connected may be dropped, but then we must replace $\mathfrak{P}$ by $\mathfrak{R}$ in the discussion. A case familiar to the reader is that of an annulus: $D = [z \mid R_1 < |z - a| < R_2]$. Here the family $H[D]$ coincides with the set of Laurent expansions in powers of $(z - a)$ convergent in the annulus in question. Such a function is obviously the limit of a sequence of elements of $\mathfrak{R}$. For the general case we refer again to Section 16.6.

Now, given an $f(z) \in H[D]$, the problem arises whether or not there exists a domain $D_1 \supset D$ such that $f(z)$ also belongs to $H[D_1]$. More precisely, we ask whether there exists a function $f_1(z) \in H[D_1]$ which coincides with $f(z)$ in D. If no such domain D_1 exists, then D is the domain of existence of $f(z)$, and the boundary of D, denoted by ∂D, is the *natural boundary* of $f(z)$. Examples of functions having the unit circle $|z| = 1$ as natural boundary are to be found in Section 5.7 of Volume I. See also Section 11.7.

If such a domain D_1 and a function $f_1(z)$ do exist, then we say that $f_1(z)$ is the *analytic continuation* of $f(z)$ in $D_1 \ominus D$. We can ask a more general question: do there exist a domain $D_1 \neq D$ such that $D_1 \cap D$ is connected and non-void and a function $f_1(z)$ such that $f_1(z) \in H[D_1]$ and $f(z) = f_1(z)$ in $D_1 \cap D$? Again, $f_1(z)$ is the analytic continuation of $f(z)$ in $D_1 \ominus (D_1 \cap D)$, and $f(z)$ is the analytic continuation of $f_1(z)$ in $D \ominus (D_1 \cap D)$. Here, $f(z)$ and $f_1(z)$ may be regarded as local representations of one and the same analytic function, one representation in $H[D]$ and the other in $H[D_1]$. The problem of describing an analytic function then involves finding its local representations. Obviously we shall need a criterion for deciding when two holomorphic functions are local representations of the same analytic function, and the latter concept has to be made precise. The notions of singular point and domain of existence also need clarification. The required definitions will be given in Section 10.3 and the following sections.

We have at our disposal a large variety of methods of analytic continuation.

We shall consider the following types and mention those sections of the book where they will be discussed.

(1) CHAINS OF REARRANGED POWER SERIES. This is the method due to Weierstrass and it is basic for the theory. The method was outlined in Section 5.6, and it will be discussed in detail in Section 10.2.

(2) THE LAW OF PERMANENCE OF FUNCTIONAL EQUATIONS. If an analytic function satisfies a functional equation in some domain, then its analytic continuation can often be obtained with the aid of the equation. We used this method in the study of the Gamma function in Section 8.8. The basis of the method will be discussed in Section 10.7.

(3) HOLOMORPHY-PRESERVING OPERATORS. There exist wide classes of linear operators which apply to holomorphic functions and give holomorphic transforms. The analytic continuation of the transform of a function is then the transform of the analytic continuation of the function. Methods of this type are discussed in Sections 11.1 and 11.2.

(4) POWER SERIES WITH ANALYTIC COEFFICIENTS. If the nth coefficient is an analytic function of the subscript, the analytic continuation of the series can usually be found by the calculus of residues or by operator methods. See Section 11.3.

(5) METHODS OF SUMMABILITY. Certain methods of summability will evaluate a power series in a starlike domain extending beyond the circle of convergence. Some of these methods will be discussed in Section 11.4.

(6) POLYNOMIAL SERIES. A function defined by a power series in z may be represented by a polynomial series valid in a region starlike with respect to $z = 0$ in which the function is holomorphic. See Section 11.5.

The methods under (3) and (4) are of somewhat special nature, but the series to which they apply are among the most important in analysis. The methods under (5) and (6) apply to all power series without exception.

10.2. Rearrangements of power series. Let us consider $f(z)$, an element of $H[\,|\,z - a\,|\,< R]$, defined by the power series

$$(10.2.1) \qquad f(z) \equiv \sum_{n=0}^{\infty} c_n (z - a)^n,$$

having the radius of convergence $R(a) = R$. We recall that

$$R(a) = [\lim_{n \to \infty} \sup |\,c_n\,|^{\frac{1}{n}}]^{-1}.$$

Let $D(a)$ be the disk $|\,z - a\,|\,< R(a)$ and suppose that $b \in D(a)$. Proceeding as in Section 5.6, we can rearrange (10.2.1) into a power series in $(z - b)$ by setting $z = (z - b) + b$. Expanding by the binomial theorem and summing the resulting double series by columns, we get

$$(10.2.2) \qquad f(z; b) = \sum_{n=0}^{\infty} c_n(b)(z - b)^n,$$

where

$$c_n(b) = \sum_{k=n}^{\infty} k(k - 1) \cdots (k - n + 1) c_k (b - a)^{k-n} = \frac{f^{(n)}(b)}{n!}.$$

We say that $f(z; b)$ is a *direct rearrangement* of $f(z)$.

Let the radius of convergence of the rearranged series be $R(b)$. It was proved in Section 5.6 that

$$(10.2.3) \qquad R(a) - |\,a - b\,| \leq R(b) \leq R(a) + |\,a - b\,|.$$

There are two possibilities according as the first sign of equality holds or not. Let us start with the second case, so that

$$R(b) > R(a) - |\,a - b\,|.$$

Let us write $D(b)$ for the circular disk $|\,z - b\,|\,< R(b)$. Then the disks $D(a)$ and $D(b)$ overlap, and $D(b) \ominus [D(b) \cap D(a)] \neq \emptyset$. We have now a holomorphic function $f(z)$ defined in $D(a)$ and a holomorphic function $f(z; b)$ defined in $D(b)$, and these two functions coincide in $D(b) \cap D(a)$, as was shown in Section 5.6. There exists, consequently, a function $F(z)$, holomorphic in $D(a) \cup D(b)$, which coincides with $f(z)$ in $D(a)$ and with $f(z; b)$ in $D(b)$. We say that $f(z; b)$ *is the analytic continuation of* $f(z)$ *in* $D(b) \ominus [D(b) \cap D(a)]$. Similarly, $f(z)$ is the analytic continuation of $f(z; b)$ in $D(a) \ominus [D(a) \cap D(b)]$.

The other possibility is that

$$(10.2.4) \qquad R(b) = R(a) - |\,b - a\,|.$$

In this case $D(b)$ is interior to $D(a)$, but their boundaries have a point in common, say $z = z_0$. We shall prove that $z = z_0$ is a *singular point* of $f(z)$ in the following sense: Let δ be a preassigned small positive number, and let D_δ be the disk $|\,z - z_0\,|\,< \delta$. Then no function $F(z)$ having the following properties can exist: (1) $F(z)$ is holomorphic in $D(a) \cup D_\delta$, and (2) $F(z) = f(z)$ in $D(a)$.

Suppose that this is false. Then $D(a) \cup D_\delta$ contains a disk $|z - b| < R_\delta$ with $R(b) < R_\delta$. Since $F(z)$ is holomorphic in this disk, Theorem 8.1.1 asserts that $F(z)$ is represented by its Taylor series in the disk:

$$F(z) = \sum_{n=0}^{\infty} \frac{F^{(n)}(b)}{n!} (z - b)^n, \quad |z - b| < R_\delta.$$

But $F^{(n)}(b) = f^{(n)}(b)$ for every n. Thus the series actually coincides with (10.2.2), which has the radius of convergence $R(b) < R_\delta$. This is a contradiction. Hence we conclude that no such function $F(z)$ can exist, no matter how small δ is.

From this we can conclude that for any b of the form

$$b = a + \alpha(z_0 - a), \quad 0 < \alpha < 1,$$

we must have

$$R(b) = R(a) - |a - b|,$$

since otherwise a contradiction would arise. This means that $z = z_0$ is also a singular point in the sense of Definition 5.7.1. In this case every disk $D(b)$ with center on the ray

$$\arg(z - a) = \arg(z_0 - a)$$

is interior to $D(a)$, and its boundary is tangent to the boundary of $D(a)$. Moreover, $R(b) \to 0$ as $b \to z_0$. This we express by saying that $z = z_0$ *is a singular point of $f(z)$ for radial approach.* But much more can be said:

No disk $D(b)$ with $|b - a| < R(a)$ can contain $z = z_0$, and $R(b) \to 0$ as $b \to z_0$ from the interior of the circle $|z - a| = R(a)$.

In fact, suppose that $z_0 \in D(b)$ for some such b. Then there would exist a function $F(z)$ holomorphic in $D(a) \cup D(b)$ such that $F(z) = f(z)$ in $D(a)$. But this open set contains a neighborhood of $z = z_0$, and we are faced with the same contradiction as above. This proves the first assertion. Actually it also proves the second one, for if $\lim \sup R(b) > 0$ when $b \to z_0$, then there would be values of b for which $D(a) \cup D(b)$ would contain a neighborhood of $z = z_0$. Thus, $z = z_0$ is a singular point of $f(z)$, no matter how we approach z_0 from the interior of the circle $|z - a| = R(a)$.

$R(b)$ *is a continuous function of b in $|b - a| < R(a)$. More precisely, if $|b - c| < \mathrm{Max}\,[R(b), R(c)]$, then*

$$(10.2.5) \qquad\qquad |R(b) - R(c)| \leqq |b - c|.$$

The continuity is obviously a consequence of this relation. Suppose that b and c are given in $D(a)$ and that, for instance, $c \in D(b)$. $D(a) \cap D(b) \cap D(c)$ is then a domain containing c. In this domain the functions $f(z)$, $f(z; b)$, and $f(z; c)$ coincide. Here $f(z)$ is defined by (10.2.1), $f(z; b)$ by (10.2.2), and $f(z; c)$ by (10.2.2) with b replaced by c. However, the series $f(z; c)$ may be obtained either by direct rearrangement of (10.2.1) or by direct rearrangement of (10.2.2). The results must be the same. But if $f(z; c)$ is thought of as a rearrangement of

$f(z; b)$, then the two radii of convergence, $R(b)$ and $R(c)$, must obey (10.2.3); that is, we have

$$R(b) - |b - c| \leq R(c) \leq R(b) + |b - c|,$$

and (10.2.5) follows. The case $b \in D(c)$ is handled in the same manner. Next we prove:

THEOREM 10.2.1. *For each point* a *in the domain of* $f(z)$ *there is at least one number* ω *such that* $0 \leq \omega < 2\pi$, *and such that we have*

$$(10.2.6) \qquad\qquad R(b) = R(a) - |b - a|$$

for any b *on the ray* $\arg(z - a) = \omega, |z - a| < R(a)$.

Proof. Our previous discussion shows that it is enough to prove that (10.2.6) holds for some b on the ray in question. Then, it is enough to show that

$$(10.2.7) \qquad\qquad \min_{\theta} R[a + \tfrac{1}{2}R(a)e^{i\theta}] = \tfrac{1}{2}R(a),$$

and this condition is necessary as well as sufficient. Let us denote the left-hand side by R_1. Then (10.2.3) shows that $R_1 \geq \tfrac{1}{2}R(a)$. Suppose that $R_1 > \tfrac{1}{2}R(a)$. Then the disks

$$D[a + \tfrac{1}{2}R(a)e^{i\theta}], \quad 0 \leq \theta < 2\pi,$$

cover the disk $D_1(a)$: $|z - a| < \tfrac{1}{2}R(a) + R_1$. Consider the corresponding power series

$$(10.2.8) \qquad\qquad f(z; a + \tfrac{1}{2}R(a)e^{i\theta}), \quad 0 \leq \theta < 2\pi.$$

Each such series is a rearrangement of (10.2.1); thus its sum is equal to $f(z)$ in $D(a) \cap D(a + \tfrac{1}{2}R(a)e^{i\theta})$. Two series corresponding to different values of θ coincide in their common domain of convergence, since the latter contains the disk $|z - a| < R_1 - \tfrac{1}{2}R(a)$ in which both series have the sum $f(z)$. It follows that the totality of series (10.2.8) defines a function $F(z)$ holomorphic in $D_1(a)$ and coinciding with $f(z)$ in $D(a)$. Then, on one hand, the Taylor series of $F(z)$ must converge in $D_1(a)$. On the other, that Taylor series must coincide with (10.2.1), which converges only in $D(a)$. This contradiction proves (10.2.7) and, hence, also (10.2.6).

COROLLARY. *There is at least one singular point of the function defined by the power series on the circle of convergence.*

For we have seen that if (10.2.6) holds, then the point $z_0 = R(a)e^{i\omega}$ is singular.

The geometric series shows that there need not be more than one singular point on the circle of convergence. In this special case, the point $z = 1$ is the only singularity in the extended plane.

The opposite extreme is that in which every point on the circle is a singular point, so that the circle is the natural boundary of the function defined by the power series. Formula (10.2.7) shows that *a necessary and sufficient condition for the circle* $|z - a| = R(a)$ *to be the natural boundary of* $f(z)$ *is that*

(10.2.9)
$$R[a + \tfrac{1}{2}R(a)e^{i\theta}] \equiv \tfrac{1}{2}R(a).$$

Here we can express the left-hand side in terms of the coefficients c_n and obtain after some manipulation the condition

(10.2.10)
$$\limsup_{k \to \infty} \left| \sum_{n=k}^{\infty} \binom{n}{k} c_n [\tfrac{1}{2}R(a)e^{i\theta}]^n \right|^{\frac{1}{k}} \equiv 1.$$

Suppose now that the circle of convergence of (10.2.1) is not a natural boundary, and consider the set of all power series which are obtained by direct rearrangement of (10.2.1) about points b in $D(a)$. As above, let $D(b)$ be the interior of the circle of convergence of $f(z; b)$. If b_1 and b_2 are two points of $D(a)$, suppose that $D(b_1) \cap D(b_2) \neq \emptyset$. Then we have also

$$[D(b_1) \cap D(b_2)] \cap D(a) \neq \emptyset,$$

and in this domain $f(z; b_1) = f(z)$ and $f(z; b_2) = f(z)$. We conclude that $f(z; b_1) = f(z; b_2)$ in $D(b_1) \cap D(b_2)$. See Figure 1. It follows that there exists a function $F(z)$ uniquely defined and holomorphic in $D_1(a)$: $\cup\{D(b) \mid b \in D(a)\}$, and $F(z)$ coincides with $f(z)$ in $D(a)$. We note that $D_1(a)$ is a simply-connected domain containing, but not coinciding with, $D(a)$. See Figure 2, which depicts $D_1(0)$ corresponding to $f(z) = (1 - z)^{-1}$. $D_1(a)$ and $D(a)$ have at least one boundary point in common. It is possible that $\partial D_1(a)$ is the natural boundary of $F(z)$.

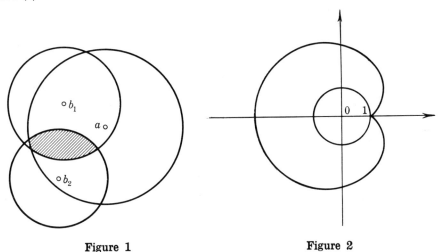

Figure 1　　　　　　　**Figure 2**

Suppose that this is not the case and suppose that $c \in D(b)$ where b is exterior to $D(a)$ but, of course, in $D_1(a)$. Then we can rearrange $f(z; b)$ in powers of $z - c$, obtaining a power series $f(z; b, c)$ which we can think of as a secondary rearrangement of $f(z)$. The series converges in a disk $D(c)$ of radius $R(c)$ where

(10.2.5) holds. The interesting case is that in which $D(c)$ is not wholly contained in $D_1(a)$. See Figure 3. In this case $f(z; b, c)$ defines an analytic continuation of $F(z)$ in $D(c) \ominus [D(c) \cap D_1(a)]$. Let us imagine that all such secondary rearrangements $f(z; b, c)$ have been found and let us adjoin the set $\{f(z; b, c)\}$ to the set $f(z) \cup \{f(z; b)\}$. We have then obtained local representations of $F(z)$ in a domain $D_2(a) = \cup \{D(c) \mid c \in D_1(a)\}$. Here we note that if $c \in D(a)$ then, since $b \in D(a)$, we have $f(z; b, c) = f(z; c)$, that is, the secondary rearrangement about c agrees with the primary, direct rearrangement of $f(z)$.

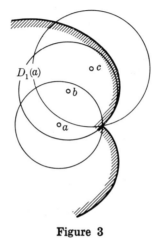

Figure 3

Suppose now that $D(c_1) \cap D(c_2) \neq \emptyset$ and consider the corresponding power series $f(z; b_1, c_1)$ and $f(z; b_2, c_2)$. If $b_1 = b_2$, then $f(z; b_1, c_1)$ and $f(z; b_2, c_2)$ must agree in $D(c_1) \cap D(c_2)$, but if $b_1 \neq b_2$, then no such claim can be made and it is actually possible for the two series to represent different branches of the analytic continuation. It should be observed that if $f(z; b_1, c)$ and $f(z; b_2, c)$ both exist, then they must be identical since $f(z; b_1)$ and $f(z; b_2)$ must agree in the set $D(b_1) \cap D(b_2)$ and hence must agree at $z = c$. Thus secondary rearrangements are uniquely determined by c in $D_1(a)$.

We can now proceed to tertiary rearrangements, which we shall denote by $f(z; a_1, a_2, a_3)$ where $a_1 \in D(a)$, $a_2 \in D(a_1)$, $a_3 \in D(a_2)$. Here we have to note that there may be several distinct power series in $z - a_3$ which are tertiary rearrangements of $f(z)$. We have to know how we get to $z = a_3$, not merely that it is possible to get there in three steps. At this stage of the continuation process it may happen that a power series is obtained whose region of convergence intersects $D(a)$ and whose values do not agree with those of $f(z)$ in the intersection. Rearranging about a point $z = a_4$ in the intersection, we would then get a power series $f(z; a_1, a_2, a_3, a_4)$ with $a_4 \in D(a)$ which is distinct from the direct rearrangement $f(z; a_4)$. It may even happen that $f(z; a_1, a_2, a_3, a_4)$ has a singular point in $D(a)$, whereas $f(z; a_4)$ has its singularities either exterior to

$D(a)$ or on its boundary. It is clear that such phenomena can arise only in the case of multivalued functions.

EXERCISE 10.2

1. Given the power series

$$\sum_{n=2}^{\infty} \frac{z^n}{n(n-1)}.$$

Sum the series in closed form and determine the singular points of the function defined by the series.

2. Show that for the series in the preceding problem the domain D_1 is the interior of a certain cardioid.

3. For the same series it is possible to choose a_1 with $|a_1| < 1$ and $a_2 \in D_1$ in such a manner that the secondary rearrangement $f(z; a_1, a_2)$ converges in some interval $(1, \alpha)$, $1 < \alpha$, of the real axis. What is the sum of the series at a point of this interval if a_1 and a_2 are (i) in the upper half-plane? (ii) in the lower half-plane?

4. Discuss the singular points of the function $\log \log z$.

10.3. Analytic functions. The reader should now have a fairly clear idea of the process of analytic continuation based on repeated rearrangements of power series. We shall base our definition of analytic functions on this process.

DEFINITION 10.3.1. *A finite or infinite sequence of power series*

$$(10.3.1) \quad f_n(z) = \sum_{k=0}^{\infty} c_{k,\,n}(z - a_n)^k, \quad |z - a_n| < R_n, \quad n = 1, 2, 3, \cdots,$$

is said to form a chain if for each $n > 1$ we have $|a_n - a_{n-1}| < R_{n-1}$, and if $f_n(z)$ is a direct rearrangement of $f_{n-1}(z)$. The chain joins $f_j(z)$ with $f_n(z)$ if $1 \leq j < n$.

Thus, the relation between $f_{n-1}(z)$ and $f_n(z)$ is

$$(10.3.2) \qquad\qquad f_n(z) = \sum_{k=0}^{\infty} \frac{1}{k!} f_{n-1}^{(k)}(a_n)(z - a_n)^k.$$

A finite chain can always be completed by intercalating additional elements so that it also joins its last element with the first.

DEFINITION 10.3.2. *Two power series,*

$$f(z) = \sum_{n=0}^{\infty} f_n(z - a)^n \quad and \quad g(z) = \sum_{n=0}^{\infty} g_n(z - b)^n$$

are said to be equivalent, written

$$(10.3.3) \qquad\qquad\qquad f(z) \sim g(z),$$

if there exists a finite chain of power series joining $f(z)$ with $g(z)$.

This relation obviously has the usual properties of an equivalence relation in that it is

(1) *reflexive*, $f \sim f$;
(2) *symmetric*, $f \sim g$ implies $g \sim f$;
(3) *transitive*, $f \sim g$ and $g \sim h$ implies $f \sim h$.

With respect to this equivalence relation the set of all power series breaks up into equivalence classes. Each class is the union of all power series equivalent to any one of its members. Such a member may be taken as a representative of its equivalence class, and it determines the latter uniquely.

DEFINITION 10.3.3. *An equivalence class of power series is called an analytic function.*

This definition is analogous to the tacit working definition used in Section 10.1, where we regarded an analytic function as the union of its local representations. In Definition 10.3.3 we restrict ourselves to local representations defined by power series. For some purposes this choice is too restrictive. It is desirable to make additional provisions for the point at infinity and for poles and algebraic singularities which belong to the domain of definition of the function. But for the time being Definition 10.3.3 will serve our needs.

We shall use $f(z)$ and similar functional notation for an equivalence class of power series, that is, for an analytic function, and the individual power series are referred to as *regular elements* of the analytic function $f(z)$.

We come next to the notion of *the analytic continuation of an element of an analytic function $f(z)$ along a rectifiable path C* leading from the center of the element, $z = a$, to some point $z = b$. It is understood that C is not completely confined to the interior of the circle of convergence, but we admit the possibility that $b = a$. Suppose that C is given by $z = p(t)$, $0 \leq t \leq 1$, where $p(0) = a$, $p(1) = b$. We speak of an analytic continuation along C if for each t, $0 \leq t < 1$, there is an element with center at $p(t)$, say $f(z; p(t))$, such that (i) $f(z; p(0))$ is the original element, (ii) each $f(z; p(t))$ is an element of $f(z)$, and (iii) for each t_1, $0 \leq t_1 < 1$, there is a $\delta = \delta(t_1)$ such that $f(z; p(t_1))$ and $f(z; p(t_2))$ are direct rearrangements of each other for $|t_1 - t_2| < \delta$. We can say that the power series $\{f(z; p(t)) \mid 0 \leq t < 1\}$ form a *continuous chain*. It is clear that by a suitable choice of points a_n on C we can obtain a finite or countable chain which serves the same purpose. This implies that for purposes of analytic continuation we can always replace a rectifiable path C by a polygonal line approximating C and having its vertices on C. Moreover, we do not need to consider all elements on the broken line; if the vertices are sufficiently dense, it will be sufficient to use only the elements whose centers are at the vertices, otherwise we may have to interpolate intermediary points.

We shall need a more restricted type of equivalence in the following.

DEFINITION 10.3.4.   *Let D be a domain in the complex plane containing the two points a and b. Two power series*

$$f(z) = \sum_{n=0}^{\infty} f_n(z-a)^n \quad and \quad g(z) = \sum_{n=0}^{\infty} g_n(z-b)^n$$

are said to be equivalent modulo D, written

(10.3.4) $f(z) \sim g(z) \quad (\text{mod } D),$

if $f(z)$ and $g(z)$ can be joined by a chain of power series having their centers in D.

In other words, $f \sim g$ (mod D) if and only if it is possible to pass from $f(z)$ to $g(z)$ by analytic continuation along a path which lies entirely in D. It is a simple matter to verify that the relation $f \sim g$ (mod D) is also reflexive, symmetric, and transitive. Under this relation, the set of all convergent power series with centers in D is subdivided into equivalence classes, and this is a finer subdivision than that defined by $f \sim g$. We shall need a suitable terminology.

DEFINITION 10.3.5.   *If $[f; D]$ is an equivalence class of power series with centers in D under the relation $f_1 \sim f_2$ (mod D), and if $[f; D]$ is defined everywhere in D, then $[f; D]$ is said to be analytic in D.*

Here the requirement that $[f; D]$ be defined everywhere in D is understood to mean that for each point a in D there is at least one power series in $(z-a)$ which belongs to $[f; D]$. It is not required that there be only one such series, for we distinguish between "being holomorphic in D" and "being analytic in D." A single-valued differentiable function $f(z)$, defined everywhere in a domain D, is holomorphic in D. Its various expansions in Taylor's series about the points of D form the equivalence class $[f; D]$ which is analytic in D by definition. But the converse is not necessarily true: if an equivalence class $[f; D]$ is analytic in D, then there need not exist a corresponding function $f(z)$ which is holomorphic in D. An example will clarify the terminology.

Consider the power series

$$-\sum_{n=1}^{\infty} \frac{1}{n}(1-z)^n, \quad |z-1| < 1.$$

Its sum is the principal determination of log z restricted to the circle of convergence. Now consider the disk $D_1 = [z \mid |z-1| < 1]$ and the equivalence class $[f; D_1]$ defined by the series in D_1. There is one and only one power series at each point a of D_1, and these power series together, that is, $[f; D_1]$, define a holomorphic function of z interior to the cardioid

$$z = (1 + e^{it})^2, \quad 0 \le t < 2\pi,$$

namely, the restriction of the principal determination of log z to this region. We now take a different domain, $D_2 = [z \mid 0 < |z| < 2]$, and consider the equivalence class $[f; D_2]$ determined by the same series. This can be described as the restriction of the infinitely many-valued analytic function $\log z$ to the domain $0 < |z| < 4$. Note that in general $[f; D]$ is not the restriction of an analytic function $f(z)$ to the domain D. For one thing, some elements of $[f; D]$ are apt to converge outside D. For another, the elements of $[f; D]$ usually form a proper subset of all the power series with centers in D which belong to the equivalence class defining the analytic function $f(z)$.

There is one case in which "analytic in D" and "holomorphic in D" mean the same thing. We shall state and prove this theorem, due to Weierstrass:

THEOREM 10.3.1. [THEOREM OF MONODROMY.] *If D is a simply-connected domain and if $[f; D]$ is analytic in D, then $[f; D]$ defines a function holomorphic in D.*

This theorem is usually stated as follows:

If D is a simply-connected domain, if $a \in D$, and if an element $f(z; a)$ can be continued analytically along every path in D, then the continuations define a holomorphic function in D.

Proof. We shall use an argument analogous to that used in proving Cauchy's theorem [Theorem 7.2.1]. We have to prove that at each point of D there is exactly one element belonging to $[f; D]$ and having this point as its center. In order to prove this, it is enough to show that analytic continuation along a closed path in D carries an element back into itself. We shall prove this for triangular paths; the extension to general rectifiable paths is then a simple matter.

Suppose that $z_1 \in D$, that $f(z; z_1)$ is an element of $[f; D]$, and that $\triangle = [z_1, z_2, z_3, z_1]$ is a positively oriented triangle which, together with its interior, lies in D. By assumption, $f(z; z_1)$ can be continued analytically along $\triangle$ from z_1 and back. Suppose that in so doing we arrive at an element $f_1(z; z_1) \neq f(z; z_1)$. We shall show that this leads to a contradiction. As in the proof of Theorem 7.2.1, we subdivide the triangle $\triangle$ into four subtriangles by joining the midpoints of the sides. See Figure 4. Let z_{jk} denote the midpoint of the side $[z_j, z_k]$.

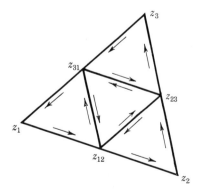

Figure 4

We start at $z = z_1$ with an element $f(z; z_1) \equiv E_1$ and proceed to $z = z_{12}$, where we encounter the element E_{12}, proceed to z_2 and the element E_2, and so on. Finally we return to z_1 with a new element, E_1^* say. Let us now consider what happens when we describe the subtriangles $\triangle_1$, $\triangle_2$, $\triangle_3$, and $\triangle_4$. Here $\triangle_4$ contains the center of gravity of $\triangle$, and $\triangle_j$ has the vertex $z_j, j = 1, 2, 3$. Let us consider analytic continuation along the sides of $\triangle_4$. Specifically we ask whether the following relation is true:

$$(10.3.5) \qquad\qquad E_{12} \to E_{23} \to E_{31} \to E_{12}.$$

In other words, we ask whether, starting with the element E_{12} at $z = z_{12}$ and describing $\triangle_4$ in the positive sense, we reach each of its vertices with the same determination as we did when we described $\triangle$. If this were true, then we see that describing $\triangle_1$ would lead to the diagram

$$E_1 \to E_{12} \to E_{31} \to E_1^* \neq E_1;$$

that is, we would have a smaller triangle than $\triangle$ along which an element of $[f; D]$ is carried into a different element. There is no reason, however, for assuming that (10.3.5) really holds. Suppose it is false because the first passage $E_{12} \to E_{23}$ as z traverses $[z_{12}, z_{23}]$ is not true. But then we see that E_{12} is not carried back into itself when z describes $\triangle_2$. If it is $E_{23} \to E_{31}$ that breaks down instead, then we see that E_{23} is not carried into itself by the circuit $\triangle_3$. Finally, if the first two steps in (10.3.5) hold, but the third breaks down, then E_{31} is not carried into itself by $\triangle_4$. In any case, we see that the assumption that E_1 is not carried into itself by $\triangle$ implies the existence of a subtriangle of $\triangle$ and of an element of $[f; D]$ which is not returned to its initial value when the subtriangle is described.

    This means that we can start a nesting process. There exists a sequence of nested triangles $\{\triangle_n\}$ in D and a corresponding sequence of elements $\{E_n\}$ belonging to $[f; D]$ such that E_n is not carried into itself when z describes $\triangle_n$. But there is a single point z_0 which lies interior to or on each of the triangles $\triangle_n$, and, if a_n is the center of E_n, then $a_n \to z_0$. Now, the radius of convergence of E_n is the distance of a_n from the boundary of D, and this distance tends to a positive limit as $a_n \to z_0$. This limit is r_0, the distance of z_0 from the boundary of D. But that means that there exists an N with the following properties: for each $n > N$ the triangle $\triangle_n$ is interior to the circle $|z - z_0| < r_0$, and E_{n+1} is a direct rearrangement of E_n. Further, $\triangle_{n+1}$ must be interior to the circle of convergence of E_n. Consequently, as we describe $\triangle_{n+1}$, the terminal value of E_{n+1} must coincide with the initial value. From this contradiction we conclude that the original hypothesis must be false, that is, analytic continuation of an element along the sides of a triangle in D must lead back to the original element.

    From the fact that analytic continuation along the perimeter of a triangle in D leaves the elements invariant, we conclude that the same holds for simple closed polygons. By Theorem B.2.1 of Appendix B, Volume I, each such polygon may be triangulated. Using this result, it is not hard to see that the assumption that an element is changed when z describes a polygon of n sides implies that some element is changed when z describes a polygon of $(n - 1)$ sides, and this ultimately leads to a contradiction. Since an arbitrary polygon is the union of simple closed polygons and multiple line segments, the result also holds for such a path. Finally, any simple closed rectifiable curve may be approximated arbitrarily closely by "inscribed polygons." It follows that analytic continuation along any closed rectifiable path in D must leave the elements unchanged. Hence, at each point of D there is at most one (and hence exactly one) element having that point as its center. Thus, $[f; D]$ defines a single-valued and, hence, a holomorphic function. This completes the proof.

DEFINITION 10.3.6. *An equivalence class of power series containing at most one element for each center c defines a single-valued function $f(z)$. The set $\{c\}$ of centers actually represented by elements in the class (plus possibly the point at infinity) is the domain of holomorphism D of $f(z)$. The domain of existence of $f(z)$ is obtained by adding to D the set of poles, if any.*

From its definition, D is open and connected and hence actually a domain. The poles form an isolated countable set.

It is natural to ask how many different elements of an analytic function may have the same center. The function $\log z$ shows that the number may actually be countably infinite. The theorem of Poincaré and Volterra[1] shows that no higher cardinal number may arise.

THEOREM 10.3.2. *The number of different power series in $(z - a)$, a fixed, which belong to an equivalence class, is finite or countably infinite.*

Proof. Suppose that we start with an element $f(z; a)$ and let $z = b$ be a point where $f(z)$ is defined. Here b may be distinct from or equal to a. The various elements of $f(z)$ at $z = b$ are obtainable from $f(z; a)$ by analytic continuation along suitably chosen paths. With each such path we associate a chain

$$f(z; a_k), \quad k = 0, 1, 2, \cdots, n, \quad a_0 = a, \quad a_n = b.$$

Without restricting the generality, we may suppose that the points a_k have

---

[1] Henri Poincaré (1854–1912), a French star of the first magnitude, was a prolific writer whose 500-odd papers and 30 books covered most fields of pure and applied mathematics including celestial mechanics and fluid dynamics, with occasional excursions into cosmogony, epistemology, and logic. He did pioneering work in asymptotic series, infinite determinants, and topology (analysis situs in his terminology); he created the theory of automorphic functions (in competition with Felix Klein (1849–1925)) and of uniformization; and he enriched the theory of differential equations, real and complex. In wealth of ideas, versatility, and power of analysis he has no compeer in the annals of mathematics. He was born in Nancy, where the local *lycée* carries his name. So also does the mathematical institute of the Sorbonne, built after his premature death.

Vito Volterra (1860–1940) exercised a strong influence on the development of analysis through his work on integral equations, partial differential equations, functions of lines, and functions of composition.

rational coordinates. If we can show that the number of such chains is countable, then it follows a fortiori that the number of distinct elements at $z = b$ is also countable. In order to get from $z = a$ to $z = b$ we shall normally need intermediary points, that is, the integer n has a minimum value, m say, where $m \geqq 1$. We refer to n as the length of the chain. Now the number of chains of length n is countable, for we have at our disposal for the choice of $a_1, a_2, \cdots, a_n$ certain subsets of points of rational coordinates. If we let $a_1, a_2, \cdots, a_n$ range independently of one another over all points of rational coordinates, the set $(a_1, a_2, \cdots, a_n)$ is in one-to-one correspondence with the set of points in a real Euclidean $(2n)$-space having rational coordinates, and this set is countable. It follows that the set of chains of length n is at most countable. Here $n = m$, $m + 1, m + 2, \cdots$, and a countable set of countable sets is also countable. This proves the theorem.

EXERCISE 10.3

1. Prove the following generalization of the theorem of monodromy: If an element of an analytic function can be continued along every path in a simply-connected domain D without encountering any singularities other than poles, then $[f; D]$ is meromorphic in D.

2. Show that there exists a countable subset of the elements defining an analytic function $f(z)$ such that every element of $f(z)$ is either in the subset or may be obtained by direct rearrangement of one of the elements in the subset.

3. Consider the set of all series

$$\sum_{n=1}^{\infty} 2^{-n-1} \left\{ 1 - \left[\left(1 - \frac{z}{2n}\right)\left(1 - \frac{z}{2n+1}\right) \right]^{\frac{1}{2}} \right\},$$

where the choice of each individual square root at $z = 0$ is arbitrary. The sum of the series at $z = 0$ is $\sum \varepsilon_n 2^{-n}$ where $\varepsilon_n = 0$ or 1 according as the nth square root is $+1$ or -1 at $z = 0$. Thus, the set of all such series is in a one-to-one correspondence with the set of real numbers between 0 and 1. How do these series group themselves into equivalence classes with respect to analytic continuation? Show in particular that two series belong to the same class if and only if in the corresponding dyadic fractions $\sum \varepsilon_n 2^{-n}$ and $\sum \eta_n 2^{-n}$ we have $\varepsilon_n = \eta_n$ for all large n.

10.4. Singularities. We have encountered several classes of singular points such as poles, isolated essential singularities, and algebraic and logarithmic branch points. We have defined singular points on the circle of convergence of a power series. We have proved that there is at least one singularity on the circle, and we have found a criterion to ensure that the circle of convergence is a singular line. The time has come to give a general definition of singular points. This we attach to the notion of *singular chains*.

Consider an analytic function $f(z)$ defined by the equivalence class $\{f_\alpha\}$ of its elements, and take an infinite subset of $\{f_\alpha\}$:

(10.4.1) S: $\{f(z; a_n), \quad |z - a_n| < R_n, \quad n = 1, 2, 3, \cdots\}.$

DEFINITION 10.4.1.   *S is said to be a singular chain of $f(z)$ if (1) S is a chain, (2) $\lim a_n = a_0$ exists, and (3) $\lim R_n = 0$. We call a_0 the endpoint of S.*

Let $\varepsilon > 0$ be given and consider the disk D_ε: $|z - a_0| < \varepsilon$. Let S_ε be the subset of S whose centers lie in D_ε, and let $[f; D_\varepsilon]$ be the equivalence class modulo D_ε which contains S_ε. Consider another singular chain S_1: $\{g(z; b_n)\}$ having the same endpoint a_0, and construct for it the corresponding equivalence class modulo D_ε, $[g; D_\varepsilon]$ say. *Two singular chains are said to be equivalent if and only if the corresponding classes $[f; D_\varepsilon]$ and $[g; D_\varepsilon]$ are identical for sufficiently small values of ε.*

DEFINITION 10.4.2.   *A finite point $z = a$ is a singular point of an analytic function $f(z)$ if and only if it is the endpoint of a singular chain made up of elements of $f(z)$. It is then said to be a singular point for approach to $z = a$ along the chain in question. Two singular chains having the same endpoint determine the same singular point if and only if the chains are equivalent.*

In order to discuss the point at infinity, we make the usual transformation and consider $f(1/z)$ at $z = 0$.

The above definition has several noteworthy features. If $f(z)$ is a single-valued function, its singular points are among the boundary points of its domain of holomorphism. Not every such boundary point can be the endpoint of a singular chain, however; for this to be the case it must be *accessible*. A *boundary point $z = a$ of a domain D is said to be accessible from D, if, given a point $z = b$ in D, it is possible to join a and b by a Jordan arc in D.* If such a path exists, then analytic continuation of $f(z)$ along the path is possible; a singular chain having its centers on the path is then easily constructed. The existence of such a path is clearly also necessary for the existence of a singular chain. Thus we have:

THEOREM 10.4.1.   *The singular points of a single-valued analytic function are the accessible boundary points of its domain of holomorphism.*

If $f(z)$ is not single-valued, then our definition foresees the possibility that the same point $z = a$ can be the carrier of several, even of infinitely many, singular points. This does not exclude the possibility that the same point may also be the center of regular elements. The following examples illustrate these possibilities.

As a first example we take the function

(10.4.2) $[1 - (1 - z)^{\frac{1}{2}}]^{\frac{1}{2}}.$

It has algebraic branch points at $z = 0$, 1, and ∞. At $z = 0$ there are two regular elements and a branch point of order one; that is, a small circuit about $z = 0$ will commute two branches and leave the other two invariant. At $z = 1$ there are two distinct singularities, both of which are branch points of order one; the four branches form two cycles, each consisting of two commuting branches. At infinity, all four branches form a single cycle, and there is a branch point of order three.

Next, we observe that the analytic function

(10.4.3)   arc sin z

has singularities located at $z = -1, +1$, and ∞. The point $z = -1$ is the carrier of infinitely many algebraic branch points each of order one, and there are no regular elements. The same is true at $z = +1$, and at $z = \infty$ we have a logarithmic branch point and again no regular elements. Compare Section 6.5.

Finally, we take

(10.4.4)   log log z

with singularities at $z = 0, 1, \infty$. At $z = 1$ we have infinitely many regular elements and one logarithmic branch point, whereas there are infinitely many logarithmic branch points at $z = 0$ and ∞ and no regular elements.

We return to the singular points of a single-valued analytic function. We have seen that they are formed by the accessible boundary points of the domain of holomorphism of the function. The structure of such a domain was completely settled by a result proved by G. Mittag-Leffler [*Acta Mathematica*, Vol. 4 (1884)]:

THEOREM 10.4.2.   *Given any domain D in the complex plane, then there exists a function $f(z)$ having D as its domain of holomorphism.*

Proof.   There are certain trivial cases in which the construction of $f(z)$ is obvious. Thus, if D is the extended plane, then $f(z)$ is necessarily a constant. If the complement of D is finite, then a rational function having poles at the points of $\mathbf{C}[D]$ is a solution of the existence problem. If $\mathbf{C}[D]$ is a countable set with the single limit point a, then a meromorphic function of $(z - a)^{-1}$ or of z, according as a is finite or infinite, can be constructed with the desired properties. The extension to the case in which $\mathbf{C}[D]$ has a finite or countably infinite set of limit points does not involve serious difficulties. We split $\mathbf{C}[D]$ into countable subsets, each having a single limit point. For each of these subsets we construct a meromorphic function whose poles are the points of the subset, and by using suitable constant multipliers we can then obtain an absolutely convergent series of these functions and, thus, a solution to the construction problem.

For the general case we shall use an argument due to Jean Besse ["Sur le domaine d'existence d'une fonction analytique," *Commentarii Mathematici Helvetici*, Vol. 10 (1937-38), pp. 302–305]. We may suppose that the point at infinity is an interior point of D. This can always be achieved by a fractional linear transformation. It follows that the boundary of D, the set ∂D, is bounded. We start by selecting a distinguished countable set of boundary points, called "well visible" by Besse. For this purpose, we consider an enumeration of the points of D which have rational coordinates. Let $\{\alpha_n\}$ be this set. For each point α_n we determine the nearest point β_n of ∂D, where, if there are several contenders, we choose β_n so that arg $(\beta_n - \alpha_n)$ has its least positive value. This gives a sequence $\{\beta_n\}$ of boundary points of D, and each β_n is "well visible" in the sense that it lies on the circumference of a circle the interior of which belongs to D. The sequence $\{\beta_n\}$ contains many repetitions. We let $\{b_n\}$ be the subset of distinct elements of $\{\beta_n\}$. We note that every b_n is a "well visible" boundary point of D. Further, every isolated point of ∂D is in the set $\{b_n\}$, and every accessible point of ∂D is either a b_n or the limit of such points.

Next, we choose a convergent series with positive terms a_n such that

$$(10.4.5) \qquad a_n > 2 \sum_{k=n+1}^{\infty} a_k$$

for each n. We may take, for instance, $a_n = a^{-n}$ with $a > 3$. With these two sequences $\{a_n\}$ and $\{b_n\}$ we form the series

$$(10.4.6) \qquad \sum_{n=1}^{\infty} \frac{a_n}{z - b_n}.$$

Let D_δ be the subset of D whose points have a distance from $\{b_n\}$ which is $\geq \delta$. Then the series converges uniformly with respect to z in D_δ. It follows that the series converges in D, and that its sum is a holomorphic function in D. We denote this function by $f(z)$.

Next, let us consider the behavior of $f(z)$ when z approaches one of the points b_k. The latter being "well visible" from D, there exists a circle C_k, passing through b_k, the interior of which lies in D. We approach b_k along the radius of C_k. For z on this radius we have

$$|z - b_k| < |z - b_n|, \quad n \neq k.$$

We write

$$f(z) = \sum_{n=1}^{k-1} \frac{a_n}{z - b_n} + \frac{a_k}{z - b_k} + \sum_{n=k+1}^{\infty} \frac{a_n}{z - b_n}.$$

Here, the first finite sum is holomorphic in C_k, and it has a finite upper bound B on the radius under consideration, since the b_n's are distinct. Using (10.4.5), we see that

$$\sum_{n=k+1}^{\infty} \frac{a_n}{|z - b_n|} < \frac{1}{2} \frac{a_k}{|z - b_k|}.$$

It follows that on the radius

$$|f(z)| > \frac{1}{2} \frac{a_k}{|z - b_k|} - B,$$

that is, $|f(z)|$ becomes infinite when $z \to b_k$ along the radius. Thus, $z = b_k$ is a singular point of $f(z)$ for each k.

From this follows the stronger assertion that $f(z)$ cannot be continued analytically outside of D. Such a continuation would involve a chain of regular elements of $f(z)$, and it is clearly no restriction to assume that the centers of these elements are at points having rational coordinates (see Problem 2 of Exercise 10.3). By the construction of the set $\{b_n\}$, each of these elements has on its circle of convergence one of the points b_n. Now, if an element of the chain converged at a point of ∂D which is not a b_n, then it would also have to converge at the nearest point of ∂D, counting from the center of the element. But this is a point b_k, and we have seen above that no such point can be interior to the circle of convergence of an element of $f(z)$ from the center of which b_k can be approached radially. This shows that it is impossible to continue $f(z)$ analytically beyond ∂D. Thus, D is the exact domain of existence of this function $f(z)$. The construction problem has been solved, and the theorem is proved.

It remains to say something about the nature of the singularities of a single-valued analytic function. We know that an isolated singularity is either a pole or an essential singular point. The function is uniformly large in the neighborhood of a pole, and its behavior at an essential singularity is governed by Picard's theorem. This asserts that $f(z)$ takes on every finite value infinitely often with at most one exception in any neighborhood of the point.

Let us now turn to the case in which $z = a$ is a limit point of singular points. Here the simplest case is that in which $z = a$ is the limit point of a sequence of poles. Picard's theorem holds, but now two finite values may be omitted, since the value ∞ is taken on infinitely often. The function

$$\sin\left(\cot\frac{1}{z}\right)$$

is an example of a function having infinitely many essential singularities with a limit point at $z = 0$. It is an easy matter to give examples of "elementary" functions whose singularities have a countable number of limit points.

More interesting situations arise when the singularities have the power of the continuum. We may assume that the singular points form a perfect set, since isolated points have already been treated. Such a set may be totally disconnected, or it may contain one or more continua. It now becomes possible for a function to exist in a domain D bounded by a perfect set S, the function being holomorphic and bounded in D, with each point of S being singular. It is even possible for the function to be continuous in $D \cup S$. We have already encountered such a function in Section 5.7, where we studied the power series

$$\sum_{n=0}^{\infty} \frac{1}{n!} z^{2^n},$$

which is continuous together with its derivatives of all orders in the closed unit disk. Here S is the circle $\mid z \mid = 1$, which is the natural boundary of the function. But such phenomena can arise even if S is a *dyadic discontinuum* (generalization of the Cantor set; compare Problem 6 below). The Rumanian mathematician Dimitrie Pompeiu (1873–1954) has constructed analytic functions which are continuous in the extended plane and have their singularities in a dyadic discontinuum.

Singularities of multiple-valued functions will be discussed in Section 10.6.

EXERCISE 10.4

1. Show that the following construction leads to a domain D whose boundary is partly inaccessible. Starting with the open unit square $0 < x < 1, 0 < y < 1$, we remove the line segments $y = 2^{-2k+1}, \ 0 < x \leq 1 - y; \ y = 2^{-2k}, \ y \leq x < 1$ for $k = 1, 2, 3, \cdots$. D is the remaining point set.

2. Verify the statements made in the text concerning the nature of the singularities of the functions (10.4.2), (10.4.3), and (10.4.4).

3. Construct a function having a simple pole at each of the points

$$\frac{1}{m} + \frac{1}{n} i, \quad m, n = 1, 2, 3, \cdots.$$

The residues are at your disposal.

4. The function

$$F(z) = \sum_{n=0}^{\infty} \frac{(-1)^n}{n!} \frac{1}{1 + a^{2n} z^2} = \frac{1}{2} \sum_{n=0}^{\infty} \frac{(-1)^n}{n!} \left(\frac{1}{1 - i a^n z} + \frac{1}{1 + i a^n z} \right),$$

where $a > 1$, has simple poles at the points $\pm i a^{-n}, n = 1, 2, 3, \cdots$, which have a limit point at the origin. The series, as well as all its derived series, converges for all values of z except for the poles. In particular, these series converge at $z = 0$. Compute $F^{(k)}(0)$ in closed form and show that the corresponding Maclaurin series

$$\sum_{k=0}^{\infty} F^{(k)}(0) \frac{z^k}{k!}$$

defines an entire function and thus cannot represent $F(z)$ in any neighborhood of the origin.

5. Given

$$f(z) = \sum_{n=1}^{\infty} e^{-n^2} \sum_{m=1}^{n-1} {}' \left(z - \frac{m}{n} \right)^{-1},$$

where the prime indicates that m and n are relatively prime. Show that $f(z)$ is holomorphic in the z-plane cut along the line segment $[0, 1]$. Let p and q be positive, relatively prime integers with $p < q$. Show that $\lim_{y \to 0} \left| f\left(\frac{p}{q} + iy \right) \right| = \infty$

and hence that the line segment is made up of singular points. (*Hint:* Prove an estimate of the type

$$\left| f\left(\frac{p}{q} + iy\right) + \frac{i}{y} e^{-q^2} \right| < q \sum_{n=1}^{\infty} n^2 e^{-n^2} \Big).$$

6. A two-dimensional Cantor set is obtained as follows: Divide the unit square $0 \leq x \leq 1, 0 \leq y \leq 1$, into nine equal squares and omit the five squares forming a cross in which either x or y or both lie between $\frac{1}{3}$ and $\frac{2}{3}$, limits excluded. The four remaining squares are treated in the same manner, and so on. After n subdivisions and omissions there are left 4^n squares of total area $(\frac{4}{9})^n$. The Cantor set C is the perfect set of points which is common to all "remaining" closed squares. Denote the centers of the squares remaining at the nth stage by $z_{n, k}, k = 1, 2, \cdots, 4^n$, and define

$$f_n(z) = 4^{-n} \sum_{k=1}^{4^n} \frac{1}{z - z_{n, k}}.$$

Then $\lim_{n \to \infty} f_n(z) \equiv f(z)$ exists for all z. $f(z)$ is continuous in the extended plane and holomorphic outside of C. $f(\infty) = 0$, but $f(z)$ is not identically zero since

$$\int_{|z|=2} f(z)\, dz = 2\pi i.$$ See P. Urysohn, *Fundamenta Mathematicæ*, Vol. 4 (1923), pp. 144–150.

10.5. Borel monogenic functions. The following series expansions exhibit phenomena discovered by Émile Borel. We start with

(10.5.1) $$F(z) = \sum_{n=2}^{\infty} e^{-n^2} \sum_{m=1}^{n-1}{}' (z - e^{2\pi im/n})^{-1},$$

where the prime indicates that m is restricted to values which are relatively prime to n. This series defines two holomorphic functions: one, $f_1(z)$, inside the unit circle, and the other, $f_2(z)$, outside. The unit circle is a natural boundary for both functions, since the sum of the series becomes infinite when z approaches radially any one of the points $e^{2\pi im/n}$. Nevertheless, there is very strong coherence between $f_1(z)$ and $f_2(z)$, as is shown by the following observation: Suppose that α is an irrational algebraic number between 0 and 1; that is, α is a root of an algebraic equation with integral coefficients and irreducible in the rational field. Then the series (10.5.1) converges on the ray

$$\arg z = 2\pi\alpha$$

uniformly with respect to z, $0 \leq |z| < \infty$. Moreover, each of the derived series has the same property. This implies that we can cross the unit circle at each of the points

$$e^{2\pi i\alpha}$$

which are dense on the circle (actually the choice $\alpha = j\sqrt{2} - k$, where j and k run through the positive integers, will give a dense set). At these points the radial limits of $f_1(z)$ and of $f_2(z)$ coincide, and the same is true for each of their derivatives. Thus, while $f_2(z)$ is not the analytic continuation of $f_1(z)$ in the sense of Weierstrass, there is a dense set of rays on which one can pass from one function to the other and preserve the continuity of all derivatives.

We shall verify this for the choice

$$\alpha = j\sqrt{2} - k,$$

where j and k are positive integers, and $0 < \alpha < 1$. On such a ray the expression $|z - \exp(2\pi i m/n)|$ does not vanish, but it is not bounded away from zero. We need a lower bound in terms of n for this expression, in order to establish the uniform convergence. Such a bound is obtained by the following considerations: For arbitrary real values of θ and γ, and for $r \geq 0$, we have

$$|re^{i\theta} - e^{i\gamma}| = |r - e^{i(\gamma - \theta)}| \geq |\sin(\theta - \gamma)|,$$

for the last member is the minimum value of the second one. Now for $0 < t < \frac{1}{2}\pi$, we have

$$\sin t > \frac{2}{\pi} t.$$

Hence, if we denote the distance from the real number u to the nearest integer by $\{u\}$, we see that

$$|\sin(\theta - \gamma)| \geq 2\left\{\frac{1}{\pi}(\theta - \gamma)\right\}.$$

We set

$$\theta = 2\pi\alpha = 2\pi(j\sqrt{2} - k), \quad \gamma = 2\pi\frac{m}{n},$$

and we obtain

$$|re^{2\pi i\alpha} - e^{2\pi i m/n}| \geq 2\left\{2\left(j\sqrt{2} - k - \frac{m}{n}\right)\right\}.$$

Here the quantity between the braces lies between -2 and $+2$ by the assumptions on m, n, and α. It is desirable that it does not get too close to any one of the integers -2, -1, 0, 1, and 2 for which $\{u\}$ is zero. We give here a lower bound for the distance from 0. The other integers are handled in the same manner and lead to similar estimates; only the factor depending upon k is changed. Now

$$j\sqrt{2} - k - \frac{m}{n} = \frac{2(nj)^2 - (m + nk)^2}{n(nj\sqrt{2} + m + nk)}.$$

Here the numerator is an integer, positive or negative, but not zero. In the denominator,

$$nj\sqrt{2} + m + nk < n(k + 1) + m + nk < 2n(k + 1),$$

since $j\sqrt{2} < k + 1$ and $m < n$. It follows that

$$2 \left| j\sqrt{2} - k - \frac{m}{n} \right| > \frac{1}{(k+1)n^2} .$$

Similar estimates are obtained for the other distances. The most unfavorable estimate is obtained for the distance from $+1$, where $k + 1$ is replaced by $4k + 5$. This estimate also gives a safe lower bound for the brace, as well as the final inequality

$$\left| e^{2\pi i \alpha} - e^{2\pi i m/n} \right| > \frac{2}{(4k+5)n^2} .$$

Since

$$F^{(p)}(z) = (-1)^p p! \sum_{n=2}^{\infty} e^{-n^2} \sum_{m=1}^{n-1}{}' \, (z - e^{2\pi i m/n})^{-p-1},$$

we see that on the ray $\arg z = 2\pi(j\sqrt{2} - k)$ the terms of the pth derived series are dominated by those of the convergent series

$$p! \, [\tfrac{1}{2}(4k + 5)]^{p+1} \sum_{n=2}^{\infty} n^{2p+3} e^{-n^2}.$$

This proves the uniform convergence.

As the second example we take the series

(10.5.2) $$F(z) = \sum_{n,q} e^{-n^2 - q^2} \sum_{m,p} \left(z - \frac{m}{n} - \frac{p}{q} i \right)^{-1},$$

where n and q range over the positive integers, $0 \leqq m \leqq n$ and $0 \leqq p \leqq q$. This function is clearly holomorphic outside the unit square $0 \leqq x \leqq 1$, $0 \leqq y \leqq 1$, $z = x + iy$, and the perimeter of this square is a singular line, since $F(z)$ becomes infinite if z approaches the boundary along lines $x = r$ or $y = r$ where r is rational. Nevertheless, there is a function defined in the square which is a continuation of the holomorphic function. With the argument used above we can prove that the series (10.5.2) converges uniformly on lines

$$x = j\sqrt{2} - k \quad \text{and} \quad y = j\sqrt{2} - k,$$

and the same applies for each of the derived series. These lines are dense in the square. In the present case we can also get two-dimensional results.

Let $\varepsilon > 0$ be given and delete from the plane all disks

$$\left| z - \frac{m}{n} - \frac{p}{q} i \right| \leqq \varepsilon(n + q)^{-3},$$

where m, n, p, q satisfy the restrictions stated above. The total area of these disks is $\pi \varepsilon^2$ times the sum of the series

$$\sum\sum (n + 1)(q + 1)(n + q)^{-6}.$$

The double series is convergent, and its sum is $< \frac{1}{5}$, so the total area is $< \varepsilon^2$, that is, as small as we please. Denote the subset of the unit square outside these disks by D_ε. The series (10.5.2) converges uniformly in D_ε because the series is dominated by the convergent double series

$$\varepsilon^{-1} \sum\sum (n+1)(q+1)(n+q)^3 \, e^{-n^2-q^2}.$$

Thus, $F(z)$ is continuous in D_ε. Moreover, each of the derived series

$$F_k(z) \equiv (-1)^k k! \sum e^{-n^2-q^2} \sum \left(z - \frac{m}{n} - \frac{p}{q} i \right)^{-k-1}$$

converges uniformly in D_ε, since it is dominated by a series of the form

$$\varepsilon^{-k-1} k! \sum\sum (n+1)(q+1)(n+q)^{3(k+1)} \, e^{-n^2-q^2}.$$

Since $F(z)$ is not holomorphic in the unit square, and since D_ε is not a domain, $F_1(z)$ is not the derivative of $F(z)$ in the sense that we have previously used this term. But if $z_0 \in D_\varepsilon$ and if $h \to 0$ in such a manner that $z_0 + h \in D_\varepsilon$, then

$$\lim_{h \to 0} \frac{1}{h} [F(z_0 + h) - F(z_0)] = F_1(z_0),$$

and in the same sense $F_k(z)$ is the derivative of $F_{k-1}(z)$.

Borel has created an extensive theory of what he calls *monogenic non-analytic functions* defined in *Cauchy domains* where they are differentiable. Such functions admit also a generalization of Cauchy's integral theorem. We restrict ourselves to the examples given above.

10.6. Multivalued functions and Riemann surfaces.

The classification of singularities of multivalued functions, even of isolated singularities, is much less developed than in the single-valued case. We shall restrict ourselves to pointing out a few important cases.

The three functions

$$(10.6.1) \qquad z^{\frac{1}{n}}, \quad \log z, \quad \text{and} \quad z^{\sqrt{2}}$$

exemplify three different types of singularities, which are known as *algebraic branch points, logarithmic branch points,* and *transcendental critical points,* respectively. In the first type of singularity only, we include the point $z = 0$ in the domain of existence of the function. A more complex type of algebraic branch point is furnished by the expansion

$$(10.6.2) \qquad A(z) \equiv z^{\frac{p}{n}} \sum_{m=0}^{\infty} c_m z^{\frac{m}{n}},$$

where n and p are integers, $n \geq 2$, $p \gtrless 0$. This function has a zero at $z = 0$ if $p > 0$ but a pole if $p < 0$. Such a series is called an *algebraic element* of the function in question, and such elements are to be adjoined to the regular elements defining the function.

We observe that it is possible to find a simple "uniformizing" transformation $z = g(t)$ which carries the first function under (10.6.1) and the function (10.6.2) into single-valued functions of t. Clearly, we may take $g(t) = t^n$, and then we obtain

$$z^{\frac{1}{n}} = t \quad \text{and} \quad A(z) = t^p \sum_{m=0}^{\infty} c_m t^m$$

respectively. We note that at the origin $A(t^n)$ has a zero or a pole of order $|p|$ according as $p > 0$ or $p < 0$.

It is also possible to uniformize the second and third functions under (10.6.1) by taking $g(t) = e^t$. We obtain

$$\log z = t \quad \text{and} \quad z^{\sqrt{2}} = e^{\sqrt{2}t}.$$

Functions like

(10.6.3) $$(\log z)^{\sqrt{3}}, \quad \log \log z, \quad z^{\sqrt{2}}(\log z)^{\sqrt{3}}$$

have singularities at the origin of a higher degree of complexity than those of formula (10.6.1). Here

$$z = e^{e^t}$$

is a uniformizing transformation, and it leads to entire functions of t

$$e^{\sqrt{3}t}, \quad t, \quad \exp(\sqrt{2}e^t + \sqrt{3}t).$$

The difference in the character of the three uniformizing transformations is rather striking. In the first case, $z = t^n$, the point $z = 0$ is the image of $t = 0$, and a neighborhood of $z = 0$ is obtained as the image of (partial) neighborhoods of $t = 0$, a point where t^n is holomorphic. In the second case, $z = e^t$, there is no point t which maps into $z = 0$, and in order to get punctured disks $0 < |z| < \varepsilon$, we have to use strips in the t-plane extending out to infinity, which is an essential singularity of the uniformizing function. The same is true in the third case, but the distribution of the strips in the t-plane follows a more complicated pattern.

These indications suffice to show both the intrinsic complexity of the situation and the use of uniformizing transformations to bring some order into the chaos.

The series

(10.6.4) $$f(z) = \sum_{n=1}^{\infty} 3^{-n} \left[\left(\frac{1}{2n} - z \right) \left(\frac{1}{2n+1} - z \right) \right]^{\frac{1}{2}}$$

gives an example of a function for which the origin is a limit point of algebraic branch points. To fix the ideas, we suppose that all square roots are positive at the origin. The function is single-valued in the plane cut along the line segments $\left[\dfrac{1}{2n+1}, \dfrac{1}{2n} \right]$, $n = 1, 2, 3, \cdots$, of the real axis. Crossing one of these segments changes the determination of the corresponding square root. There is clearly no closed path in the plane which surrounds the origin without also surrounding infinitely many of the algebraic branch points. A singular point of this type is sometimes known as an *indirectly critical point*. We note that the series for $f(z)$, as well as all its derived series, converges at $z = 0$. The functions $f^{(k)}(z)$ tend to finite limits as $z \to 0$ in any sector omitting the positive real axis. It is obvious how we can use the same method to condense more complicated singularities than first-order branch points at a desired limit point. We shall not pursue any further, however, the study of singularities per se.

An analytic function is an equivalence class of *regular elements*, power series in a variable $z - a$, where a ranges over a set E_0 of the finite plane. E_0 is the (restricted) domain of the function. To the set of regular elements we adjoin the following classes of elements in case they exist:

(1) *polar elements* of the form

$$(z - a)^{-p} \text{ times power series in } (z - a), \quad p \geq 1,$$

corresponding to the *finite poles* of $f(z)$;

(2) *algebraic elements* of the form

$$(z - a)^{\frac{p}{n}} \text{ times power series in } (z - a)^{\frac{1}{n}}, \quad n > 1,$$

corresponding to the *finite algebraic branch points*; and

(3) *infinitary elements* of the form

$$z^{\frac{p}{n}} \text{ times power series in } z^{-\frac{1}{n}}, \quad n \geq 1,$$

corresponding to the point at infinity.

After adding these elements to the regular ones and adjoining the corresponding centers a to the set E_0, we obtain the set E, the *domain of definition* of $f(z)$.

The set E gives fairly vague information concerning $f(z)$. We get a much more satisfactory idea of $f(z)$ by considering its *graph* $\mathfrak{G}[f]$. This is the set of ordered pairs

$$\mathfrak{p} \equiv (a, f_\alpha(z; a)),$$

where a ranges over E and $f_\alpha(z; a)$ ranges over the elements of $f(z)$ having center at a. By Theorem 10.3.2, the set $\{\alpha\}$ is finite or countable. We refer to $\mathfrak{p}$ as a *place* on the graph. To complete the description of the graph, we have to know what places $\mathfrak{p}$ have the property of being "near" to a given place $\mathfrak{p}_0$. In other words, we have to define *neighborhoods* on the graph.

Now $\mathfrak{G}[f]$ is a *Cartesian product space*, that is, a set of ordered pairs $[x, y]$ where x belongs to a subset of a space X and y belongs to a subset of a space Y. In our case, X is the set of complex numbers and Y is the set of all elements: regular, polar, algebraic, and infinitary. In each of these spaces we have a natural notion of nearness which we can use to define neighborhoods. In the space of complex numbers these are the regular ε-neighborhoods, the set $N(a, \varepsilon)$ of all points z such that $|z - a| < \varepsilon$. By extension, any domain in the plane containing the point a may be regarded as a neighborhood of a. In the space of elements, an element $f(z; b)$ may be regarded as near to an element $f(z; a)$ if the former element is an immediate rearrangement of the latter. The set of all immediate rearrangements of $f(z; a)$ is a neighborhood of $f(z; a)$. We can get narrower or wider neighborhoods of $f(z; a)$, as desired, by including in the neighborhood all elements $f(z; b)$ which are equivalent to $f(z; a)$ modulo D, where D is some domain containing $z = a$.

Before going any farther, let us observe that the process of direct rearrangement applies also to algebraic elements. Given a series

$$(10.6.5) \qquad \sum_{m=0}^{\infty} c_m(z - a)^{\frac{p+m}{n}}, \quad |z - a| < R,$$

where $c_m \neq 0$ for at least one value of m such that n is relatively prime to $p + m$. This series represents n branches of the multivalued function $f(z)$. We can obtain the Taylor series for each of these branches at any point b inside the circle of convergence by substituting $z = (z - b) + b$ and by using the binomial series to expand

$$(z - a)^{\frac{p+m}{n}} = (b - a)^{\frac{p+m}{n}} \left\{ 1 + \frac{z - b}{b - a} \right\}^{\frac{p+m}{n}}.$$

Here, the first factor on the right is taken as the $(p + m)$th power of $(b - a)^{1/n}$, the determination of which fixes the branch of $f(z)$ under consideration. The binomial series gives the principal value of the second factor. The resulting double series

$$(10.6.6)$$
$$\sum_{m=0}^{\infty} c_m(b - a)^{\frac{p+m}{n}} \sum_{k=0}^{\infty} \frac{(p + m)(p + m - n) \cdots (p + m - (k - 1)n)}{n^k k!} \left(\frac{z - b}{b - a} \right)^k$$

is absolutely convergent for

$$|z - b| < \min \left[|b - a|, R - |b - a| \right].$$

For such values of z we may interchange the order of summation and obtain the desired power series in $(z - b)$ which represents the branch. We note that there exists an R_1, $R_1 \leq R$, such that for $0 < |b - a| < R_1$ distinct choices of $(b - a)^{1/n}$ give distinct branches of $f(z)$. The rearrangement procedure evidently applies also for polar elements; we have merely to set $n = 1$ and take $p < 0$ in the formulas. Infinitary elements can be handled in the same manner.

After this digression, let us return to the neighborhoods.

DEFINITION 10.6.1.   *A neighborhood of the place* $p_0 = (a, f(z; a))$ *is the set of all places* $p = (b, f_\alpha(z; b))$ *such that* $b \in N(a) \cap E$, *where* $N(a)$ *is a neighborhood of* a, E *is the domain of definition of* $f(z)$, *and* $f_\alpha(z; b)$ *is equivalent to* $f(z; a)$ *modulo* $N(a)$.

DEFINITION 10.6.2.   *A set X of points x and a collection Σ of subsets N of X, called neighborhoods, defines a Hausdorff space[1] if*

H$_1$.   *For any two distinct points x and y of X there exist disjoint sets N_1 and N_2 of Σ such that $x \in N_1$ and $y \in N_2$.*

H$_2$.   *For any two sets N_1 and N_2 which contain the point x, there exists a set N_3 of Σ such that $x \in N_3 \subset N_1 \cap N_2$.*

THEOREM 10.6.1. $\mathfrak{G}[f]$ *is a Hausdorff space if neighborhoods are introduced as in Definition 10.6.1.*

In verifying conditions H$_1$ and H$_2$ we use, on the one hand, the fact that the complex plane is a Hausdorff space under the usual neighborhood topology. On the other hand, we use the fact that the elements $f_\alpha(z; b)$ which are equivalent to $f(z; a)$ modulo $N_3(a)$ form a non-void subset of those which are equivalent modulo $N_1(a)$ or $N_2(a)$ if $N_3(a) \subset N_1(a) \cap N_2(a)$.

For any given place $\mathfrak{p}_0$ on $\mathfrak{G}[f]$, there exist a positive integer n and a positive δ such that for each $b \neq a$ the neighborhood

(10.6.7) $\{(b, f_\alpha(z; b)) \mid b \in N(a) \cap E, f_\alpha(z; b) \sim f(z; a) \text{ modulo } N(a)\}$

contains exactly n places $\mathfrak{p}$ having first coordinate b, provided the diameter of $N(a)$ does not exceed δ. Here $n = 1$ if $f(z; a)$ is regular or polar, but $n > 1$ if $f(z; a)$ is an algebraic element. This follows from (10.6.5). If, at the outset, $N(a)$ is confined to the disk $\mid z - a \mid < R$, then $f_\alpha(z; b) \sim f(z; a)$ modulo $N(a)$ means that $f_\alpha(z; b)$ is obtained by direct rearrangement of $f(z; a)$ at $z = b$. We saw above that there exists an R_1, $R_1 \leq R$, such that for $\mid b - a \mid < R_1$ the rearrangement process leads to the n distinct elements

$$f_1(z; b), f_2(z; b), \cdots, f_n(z; b)$$

corresponding to the n determinations of $(b - a)^{1/n}$. It follows that if $N(a)$ lies in the disk $\mid z - a \mid < R_1$, then the neighborhood (10.6.7) has the desired property.

This means that if we take for $N(a)$ the disk $\mid z - a \mid < R_1$, then we have a one-to-one correspondence between the neighborhood (10.6.7) and the open unit disk of the complex t-plane, D say. If $t_1 \in D$ and if

$$(j - 1) \frac{2\pi}{n} \leq \arg t_1 < j \frac{2\pi}{n}, \quad j = 1, 2, \cdots, n,$$

then we order to $t = t_1$ the place $(b_1, f_j(z; b_1))$, where

$$b_1 = a + (R_1 t_1)^n,$$

(10.6.8)

$$f_j(z; b_1) = \sum_{k=0}^{\infty} \frac{(R_1 t_1)^{-nk}}{k!} (z - b_1)^k \sum_{m=0}^{\infty} c_m (p + m) \cdots (p + m - (k - 1)n)(R_1 t_1)^{p+m}.$$

---

[1] The *Grundzüge der Mengenlehre* (Veit & Co., Leipzig, 1914) of Felix Hausdorff (1868–1942) was epoch making and strongly influenced the development of point-set topology and abstract spaces. Hausdorff also wrote important papers on Fourier series, methods of summability, and moment problems.

The latter series is what we obtain from (10.6.6) by changing the order of summation, replacing b by b_1 and $(b - a)^{1/n}$ by $R_1 t_1$. The correspondence is not merely one-to-one, it is also continuous in both directions. Here, continuity must be understood in the sense induced by the Hausdorff topology. Thus, the mapping $t \to \mathfrak{p}$ is continuous at $t = t_1$ if, for every neighborhood $N(\mathfrak{p}_1)$ of $\mathfrak{p}_1$, there is a neighborhood $N(t_1)$ of t_1 such that $\mathfrak{p}[N(t_1)] \subset N(\mathfrak{p}_1)$. In other words, all the places $\mathfrak{p}$ which are images of points t in $N(t_1)$ belong to $N(\mathfrak{p}_1)$. In the opposite direction it is required that for any given $N(t_1)$ there is an $N(\mathfrak{p}_1)$ such that $t[N(\mathfrak{p}_1)] \subset N(t_1)$. The reader is urged to convince himself that both statements are true and that they follow from our definitions. *Such a one-to-one and bicontinuous mapping is known as a homeomorphism.* Thus, we have proved:

THEOREM 10.6.2. *Every place $\mathfrak{p}_0$ of $\mathfrak{G}[f]$ has at least one neighborhood $N(\mathfrak{p}_0)$ which is the homeomorphic image of the unit disk D: $| t | < 1$ of the complex t-plane.*

Formula (10.6.8) gives a parametric representation of the places $\mathfrak{p}$ in the neighborhood $N(\mathfrak{p}_0)$, and t is known as the *uniformizing parameter*. Sometimes it is desirable to consider more general representations where the first coordinate is also given by an infinite power series in a parameter t. It is clear that the choice of the parameter is highly arbitrary, as long as it gives the homeomorphic map of D onto a neighborhood of the place $\mathfrak{p}_0$ under consideration. If two choices of parameters, s and t say, lead to different neighborhoods $N_1(\mathfrak{p}_0)$ and $N_2(\mathfrak{p}_0)$, then there exist a neighborhood $N_3(\mathfrak{p}_0) \subset N_1(\mathfrak{p}_0) \cap N_2(\mathfrak{p}_0)$ and a parameter u which maps D onto $N_3(\mathfrak{p}_0)$. In this sense all possible parametrizations of neighborhoods of $\mathfrak{p}_0$ are equivalent.

Suppose now that we have two places $\mathfrak{p}_1$ and $\mathfrak{p}_2$ with corresponding neighborhoods $N(\mathfrak{p}_1)$ and $N(\mathfrak{p}_2)$, each of which is a homeomorphic image of D. Thus, we have two transformations h_1 and h_2 such that

$$N(\mathfrak{p}_1) = h_1[D], \quad N(\mathfrak{p}_2) = h_2[D],$$

while, conversely,

$$D = h_1^{-1}[N(\mathfrak{p}_1)], \quad D = h_2^{-1}[N(\mathfrak{p}_2)].$$

All four mappings involved are homeomorphisms. There are some interesting composite mappings. In particular, we have

$$N(\mathfrak{p}_2) = h_2\{h_1^{-1}[N(\mathfrak{p}_1)]\},$$

that is, we can map $N(\mathfrak{p}_1)$ onto $N(\mathfrak{p}_2)$ via D, and this mapping is also a homeomorphism.

Suppose next that $N(\mathfrak{p}_1)$ and $N(\mathfrak{p}_2)$ intersect and that $\mathfrak{p}_0 \in N(\mathfrak{p}_1) \cap N(\mathfrak{p}_2)$. We have then

$$\mathfrak{p}_0 = h_1(t_1) = h_2(t_2),$$

where t_1 is the point of D which maps into $\mathfrak{p}_0$ under h_1 and t_2 is the point which goes into $\mathfrak{p}_0$ under h_2. These values of t satisfy

$$t_2 = h_2^{-1}[h_1(t_1)], \quad t_1 = h_1^{-1}[h_2(t_2)].$$

We shall look a little more closely at this mapping. If the traces of $N(\mathfrak{p}_1)$ and $N(\mathfrak{p}_2)$ in the complex z-plane, that is, the sets $N(b_1)$ and $N(b_2)$ of first coordinates, are circular disks, then $N(\mathfrak{p}_1) \cap N(\mathfrak{p}_2)$ is a connected open set, in other words a domain, Π say. We set

$$D_1 = h_1^{-1}[\Pi], \quad D_2 = h_2^{-1}[\Pi].$$

Then D_1 and D_2 are two subdomains of D, and

$$D_2 = h_2^{-1}\{h_1[D_1]\}, \quad D_1 = h_1^{-1}\{h_2[D_2]\}.$$

These mappings are also homeomorphisms. But if $N(\mathfrak{p}_1)$ and $N(\mathfrak{p}_2)$ have parametric representations of the type (10.6.8), then

$$b_0 = b_1 + (R_1 t_1)^{n_1} = b_2 + (R_2 t_2)^{n_2},$$

where n_1 and n_2 are positive integers. This shows that t_2 is an analytic function of t_1 and, since the correspondence is one-to-one, we see that the mappings $h_2^{-1}h_1$ and $h_1^{-1}h_2$ are (directly) conformal. We restate this as a theorem.

THEOREM 10.6.3.   *If the homeomorphisms h_1 and h_2 map the unit disk D onto neighborhoods $N(\mathfrak{p}_1)$ and $N(\mathfrak{p}_2)$, and if $N(\mathfrak{p}_1) \cap N(\mathfrak{p}_2)$ is non-void and connected, then the composite transformation $h_2^{-1}h_1$ maps a subdomain D_1 of D conformally onto another subdomain D_2.*

One more definition, and we are at our goal!

DEFINITION 10.6.3.   *An abstract Riemann surface $\mathfrak{R}$ is a connected Hausdorff space with a set of neighborhoods $\{N\}$ and associated mappings $\{h\}$ having the following properties*:

(1) *Each N is the homeomorphic image of the unit disk $D = [t \mid \mid t \mid < 1]$ under the corresponding mapping $h = h_N$.*

(2) *If $N_1 \cap N_2$ is non-void and connected, then the composite mapping $t_2 = h_2^{-1}\{h_1(t_1)\}$ is conformal.*

We have then

THEOREM 10.6.4.   *The graph $\mathfrak{G}[f]$ is an abstract Riemann surface.*

We can say that $\mathfrak{G}[f]$ is a Riemann surface associated with the analytic function $f(z)$. It is fairly obvious that it is not the only surface for which such a claim can be made. In the above construction we have, obviously, considerable liberty in the choice of the neighborhoods and of the associated mappings. But the points themselves are not necessarily chosen as we have done. In particular, following H. Weyl[1] one could define points on the surface $\mathfrak{R}$ as

---

[1] The modern theory of Riemann surfaces goes back to the treatise *Die Idee der Riemannschen Fläche* (B. G. Teubner, Leipzig and Berlin, 1913) by Hermann Weyl (1885–1955). Weyl made profound contributions to many fields of mathematics, among them singular boundary-value problems, group representations, quantum mechanics, and relativity.

ordered pairs $(P(t), Q(t))$, where $P(t)$ and $Q(t)$ are Laurent series in t each involving only a finite number of negative powers. To the place $(a, f(z; a))$ in our notation would correspond the point

$$\left(a + t^n, \sum_{m=0}^{\infty} c_m t^{p+m} \right)$$

in the notation of formula (10.6.5). Again, the same point may have many representations, and it is necessary to have a definition of equivalence to decide when two symbols

$$(P_1(s), Q_1(s)) \quad \text{and} \quad (P_2(t), Q_2(t))$$

represent the same point on the surface.

Given an abstract Riemann surface $\Re$, an important question is whether or not it can be realized as the graph of an analytic function $f(z)$. Another question of importance is the following: Given a surface $\mathfrak{G}[f]$, what functions are single-valued on the surface? It is clear that $f(z)$ has this property, as well as any rational function of $f(z)$ or the derivative of such a function. We shall not pursue these questions further.

EXERCISE 10.6

1. Describe the places p on the Riemann surface $\mathfrak{G}[\sqrt{z}]$. Find a neighborhood of a place whose first coordinate is $+1$.

2. Describe the places of first coordinate $+1$ on the surface $\mathfrak{G}[\arcsin z]$. Find a neighborhood of one of these places.

3. Consider the function defined by (10.4.2) and find the infinitary element.

10.7. Law of permanence of functional equations. In analysis, we frequently encounter analytic functions which satisfy a *functional equation* in some domain. Usually this relationship is a differential equation, but other varieties occur also. Such a functional equation dominates the properties of the function in its whole domain of existence. We shall start with a simple case, and then we shall indicate some applications and generalizations.

Suppose we are given a function $F(z, w_1, w_2)$ of three variables, analytic in each of them. More precisely, we assume the existence of three domains D_0, D_1, D_2, such that for $z \in D_0$, $w_1 \in D_1$, $w_2 \in D_2$, the function $F(z, w_1, w_2)$ and its first-order partial derivatives exist and are continuous. Let $a \in D_0$, and suppose there is a circular disk $D: [z \mid \ |z - a| < R]$, also in D_0, and two analytic functions $f_1(z)$ and $f_2(z)$ which are holomorphic in D and there take on values confined to D_1 and D_2 respectively:

$$f_1(D) \subset D_1, \quad f_2(D) \subset D_2.$$

Then

$$F(z, f_1(z), f_2(z))$$

is a holomorphic function of z in D, for this function is well defined there, and it has a unique derivative since $F(z, w_1, w_2)$ has partial derivatives and $f_1(z)$ and $f_2(z)$ are differentiable functions. We assume further that

(10.7.1) $$F(z, f_1(z), f_2(z)) \equiv 0, \quad z \in D.$$

We have then the following:

THEOREM 10.7.1. *Suppose that $f_1(z)$ and $f_2(z)$ can be continued analytically along a rectifiable path $z = p(t)$, $0 \leq t \leq 1$, leading from $z = a$ to $z = b$. Suppose that for each such t there is a disk $D(t)$: $[z \mid \mid z - p(t) \mid < R(t)]$ with the properties (i) $D(t) \subset D_0$, (ii) $f_j(z; p(t))$ is holomorphic in $D(t)$, $j = 1, 2$, and (iii) for z restricted to $D(t)$ the values of $f_j(z; p(t))$ belong to D_j, $j = 1, 2$. Then*

(10.7.2) $$F(z, f_1(z; p(t)), f_2(z; p(t))) \equiv 0, \quad z \in D(t), 0 \leq t \leq 1.$$

Proof. We can find a finite sequence of t-values

$$0 = t_0 < t_1 < t_2 < \cdots < t_n = 1$$

such that

$$p(t) \in D(t_{k-1}), \quad t_{k-1} \leq t \leq t_k, k = 1, 2, \cdots, n.$$

This implies that $f_j(z; p(t))$ is a direct rearrangement of $f_j(z; p(t_{k-1}))$ for such values of t. Now, the assumptions imply that

$$F(z, f_1(z; p(t)), f_2(z; p(t)))$$

is a holomorphic function of z in $D(t)$. For $0 \leq t \leq t_1$ this function coincides in $D(t_0) \cap D(t) = D \cap D(t)$ with

$$F(z, f_1(z; p(t_0)), f_2(z; p(t_0))) = F(z, f_1(z), f_2(z)) \equiv 0.$$

It must then be identically zero also in all of $D(t)$. Thus (10.7.2) holds for $0 \leq t \leq t_1$. We can then proceed step by step from one interval to the next, and we conclude that (10.7.2) holds throughout.

The relation (10.7.1) is a functional equation satisfied by the two functions $f_1(z)$ and $f_2(z)$. Theorem 10.7.1 states that *the analytic continuations of the solutions of the functional equation are solutions of the analytic continuation of the equation.* This principle is known as the *law of permanence of functional equations.* The same argument clearly applies to the simpler equation

(10.7.3) $$F(z, w) = 0,$$

as well as to equations of the form

(10.7.4) $$F(z, w_1, w_2, \cdots, w_n) = 0.$$

It also applies to systems of equations. We shall give several applications of this principle. We start with:

The derivative of the analytic continuation of a function equals the analytic continuation of the derivative of the function.

This is a special case of Theorem 10.7.1 with

$$F(z, w_1, w_2) = w_2 - w_1, \quad f_1(z) = f(z), \quad f_2(z) = g'(z),$$

D_0, D_1, D_2 each being the finite plane. The observation of course extends to higher-order derivatives. This fact is basic in the theory of differential equations. Given a first-order equation

$$(10.7.5) \qquad F(z, w, w') = 0,$$

where the function $F(z, w_1, w_2)$ satisfies the assumptions made above. Suppose that we have found a function $f(z)$, holomorphic in a disk $D \subset D_0$, where its values belong to the domain D_1, whereas those of its derivative belong to D_2, and suppose that $w = f(z)$ is a solution of (10.7.5) in D. Then by Theorem 10.7.1 any analytic continuation of $f(z)$ by elements $f(z; p(t))$ such that

$$(10.7.6) \qquad F(z, f(z; p(t)), f'(z; p(t)))$$

remains holomorphic in a neighborhood of each point $z = p(t)$ of the path, will also be a solution of (10.7.5) for z in that neighborhood. This evidently extends immediately to equations of higher order and to systems of differential equations.

There is one case in which the requirement that (10.7.6) or a similar expression shall remain holomorphic when $f(z)$ is continued, is automatically satisfied. Suppose that $F(z, w_1, w_2, \cdots, w_{n+1})$ is a polynomial in $w_1, w_2, \cdots, w_{n+1}$ with coefficients which are entire functions of z. Then the differential equation

$$(10.7.7) \qquad F(z, w, w', \cdots, w^{(n)}) = 0$$

has the property that any regular element of the analytic function determined by a local solution is also a solution. For if we substitute a holomorphic function of z for w in the left member of (10.7.7), the result is always a holomorphic function. In the case of a *linear differential equation* a sharper statement can be made. Here the singularities of the solutions are restricted to the zeros of the coefficient of the highest-order derivative, so that any point a which does not belong to the set of zeros is the center of regular elements only. Thus, *any solution can be continued along any path that avoids the fixed singularities.*

Our considerations apply also to difference equations. For the sake of simplicity we take

$$(10.7.8) \quad p_0(z)\, w(z + n) + p_1(z)\, w(z + n - 1) + \cdots + p_n(z)\, w(z) = 0,$$

where the coefficients $p_k(z)$ are polynomials in z. This is a linear difference equation of order n. We have

$$F(z, w_1, \cdots, w_{n+1}) = p_0(z)\, w_{n+1} + p_1(z)\, w_n + \cdots + p_n(z)\, w_1,$$

and we conclude that any continuation of a solution remains a solution. In this case we can actually use the equation to obtain the analytic continuation. Compare the discussion of the Gamma function in Section 8.8. The structure of the equation is such that it is natural to assume the existence of a solution which is holomorphic in a right or a left half-plane. Take the first alternative, and assume that $w(z)$ is holomorphic for $\Re(z) > b$ and that $p_n(z) \neq 0$ in this half-plane. Then for $\Re(z) > b$

$$w(z) = - \frac{1}{p_n(z)} [p_{n-1}(z)\, w(z+1) + \cdots + p_0(z)\, w(z+n)].$$

But here the expression in brackets is well defined for $\Re(z) > b - 1$ and can be used to define $w(z)$ throughout the larger half-plane. This argument can be repeated, and we find that $w(z)$ can be extended to the whole plane as a meromorphic function of z having poles at the points

$$\alpha_j - m, \quad j = 1, 2, \cdots, k, \quad m = 0, 1, 2, \cdots,$$

where $p_n(\alpha_j) = 0$. In a similar manner we can extend a solution which is holomorphic in a left half-plane $\Re(z) < a$ in which $p_0(z - n) \neq 0$. The extension is meromorphic and has poles at the points

$$\beta_j + n + m, \quad j = 1, 2, \cdots, l, \quad m = 0, 1, 2, \cdots,$$

where $p_0(\beta_j) = 0$.

There are also important applications of the law of permanence to the theory of implicit functions. In Theorem 9.4.4 we proved the existence of a function $f(z)$, holomorphic in a disk $|z| < \rho$, such that

$$F(z, f(z)) \equiv 0, \quad f(0) = 0.$$

Here $F(z, w)$ is holomorphic in the di-cylinder $|z| < R_1, |w| < R_2$, and $F(0, 0) = 0$, while $F_w(0, 0) \neq 0$. We got certain estimates for ρ in Theorems 9.4.5 and 9.4.6, but in general $f(z)$ will also exist outside of the circle $|z| = \rho$. Suppose that it is possible to continue the function $f(z)$ analytically along a path $z = p(t)$ by elements $f(z; p(t))$. As long as $|p(t)| < R_1$ and $|f(z; p(t))| < R_2$, we have

$$F(z, f(z; p(t))) \equiv 0$$

by the previous argument. But if the path gets out of the disk $|z| < R_1$, then it does not necessarily follow that $F(z, f(z; p(t))$ has a meaning. This is illustrated by the following example: Take

$$F(z, w) = (w - z)(w - g(z)),$$

where $g(z)$ is a given function, holomorphic in $|z| < 1$ but having $|z| = 1$ as its natural boundary. If $g(0) \neq 0$, the initial condition $f(0) = 0$ selects $f(z) = z$. This function exists in the whole plane, whereas $F(z, w)$ exists only for $|z| < 1$.

The situation is simpler and more favorable in the case of inverse functions. Here we have

THEOREM 10.7.2.   *Let $G(w)$ be an analytic function. Then there exists an analytic function $f(z)$ with the following properties: For every point w_0 where there exists a regular element of $G(w)$, $G(w; w_0)$ say, such that $G(w_0; w_0) = z_0$ and $G'(w_0; w_0) \neq 0$, there exists a regular element of $f(z), f(z; z_0)$ say, such that $f(z_0; z_0) = w_0, f'(z_0; z_0) \neq 0$, and*

$$(10.7.9) \qquad\qquad z \equiv G(f(z; z_0); w_0).$$

Conversely, for every point z_0 where there exists a regular element $f(z; z_0)$ of $f(z)$ such that $f(z_0; z_0) = w_0$ and $f'(z_0; z_0) \neq 0$, there exists a regular element $G(w; w_0)$ of $G(w)$ such that $G(w_0; w_0) = z_0$, $G'(w_0; w_0) \neq 0$, and

$$(10.7.10) \qquad\qquad w \equiv f(G(w; w_0); z_0).$$

Proof.   The existence of functions $f(z; z_0)$ and $G(w; w_0)$ having the stated local properties is a consequence of Theorem 4.5.1. What should be proved, however, is that the functions $f(z; z_0)$ are all regular elements of the same analytic function $f(z)$ and that a similar statement holds for the functions $G(w; w_0)$ of the second half of the theorem. To see this, we argue as follows: We start with a regular element $G(w; w_0)$ of $G(w)$ such that $G(w_0; w_0) = z_0$ and $G'(w_0; w_0) \neq 0$. We have then a corresponding function $f(z; z_0)$ satisfying (10.7.9) such that $f(z_0; z_0) = w_0$ and $f'(z_0; z_0) \neq 0$. Let us now continue $G(w; w_0)$ along an arc $C: w = p(t), 0 \leq t \leq 1, p(0) = w_0$, such that at every point of C the continuation $G(w; p(t))$ is holomorphic and $G'(p(t); p(t)) \neq 0$. The second condition can always be satisfied: if we can continue $G(w)$ analytically, we can avoid the isolated points where its derivative vanishes. At each point of C we have then a corresponding uniquely defined inverse function, $f(z; z(t))$ say, such that $f(z(t); z(t)) = p(t), f'(z(t), z(t)) \neq 0$, and

$$(10.7.11) \qquad\qquad z \equiv G(f(z; z(t)); p(t)), \quad z(t) = G(p(t); p(t)).$$

For $t = 0$ we have $f(z; z(0)) = f(z; z_0)$. Since $G(w; w_0)$ is holomorphic in some neighborhood of $w = w_0$, we conclude from the analogue of Theorem 10.7.1 for equations of the form $F(z, w) = 0$ that the following must be true: There exists an interval $[0, t_1]$ such that the analytic continuation of $f(z; z_0)$ along the arc $z = z(t), 0 \leq t \leq t_1$, will satisfy

$$z \equiv G(f^*(z; z(t)); p(t)),$$

where we have denoted the analytic continuation of $f(z; z_0)$ by $f^*(z; z(t))$ for the time being. In particular, this holds at $z = z(t)$. But $G(w; p(t))$ takes on the value $z(t)$ at $w = p(t)$, and $G(w; p(t)) \neq z(t)$ in some neighborhood of $p(t)$ not including $p(t)$. If t_1 is sufficiently small, we must then have $f^*(p(t); p(t)) = z(t)$.

But the inverse function, if it exists, is unique. It follows that

(10.7.12) $f^*(z; z(t)) \equiv f(z; z(t)), \quad 0 \leq t \leq t_1.$

We can then apply the same argument at the point $t = t_1$ and see that there must exist a larger t-interval in which (10.7.12) holds. It is then not difficult to see that (10.7.12) must hold everywhere in $[0, 1]$. Since we can reach every regular element of $G(w)$, where $G'(w) \neq 0$, by analytic continuation from $w = w_0$, starting with the element $G(w; w_0)$, we conclude that the local inverse, which exists everywhere, is actually a regular element of one and the same analytic function $f(z)$, the inverse of $G(w)$. Thus, *the inverse of the continuation is the continuation of the inverse.* Interchanging the roles of z and w in this argument, we arrive at a proof of the second part of the theorem.

EXERCISE 10.7

1. Show that if $f(z)$ is a meromorphic solution of (10.7.8), and if $\omega(z)$ is an entire function of period 1, then $f(z)\omega(z)$ is also a solution. Suppose that $f(z)$ had no poles in a right half-plane, how could you choose $\omega(z)$ so that the resulting solution would be an entire function?

2. Find a solution of $f(z + 1) = zf(z)$ which is an entire function.

3. A function $f(z)$ is holomorphic in a neighborhood of the origin and satisfies the functional equation

$$f(2z) = 2f(z)f'(z).$$

Show that $f(z)$ can be extended to the finite plane as an entire function. Give an example of such a function.

4. A function $g(z)$ is holomorphic in a neighborhood of the origin and satisfies the functional equation

$$g(2z) = \frac{2g(z)}{1 - g^2(z)}.$$

Show that $g(z)$ can be extended to the finite plane as a meromorphic function. Give an example of such a function.

5. Given an integral equation of the Volterra type

$$f(z) = g(z) + \int_0^z K(z - w)f(w)\, dw,$$

where $g(z)$ and $K(z)$ are given entire functions, and given a solution $f(z)$ holomorphic in some neighborhood of the origin, show that $f(z)$ can be extended to the finite plane as an entire function.

6. Are there any regular elements of arc sin z which do not have a regular element as an inverse? Same question for sin z.

COLLATERAL READING

As general references see

AHLFORS, L. V. *Complex Analysis. An Introduction to the Theory of Analytic Functions of One Complex Variable*, Chap. 6. McGraw-Hill Book Company, Inc., New York, 1953.

BEHNKE, H., and SOMMER, F. *Theorie der analytischen Funktionen einer komplexen Veränderlichen.* Springer-Verlag, Berlin, 1955.

SAKS, S., and ZYGMUND, A. *Analytic Functions*, Chap. 6. Monografie Matematyczne. Vol. 28, Warsaw and Wroclaw, 1952.

For Section 10.5 see

BOREL, ÉMILE. *Leçons sur les Fonctions Monogènes Uniformes d'une Variable Complexe.* Gauthier-Villars, Paris, 1917.

For Section 10.6 consult also

AHLFORS, L. V., and SARIO, L. *Riemann Surfaces.* Princeton University Press, Princeton, New Jersey, 1960.

PFLUGER, A. *Theorie der Riemannschen Flächen.* Springer-Verlag, Berlin, 1957.

SPRINGER, G. *Introduction to Riemann Surfaces.* Addison-Wesley Publishing Company, Inc., Reading, Massachusetts, 1957.

WEYL, H. *Die Idee der Riemannschen Fläche*, Third Edition. B. G. Teubner, Stuttgart, 1955.

SINGULARITIES AND REPRESENTATION
OF ANALYTIC FUNCTIONS

11.1. Holomorphy-preserving transformations: I. Integral operators. In the present chapter we shall be concerned with various investigations dealing with the nature and the distribution of the singularities of an analytic function. If the latter is defined by a regular element at the origin,

$$(11.1.1) \qquad f(z; 0) \equiv \sum_{n=0}^{\infty} c_n z^n, \quad |z| < R,$$

then all the properties of $f(z)$ are determined somehow by properties of the infinite sequence $\{c_n\}$. In particular, the singularities of $f(z)$ are determined by $\{c_n\}$, and it becomes a fascinating problem to try to discern how properties of the sequence are reflected in properties of the singularities. There are obviously close relations between this problem and that of analytic continuation of the element in question; and these aspects will be a recurrent theme in this chapter.

Since continuation along a ray plays an important part in this work, we shall be concerned with regions starlike with respect to the origin which are associated with the given regular element $f(z; 0)$. The most important of these is the *principal* or *Mittag-Leffler star*, $\mathbf{A} = \mathbf{A}[f]$. This is the set of all points a such that the function $f(z)$ defined by $f(z; 0)$ can be continued analytically along the line segment $[0, a]$. Every finite boundary point of $\mathbf{A}[f]$ is a singular point of $f(z)$ if accessible by radial approach. Such a point is usually called a *vertex* of the star.

We shall start the investigation by discussing some operations, applicable to a function $f(z)$ holomorphic in a domain D, such that the result is also a function holomorphic in D. The reader will be familiar with three such operations: (i) multiplication by a fixed function holomorphic in the given domain D, (ii) differentiation with respect to z, and (iii) integration with respect to z. In the last case we must assume that D is simply-connected, since otherwise the result is not necessarily single-valued as shown by the function $1/z$.

We have no comments on case (i). Case (ii) will be elaborated in Section 11.2, and various generalizations of case (iii) will be considered in the present section. We start with the idea of a *moment sequence*. Consider an interval $[a, b]$ and let $q(u) \in BV[a, b]$, that is, $q(u)$ is of bounded variation in $[a, b]$. If $b - a < \infty$, the integrals

$$(11.1.2) \qquad \int_a^b u^n \, dq(u) \equiv \mu_n, \quad n = 0, 1, 2, \cdots,$$

exist, and $\{\mu_n\}$ is called a moment sequence corresponding to the interval $[a, b]$. If the interval is infinite, the existence of the *moments* μ_n imposes a severe restriction on $q(u)$, and we shall not have occasion to consider this case here.

We specialize and take $[a, b] = [0, 1]$, so that

$$(11.1.3) \qquad\qquad \mu_n = \int_0^1 u^n \, dq(u).$$

We suppose that $f(z)$ is holomorphic in some neighborhood of $z = 0$ and form the transform

$$(11.1.4) \qquad\qquad Q[f] \equiv \int_0^1 f(zu) \, dq(u).$$

Here $f(zu)$ is a continuous function of u for any fixed $z \in \mathbf{A}[f]$. Thus, the integral exists for $z \in \mathbf{A}[f]$, and $Q[f](z)$ is a holomorphic function of z in $\mathbf{A}[f]$, since we can obviously differentiate under the sign of integration, obtaining

$$\frac{d}{dz} Q[f](z) = \int_0^1 uf'(zu) \, dq(u).$$

Further we have

$$(11.1.5) \qquad\qquad Q[f](z) = \sum_{n=0}^{\infty} \mu_n c_n z^n, \quad |z| < R,$$

if the regular element of $f(z)$ at the origin is the series (11.1.1).

DEFINITION 11.1.1. *A sequence* $\{\alpha_n \mid n = 0, 1, 2, \cdots\}$ *is a holomorphy-preserving factor sequence if, for every function $f(z)$ defined by an element*

$$f(z; 0) = \sum_{n=0}^{\infty} c_n z^n,$$

the principal star of the transform $T[f]$ of $f(z)$ defined by the element

$$(11.1.6) \qquad\qquad \sum_{n=0}^{\infty} \alpha_n c_n z^n$$

contains the principal star of $f(z)$:

$$(11.1.7) \qquad\qquad \mathbf{A}\{T[f]\} \supset \mathbf{A}[f].$$

Thus we see that *a moment sequence of the type* (11.1.3) *is a holomorphy-preserving factor sequence.* For the following theorem consult Section 4.7 and Theorem 7.10.2 of Volume I.

THEOREM 11.1.1. *Let $HB[D]$ be the Banach space of all functions holomorphic and bounded in a domain D which is starlike with respect to $z = 0$. Then formula* (11.1.4) *defines a linear bounded transformation on $HB[D]$ into itself, and the norm of the operator Q does not exceed V, the total variation of $q(u)$ in $[0, 1]$:*

$$(11.1.8) \qquad\qquad \| Q \| \leq V.$$

Proof. $Q[f]$ is well defined for any $f \in HB[D]$ and is a holomorphic function of z in D. Here it is essential that D is starlike. Further,

$$| Q[f](z) | \leq \int_0^1 | f(zu) | \, | dq(u) | \leq \| f \| \int_0^1 | dq(u) | = V \| f \|$$

by the properties of the sup-norm and of the Stieltjes integral. For the latter, see Section C.3 of Volume I. Thus $Q[f] \in HB[D]$ and

(11.1.9) $$\| Q[f] \| \leq V \| f \|.$$

But this is what we mean by saying that Q is a bounded transformation on $HB[D]$ to itself. The norm of the operator Q is by definition

$$\| Q \| = \sup \, [\| Q[f] \| \, | \, \| f \| = 1],$$

and formula (11.1.9) then implies (11.1.8). Actually, this is the best possible estimate of $\| Q \|$, for, if $q(u)$ is monotone increasing and $f(z) \equiv 1$, we have

$$Q[1] = q(1) - q(0) = V.$$

Finally, we recall that a transformation T is said to be *linear* if for all constants γ_1, γ_2 and all elements f_1 and f_2 of the domain of T we have

$$T[\gamma_1 f_1 + \gamma_2 f_2] = \gamma_1 T[f_1] + \gamma_2 T[f_2].$$

Since this relation is obviously satisfied by Q, the theorem is proved.

We define the powers of Q as follows:

(11.1.10) $$Q^0 = I, Q^1 = Q, Q^n[f] = Q\{Q^{n-1}[f]\}, \quad n = 2, 3, 4, \cdots,$$

where I is the *identical transformation*, that is, $I[f] = f$. Since Q is a bounded transformation on $HB[D]$ into itself, all these transformations exist and are linear and bounded. We have

(11.1.11) $$\| Q^n \| \leq \| Q \|^n \leq V^n,$$

and for $q(u)$ monotone increasing we have equality in both places. Having defined arbitrary powers Q^n, we can also define polynomials in Q as well as power series. Thus if

$$G_n(\lambda) \equiv \sum_{p=0}^n g_p \lambda^p$$

is a polynomial in the complex variable λ, we set

$$G_n(Q)[f](z) = g_0 f(z) + g_1 Q[f](z) + \cdots + g_n Q^n[f](z),$$

and this is a linear bounded transformation on $HB[D]$ into itself. Under certain circumstances we may let $n \to \infty$. In this way we obtain the following theorem, the proof of which is left to the reader:

THEOREM 11.1.2.   *Let*

$$(11.1.12) \qquad\qquad G(\lambda) = \sum_{p=0}^{\infty} g_p \lambda^p$$

have a positive radius of convergence ρ. *Suppose that* $V < \rho$. *Then*

$$(11.1.13) \qquad G(Q)[f](z) \equiv g_0\, f(z) + \sum_{p=1}^{\infty} g_p\, Q^p[f](z)$$

is a linear bounded transformation on $HB[D]$ *into itself.   Further,* $G(Q)[f](z)$ *is holomorphic in* $\mathbf{A}[f]$, *and the sequence* $\{G(\mu_n)\}$ *is a holomorphy-preserving factor sequence.*

We note that

$$(11.1.14) \qquad\qquad \| \, G(Q) \, \| \leq \sum_{p=0}^{\infty} | \, g_p \, | \, V^p,$$

and that this is the best possible estimate if $q(u)$ is increasing and the coefficients g_p are non-negative.

At this juncture we make two observations which have a bearing on our problem.   The beginning of a power series is of no importance for the distribution in the finite plane of the singularities of the function defined by the series, and it may be omitted if desired.   Likewise, the beginning of a factor sequence is immaterial for its holomorphy-preserving character and, thus, may be changed ad lib.   These considerations lead to

THEOREM 11.1.3.   *If* ρ *is the radius of convergence of* (11.1.12), *if*

$$| \, q(1) - q(1 - 0) \, | < \rho,$$

and if D *is bounded, then there exists an integer* m *with the following properties:* (i) $G(\mu_n)$ *is defined for* $n \geq m$, (ii) *the sequence* $\{0, 0, \cdots, G(\mu_m), G(\mu_{m+1}), \cdots\}$ *is a holomorphy-preserving factor sequence, and* (iii) *the operator* $G(Q)$ *is bounded when restricted to that subset of* $HB[D]$ *the elements of which satisfy the conditions*

$$(11.1.15) \qquad f(0) = f'(0) = \cdots = f^{(m-1)}(0) = 0.$$

Proof.   By assumption we can find a value δ, $0 < \delta < 1$, such that $V_\delta^1[q] \leq \rho_1 < \rho$, where $V_\delta^1[q]$ is the total variation of $q(u)$ on $[\delta, 1]$.   Thus

$$| \, \mu_n \, | \leq \int_0^1 u^n \, | \, dq(u) \, | = \left\{ \int_0^\delta + \int_\delta^1 \right\} u^n \, | \, dq(u) \, |$$

$$\leq \delta^n V + \rho_1.$$

This expression is clearly $< \rho$ for all large values of n.   Let m be the least integer n such that $\int_0^1 u^n \, | \, dq(u) \, | < \rho$ and set

$$\int_0^1 u^m \, | \, dq(u) \, | = \rho_0.$$

Then $G(\mu_n)$ is defined for $n \geq m$. With this value of m, suppose that $f(z)$ satisfies conditions (11.1.15). We have then $f(z) = z^m h_0(z)$, where $h_0(z)$ is holomorphic in D. Actually $h_0(z) \in HB[D]$ and, if d is the distance of ∂D from the origin, then

$$\| h_0 \| \leq d^{-m} \| f \|.$$

We have now

$$Q[f](z) = z^m \int_0^1 h_0(zu) u^m \, dq(u) \equiv z^m h_1(z),$$

where $h_1(z) \in HB[D]$ and

$$\| h_1 \| \leq \rho_0 \| h_0 \| \leq \rho_0 d^{-m} \| f \|.$$

By complete induction one proves that

$$Q^p[f](z) = z^m h_p(z), \quad h_p(z) \in HB[D], \quad \| h_p \| \leq \rho_0{}^p d^{-m} \| f \|.$$

Hence

$$G(Q)[f](z) = z^m \sum_{p=0}^{\infty} g_p h_p(z) \equiv z^m H(z),$$

where

$$| H(z) | \leq \| f \| d^{-m} \sum_{p=0}^{\infty} | g_p | \rho_0{}^p, \quad z \in D.$$

It follows that $H(z) \in HB[D]$ and, since D is bounded, that $G(Q)[f] \in HB[D]$. Let us denote by $HB_m[D]$ the subset of $HB[D]$ the elements of which satisfy (11.1.15). We now see that $G(Q)$ is a linear bounded transformation on $HB_m[D]$ into itself. By the usual argument we show that $G(Q)[f](z)$ is holomorphic in $\mathbf{A}[f]$. Finally, we have

(11.1.16) $$G(Q)[f](z) = \sum_{n=m}^{\infty} G(\mu_n) c_n z^n, \quad | z | < R.$$

This completes the proof.

COROLLARY. *The function $g(z)$ defined by the series*

(11.1.17) $$\sum_{n=m}^{\infty} G(\mu_n) z^n$$

is holomorphic in the domain obtained by omitting the line segment $[+1, +\infty)$ from the complex plane.

For the series (11.1.17) is the $G(Q)$-transform of the function $z^m(1 - z)^{-1}$ and as such $g(z)$ is holomorphic in the principal star of the original function. We denote this star simply by $\mathbf{A}$.

To illustrate the Corollary, we start with the case

$$q(u) = \frac{1}{\Gamma(\alpha)} \int_0^u \left(\log \frac{1}{s}\right)^{\alpha-1} ds, \quad \Re(\alpha) > 0.$$

A simple calculation shows that the corresponding moments are

$$\mu_n = (n + 1)^{-\alpha}.$$

Since $q(u)$ is continuous at $u = 1$, no condition need be imposed on the radius of convergence of $G(\lambda)$ beyond its being positive. Thus, if $G(\lambda)$ is holomorphic at the origin,

$$(11.1.18) \qquad \sum_{n=m}^{\infty} G((n + 1)^{-\alpha}) z^n$$

defines a function holomorphic in **A**. For $\alpha = 1$, this is known as the theorem of Leau (Léopold Leau, 1868–1943).

Actually the line segment $[1, \infty)$ is not a singular line for (11.1.18); only the endpoints are necessarily singular. On the other hand, if we cross the segment and then return to the origin, the latter is ordinarily found to be a singular point. This happens even in the simplest cases. Thus, if

$$\alpha = 1, \quad G(\lambda) = \lambda^2, \quad g(z) = \sum_{n=0}^{\infty} \frac{z^n}{(n + 1)^2},$$

we find that

$$\frac{d}{dz}[z\, g(z)] = \frac{1}{z} \log \frac{1}{1 - z},$$

and here the right-hand side has a singular point at $z = 0$ for every determination of the logarithm except for the principal branch. We shall prove the following theorem, which is a special case of results due to Alexander Ostrowski (1933).

THEOREM 11.1.4. *The function $g(z)$ defined by (11.1.18) can be continued analytically along any path C such that* (i) *C starts at $z = 0$ but does not return to the origin,* (ii) *C is bounded,* (iii) *C does not pass through $z = 1$, and* (iv) *C does not separate $z = 1$ from $z = \infty$.*

REMARK. Condition (iv) holds, in particular, if C does not intersect itself. It may be omitted if α is an integer.

Proof. By definition

$$(11.1.19) \qquad Q\!\left\{\frac{z^m}{1 - z}\right\} = \frac{z^m}{\Gamma(\alpha)} \int_0^1 \left(\log \frac{1}{u}\right)^{\alpha-1} \frac{u^m\, du}{1 - zu} = \sum_{n=m}^{\infty} \frac{z^n}{(n + 1)^\alpha},$$

where the series converges for $|z| < 1$. It follows that the effect of replacing Q by Q^p is to replace α by αp in the second and third members of (11.1.19). Hence

$$(11.1.20) \qquad G(Q)\!\left\{\frac{z^m}{1 - z}\right\} = g_0 \frac{z^m}{1 - z} + z^m \int_0^1 G_\alpha(u) \frac{u^m\, du}{1 - zu},$$

where

$$(11.1.21) \qquad G_\alpha(u) = \sum_{p=1}^{\infty} \frac{g_p}{\Gamma(\alpha p)} \left(\log \frac{1}{u}\right)^{\alpha p - 1}.$$

This kernel is an entire function of $\left(\log \dfrac{1}{u}\right)^\alpha$, save for the factor $\left(\log \dfrac{1}{u}\right)^{-1}$, and m

is so chosen that u^m times the sum of the absolute values of the terms of (11.1.21) is integrable over (0, 1).

So far z has been restricted to the star **A** and the integral has been taken along the real axis. Since $G_\alpha(u)$ is an analytic function of u, we may, however, deform the path of integration, using Cauchy's theorem, in order to allow z to cross the cut from $+1$ to $+\infty$. Let C be a path in the z-plane joining $z = 0$ with $z = a$ and satisfying the conditions of Theorem 11.1.4. Let C^* be the curve described by $1/z$ when z describes C. Then C^* starts at $z = \infty$ but does not return there, is bounded away from $z = 0$ and $z = 1$, and does not separate these points. We can then find a path Γ, joining $u = 0$ with $u = 1$, such that Γ coincides with the real axis on two short intervals $(0, \delta)$ and $(1 - \delta, 1)$, and Γ does not intersect C^*. It follows that

$$g_0 \frac{z^m}{1 - z} + z^m \int_\Gamma G_\alpha(u) \frac{u^m\, du}{1 - zu}$$

exists and defines a holomorphic function of z along the curve C. Since this function coincides with $g(z)$ for small values of z, it must give the analytic continuation of $g(z)$ along C. This completes the proof.

An analogue of Theorem 11.1.4 holds for analytic continuation of $G(Q)[f]$ for arbitrary functions $f(z)$. See Problem 3 of Exercise 11.1. Theorem 11.1.4 holds for more general moment sequences than $\{(n + 1)^{-\alpha}\}$ since the proof depends essentially only upon the fact that the corresponding mass function $q(u)$ is analytic save for singular points at $u = 0, 1, \infty$. It should be observed, however, that the theorem is not true for arbitrary functions $q(u) \in BV[0, 1]$. A counterexample is found in Problem 2 below.

As a transition to the subject matter of the next section we consider briefly an integral transform of a different nature. The following theorem is due to A. Ostrowski (1923), who obtained his result as a generalization of results due to H. Cramér (1918). For the latter see Problem 5 below.

THEOREM 11.1.5.   *Let*

$$(11.1.22) \qquad g(z) = \sum_{k=0}^{\infty} g_k\, z^{-k-1}$$

be an entire function of z^{-1}. Suppose that $f(z)$ is holomorphic in a domain D and let Δ be a compact subset of D whose distance from ∂D equals 2ρ. Form

$$(11.1.23) \qquad f_g(z) \equiv \frac{1}{2\pi i} \int_{|u| = \rho} f(z + u)\, g(u)\, du, \quad z \in \Delta.$$

Then $f_g(z)$ is holomorphic in Δ and may be represented there by the absolutely and uniformly convergent series

$$(11.1.24) \qquad f_g(z) = \sum_{p=0}^{\infty} \frac{g_p}{p!} f^{(p)}(z).$$

Proof.   For $z \in \Delta$, $|u| = \rho$, we have

$$f(z + u)\, g(u) = \sum_{n=0}^{\infty} \frac{u^n}{n!} f^{(n)}(z) \sum_{k=0}^{\infty} g_k u^{-k-1},$$

where the two series in the product are absolutely and uniformly convergent with respect to z and u. We can then multiply out and integrate term by term. The result is the absolutely and uniformly convergent series (11.1.24), whose sum then must be a holomorphic function. This completes the proof.

EXERCISE 11.1

1. Prove: A necessary condition that $\{\alpha_n\}$ be a holomorphy-preserving factor sequence is that the function defined by the series

$$\sum_{n=0}^{\infty} \alpha_n z^n$$

be holomorphic in the star **A** ($=$ the star of the geometric series).

2. Let $q(u) = 0$ or 1 according as $0 \leq u < \alpha$ or $\alpha \leq u \leq 1$, and define $G(Q)[f](z)$ by (11.1.13). What form does the series take in the present case? Choose $f(z) = z^m(1 - z)^{-1}$ and determine the nature and location of the singularities of $G(Q)[f]$. Compute $G(\mu_n)$.

3. Prove the following generalization of Theorem 11.1.4: The function defined by the element

$$\sum_{n=m}^{\infty} G((n + 1)^{-\alpha})\, c_n z^n$$

of $G(Q)[f](z)$ can be continued analytically along any path C such that (i) C starts at $z = 0$ but does not return to the origin, (ii) C is bounded, (iii) $f(z)$ can be continued along C, and (iv) C does not intersect itself.

*4. Show that $\{[\log (n + 2)]^{-\alpha}\}$, $\alpha > 0$, is a moment sequence and that Theorem 11.1.4 holds for the series

$$\sum_{n=m}^{\infty} G([\log (n + 2)]^{-\alpha})\, z^n.$$

(For the first and basic part of the problem, see F. Hausdorff, "Summationsmethoden und Momentfolgen. I," *Mathematische Zeitschrift*, Vol. 9 (1921), pp. 74–109.)

5. Prove the following special case of Cramér's theorem: Let $g(z)$ be defined by (11.1.22) and be an entire function of $1/z$. Let

$$G(z) = \sum_{k=0}^{\infty} \frac{g_k}{k!} z^k.$$

Let $0 < \lambda_n < \lambda_{n+1}$, $\lambda_n \to \infty$, and let the Dirichlet series

$$f(z) = \sum_{n=1}^{\infty} a_n e^{-\lambda_n z}$$

be uniformly convergent for $\Re(z) \geq \delta$. Form the function $f_g(z)$ of Theorem 11.1.5. Then for $\Re(z) > \delta$ we have

$$f_g(z) = \sum_{n=1}^{\infty} a_n G(-\lambda_n) e^{-\lambda_n z}.$$

11.2. Holomorphy-preserving transformations: II. Differential operators.

At the end of the preceding section we encountered a holomorphy-preserving differential operator of infinite order, namely,

$$(11.2.1) \qquad G\left(\frac{d}{dz}\right)[f] = \sum_{k=0}^{\infty} \frac{g_k}{k!} f^{(k)}(z),$$

where

$$(11.2.2) \qquad G(w) = \sum_{k=0}^{\infty} \frac{g_k}{k!} w^k$$

is an entire function such that

$$(11.2.3) \qquad \lim_{k \to \infty} |g_k|^{\frac{1}{k}} = 0.$$

In Chapter 14 we shall learn to describe such a function as being at most of *order one and minimal type*. For the time being, (11.2.3) serves as definition of the class under consideration. We note that if (11.2.3) holds, then for any given $\varepsilon > 0$ there exists an $M(\varepsilon)$ such that

$$(11.2.4) \qquad |g_k| \leq M(\varepsilon)\varepsilon^k, \quad k = 0, 1, 2, \cdots,$$

and

$$(11.2.5) \qquad |G(w)| \leq M(\varepsilon) e^{\varepsilon|w|}.$$

It is clear that (11.2.4) implies (11.2.5). Conversely, using Theorem 8.2.1 with $a = 0$, $r = k/\varepsilon$, we see that (11.2.5) implies the validity of (11.2.3) and, hence, of (11.2.4). Thus, if either (11.2.4) or (11.2.5) holds for every $\varepsilon > 0$, then (11.2.3) also holds.

The operator d/dz is the natural one to consider in connection with Dirichlet series since

$$(11.2.6) \qquad \frac{d}{dz} e^{az} = \alpha e^{az}.$$

With a terminology inspired by operator theory, we may say that e^{az} is a *characteristic function* of the operator d/dz.

In the theory of power series another differential operator plays the leading role. Setting

(11.2.7)
$$\vartheta \equiv z \frac{d}{dz},$$

we note that

(11.2.8)
$$\vartheta z^{\alpha} = \alpha z^{\alpha},$$

that is, every power of z is a characteristic function of the operator ϑ. We shall study this operator in some detail. We start with the usual definition of powers of the operator. We set

(11.2.9) $\vartheta^0[f] = f, \quad \vartheta^1[f] = \vartheta[f], \quad \vartheta^n[f] = \vartheta\{\vartheta^{n-1}[f]\}, \quad n = 2, 3, 4, \cdots.$

We need information concerning the functions $\vartheta^n[(a - z)^{-1}]$.

LEMMA 11.2.1.   *The equations*

$$\vartheta^n \frac{1}{a - z} = \frac{P_n(z, a)}{(a - z)^{n+1}}, \quad n = 0, 1, 2, \cdots,$$

define functions $P_n(z, a)$ with the following properties:

(i) $P_n(z, a)$ *is a polynomial of degree n in z and $n - 1$ in a $(n = 1, 2, 3, \cdots)$:*

$$P_n(z, a) = z \sum_{k=0}^{n-1} \alpha_{k,n} a^{n-k-1} z^k,$$

(ii) *the coefficients $\alpha_{k,n}$ are positive integers such that*

$$\sum_{k=0}^{n-1} \alpha_{k,n} = n!.$$

Proof.   By definition $P_1(z, a) \equiv z$ and

$$P_n(z, a) = z [nP_{n-1}(z, a) + (a - z)P'_{n-1}(z, a)],$$

where the prime denotes differentiation with respect to z. Hence, by induction we deduce the statement about the polynomial character of P_n and the degrees. We also obtain the following recurrence relations for the α's:

$$\alpha_{0,n} = \alpha_{0,n-1},$$
$$\alpha_{k,n} = (k + 1)\alpha_{k,n-1} + (n - k)\alpha_{k-1,n-1}, \quad k = 1, 2, \cdots, n - 2,$$
$$\alpha_{n-1,n} = \alpha_{n-2,n-1}.$$

Since $\alpha_{0,1} = 1$, induction also yields

$$\alpha_{0,n} = \alpha_{n-1,n} = 1, \quad n = 1, 2, 3, \cdots,$$

and we see that all the coefficients are positive integers. Finally, we have

$$P_n(a, a) = naP_{n-1}(a, a) = \cdots = n! \, a^{n-1}P_1(a, a) = n! \, a^n,$$

and thus $\sum_{k=0}^{n-1} \alpha_{k,n} = n!$, as asserted. This completes the proof.

COROLLARY.   *If $r = \max(|z|, |a|)$, we have*

$$(11.2.10) \qquad\qquad |P_n(z, a)| \leq n!\, r^n.$$

We now take $G(w)$ as a function satisfying (11.2.2) and (11.2.3). Suppose that $f(z)$ is holomorphic in some domain D. We say that *the differential operator $G(\vartheta)$ applies to $f(z)$ in D* if

$$(11.2.11) \qquad\qquad G(\vartheta)[f] = \lim_{n \to \infty} \sum_{k=0}^{n} \frac{g_k}{k!}\, \vartheta^k[f]$$

exists for z in D, uniformly with respect to z on compact subsets of D. The general question of applicability will be discussed below. Here we want to show that $G(\vartheta)$ applies to $(1 - z)^{-1}$ in the domain $z \neq 1$.

THEOREM 11.2.1.   *If $G(w)$ satisfies (11.2.2) and (11.2.3), then $G(\vartheta)$ applies to $(1 - z)^{-1}$ for $z \neq 1$. The result*

$$(11.2.12) \qquad\qquad G(\vartheta)\, \frac{1}{1 - z} \equiv F(z)$$

is an entire function of $(1 - z)^{-1}$ which vanishes at $z = \infty$. Further,

$$(11.2.13) \qquad\qquad F(z) = \sum_{n=0}^{\infty} G(n) z^n, \quad |z| < 1,$$

and

$$(11.2.14) \qquad\qquad F(z) = - \sum_{n=1}^{\infty} G(-n) z^{-n}, \quad |z| > 1.$$

Proof.   We have

$$(11.2.15) \qquad\qquad \sum_{k=m}^{n} \frac{g_k}{k!}\, \vartheta^k\, \frac{1}{1 - z} = \sum_{k=m}^{n} \frac{g_k}{k!}\, \frac{P_k(z, 1)}{(1 - z)^{k+1}}\,.$$

This expression can be estimated with the aid of (11.2.4) and (11.2.10). Its absolute value is found not to exceed

$$M(\varepsilon) \sum_{k=m}^{n} \frac{\varepsilon^k}{|1 - z|^{k+1}} \quad \text{for} \quad |z| \leq 1,$$

$$M(\varepsilon) \sum_{k=m}^{n} \frac{(\varepsilon |z|)^k}{|1 - z|^{k+1}} \quad \text{for} \quad |z| > 1.$$

Suppose now that $|1 - z| > 2\varepsilon$, where $\varepsilon < \tfrac{1}{4}$. Then

$$\frac{\varepsilon}{|1 - z|} \leq \tfrac{1}{2} \quad \text{and} \quad \frac{\varepsilon |z|}{|1 - z|} \leq \tfrac{1}{2} + \varepsilon < \tfrac{3}{4}$$

for such values of z. The absolute value of the right member of (11.2.15) consequently does not exceed $M(\varepsilon)(2\varepsilon)^{-1}(\tfrac{3}{4})^{m-1}$, which tends to zero as $m \to \infty$.

It follows that $G(\vartheta)$ applies to $(1 - z)^{-1}$ for $z \neq 1$ and that the series

$$(11.2.16) \qquad G(\vartheta) \frac{1}{1 - z} = \sum_{k=0}^{\infty} \frac{g_k}{k!} \vartheta^k \frac{1}{1 - z} \equiv F(z)$$

converges uniformly in any region $| 1 - z | \geq \delta > 0$ to a holomorphic function of z. Since $F(z)$ is single-valued, it is an entire function of $(1 - z)^{-1}$. At the end of the section it will be shown that $F(z)$ does not vanish identically. It does vanish at $z = \infty$, since every term of the series (11.2.16) has this property.

To get the series expansions, we note that for $| z | < 1$

$$(11.2.17) \qquad \vartheta^k \frac{1}{1 - z} = \sum_{n=0}^{\infty} n^k z^n$$

by (11.2.8). Hence

$$F(z) = \sum_{k=0}^{\infty} \frac{g_k}{k!} \sum_{n=0}^{\infty} n^k z^n = \sum_{n=0}^{\infty} z^n \sum_{k=0}^{\infty} \frac{g_k}{k!} n^k = \sum_{n=0}^{\infty} G(n) z^n,$$

as asserted. The change of the order of summation is justified by the absolute convergence of the double series. For $| z | > 1$ we have

$$\frac{1}{1 - z} = - \sum_{n=1}^{\infty} z^{-n}$$

and

$$\vartheta^k \frac{1}{1 - z} = - \sum_{n=1}^{\infty} (-n)^k z^{-n},$$

again by (11.2.8). Formula (11.2.14) is an immediate consequence of this relation. This completes the proof.

Theorem 11.2.1 admits of an interesting converse.

THEOREM 11.2.2.   *If $F(z)$ is an entire function of $(1 - z)^{-1}$ which vanishes at $z = \infty$, then there exists an entire function $G(w)$, satisfying (11.2.3), such that $F(z)$ is represented by the series (11.2.13) for $| z | < 1$ and by (11.2.14) for $| z | > 1$.*

Proof.   Suppose that

$$(11.2.18) \qquad F(z) = \sum_{k=1}^{\infty} \frac{c_k}{(1 - z)^k} \quad \text{where} \quad \lim_{k \to \infty} | c_k |^{\frac{1}{k}} = 0.$$

Here the right-hand side is the general expression for an entire function of $(1 - z)^{-1}$ which vanishes at ∞. We have

$$F(z) = \sum_{k=1}^{\infty} c_k \sum_{n=0}^{\infty} \binom{-k}{n} (-z)^n, \qquad | z | < 1,$$

$$F(z) = \sum_{k=1}^{\infty} c_k \sum_{m=0}^{\infty} \binom{-k}{m} \left(-\frac{1}{z}\right)^{m+k}, \qquad | z | > 1.$$

Here we can interchange the order of summation since the double series involved are absolutely convergent. In the first case we find that the coefficient of z^n, $n = 0, 1, 2, \cdots$, is given by

$$a_n \equiv c_1 + c_2 \frac{n+1}{1!} + c_3 \frac{(n+1)(n+2)}{2!} + \cdots$$

$$+ c_{k+1} \frac{(n+1)(n+2)\cdots(n+k-1)}{k!} + \cdots.$$

In the second case the coefficient of z^{-n}, $n = 1, 2, 3, \cdots$, is

$$b_n \equiv -c_1 - c_2 \frac{1-n}{1!} - c_3 \frac{(1-n)(2-n)}{2!} - \cdots$$

$$- c_{k+1} \frac{(1-n)(2-n)\cdots(k-1-n)}{k!} - \cdots,$$

which, of course, breaks off after the nth term. Thus we see that if

$$(11.2.19) \qquad G(w) \equiv c_1 + \sum_{k=1}^{\infty} c_{k+1} \frac{(w+1)(w+2)\cdots(w+k-1)}{k!}$$

is an entire function of w with the desired properties, then we have

$$a_n = G(n), \quad b_n = -G(-n),$$

and the two expansions for $F(z)$ coincide with (11.2.13) and (11.2.14).

Here $G(w)$ is defined by a binomial series in the variable $-w$ (see Section 8.1). To prove the convergence of the series and to estimate the rate of growth of $G(w)$, we use the information available about c_k, namely, that $|c_k|^{1/k} \to 0$. This implies that for any $\varepsilon > 0$ we can find an $M(\varepsilon)$ such that

$$|c_k| \leq M(\varepsilon)\varepsilon^{k-1}$$

for all k. It follows that the series for $G(w)$ is absolutely convergent if the dominating series

$$M(\varepsilon)\left\{1 + \sum_{k=1}^{\infty} \varepsilon^k \frac{(|w|+1)(|w|+2)\cdots(|w|+k-1)}{k!}\right\}$$

has this property. By formula (6.3.11) this binomial series has the sum

$$M(\varepsilon)(1-\varepsilon)^{-1-|w|}.$$

Hence the series (11.2.19) is convergent for every finite w. It then defines an entire function of w and

$$|G(w)| \leq M(\varepsilon) \exp\left\{(1+|w|) \log \frac{1}{1-\varepsilon}\right\}.$$

If $\varepsilon < \frac{1}{2}$, the logarithm does not exceed $\varepsilon \log 4$, so that

$$| \, G(w) \, | \leq 2M(\varepsilon) \exp \, [\varepsilon \, | \, w \, | \, \log 4].$$

Since this inequality holds for every ε with $0 < \varepsilon < \frac{1}{2}$, we conclude that $G(w)$ has a series expansion of type (11.2.2) satisfying condition (11.2.3). It follows that the coefficients a_n and b_n actually have the desired representation. This completes the proof.

These two theorems together are known as the theorem of Wigert (1900) (Severin Wigert, 1871–1941).

We turn now to the general question of the applicability of the operator $G(\vartheta)$ to holomorphic functions.

THEOREM 11.2.3.   *If $f(z)$ is holomorphic in the domain D, then $G(\vartheta)$ applies to $f(z)$ in D, and $G(\vartheta)[f]$ is holomorphic in D. In particular, if $f(z)$ is a single-valued analytic function, then so is $G(\vartheta)[f](z)$, and the domain of holomorphy of $G(\vartheta)[f]$ contains that of $f(z)$. Finally, $\{G(n)\}$ is a holomorphy-preserving factor sequence.*

Proof.   Suppose that Δ is a compact subset of D and that C is a contour in D made up of a finite number of simple closed rectifiable oriented curves which do not intersect each other, so that

$$f(z) = \frac{1}{2\pi i} \int_C \frac{f(t) \, dt}{t - z}$$

for every z in Δ. Suppose that the distance of Δ from C is $\delta > 0$. We can apply the operator ϑ under the sign of integration as often as we please, for we saw in Section 7.5 that differentiation could be carried out under the sign of integration in Cauchy's integral and that the formal result so obtained was correct. Hence by Lemma 11.2.1

$$\vartheta^n f(z) = \frac{1}{2\pi i} \int_C \frac{P_n(z, t) f(t)}{(t - z)^{n+1}} \, dt$$

and

$$(11.2.20) \qquad \sum_{k=m}^{n} \frac{g_k}{k!} \vartheta^k f(z) = \frac{1}{2\pi i} \int_C \left\{ \sum_{k=m}^{n} \frac{g_k}{k!} \frac{P_k(z, t)}{(t - z)^k} \right\} \frac{f(t) \, dt}{t - z} .$$

Since Δ is compact, we can find a finite positive r such that

$$\max \, (| \, z \, |, | \, t \, |) \leq r, \quad z \in \Delta, t \in C.$$

With this value of r fixed, we can use condition (11.2.3) to choose a value $M = M(\delta, r)$ such that for all k

$$| \, g_k \, | < M \left(\frac{\delta}{2r} \right)^k.$$

Using this inequality and (11.2.10) we find that (11.2.20) does not exceed

$$\frac{M}{2\pi\delta} \, l(C) \, M(f; C) \, 2^{1-m},$$

where $l(C)$ is the length of C and $M(f; C)$ is the maximum of $|f(z)|$ on C. This upper bound tends to zero as $m \to \infty$. This shows that the limit in (11.2.11) exists uniformly with respect to z in Δ. It follows that $G(\vartheta)$ applies to $f(z)$ in D and that $G(\vartheta)[f](z)$ is a holomorphic function of z in D. If $f(z)$ is a single-valued analytic function, then we can identify D with the domain of holomorphy of $f(z)$ and conclude that the transform is holomorphic in the same domain.

In particular, if

$$(11.2.21) \qquad\qquad f(z) = \sum_{n=0}^{\infty} c_n z^n, \quad |z| < R,$$

then

$$(11.2.22) \qquad\qquad G(\vartheta)[f](z) = \sum_{n=0}^{\infty} c_n G(n) \, z^n, \quad |z| < R,$$

whence it follows that $\{G(n)\}$ is a holomorphy-preserving factor sequence. This completes the proof.

We shall find the factor sequences $\{G(n)\}$ quite useful in the discussion of singularities. For the applications we have to have at our disposal some special functions $G(w)$ having the desired growth properties. We list without proof the following two types:

$$(11.2.23) \qquad\qquad G(w) = \prod_{n=1}^{\infty} \left\{1 - \frac{w}{a_k}\right\}, \quad \sum \frac{1}{|a_k|} < \infty,$$

$$(11.2.24) \qquad\qquad G(w) = \prod_{n=1}^{\infty} \left\{1 - \frac{w^2}{a_k^2}\right\}, \quad \frac{a_k}{k} \to \infty.$$

The discussion of such products will be postponed until Chapter 14. At this point, however, we shall insert some remarks on the sequence $\{G(n)\}$. For this purpose we need the following:

LEMMA 11.2.2. *For an entire function $G(w)$ that satisfies (11.2.3), the number of zeros in the disk $|w| < r$ is $o(r)$ as $r \to \infty$. In particular, the number of integers n such that $n \leq N$ and $G(n) = 0$ is $o(N)$ as $N \to \infty$.*

Proof. We use Theorem 9.2.5. We may assume $G(0) \neq 0$. Then

$$\frac{1}{2\pi} \int_0^{2\pi} \log |G(re^{i\theta})| \, d\theta = \log |G(0)| + \sum \log \frac{r}{|a_j|},$$

where the a_j's run through the zeros of $G(w)$ in the disk $|w| < r$. By (11.2.5) the left-hand side does not exceed $\varepsilon r + \log M(\varepsilon)$. Suppose that on the right-hand side we restrict the summation to those values of j for which $|a_j| < \frac{1}{2}r$ and suppose that the number of such zeros is $n(\frac{1}{2}r)$. We then have

$$\varepsilon r + \log M(\varepsilon) > \log |G(0)| + n(\tfrac{1}{2}r) \log 2.$$

Replacing r by $2r$ we get

$$\limsup_{r \to \infty} \frac{n(r)}{r} < 3\varepsilon.$$

Since $\varepsilon > 0$ is arbitrary, this gives

$$\lim_{r \to \infty} \frac{n(r)}{r} = 0,$$

as asserted.

From this we see in particular that $G(n) \neq 0$ for infinitely many values of n. This implies that the function $F(z)$ of Theorem 11.2.1 does not reduce to a polynomial in z. Thus, $z = 1$ is actually a singular point of $F(z)$, that is, the radius of convergence of the series (11.2.13) is equal to 1, and

$$(11.2.25) \qquad \limsup_{n \to \infty} |G(n)|^{\frac{1}{n}} = 1.$$

We shall use these facts to prove a generalization of Theorem 5.7.1.

THEOREM 11.2.4.   *If the series (11.2.21) has real coefficients such that*

$$(11.2.26) \quad (-1)^{k-1}c_n > 0 \quad \text{for} \quad n_{k-1} \leqq n < n_k, \quad k = 1, 2, 3, \cdots,$$

where $n_0 = 0$ and $k/n_k \to 0$ as $k \to \infty$, then $z = R$ is a singular point of $f(z)$.

Proof.   We form the entire function

$$G(w) = \prod_{k=1}^{\infty} \left\{ 1 - \frac{w^2}{(n_k - \frac{1}{2})^2} \right\},$$

which is of type (11.2.24). $G(w)$ is real on the real axis and its sign is $(-1)^{k-1}$ for $n_{k-1} \leqq u \leqq n_k - 1$. Further, if w tends to infinity along the positive real axis, omitting the intervals $(n_k - 1, n_k)$, then, as will be shown in Section 14.3, we have

$$(11.2.27) \qquad \lim_{w \to \infty} \frac{1}{w} \log |G(w)| = 0.$$

The corresponding transformed series

$$G(\vartheta)[f](z) = \sum_{n=0}^{\infty} c_n G(n) z^n$$

has positive coefficients, and by (11.2.25) and (11.2.27) its radius of convergence also equals R. By Theorem 5.7.1, $z = R$ is a singular point of the transform, whence it follows by Theorem 11.2.3 that $z = R$ is also singular for $f(z)$. This completes the proof.

EXERCISE 11.2

1. Find the Maclaurin series of $G(\vartheta)[(a - z)^{-p}]$, where $a \neq 0$ and p is a positive integer.

2. Show that the operator $G(\vartheta)$ cannot annihilate a pole. Use the result of the preceding problem to obtain the effect of $G(\vartheta)$ on the principal part and show that the Maclaurin series cannot reduce to a polynomial.

3. Show that an operator $G(\vartheta)$ may annihilate a function with a natural boundary by considering the lacunary series (5.7.5) and taking

$$G(w) = \prod_{n=0}^{\infty}(1 - w\, 2^{-n}).$$

The latter function is of type (11.2.23).

4. If $G(w)$ is chosen as in the preceding example, what remarkable fact can you state regarding the solutions of the differential equation of infinite order

$$G(\vartheta)[w] = 0?$$

11.3. Power series with analytic coefficients. In the preceding sections we have encountered several classes of power series whose coefficients are analytic functions of the index. Under the special assumptions made above, the singularities of the functions defined by the series showed a preference for the line segment $[1, \infty)$ of the real axis and, in particular, for the points 1 and ∞. We shall now study another class of series whose singularities have a wider distribution. This class has been the object of much research starting with E. Le Roy (1870–1954), E. Lindelöf, and Walter B. Ford at the turn of the century. We shall use calculus of residues rather than operator methods for this study.

The coefficients of the power series will be determined by a function $g(w)$, holomorphic and of *exponential type* in a right half-plane, that is, $g(w)$ is holomorphic for $\Re(w) \geq \beta$ and there exists a constant a such that

(11.3.1) $|g(w)| \leq e^{a|w|}, \quad R \leq |w|,$

in this half-plane. For such a function it is advantageous to consider the *growth indicator*

(11.3.2) $h(\varphi) = h(\varphi; g) = \limsup_{r \to \infty} \frac{1}{r} \log |g(\beta + re^{i\varphi})|, \quad -\frac{\pi}{2} \leq \varphi \leq \frac{\pi}{2}.$

Such growth-measuring functions were introduced by E. Phragmén (1863–1937) and E. Lindelöf in 1908; the properties of $h(\varphi; g)$ will be studied in Section 19.3. Here we merely state as a lemma without proof the properties that will be needed in the discussion which follows.

LEMMA 11.3.1.   *Either $h(\varphi; g) \equiv -\infty$ for $-\dfrac{\pi}{2} < \varphi < \dfrac{\pi}{2}$ or this function is continuous in $\left(-\dfrac{\pi}{2}, \dfrac{\pi}{2}\right)$ and has finite left- and right-hand derivatives there. In the second case, $h(\varphi; g)$ is the function of support* (in the sense of Section 2.3) *of an unbounded closed convex set $\bar{K}$ known as the indicator diagram of $g(w)$.*

The conjugate set $K = [w \mid \bar{w} \in \bar{K}]$ is known as the *conjugate diagram.* This terminology is due to G. Pólya, who has contributed much to the exploration of functions of exponential type.

$\bar{K}$ is located in the strip

(11.3.3) $u \leq h(0), \quad -h\left(-\dfrac{\pi}{2}\right) \leq v \leq h\left(\dfrac{\pi}{2}\right), \quad w = u + iv,$

and either the lines

$$v = -h\left(-\dfrac{\pi}{2}\right) \quad \text{and} \quad v = h\left(\dfrac{\pi}{2}\right)$$

are asymptotes of $\partial\bar{K}$, the boundary of $\bar{K}$, or certain infinite segments of these lines form part of $\partial\bar{K}$. We note in particular that

(11.3.4) $\omega(\bar{K}) \equiv h\left(\dfrac{\pi}{2}\right) + h\left(-\dfrac{\pi}{2}\right) \geq 0.$

This quantity is called the *width* of $\bar{K}$.

If m is an integer and $g_m(w) = e^{-2\pi i m w} g(w)$, then $g_m(n) = g(n)$ for every integer n, while

(11.3.5) $h(\varphi; g_m) = h(\varphi; g) + 2m\pi \sin \varphi,$

whence it follows that

$$\bar{K}(g_m) = \bar{K}(g) + 2m\pi i$$

with obvious notation.

We can, consequently, move $\bar{K}$ up or down by multiples of $2\pi i$ without affecting the values of $g(w)$ at the integers. Hence, it is no restriction to assume that $\bar{K}(g)$ straddles one of the strips

$$-\pi \leq v \leq \pi \quad \text{or} \quad 0 \leq v \leq 2\pi$$

if $\omega(\bar{K}) > 2\pi$. If, on the other hand, $\omega(\bar{K}) \leq 2\pi$, we may assume that $\bar{K}(g)$ is confined to one of these strips. To achieve this we may have to replace z by $-z$, which moves $\bar{K}(g)$ up or down by πi.

We shall also have occasion to consider entire functions $g(w)$ of exponential type. In this case (11.3.1) holds for all large $\mid w \mid$, the indicator $h(\varphi; g)$ is defined as a continuous function for $-\pi \leq \varphi \leq \pi$ with $h(-\pi; g) = h(\pi; g)$, and $\bar{K}$ is a bounded closed convex set. The formula (11.3.4) now generalizes to

(11.3.6) $0 \leq h\left(\varphi + \dfrac{\pi}{2}\right) + h\left(\varphi - \dfrac{\pi}{2}\right), \quad -\dfrac{\pi}{2} \leq \varphi \leq \dfrac{\pi}{2}.$

This expresses that the width of $\bar{K}$ is not negative in any direction. Note that $\bar{K}$ may very well reduce to a point or to a line segment.

In defining the growth indicator, it is not essential that the rays emanate from the point $w = \beta$; we could use any other center $w = \gamma$ with $\Re(\gamma) \geq \beta$ and we would get the same result. Thus, if

$$h_\gamma(\varphi; g) \equiv \limsup_{r \to \infty} \frac{1}{r} \log | g(\gamma + re^{i\varphi}) |,$$

then

(11.3.7) $$h_\gamma(\varphi; g) = h_\beta(\varphi; g) = h(\varphi; g).$$

THEOREM 11.3.1.   *Let $g(w)$ be holomorphic and of exponential type in $\Re(w) \geq \beta$ where $\beta < 0$, let $h(\varphi; g)$ be defined by (11.3.2), let $h(\varphi; g)$ be bounded, and let $\bar{K}$ be the indicator diagram of $g(w)$, where the position of $\bar{K}$ satisfies the restrictions stated above. The power series*

(11.3.8) $$\sum_{n=0}^{\infty} g(n)\, z^n$$

converges at least for

$$| z | < e^{-h(0)}$$

and defines a function $f(z)$, holomorphic in the domain

(11.3.9) $$D = \mathbf{C}[e^{-K}],$$

which is starlike with respect to the origin. If $\omega(\bar{K}) < 2\pi$, D extends to infinity and contains the sector

(11.3.10) $$S: \quad h\left(\frac{\pi}{2}\right) < \arg z < 2\pi - h\left(-\frac{\pi}{2}\right).$$

If $\gamma = \min(-\beta, 1)$, then $z^\gamma f(z)$ is bounded in every fixed interior subsector of S for $| z | > 1$. Moreover, if $-m - 1 < \beta \leq -m$, then in the interior of S we have

(11.3.11) $$f(z) = -\frac{g(-1)}{z} - \frac{g(-2)}{z^2} - \cdots - \frac{g(-m)}{z^m} + O(| z |^\beta).$$

REMARKS.   The last two statements may be strengthened. One can show that $z^\gamma f(z) \to 0$ as $z \to \infty$ in the interior of S and that in (11.3.11) the remainder term is $o(| z |^\beta)$. The formula (11.3.9) means that D is the complement of the set $[e^{-w} \mid w \in K]$. Finally, if $h(\varphi; g) \equiv -\infty$, $-\frac{\pi}{2} < \varphi < \frac{\pi}{2}$, then $f(z)$ is an entire function. We shall consider this case later.

Proof.   Since

$$\limsup_{n \to \infty} \frac{1}{n} \log | g(n) | \leq \limsup_{r \to \infty} \frac{1}{r} \log | g(r) | = h(0),$$

we see that the radius of convergence of (11.3.8) is at least $e^{-h(0)}$. In order to extend $f(z)$ outside of the circle of convergence, we shall use the methods of Section 9.3. We see that

$$(11.3.12) \qquad \sum_{k=0}^{n} g(k) \, z^k = \int_{\Gamma_n} \frac{g(w) \, z^w}{e^{2\pi i w} - 1} \, dw,$$

where Γ_n is a closed contour surrounding the points $w = 0, 1, \cdots, n$. We shall take Γ_n as a triangle with vertices at the points

$$\alpha = \max \, (\beta, -\tfrac{1}{2}), \; w_+(n) = (n + \tfrac{1}{2})[1 + iv_+(n)], \; w_-(n) = (n + \tfrac{1}{2})[1 - iv_-(n)],$$

where $v_+(n)$ and $v_-(n)$ are so chosen that for all n

$$\arg \, [w_+(n) - \alpha] = \varphi_+, \quad \arg \, [w_-(n) - \alpha] = \varphi_-$$

have fixed values such that

$$-\frac{\pi}{2} < \varphi_- < 0 < \varphi_+ < \frac{\pi}{2}.$$

These arcs φ_- and φ_+ will be disposed of later.

We return to (11.3.12) and plan to show that the integral along the vertical boundary tends to zero as $n \to \infty$, provided z is suitably restricted. We carry through the argument under the assumption that there exists a σ such that for a given ε, $0 < \varepsilon$, we have

$$(11.3.13) \qquad h(\varphi; g) \leqq \sigma - \varepsilon < \sigma < \pi, \quad -\frac{\pi}{2} < \varphi < \frac{\pi}{2}.$$

As we shall see in a moment, the representation

$$(11.3.14) \qquad f(z) = -\int_{L_+} \frac{g(w) \, z^w}{e^{2\pi i w} - 1} \, dw + \int_{L_-} \frac{g(w) \, z^w}{e^{2\pi i w} - 1} \, dw$$

with

$$L_+: \; \arg \, (w - \alpha) = \varphi_+, \qquad L_-: \; \arg \, (w - \alpha) = \varphi_-$$

is then valid for

$$(11.3.15) \qquad |z| < e^{-\sigma}, \quad \sigma < \arg z < 2\pi - \sigma,$$

and arbitrary choices of φ_+ and φ_-. The representation remains valid without the stated restrictions on σ and on $\arg z$, provided $|z|$ is subjected to a stronger restriction.

Suppose then that (11.3.13) holds and consider the integral on the line segment from $w = n + \tfrac{1}{2}$ to $w = w_+(n)$. Here the denominator of the integrand in (11.3.12) is > 1 in absolute value. Further, if $z = re^{i\theta}$, then

$$|z^w| = r^{n+\frac{1}{2}} e^{-\theta v}, \quad |g(w)| < \exp \sigma(n + \tfrac{1}{2} + |v|),$$

so that this part of the integral is dominated by a constant multiple of

$$(e^\sigma r)^{n+\frac{1}{2}} \int_0^\infty e^{-(\theta - \sigma)v} \, dv.$$

This exists for $\sigma < \theta$ and tends to zero as $n \to \infty$ provided $r < e^{-\sigma}$. Similarly, the integral from $w_-(n)$ to $n + \frac{1}{2}$ is dominated by a constant multiple of

$$(e^\sigma r)^{n+\frac{1}{2}} \int_0^\infty e^{-(2\pi - \theta - \sigma)v}\, dv.$$

This majorant exists if $\theta < 2\pi - \sigma$. Thus, (11.3.15) implies the validity of (11.3.14). If $\pi \leq \sigma$ or if θ lies outside of the range $(\sigma, 2\pi - \sigma)$, we remark that more favorable majorants are obtained by replacing the infinite upper limit in the integrals by $(n + \frac{1}{2} - \alpha)$ sec φ, where $\varphi = \varphi_+$ in the first integral and $\varphi = \varphi_-$ in the second. It is clear that we can still force the majorants to tend to zero as $n \to \infty$ by making r sufficiently small. The details are left to the reader.

We come now to the problem of analytic continuation of $f(z)$ by means of formula (11.3.14). Since z^w is an entire function of $\log z$, it is sufficient to show that if $z_0 \in D = \mathbf{C}[e^{-K}]$ it is possible to choose φ_+ and φ_- in such a manner that the integrals in (11.3.14) converge uniformly with respect to z in some neighborhood of z_0. Let us first see how one arrives at this description of the domain of holomorphism of $f(z)$. It is of course understood that D may be a proper subset of the actual domain of holomorphism.

On L_+ we have

$$w = \alpha + se^{i\varphi_+}, \quad 0 \leq s < \infty,$$

and, taking (11.3.7) into account, we see that we can find an $M(\varepsilon)$ such that

$$|\,g(w)\,| < M(\varepsilon) \exp\{[h(\varphi_+) + \varepsilon]s\}.$$

Further,

$$|\,z^w\,| = r^\alpha \exp\{[\log r \cos \varphi_+ - \theta \sin \varphi_+]s\},$$

while the denominator is bounded away from 0 and ∞. Hence

$$\left|\int_{L_+}\right| \leq M_1(\varepsilon) r^\alpha \int_0^\infty \exp\{[h(\varphi_+) + \varepsilon + \log r \cos \varphi_+ - \theta \sin \varphi_+]s\}\, ds.$$

Replacing φ_+ by φ_- and $-\theta$ by $2\pi - \theta$, we can estimate the integral along L_-.

It is consequently required to choose φ_+ and φ_- in such a manner that for the values of z under consideration

(11.3.16)
$$-\log r \cos \varphi_+ + \theta \sin \varphi_+ \geq h(\varphi_+) + 2\varepsilon,$$
$$-\log r \cos \varphi_- + (\theta - 2\pi) \sin \varphi_- \geq h(\varphi_-) + 2\varepsilon.$$

For all points $w = u + iv$ of $\bar{K}$ we have

$$u \cos \varphi + v \sin \varphi \leq h(\varphi), \quad -\frac{\pi}{2} \leq \varphi \leq \frac{\pi}{2}.$$

Hence, conditions (11.3.16) express that the points

$$-\log r + i\theta \quad \text{and} \quad -\log r + i(\theta - 2\pi)$$

are outside of $\bar{K}$. This means that z belongs to the set D of (11.3.9).

That D is starlike with respect to the origin follows from the fact that K is convex and that a horizontal line $v = b$ is mapped onto a ray $\arg z = b$ by the function $z = e^{-w}$. The character of D depends essentially upon $\omega(\bar{K})$. If $\omega(\bar{K}) > 2\pi$, then the boundary ∂D of D is a simple closed curve which is the image under the mapping $z = e^{-w}$ of an arc of ∂K whose endpoints have a vertical distance 2π. This arc is uniquely determined unless a segment of $u = h(0)$ of length $> 2\pi$ forms a part of ∂K. In the latter case D is the disk

$$| z | < e^{-h(0)},$$

and the integral (11.3.14) fails to give any information about $f(z)$ beyond that furnished by the series (11.3.8). If $\omega(\bar{K}) < 2\pi$, then D extends to infinity and contains the sector S of (11.3.10). For $\omega(\bar{K}) = 2\pi$ there are two alternatives. D is bounded if segments of the lines

$$v = h\left(\frac{\pi}{2}\right) \quad \text{and} \quad v = -h\left(-\frac{\pi}{2}\right) = h\left(\frac{\pi}{2}\right) - 2\pi$$

form parts of $\partial \bar{K}$. If at least one of these lines is an asymptote of $\bar{K}$, then D exhibits a "horn" extending to infinity in the direction $\arg z = h\left(\frac{\pi}{2}\right)$.

It remains to verify the possibility of satisfying (11.3.16) when $z \in D$. Consider a truncated sector T located in D:

$$T: \quad 0 < r_1 < | z | < r_2 < \infty, \qquad \theta_1 < \arg z < \theta_2.$$

Here T is the image under the mapping $z = e^{-w}$ of any one of the rectangles

$$R_n: \quad -\log r_2 < u < -\log r_1, \qquad -\theta_2 < v - 2n\pi < -\theta_1,$$

which have no points in common with K by the definition of D. But then the conjugate rectangles

$$\bar{R}_n: \quad -\log r_2 < u < -\log r_1, \qquad \theta_1 < v + 2n\pi < \theta_2$$

do not overlap $\bar{K}$. To fix the ideas, let us consider the case illustrated by Figures 5 and 6. Here Figure 5 shows the region $\bar{K}$ and the two rectangles $\bar{R}_0$ and $\bar{R}_1$, and Figure 6 depicts the corresponding region $\exp(-K)$ and the sector T.

For the purpose of the drawings, we have taken a special case

$$(11.3.17) \qquad h(\varphi) = \frac{3}{2} \cos \varphi \log (2 \cos \varphi) + \left(\frac{3}{2} \varphi - \frac{\pi}{12}\right) \sin \varphi.$$

Here

$$h\left(-\frac{\pi}{2}\right) = \frac{5\pi}{6}, \quad h(0) = \frac{3}{2} \log 2, \quad h\left(\frac{\pi}{2}\right) = \frac{2\pi}{3},$$

and

$$\partial D: \quad r\left[2 \cos \frac{2}{3}\left(\theta - \frac{\pi}{12}\right)\right]^{\frac{3}{2}} = 1.$$

Further,

$$\theta_1 = \frac{5\pi}{6}, \quad \theta_2 = \frac{4\pi}{3}.$$

The function $h(\varphi)$ does not determine $g(w)$ uniquely by any means. Among the possible functions $g(w)$ such that $h(\varphi; g)$ equals the function $h(\varphi)$ of (11.3.17) we mention

$$g(w) = \exp\left(\frac{\pi}{12} iw\right) \sum_{k=1}^{\infty} \binom{\frac{3}{2}w + 1}{k^2},$$

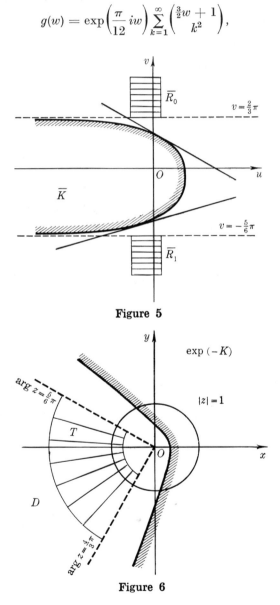

Figure 5

Figure 6

where the series is a lacunary binomial series.[1] The sum of the series is an integer for even integral values of w.

Referring to Figure 5 we see that there may be drawn to $\bar{K}$ a line of support

$$u \cos \varphi_+ + v \sin \varphi_+ = h(\varphi_+) \quad \text{with} \quad 0 < \varphi_+ < \frac{\pi}{2}$$

which separates $\bar{K}$ and $\bar{R}_0$, as well as a line of support

$$u \cos \varphi_- + v \sin \varphi_- = h(\varphi_-) \quad \text{with} \quad -\frac{\pi}{2} < \varphi_- < 0$$

which separates $\bar{K}$ and $\bar{R}_1$. Thus, for any choice of z in T there exist fixed arcs φ_- and φ_+ such that (11.3.16) holds for some small fixed positive ε uniformly with respect to z.

To prove that this implies that $f(z)$ is holomorphic in T, we argue as follows: It is enough to consider the integral along L_+. Let

$$c_1 = \alpha + \rho_1 e^{i\varphi_+}, \quad c_2 = \alpha + \rho_2 e^{i\varphi_+}, \quad \rho_1 < \rho_2,$$

be two distant points on L_+. Then, by the now familiar estimates,

$$\left| \int_{c_1}^{c_2} \right| \leq M_1(\varepsilon) r^\alpha \int_{\rho_1}^{\rho_2} \exp \{[h(\varphi_+) + \varepsilon + \log r \cos \varphi_+ - \theta \sin \varphi_+] s\} \, ds$$

$$< M_1(\varepsilon) r^\alpha \int_{\rho_1}^\infty e^{-\varepsilon s} \, ds = \varepsilon^{-1} M_1(\varepsilon) r^\alpha e^{-\varepsilon \rho_1}.$$

This tends to zero as $\rho_1 \to \infty$, uniformly for z in T. On the other hand, for fixed c the series expansion

$$\int_\alpha^c = \sum_{n=0}^\infty \frac{1}{n!} (\log z)^n \int_\alpha^c \frac{g(w) w^n}{e^{2\pi i w} - 1} \, dw$$

obviously converges uniformly for z in T and, hence, defines a holomorphic function of z in T. Thus, the integral along L_+ is the limit of functions holomorphic in T, and the limit exists uniformly with respect to z in T. It follows that the integral is holomorphic in T. The integral along L_- is handled in the same way. Hence, $f(z)$ is holomorphic in T. But we can cover domain D by the disk

$$|z| < e^{-h(0)}$$

and a countable number of sectors of the type T. Since $f(z)$ is holomorphic in each of these subdomains and since their union is simply-connected, we conclude that $f(z)$ is holomorphic in D as asserted.

---

[1] For this choice of $g(w)$, see F. Carlson, "Über ganzwertige Funktionen," *Mathematische Zeitschrift*, Vol. 11 (1921), pp. 1–23, especially pp. 21–22. F. Carlson (1888–1952) was one of the founders of the theory of binomial series. The growth function

$$\mu(\varphi) \equiv \cos \varphi \log (2 \cos \varphi) + \varphi \sin \varphi$$

is the upper bound for all functions $h(\varphi; g)$ such that $g(w)$ is representable as a convergent binomial series in the variable w in a right half-plane. Theorem 11.3.1 is taken from Carlson's Uppsala dissertation of 1914: *Sur une classe de séries de Taylor.*

We still have to discuss the asymptotic properties of $f(z)$ in the sector S of (11.3.10). Now S corresponds to the strip

$$h\left(\frac{\pi}{2}\right) < v < 2\pi - h\left(-\frac{\pi}{2}\right)$$

under the mapping $z = e^{-w}$. Referring to (11.3.16) once more, we see that for z in S we can actually take

$$\varphi_- = -\frac{\pi}{2}, \quad \varphi_+ = \frac{\pi}{2}$$

in (11.3.14), so that

(11.3.18) $$f(z) = -\int_{\alpha - i\infty}^{\alpha + i\infty} \frac{g(w)\, z^w}{e^{2\pi i w} - 1}\, dw, \quad z \in S.$$

Here we set $w = \alpha + iv$ and assume that

$$1 \leqq r, \quad h\left(\frac{\pi}{2}\right) + 2\varepsilon \leqq \arg z \leqq 2\pi - h\left(-\frac{\pi}{2}\right) - 2\varepsilon.$$

Hence

$$f(re^{i\theta}) = -i\, e^{i\alpha\theta}\, r^\alpha \int_{-\infty}^{\infty} \frac{g(\alpha + iv)\, e^{-\theta v}\, e^{iv \log r}}{e^{2\pi(i\alpha - v)} - 1}\, dv.$$

The absolute value of the integrand does not exceed

$$M(\alpha, \varepsilon)\, e^{-\varepsilon |v|}$$

uniformly with respect to r and θ. It follows that the integral is uniformly bounded. Actually a stronger statement can be made. The integrand with the factor $e^{iv \log r}$ omitted is a continuous function $H(v)$ which is integrable over $(-\infty, +\infty)$. By the Riemann-Lebesgue theorem (see Problem 4 of Exercise 11.3) its Fourier transform $\hat{H}(s)$ has the property

(11.3.19) $$\hat{H}(s) = (2\pi)^{-\frac{1}{2}} \int_{-\infty}^{\infty} H(v)\, e^{-isv}\, dv \to 0 \quad \text{as } s \to \pm\infty.$$

Setting $s = \log r$ we see that

(11.3.20) $$\lim_{r \to \infty} r^{-\alpha} f(re^{i\theta}) = 0$$

uniformly with respect to θ. By assumption $\alpha = \max\left(-\frac{1}{2}, \beta\right)$ so that this is the desired relation when $-\frac{1}{2} \leqq \beta < 0$. If, however, $-1 < \beta < -\frac{1}{2}$, we can replace the line of integration in (11.3.18) by $u = \beta$. This is easily justified by a contour integration argument. Hence we can replace α by β in the preceding argument and also in the conclusion. If $\beta = -1$ we note that

$$f(z) = \frac{g(-1)}{1 - z} - \int_{-1 - i\infty}^{-1 + i\infty} \frac{g(w) - g(-1)}{e^{2\pi i w} - 1}\, z^w\, dw,$$

whence

$$f(z) = -\frac{g(-1)}{z} + o\left(\frac{1}{|z|}\right), \quad z \to \infty.$$

The extension to the case in which $-m - 1 < \beta < -m, 1 \le m$ is immediate. If β is not an integer, then (11.3.18) with α replaced by β represents

$$f(z) + \sum_{k=1}^{m} \frac{g(-k)}{z^k},$$

and the integral is $o(|z|^\beta)$ by the previous argument. If $\beta = -m$ we have

$$f(z) + \sum_{k=1}^{m-1} \frac{g(-k)}{z^k} = \frac{g(-m)}{(1-z)z^{m-1}} - \int_{-m-i\infty}^{-m+i\infty} \frac{g(w) - g(-m)}{e^{2\pi i w} - 1} z^w \, dw,$$

and here the right-hand side is

$$-\frac{g(-m)}{z^m} + o\left(\frac{1}{|z|^m}\right), \quad z \to \infty.$$

This completes the proof of Theorem 11.3.1.

So far we have excluded the case in which $h(\varphi; g)$ is not bounded below. We state the result without proof. The reader will have no difficulties in adapting the methods used above to the new situation.

THEOREM 11.3.2. *Let $g(w)$ be holomorphic and of exponential type in $\Re(w) \ge \beta$ where $\beta < 0$ and let*

$$h(\varphi; g) = -\infty, \quad -\frac{\pi}{2} < \varphi < \frac{\pi}{2}.$$

Then the series (11.3.8) defines an entire function. If

$$h\left(\frac{\pi}{2}\right) + h\left(-\frac{\pi}{2}\right) < 2\pi$$

and if $-1 < \beta < 0$, then $z^{-\beta} f(z) \to 0$ as $z \to \infty$ in the interior of the sector S of (11.3.10), and if $-m - 1 < \beta \le -m, 1 \le m$, then (11.3.11) holds.

In connection with this theorem we note that an entire function may be of exponential type in every right half-plane without being an entire function of exponential type. An example is given by

(11.3.21) $$g(w) = \frac{1}{\Gamma(1 + \alpha w)}, \quad \alpha > 0.$$

By Stirling's formula (8.8.24) we have

$$\log |g(re^{i\varphi})| = -(\alpha r \cos \varphi + \tfrac{1}{2}) \tfrac{1}{2} \log (\alpha^2 r^2 + 2\alpha r \cos \varphi + 1)$$

$$+ \alpha r \sin \varphi \arctan \frac{\alpha r \sin \varphi}{\alpha r \cos \varphi + 1} + \alpha r \cos \varphi + O(1).$$

It follows that

$$h\left(\frac{\pi}{2}\right) = h\left(-\frac{\pi}{2}\right) = \alpha\,\frac{\pi}{2}, \quad h(\varphi) = -\infty, \quad -\frac{\pi}{2} < \varphi < \frac{\pi}{2},$$

and for $\dfrac{\pi}{2} < |\varphi| < \pi$

$$\frac{1}{r}\log |\,g(re^{i\varphi})\,| = -\alpha \cos \varphi \log r + O(1).$$

This expression tends to $+\infty$ as $r \to \infty$, so $g(w)$ is not an entire function of exponential type.

Let us observe what Theorem 11.3.2 asserts concerning the corresponding Mittag-Leffler function $E_\alpha(z)$ defined by

$$(11.3.22) \qquad E_\alpha(z) = \sum_{n=0}^{\infty} \frac{z^n}{\Gamma(1+n\alpha)}.$$

For $0 < \alpha < 2$ we find that

$$(11.3.23) \quad E_\alpha(z) = -\sum_{k=1}^{n} \frac{1}{\Gamma(1-k\alpha)z^k} + o(|\,z\,|^{-n}), \quad \alpha\,\frac{\pi}{2} < \arg z < 2\pi - \alpha\,\frac{\pi}{2}.$$

In particular, for $\alpha = 1$ we note that $g(-n) = 0$ for $n = 1, 2, 3, \cdots$ and we obtain a well-known property of the function e^z. For another example of an entire function whose asymptotic behavior is similar to that of the exponential function, see Problem 8 in Exercise 11.3.

Formula (11.3.8) raises the following question: *Is it possible for the function $f(z)$ to vanish identically when $g(w)$ satisfies the conditions of Theorem 11.3.1?* This would be the case if and only if $g(n) = 0$ for $n = 0, 1, 2, \cdots$. An obvious choice is

$$g(w) = \sin \pi w \quad \text{with} \quad h(\varphi; g) = \pi \sin \varphi, \quad -\frac{\pi}{2} \leq \varphi \leq \frac{\pi}{2}.$$

Here $\bar{K}$ is the line segment $[-\pi i, \pi i]$ on the imaginary axis and $\omega(\bar{K}) = 2\pi$. Actually this is the entire function of least growth which vanishes on the positive integers. More precisely, we have the following result, which also goes back to F. Carlson:

THEOREM 11.3.3. *Let $g(w)$ be holomorphic and of exponential type in $\Re(w) \geq -\frac{1}{2}$ and such that*

    (i) $h\left(-\dfrac{\pi}{2}; g\right) \leq \pi, \quad h\left(\dfrac{\pi}{2}; g\right) \leq \pi,$

    (ii) *$\bar{K}$ lies in the open strip $-\pi < v < \pi$,*

    (iii) $\operatorname{sech} \pi v\, g(-\frac{1}{2} + iv) \in L_2(-\infty, \infty),$

    (iv) $g(n) = 0, \; n = 0, 1, 2, \cdots.$

Then

$$g(w) \equiv 0.$$

REMARKS.   Condition (i) obviously implies $\omega(\bar{K}) \leq 2\pi$. Conversely, if for a given $g(w)$ satisfying (iv) we have $\omega(\bar{K}) \leq 2\pi$, then we can choose a real α such that for $g_1(w) = e^{i\alpha w}g(w)$ we have $\bar{K}(g_1)$ located in the strip $-\pi \leq v \leq \pi$ and $g_1(n) = 0$. Thus, there is no restriction to assume (i) rather than that $\omega(\bar{K}) \leq 2\pi$, and it makes the formulation of the integrability condition somewhat simpler. The proof given below is based on the Plancherel theorem, which says that $H(v) \in L_2(-\infty, \infty)$ implies that its Fourier transform $\hat{H}(s)$ exists and belongs to $L_2(-\infty, \infty)$, and that, moreover, the mapping $H \to \hat{H}$ is an *isometry* in the sense that

$$(11.3.24) \qquad \int_{-\infty}^{\infty} |H(v)|^2 \, dv = \int_{-\infty}^{\infty} |\hat{H}(s)|^2 \, ds.$$

For such questions see Norbert Wiener, *The Fourier Integral* (Cambridge University Press, 1933).

Proof.   Let $0 < \delta < 1$ and define

$$(11.3.25) \qquad g(w, \delta) = g[-\tfrac{1}{2} + \delta(w + \tfrac{1}{2})], \quad f(z, \delta) = \sum_{n=0}^{\infty} g(n, \delta) z^n.$$

We note that $g(w, \delta)$ is holomorphic and of exponential type in $\Re(w) \geq -\tfrac{1}{2}$ and that its growth indicator

$$h(\varphi; g_\delta) = \delta h(\varphi; g).$$

By Theorem 11.3.1 the power series defines a holomorphic function of z in

$$D_\delta \equiv \mathbf{C}(e^{-\delta K}) \supset \mathbf{C}(e^{-K}) \equiv D, \quad 0 < \delta \leq 1.$$

Here D extends to infinity by virtue of condition (ii) and contains the negative real axis, which is the image of the line $v = \pi$ under the mapping $z = e^{-w}$. Suppose now that T is a closed truncated sector in D of the type considered in the proof of Theorem 11.3.1. Then a fortiori $T \subset D_\delta$, and we can find arcs φ_+ and φ_- depending on T but not on δ such that $f(z, \delta)$ is given by formula (11.3.14) with $g(w)$ replaced by $g(w, \delta)$. Here the integrals converge uniformly with respect to z on T and with respect to δ, $0 < \delta \leq 1$. It follows in particular that

$$(11.3.26) \qquad \lim_{\delta \to 1} f(z, \delta) \equiv f(z, 1) = f(z)$$

exists uniformly on T. This implies that the limit exists everywhere in D. We know a priori that $f(z)$ is holomorphic in D, but the validity of (11.3.26) is not obvious and requires a proof. On the other hand, we know that $f(z) \equiv 0$ since the power series (11.3.8) has all its coefficients equal to zero by assumption. Hence

$$\lim_{\delta \to 1} f(z, \delta) \equiv 0, \quad z \in D,$$

uniformly on compact subsets of D.

For

$$\delta h\left(\frac{\pi}{2}\right) < \arg z < 2\pi - \delta h\left(-\frac{\pi}{2}\right),$$

$f(z, \delta)$ is represented by formula (11.3.18) with $g(w)$ replaced by $g(w, \delta)$ and α by $-\frac{1}{2}$. In particular, for z real, negative, $z = -r$, we have

(11.3.27) $\qquad f(-r, \delta) = \frac{1}{2} r^{-\frac{1}{2}} \int_{-\infty}^{\infty} g(-\frac{1}{2} + i\delta v) \operatorname{sech} \pi v \, e^{iv \log r} \, dv.$

We set

$$H(v, \delta) = g(-\tfrac{1}{2} + i\delta v) \operatorname{sech} \pi v$$

and note that for $0 < \delta < 1$

$$H(v, \delta) \in L(-\infty, \infty) \cap L_2(-\infty, \infty),$$

while for $\delta = 1$ we have at least

$$H(v, 1) \in L_2(-\infty, \infty).$$

Thus, for $0 < \delta < 1$ we have

(11.3.28)
$$\hat{H}(s, \delta) = (2\pi)^{-\frac{1}{2}} \int_{-\infty}^{\infty} H(v, \delta) e^{-isv} \, dv$$
$$= 2(2\pi)^{-\frac{1}{2}} e^{-\frac{1}{2}s} f(-e^{-s}, \delta)$$

by (11.3.27). Here the integral is absolutely convergent. For $\delta = 1$ we have a similar representation, where, however, the integral is the limit in the mean of order two of the integral from $-a$ to $+a$ as $a \to +\infty$ and the validity of the second line of (11.3.28) for $\delta = 1$ remains to be proved.

By (11.3.24) we have

(11.3.29) $\qquad \int_{-\infty}^{\infty} |H(v, \delta) - H(v, 1)|^2 \, dv = \int_{-\infty}^{\infty} |\hat{H}(s, \delta) - \hat{H}(s, 1)|^2 \, ds.$

When $\delta \to 1$, the left member tends to zero. For

$$\lim_{\delta \to 1} \int_{-a}^{a} |H(v, \delta) - H(v, 1)|^2 \, dv = 0,$$

since $H(v, \delta) \to H(v, 1)$ uniformly on $(-a, a)$. Here we can choose a so large that for a given $\varepsilon > 0$ and any δ with $\frac{1}{2} < \delta < 1$ we have

$$\int_{|v| > a} |H(v, \delta)|^2 \, dv < \int_{|v| > \delta a} |H(v, 1)|^2 \, dv < \varepsilon.$$

Then the contribution from the intervals $(-\infty, -a)$ and (a, ∞) to the left member of (11.3.29) does not exceed 4ε. Since ε is arbitrary, it follows that the limit is 0 as asserted.

It follows that

$$\lim_{\delta \to 1} \int_{-\infty}^{\infty} |\hat{H}(s, \delta) - \hat{H}(s, 1)|^2 \, ds = 0.$$

But formula (11.3.28) shows that for fixed s

$$\lim_{\delta \to 1} \hat{H}(s, \delta) = 0$$

uniformly with respect to s on compact subsets of $(-\infty, 0)$. For fixed finite a we have then

$$\int_{-a}^{a} |\hat{H}(s, 1)|^2 \, ds = \lim_{\delta \to 1} \int_{-a}^{a} |\hat{H}(s, \delta) - \hat{H}(s, 1)|^2 \, ds = 0.$$

Since this holds for every a, it follows that

$$\hat{H}(s, 1) = 0$$

for almost all s. Going back to (11.3.24), we see that this implies

$$H(v, 1) \equiv 0,$$

since $H(v, 1)$ is continuous. But from

$$g(-\tfrac{1}{2} + iv) \equiv 0$$

it follows that

$$g(w) \equiv 0,$$

and the theorem is proved.

We shall return to this theorem in a more general setting in Chapter 19 (see Problems 6 and 8 of Exercise 19.2).

In conclusion, let us state the analogue of Theorem 11.3.1 for entire functions of exponential type.

THEOREM 11.3.4.   *Let $g(w)$ be an entire function of exponential type, and form*

$$D = \mathbf{C}(e^{-K}).$$

Then the series (11.3.8) *formed with this $g(w)$ defines a holomorphic function of z in the component of D which contains the origin. If $\omega(\bar{K}) < 2\pi$, then D has a single component, $f(z)$ is holomorphic at ∞, and*

(11.3.30) $$f(z) = -\sum_{k=1}^{\infty} \frac{g(-k)}{z^k}, \quad |z| > e^{-h(\pi)}.$$

EXERCISE 11.3

1. Prove (11.3.7).

2. If $h(\varphi; g) \leq \mu(\varphi) \equiv \cos \varphi \, \log (2 \cos \varphi) + \varphi \sin \varphi$ prove that $f(z)$ is holomorphic in $\Re(z) < \tfrac{1}{2}$.

3. If $h(\varphi; g) \leqq 2\mu(\varphi)$, prove that $f(z)$ is holomorphic in the interior of the parabola $r\left[2\cos\dfrac{\theta}{2}\right]^2 = 1$.

4. Prove (11.3.19) for continuous functions $H(v)$ in $L(-\infty, \infty)$. Note that

$$-2(2\pi)^{\frac{1}{2}}\hat{H}(s) = \int_{-\infty}^{\infty}\left[H\left(v + \frac{\pi}{s}\right) - H(v)\right]e^{-isv}\, dv.$$

5. Take

$$g(w) = (1 + w)^{-\alpha(1+w)}, \quad 0 < \alpha < 2,$$

and prove that the entire function

$$L_\alpha(z) = \sum_{n=1}^{\infty} n^{-\alpha n}\, z^n,$$

introduced by Lindelöf, tends to zero in the sector $\alpha\dfrac{\pi}{2} < \arg z < 2\pi - \alpha\dfrac{\pi}{2}$.

6. Take

$$g(w) = [\log (w + \alpha)]^{-w}, \quad \alpha > 2,$$

and prove that the entire function

$$M_\alpha(z) = \sum_{n=0}^{\infty} [\log (n + \alpha)]^{-n}\, z^n,$$

introduced by Johannes Malmquist (1882–1952), tends to zero when $z \to \infty$, $0 < \arg z < 2\pi$.

7. If $2 < \alpha < \beta$, prove that the entire function

$$\exp[-M_\alpha(z)] - \exp[-M_\beta(z)]$$

tends to zero along every ray $\arg z = \theta$. Can the convergence be uniform with respect to θ? Is there any contradiction with the theorem of Liouville?

8. Prove that the entire function

$$\sum_{n=0}^{\infty} \frac{z^n}{\Gamma(\frac{1}{2}n + 1)\Gamma(n + 1)}$$

tends to zero faster than any power of $1/z$ as $z \to \infty$ in the sector

$$\frac{3\pi}{4} + \varepsilon \leqq \arg z \leqq \frac{5\pi}{4} - \varepsilon.$$

9. If $g(w)$ is an entire function of exponential type and if its indicator satisfies $|h(\varphi; g)| \leqq a < \pi$, prove that the function $f(z)$ defined by (11.3.8) is holomorphic outside the curve

$$r = \exp \pm(a^2 - \theta^2)^{\frac{1}{2}}, \quad z = re^{i\theta}.$$

10. Prove Theorem 11.3.2.

11. Prove Theorem 11.3.4.

11.4. Analytic continuation in a star. We shall return to the problem of analytic continuation of a regular element defined by a power series in z. This problem may be attacked by various methods of summability. Such methods are linear in character. A consequence of this property and of the structure of Cauchy's integral is that it suffices to test the effectiveness of the method in question on the geometric series. We shall need some terminology.

Under the name "regular method of summability" one understands two different but related concepts. First, the term may refer to a linear transformation M taking a sequence $\{A_n\}$ into a sequence $M[\{A_n\}] = \{B_n\}$ such that

$$(11.4.1) \qquad \lim_{n \to \infty} B_n = \lim_{n \to \infty} A_n$$

whenever the right-hand side exists. Or such a method may involve a family of linear transformations $\{M_\alpha\}$ taking the series $\sum_{n=0}^{\infty} a_n$ into the convergent series

$$M_\alpha \left[\sum_{n=0}^{\infty} a_n \right] = \sum_{n=0}^{\infty} \mu_n(\alpha) \, a_n$$

such that

$$(11.4.2) \qquad \lim_{\alpha \to \infty} \lim_{n \to \infty} \sum_{k=0}^{n} \mu_k(\alpha) \, a_k = \lim_{n \to \infty} \sum_{k=0}^{n} a_k,$$

whenever the right-hand side exists.

Applying a method of summability to the geometric series, we obtain a family of functions $\{G_\alpha(z)\}$ so that in a region R_M of the complex plane we have

$$(11.4.3) \qquad \lim_{\alpha \to \infty} G_\alpha(z) = \frac{1}{1 - z}, \qquad z \in R_M.$$

Here R_M contains the open unit disk. In the first case, the index set $\{\alpha\}$ is the set of non-negative integers and $G_n(z)$ is normally a polynomial in z. In either case,

$$(11.4.4) \qquad G_\alpha(z) = \sum_{m=0}^{\infty} \mu_m(\alpha) \, z^m$$

with a radius of convergence ≥ 1. If the method is to be of any use in the theory of analytic continuation, R_M must extend outside the unit circle.

We shall restrict ourselves to the case in which R_M is starlike with respect to the origin. Given a domain S, starlike with respect to $z = 0$ and not containing $z = 1$, and given an element $f(z)$ regular at $z = 0$ whose principal star is $\mathbf{A}[f]$, we proceed to construct a star $\mathbf{SA}[f]$ which is a subset of $\mathbf{A}[f]$. For each θ, $0 \leq \theta < 2\pi$, let $r(\theta)$ be the upper limit of the values of r for which $z = re^{i\theta} \in \mathbf{A}[f]$. If $r(\theta)$ is finite, we set

$$(11.4.5) \qquad S(\theta) = [z \mid z = z_1 \, r(\theta) \, e^{i\theta}, \quad z_1 \in S]$$

and define

$$(11.4.6) \qquad \mathbf{SA}[f] = \cap_\theta S(\theta),$$

where only those values of θ are considered for which $r(\theta)$ is finite.

Since each $S(\theta)$ is starlike with respect to $z = 0$, the intersection **SA**[f] also has this property. Further, **SA**[f] is a subset of **A**[f]. In fact, if $z \in$ **SA**[f] and $\arg z = \theta$ with $r(\theta) = \infty$, then $z \in$ **A**[f] automatically, and if $r(\theta) < \infty$, then $z = s\,r(\theta)\,e^{i\theta}$ where $s \in S$ and $0 < s < 1$, so that again $z \in$ **A**[f]. We note that if S is convex, then $S(\theta)$ and hence also **SA**[f] will have this property.

Further, if **SA**[f] = **A**[f] for each f, then S equals the whole plane less the segment $[+1, +\infty]$ of the real axis, that is, $S =$ **A**[g] where $g(z) = (1 - z)^{-1}$. For

$$\mathbf{SA}[g] = S(0) = [z \mid z \in S] = S,$$

and if **SA**[g] = **A**[g], then we must have $S =$ **A**[g] as asserted.

The function $r(\theta)$ introduced above is normally not continuous; it can be continuous only if **A**[f] is bounded. On the other hand, it is always lower semi-continuous, that is, for every sequence $\{\theta_n\}$ converging to θ_0 we have

$$(11.4.7) \qquad\qquad \liminf_{n \to \infty} r(\theta_n) \geqq r(\theta_0).$$

This is trivially true if the left member is infinite. If the limit is finite, then we have a sequence of points a_n, $\arg a_n = \theta_n$, which are singularities of $f(z)$ for radial approach, and a subsequence converges to a finite limit a_0. Then, either a_0 is a singularity for radial approach or there exists a singular point of $f(z)$ on the ray $\arg z = \theta_0$ which is closer to the origin. This proves (11.4.7).

LEMMA 11.4.1.   *If Δ is a compact subset of* **SA**[f], *then there exists a "scroc" C_Δ in* **A**[f] *containing 0 and Δ in its interior, such that $z/t \in S$ for all $z \in \Delta$ and all $t \in C_\Delta$.*

Proof.   We recall first that "scroc" is the abbreviation which we use for the phrase "simple closed rectifiable oriented curve." Let $z_0 \in \Delta$, $z_0 \neq 0$, and consider a ray $\arg t = \varphi$ in the t-plane. For every point t on this ray we construct the set $tS = [w \mid w = tz, z \in S]$. Since the domain S contains the origin, the inclusion $z_0 \in tS$ must hold for all large values of $|t|$. Thus, there exists a quantity $\rho(z_0, \varphi)$ such that $z_0 \in tS$ for $|t| > \rho(z_0, \varphi)$ but not for any t with $|t| \leqq \rho(z_0, \varphi)$. If S is unbounded, we may have $\rho(z_0, \varphi) = 0$, but normally $\rho(z_0, \varphi) > 0$. From its definition $\rho(z_0, \varphi)$ is a bounded continuous function of (z_0, φ) for $z_0 \in \Delta$, $0 \leqq \varphi < 2\pi$. Further, if a vertex of **A**[f] lies on the ray $\arg t = \varphi$, then it follows from (11.4.5) and (11.4.6) that

$$\rho(z_0, \varphi) < r(\varphi), \quad z_0 \in \Delta.$$

Moreover, we can strengthen this inequality to

$$(11.4.8) \qquad\qquad \sup_{z_0 \in \Delta} \rho(z_0, \varphi) < r(\varphi),$$

for Δ is closed and bounded so that if the supremum were equal to $r(\varphi)$ there would be a point $z_0 \in \Delta$ such that $\rho(z_0, \varphi) = r(\varphi)$. Actually, the inequality is valid for all values of φ since the right member is $+\infty$ if there is no vertex on the ray.

This being the case, we can find a function $s(\varphi)$ such that

(11.4.9) $b(\varphi) = \max [\sup \rho(z, \varphi), \sup |z|] < s(\varphi) < r(\varphi)$,

where the first supremum refers to z in Δ and the second involves only the subset with $\arg z = \varphi$. Obviously, $re^{i\varphi} \in \Delta$ implies $r < r(\varphi)$. Without restricting the generality we may assume that the quantity $\sup |z|$ for $z \in \Delta$ and $\arg z = \varphi$ is a continuous function of φ. Then $b(\varphi)$ is continuous and $r(\varphi)$ is lower semicontinuous. Hence, $r(\varphi) - b(\varphi)$ is also lower semicontinuous and, since the difference is positive, it has a positive minimum, δ say. Let us subdivide $[0, 2\pi)$ into small intervals $I_k = (\varphi_k, \varphi_{k+1})$, $k = 0, 1, 2, \cdots, n-1$, with $\varphi_0 = 0$, $\varphi_n = 2\pi$, such that in each I_k the oscillation of $b(\varphi)$ does not exceed $\frac{1}{5}\delta$. We can then choose $s(\varphi)$ in I_k as a linear function of φ such that

$$\min_{\varphi \in I_k} b(\varphi) + \tfrac{1}{5}\delta < s(\varphi) < \max_{\varphi \in I_k} b(\varphi) + \tfrac{4}{5}\delta.$$

Since in adjacent intervals I_k and I_{k+1} the minimum values of $b(\varphi)$ differ at most by $\frac{1}{5}\delta$ and the same is true for the maxima, we see that we can adjust the line segments in each I_k in such a manner that the resulting function $s(\varphi)$ is continuous and $s(0+) = s(2\pi-)$. It follows that this function $s(\varphi)$ is of bounded variation. Hence, the curve defined by

$$C_\Delta: \quad t = s(\varphi) e^{i\varphi}, \qquad 0 \le \varphi < 2\pi,$$

is a "scroc" having the desired properties. It clearly contains 0 and Δ in its interior since $s(\varphi) > b(\varphi)$. Further, $z/t \in S$ for $z \in \Delta$ and $t \in C_\Delta$ by the definition of $\rho(z, \varphi)$ and the choice of $s(\varphi)$.

After these preparations we can state and prove

THEOREM 11.4.1.   *Suppose that a method of summability M sums the geometric series to the limit $(1 - z)^{-1}$ for every z in a domain S, starlike with respect to the origin and containing the disk $|z| < 1$, and that the limit exists uniformly with respect to z on compact subsets of S. Then M also sums the series*

(11.4.10) $$f(z) = \sum_{n=0}^{\infty} c_n z^n$$

to the function in the star **SA**[*f*], *uniformly with respect to z on compact subsets.*

Proof.   By assumption there exists a family of functions $G_\alpha(z)$ which are the transforms of the geometric series by M and which converge to $(1 - z)^{-1}$ in S. If Δ is a compact set in **SA**[*f*] and $z \in \Delta$, consider the integral

(11.4.11) $$f_\alpha(z) = \frac{1}{2\pi i} \int_{C_\Delta} G_\alpha\left(\frac{z}{t}\right) f(t) \frac{dt}{t},$$

where C_Δ is a "scroc" of the type described in Lemma 11.4.1. The integral exists since $z/t \in S$ and $t \in \mathbf{A}[f]$. For sufficiently small values of z, we can deform the contour of integration so that it lies in the circle of convergence of (11.4.10). We can then substitute the series (11.4.4) and (11.4.10) in the integral, whose value consequently equals the term independent of t in the product of $G_\alpha(z/t)$ and $f(t)$, or

$$(11.4.12) \qquad f_\alpha(z) = \sum_{n=0}^{\infty} \mu_n(\alpha) c_n z^n.$$

Thus, $f_\alpha(z)$ is the M-transform of (11.4.10). For $z \in \Delta$, $t \in C_\Delta$, we see that z/t belongs to a compact subset of S, since Δ and C_Δ both are compact sets and t is bounded away from 0 on C_Δ. If now $\alpha \to \infty$

$$G_\alpha\left(\frac{z}{t}\right) \to \left(1 - \frac{z}{t}\right)^{-1},$$

uniformly with respect to $t \in C_\Delta$ and $z \in \Delta$. It follows that

$$\lim_{\alpha \to \infty} f_\alpha(z) = \frac{1}{2\pi i} \int_{C_\Delta} \frac{f(t)\,dt}{t - z} = f(z),$$

uniformly on Δ. This completes the proof.

We shall give some applications of this theorem. An important class of sequence-to-sequence transformations is given by the methods of Hausdorff. Each such method $[H, q]$ is characterized by a function $q(u) \in BV[0, 1]$ such that $q(u)$ is continuous to the right at $u = 0$ with $q(0) = 0$ and $q(1) = 1$. To a given sequence $\{A_n\}$ there corresponds a transformed sequence $\{B_n\}$ defined by

$$(11.4.13) \qquad B_n = \sum_{m=0}^{n} \binom{m}{n} A_m \int_0^1 u^m (1 - u)^{n-m}\,dq(u).$$

If $\lim B_n \equiv B$ exists, we say that

$$(11.4.14) \qquad [H, q]\text{-}\lim A_m = B.$$

In particular, if $q(u) = u^\alpha$, we obtain $[H, q] = (C, \alpha)$, the Cesàro method of order α. The Cesàro methods are useful in summing power series on the circle of convergence, but they are not effective outside the circle. There are other Hausdorff methods, however, which can be used for the purpose of analytic continuation.

For the geometric series the partial sums are

$$A_n = \frac{1 - z^{n+1}}{1 - z},$$

and the transformed sequence is found to be

$$(11.4.15) \qquad B_n = \left\{ 1 - z \int_0^1 [1 - (1 - z)u]^n\,dq(u) \right\} \frac{1}{1 - z}.$$

Thus the method $[H, q]$ sums the geometric series to the correct limit for a given value of z if and only if

(11.4.16) $$\lim_{n \to \infty} \int_0^1 [1 - (1 - z)u]^n \, dq(u) = 0.$$

From this criterion we obtain

THEOREM 11.4.2. *If $q(u) = 1$ for $0 < a \le u \le 1$, then the geometric series is summable $[H, q]$ in the disk*

(11.4.17) $$\left| z + \frac{1}{a} - 1 \right| < \frac{1}{a}.$$

Proof. For z in the disk we have

$$a(1 - z) = 1 - \rho e^{i\varphi}, \quad 0 \le \rho < 1, \quad 0 \le \varphi < 2\pi.$$

It follows that

$$| 1 - (1 - z)u |^2 = | 1 - \frac{u}{a}(1 - \rho e^{i\varphi}) |^2$$

$$= \left(1 - \frac{u}{a}\right)^2 + 2\rho \frac{u}{a}\left(1 - \frac{u}{a}\right)\cos\varphi + \left(\rho\frac{u}{a}\right)^2 \le \left[1 - \frac{u}{a}(1 - \rho)\right]^2.$$

This expression is < 1 except for $u = 0$. In the integral (11.4.16) the interval $(a, 1]$ gives no contribution and

$$\left| \int_0^a [1 - (1 - z)u]^n \, dq(u) \right| \le V_0^\delta[q] + \left[1 - \frac{\delta}{a}(1 - \rho)\right]^n V_\delta^1[q],$$

where $V_\alpha^\beta[q]$ is the total variation of $q(u)$ on $[\alpha, \beta]$. The superior limit of the left member as $n \to \infty$ hence does not exceed $V_0^\delta[q]$, which is arbitrarily small with δ since, by assumption, $q(u)$ is continuous at $u = 0$. It follows that (11.4.16) holds.

The most important special case covered by Theorem 11.4.2 is that in which

(11.4.18) $$q(u) = \begin{cases} 1, a \le u \le 1, \\ 0, 0 \le u < a. \end{cases}$$

This choice gives rise to the so called Euler-Knopp means. Their effectiveness as an instrument of analytic continuation grows as a decreases toward 0.

We now consider methods of the factor sequence type and start with a variant of the method introduced by É. Borel. We take

(11.4.19) $$\mu_n(\alpha) = e^{-\alpha} \sum_{k=n}^{\infty} \frac{\alpha^k}{k!}, \quad n = 0, 1, 2, \cdots,$$

and find after some computation

(11.4.20) $$G_\alpha(z) = \frac{1}{1 - z}[1 - z e^{-\alpha(1 - z)}].$$

This is an entire function of z and tends to $(1 - z)^{-1}$ as $\alpha \to \infty$, uniformly with respect to z in the half-plane $\Re(z) \leq 1 - \delta, 0 < \delta$. Thus we have

$$S \equiv B = [z \mid \Re(z) < 1].$$

The corresponding star $\mathbf{BA}[f]$ is an interesting configuration known as Borel's polygon of summability. By (11.4.6), $\mathbf{BA}[f]$ is the intersection of all half-planes

$$\Re\left(\frac{z}{v_\alpha}\right) < 1,$$

where v_α runs through the vertices of $\mathbf{A}[f]$. The intersection is obviously a convex domain, and it is a polygon if, for instance, the set $\{v_\alpha\}$ is finite. Thus, we have

THEOREM 11.4.3.   *For $z \in \mathbf{BA}[f]$ we have*

$$(11.4.21) \qquad \lim_{\alpha \to \infty} e^{-\alpha} \sum_{n=0}^{\infty} c_n \left\{ \sum_{k=n}^{\infty} \frac{\alpha^k}{k!} \right\} z^n = f(z).$$

Borel also gave another interesting representation of $f(z)$ in $\mathbf{BA}[f]$. This is based on the so-called Borel transform of $f(z)$ which, by definition, is the entire function

$$(11.4.22) \qquad F(z) = \sum_{n=0}^{\infty} \frac{c_n}{n!} z^n.$$

THEOREM 11.4.4.   *For $z \in \mathbf{BA}[f]$ the integral*

$$(11.4.23) \qquad \int_0^\infty e^{-\alpha} F(\alpha z)\, d\alpha$$

converges and represents $f(z)$.

Proof.   We start by observing that

$$(11.4.24) \qquad F(z) = \frac{1}{2\pi i} \int_C e^{\frac{z}{s}} f(s) \frac{ds}{s},$$

where C is any "scroc" surrounding $s = 0$ located in $\mathbf{A}[f]$. If we choose C as a circle $\mid s \mid = \rho$, where ρ is less than the radius of convergence of (11.4.10), then we have

$$e^{\frac{z}{s}} f(s) = \sum_{k=0}^{\infty} \frac{1}{k!} \left(\frac{z}{s}\right)^k \sum_{m=0}^{\infty} c_m s^m,$$

and the value of the integral is the sum of the terms in the product series which are independent of s. This proves (11.4.24).

It follows that

$$(11.4.25) \qquad \int_0^\infty e^{-\alpha} F(\alpha z)\, d\alpha = \int_0^\infty e^{-\alpha} \left\{ \frac{1}{2\pi i} \int_C e^{\frac{\alpha z}{s}} f(s) \frac{ds}{s} \right\} d\alpha.$$

Here we restrict z to lie in a compact subset Δ of $\mathbf{BA}[f]$. By Lemma 11.4.1 we can choose C in $\mathbf{A}[f]$ in such a manner that for $z \in \Delta$ and $s \in C$ the values of z/s belong to a compact subset of B. In particular, there exists a $\delta > 0$ such that

$$\Re\left(\frac{z}{s}\right) \leqq 1 - \delta.$$

This implies that the double integral in (11.4.25) is absolutely convergent. By the Fubini theorem the order of integration may then be interchanged. This gives

$$\frac{1}{2\pi i}\int_C f(s)\left\{\int_0^\infty e^{-\alpha\left(1-\frac{z}{s}\right)}\,d\alpha\right\}\frac{ds}{s} = \frac{1}{2\pi i}\int_C \frac{f(s)}{s-z}\,ds = f(z).$$

Since Δ is an arbitrary compact subset of $\mathbf{BA}[f]$, Theorem 11.4.4 is proved.

From the point of view of the theory of analytic continuation, the most effective methods of summability are those which sum the geometric series in its principal star so that

$$\mathbf{SA}[f] = \mathbf{A}[f]$$

for every power series. A number of such methods are available in the literature. The best known are based on factor sequences $\{\mu_n(\delta)\}$, where $\mu_n(\delta)$ has one of the following three values:

$$(11.4.26) \qquad (1+n)^{-\delta(1+n)},\quad \frac{1}{\Gamma(1+\delta n)},\quad \frac{\Gamma(1+(1-\delta)n)}{\Gamma(1+n)}.$$

These methods are due to E. Lindelöf, G. Mittag-Leffler, and E. Le Roy respectively. It should be noted that the parameter δ is supposed to tend to 0 and not to ∞ as in the cases previously considered.

THEOREM 11.4.5.   *For $z \in \mathbf{A}[f]$ we have*

$$(11.4.27) \qquad \lim_{\delta\to 0}\sum_{n=0}^\infty (1+n)^{-\delta(1+n)}c_n z^n = f(z),$$

and the same result holds for the other two factor sequences of (11.4.26).

Proof.   By Theorem 11.4.1 it suffices to prove that

$$(11.4.28) \qquad \lim_{\delta\to 0} L_\delta(z) = (1-z)^{-1},$$

uniformly on compact sets in $\mathbf{A}[g]$ where $L_\delta(z)$ is the entire function of Problem 5, Exercise 11.3. This relation obviously holds on any disk $|z| < 1 - \varepsilon,\ 0 < \varepsilon$. Let Δ be a compact set in $\mathbf{A}[g]$ which omits a neighborhood of the origin and let δ_0 be so small that Δ is contained in the sector

$$\delta_0\frac{\pi}{2} < \arg z < 2\pi - \delta_0\frac{\pi}{2}.$$

Since the function

$$(1 + w)^{-\delta(1+w)}$$

satisfies the conditions of Theorem 11.3.2 in the half-plane $\Re(w) \geq -\frac{1}{2}$, we have

$$L_\delta(z) - (1 - z)^{-1} = i \int_{-\infty}^{\infty} \frac{(\frac{1}{2} + iv)^{-\delta(\frac{1}{2}+iv)} - 1}{e^{-2\pi v} + 1} z^{-\frac{1}{2}+iv} \, dv, \quad z \in \Delta, \delta < \delta_0.$$

On a fixed finite interval $(-a, a)$ the integrand tends to zero as $\delta \to 0$ uniformly with respect to v and z. For $|v| > a$ the integrand is dominated by an expression of the form $M(\varepsilon) \exp(-\varepsilon |v|)$ where ε is a lower bound of $\theta - \delta \frac{\pi}{2}$ and of $2\pi - \delta \frac{\pi}{2} - \theta$ for $z \in \Delta$ and $\delta < \delta_0$. Thus, the contributions of the intervals $(-\infty, -a)$ and (a, ∞) do not exceed $M_1(\varepsilon)e^{-\varepsilon a}$ and can be made as small as we please by a suitable choice of a. This proves (11.4.28) and, hence, also the theorem, since the same argument applies to the other cases.

These examples do not exhaust the applications of the general theory of summability to the problem of analytic continuation, but we shall not pursue the question further.

EXERCISE 11.4

1. Prove that (11.4.2) holds if the sequence $\{\mu_n(\alpha)\}$ satisfies the following conditions:

(i) $\lim\limits_{n \to \infty} \mu_n(\alpha) = 0$ for each fixed α, $\quad 0 < \alpha < \infty$,

(ii) $\sum\limits_{n=0}^{\infty} |\mu_n(\alpha) - \mu_{n+1}(\alpha)| \leq M$ uniformly in α,

(iii) $\lim\limits_{\alpha \to \infty} \mu_n(\alpha) = 1$ for each fixed n.

2. Verify (11.4.15).

3. Prove that the sequence $\{\mu_n(\alpha)\}$ of (11.4.19) satisfies the conditions of Problem 1 above.

4. Verify (11.4.20).

5. Determine **BA**$[f]$ if $f(z)$ is the sum of the series

$$\sum_{n=1}^{\infty} 2^{-n} \tan\left(e^{\frac{n\pi i}{3}} \frac{\pi z}{n}\right).$$

6. The divergent power series $\sum_0^\infty (-1)^n n! \, z^n$ can be summed by formula (11.4.23) to the sum

$$\int_0^\infty (1 + \alpha z)^{-1} e^{-\alpha} \, d\alpha \equiv f(z).$$

This function is holomorphic in the plane cut along the negative real axis. Why? Show that the nth derivative of $f(z)$ tends to $(-1)^n(n!)^2$ as $z \to 0$ through positive values.

7. Verify that the sequence $\{[\Gamma(1 + \delta n)]^{-1}\}$ satisfies the conditions of Problem 1 (with an obvious modification in (iii)) and carry through the proof of Theorem 11.4.5 for the case of Mittag-Leffler summability.

11.5. Polynomial series. The problem of continuing a power series outside its circle of convergence may be handled by methods which are not applicable to the summation of other classes of series. In this connection we wish to call attention to the polynomial series constructed by Borel, Mittag-Leffler, and Paul Painlevé (1863–1933) among others. The following example illustrates one of the methods invented by Mittag-Leffler. It has the advantage of being intuitively simple and capable of infinite variations.

We choose arbitrarily a function $A(u)$,

$$(11.5.1) \qquad A(u) = \sum_{k=1}^{\infty} a_k u^k,$$

with the following properties:

(i) $a_k \geq 0$,

(ii) $\sum_{k=1}^{\infty} a_k = 1$,

(iii) the mapping $u \to A(u)$ is one-to-one on $|u| \leq 1$.

Thus, $A(0) = 0$, $A(1) = 1$, and $v = A(u)$ maps the disk $|u| < 1$ conformally onto a certain domain D which contains the interval $[0, 1)$ and is symmetric with respect to this line segment. $A(u)$ is continuous in $|u| \leq 1$.

Consider the function $f(z)$ defined by the power series

$$(11.5.2) \qquad f(z) = \sum_{n=0}^{\infty} c_n z^n, \quad |z| < R.$$

For given values of u and z, $|u| \leq 1$, $|z| < R$, we define w by

$$(11.5.3) \qquad z = wA(u).$$

We substitute this expression in (11.5.2) and obtain

$$
\begin{aligned}
f(z) &= \sum_{n=0}^{\infty} c_n w^n [A(u)]^n \\
&= \sum_{n=0}^{\infty} c_n w^n \left[\sum_{k=1}^{\infty} a_k u^k \right]^n \\
&\equiv \sum_{n=0}^{\infty} c_n w^n \sum_{k=n}^{\infty} a_{k,n} u^k.
\end{aligned}
$$

It is clear that $a_{k,n} \geqq 0$ for all n. Further, the double series is absolutely and uniformly convergent for $|w| \leqq R - \delta$, $|u| \leqq 1$. For such values of u and w it is permitted to interchange the order of summation, so that

$$f(z) = \sum_{k=0}^{\infty} u^k \sum_{n=0}^{k} c_n a_{k,n} w^n.$$

This series is absolutely convergent for $u = 1$, $w = z$ if $|z| < R$. Hence,

$$(11.5.4) \qquad f(z) = c_0 + \sum_{k=1}^{\infty} \left[\sum_{n=1}^{k} c_n a_{k,n} z^n \right].$$

This is obviously a polynomial series. Considered as a double series it is absolutely convergent at least for $|z| < R$ and represents $f(z)$ for such values of z. Normally its region of convergence is much larger.

THEOREM 11.5.1. *The series* (11.5.4) *converges and represents* $f(z)$ *for each* z *such that* $z\bar{D} \subset \mathbf{A}[f]$ *where* $\bar{D}$ *is the closure of* D. *The series converges uniformly with respect to* z *on any compact set* Δ *such that* $\Delta \bar{D} \subset \mathbf{A}[f]$.

Proof. Consider a compact set Δ such that $\Delta \bar{D} \subset \mathbf{A}[f]$ where we have $\Delta \bar{D} = [z \mid z = z_1 z_2, z_1 \in \Delta, z_2 \in \bar{D}]$. Fix z in Δ and consider

$$(11.5.5) \qquad f(zA(u)) \equiv \sum_{k=0}^{\infty} P_k(z;f) u^k, \quad |u| < 1.$$

The left member is a holomorphic function of u in $|u| < 1$ since $z\bar{D} \subset \mathbf{A}[f]$. It is continuous in $|u| \leqq 1$. The right member is the Maclaurin series of $f(zA(u))$ as function of u. Comparison with (11.5.4) shows that

$$(11.5.6) \qquad P_0(z;f) = c_0, \quad P_k(z;f) = \sum_{n=1}^{k} c_n a_{k,n} z^n, \quad k > 0.$$

It is obvious that the polynomials $P_k(z;f)$ are uniformly bounded for $z \in \Delta$. We shall prove the stronger assertion that

$$(11.5.7) \qquad \sum_{k=1}^{\infty} k \, |P_k(z;f)|^2 \leqq M(\Delta), \quad z \in \Delta,$$

where the series is uniformly convergent on Δ.

The proof will be based upon two lemmas which we state here and prove at the end of the section. The first is known as the theorem of Ulisse Dini (1845–1918). With $(C, 1)$-summability instead of Abel summability, the second lemma is due to L. Fejér.

LEMMA 11.5.1. *A monotone increasing sequence of continuous functions which converges to a continuous limit in a bounded closed set, converges uniformly to its limit.*

LEMMA 11.5.2.   *Given a power series*

$$\sum_{n=0}^{\infty} b_n z^n \equiv g(z), \quad |z| < 1,$$

such that

$$\sum_{n=1}^{\infty} n \, | \, b_n \, |^2 \quad converges \ and \quad \lim_{r \to 1} g(r) = B.$$

Then the series converges for $z = 1$ and its sum equals B. If the hypotheses hold uniformly with respect to a parameter, so does the conclusion.

In order to prove (11.5.7) we consider the integral

$$(11.5.8) \qquad\qquad J(zD) \equiv \int\!\!\int | \, f'(\zeta) \, |^2 \, d\xi \, d\eta$$

extended over the domain zD. The geometrical interpretation of this integral follows from the discussion in Section 4.6: it is the area of the image of zD under the conformal mapping defined by $f(\zeta)$, where, however, multiply covered portions of the map have to be counted as often as they occur. There exists a constant $M_1(\Delta)$ such that

$$| \, f'(\zeta) \, | \leq M_1(\Delta) \quad \text{for} \quad \zeta \in \Delta \bar{D}.$$

Hence,

$$(11.5.9) \qquad\qquad J(zD) \leq \rho m(D)[M_1(\Delta)]^2, \quad z \in \Delta,$$

where $\rho = \max | \, z \, |, z \in \Delta$, and $m(D)$ is the area of D.

On the other hand, $J(zD)$ is also the area of the image of the disk $| \, u \, | < 1$ under the mapping defined by $f(zA(u))$. From the second interpretation it follows that

$$(11.5.10) \qquad J(zD) = \int\!\!\int \left| \frac{\partial}{\partial u} f(zA(u)) \right|^2 r \, dr \, d\theta, \quad u = re^{i\theta},$$

where the integral is taken over the unit disk. In this integral we substitute the series (11.5.5) and use the result of Problem 6, Exercise 8.2. This gives

$$(11.5.11) \qquad\qquad J(zD) = \pi \sum_{k=1}^{\infty} k \, | \, P_k(z; f) \, |^2.$$

Combining this result with (11.5.9), we see that (11.5.7) holds. The series (11.5.11) has positive terms which are continuous functions of z, and its sum $J(zD)$ is a continuous function of z by virtue of its geometric significance. Dini's theorem then shows that the series is uniformly convergent on Δ.

Finally we note that

$$\lim_{u \to 1} f(zA(u)) = f(z),$$

uniformly with respect to z in Δ. We can then apply Lemma 11.5.2 to the series (11.5.5). It follows that the series converges for $u = 1$, its sum being $f(z)$, and the convergence is uniform with respect to z in Δ. This completes the proof of Theorem 11.5.1.

Among the functions $A(u)$ that may be used in this connection the following should be mentioned:

$$(11.5.12) \qquad \frac{\alpha u}{1 - (1 - \alpha)u}, \quad \frac{(1 - \alpha)^\alpha u}{(1 - \alpha u)^\alpha}, \quad 1 - (1 - u)^\alpha,$$

where in each case $0 < \alpha < 1$.

Proof of Lemma 11.5.1. Let S be the closed bounded set, $\{U_n(z)\}$ the monotone increasing sequence of continuous functions, and $U(z)$ the continuous limit. We set

$$V_n(z) = U(z) - U_n(z).$$

Then $\{V_n(z)\}$ is a monotone decreasing sequence of continuous functions and the limit is 0 everywhere in S. Suppose that the convergence were not uniform. Then for some positive δ, we could find a sequence $\{z_n\} \subset S$ such that $V_n(z_n) > 2\delta$ for all n. We may assume that $z_n \to z_0 \in S$, for there is at least one limit point and this point is in S since S is closed. Since $V_n(z_0) \to 0$, there exists an N such that $V_n(z_0) < \delta$ for $n \geq N$. The function $V_N(z)$ is continuous, so there exists an $\varepsilon = \varepsilon(\delta)$ such that $\mid V_N(z) - V_N(z_0) \mid < \delta$ for $\mid z - z_0 \mid < \varepsilon$. But if n is large enough, we have $\mid z_n - z_0 \mid < \varepsilon$. Hence,

$$0 \leq V_N(z_n) < V_N(z_0) + \delta < 2\delta$$

for all large n. But for $n > N$

$$V_n(z_n) \leq V_N(z_n) < 2\delta,$$

and this contradicts our hypothesis. It follows that the convergence must be uniform, and Dini's theorem is proved. In particular, we see that a convergent series whose terms are positive continuous functions and whose sum is continuous must converge uniformly. This is the form of the theorem used above.

Proof of Lemma 11.5.2. If $B_m = \sum\limits_{n=0}^{m} b_n$, we have

$$B_m - g(z) = \sum_{n=1}^{m} b_n(1 - z^n) - \sum_{n=m+1}^{\infty} b_n z^n,$$

so that

$$\mid B_m - g(z) \mid \leq \mid 1 - z \mid \sum_{n=1}^{m} n \mid b_n \mid + \sum_{n=m+1}^{\infty} \mid b_n \mid \mid z \mid^n,$$

where we have used the fact that for $\mid z \mid < 1$

$$\mid 1 - z^n \mid = \mid 1 - z \mid \mid 1 + z + \cdots + z^{n-1} \mid < n \mid 1 - z \mid.$$

For a given δ, $\delta > 0$, we now choose an integer p such that

$$\sum_{n=p+1}^{\infty} n \mid b_n \mid^2 < \delta^2.$$

Hence, by Cauchy's inequality we have for $m > p$

$$\left[\sum_{p+1}^{m} n \mid b_n \mid\right]^2 \leqq \left[\sum_{p+1}^{m} n \mid b_n \mid^2\right]\left[\sum_{p+1}^{m} n\right] < \delta^2(m+1)^2.$$

Further,

$$\left[\sum_{m+1}^{\infty} \mid b_n \mid \mid z \mid^n\right]^2 < \left[\sum_{m+1}^{\infty} n \mid b_n \mid^2\right]\left[\sum_{m+1}^{\infty} \frac{\mid z \mid^{2n}}{n}\right]$$

$$< \frac{\delta^2}{m+1}(1 - \mid z \mid^2)^{-1} < \delta^2[(m+1)(1 - \mid z \mid)]^{-1}.$$

It follows that

$$\mid B_m - g(z) \mid < \mid 1 - z \mid \sum_{n=1}^{p} n \mid b_n \mid + \delta \mid 1 - z \mid(m+1) + \delta[(m+1)(1 - \mid z \mid)]^{-\frac{1}{2}}.$$

Here we set

$$z = \frac{m}{m+1}$$

and obtain

$$\left| B_m - g\left(\frac{m}{m+1}\right)\right| < \frac{1}{m+1}\sum_{n=1}^{p} n \mid b_n \mid + 2\delta.$$

As $m \to \infty$, the superior limit of the right member is 2δ. Since δ is arbitrarily small and since

$$\lim_{m \to \infty} g\left(\frac{m}{m+1}\right) = B,$$

it follows that

$$\lim_{m \to \infty} B_m = B$$

as asserted. The extension to uniform convergence in the case in which the coefficients depend upon a parameter is immediate. This completes the proof.

EXERCISE 11.5

1. What is the domain D for the first function $A(u)$ listed under (11.5.12)? Find $P_k(z;f)$.

2. Same question for the third function.

3. Prove Fejér's theorem: If Σb_n is $(C, 1)$-summable to the sum B and if $\Sigma n \mid b_n \mid^2$ is convergent, then Σb_n is convergent.

$\left(Hint:\ \text{Prove first that}\ \dfrac{B_0 + B_1 + \cdots + B_n}{n+1} = B_n - \dfrac{b_1 + 2b_2 + \cdots + nb_n}{n+1}.\right)$

11.6. Composition theorems. In the early sections of this chapter we encountered several classes of holomorphy-preserving transformations. A necessary and sufficient condition in order that $\mathbf{A}[h]$, the principal star of

$$(11.6.1) \qquad h(z) \equiv \sum_{n=0}^{\infty} a_n b_n z^n,$$

contain the principal star $\mathbf{A}[g]$ of

$$(11.6.2) \qquad g(z) \equiv \sum_{n=0}^{\infty} b_n z^n$$

was found to be that the principal star $\mathbf{A}[f]$ of

$$(11.6.3) \qquad f(z) \equiv \sum_{n=0}^{\infty} a_n z^n$$

should reduce to the whole plane less the interval $[+1, +\infty)$ of the real axis. This is a special case of Hadamard's composition theorem, to which we now turn.[1]

The formulas above define a type of *composition* or product for power series

$$(11.6.4) \qquad h = f * g.$$

This is a commutative and associative operation and the geometric series acts as a unity or neutral element. Not every $f(z)$ has an inverse with respect to this operation; a necessary condition is that $a_n \neq 0$ for every n. A necessary and sufficient condition is that the power series

$$\sum_{n=0}^{\infty} \frac{1}{a_n} z^n$$

shall exist and have a positive radius of convergence. The set of convergent power series is an algebra under the operations of addition, Hadamard composition, and multiplication by constants. In this algebra there are divisors of zero since, obviously, $h(z)$ may be identically zero without either $f(z)$ or $g(z)$ having this property.

Suppose now that $\mathbf{A}[f]$, $\mathbf{A}[g]$, and $\mathbf{A}[h]$ are the principal stars of the three functions figuring in (11.6.4). We shall define also a composition of principal

---

[1] Jacques Hadamard, born in 1865, the revered Nestor of French mathematicians, gave the impetus to much of the later research on power series and their singularities by his dissertation "Essai sur l'étude des fonctions données par leur dévelopment de Taylor," *Journal des mathématiques pures et appliquées*, Series 4, Vol. 8 (1892), pp. 101–186. This was followed by important papers on the theory of entire functions with applications to the Riemann zeta function. His lifework was devoted to Cauchy's problem in the theory of partial differential equations of the hyperbolic type, where his contributions are basic.

stars somewhat analogous to the definition of **SA**[f] in Section 11.4. We define

(11.6.5) $\mathbf{A}[f] * \mathbf{A}[g] = \mathbf{C}\{\mathbf{C}(\mathbf{A}[f]) \cdot \mathbf{C}(\mathbf{A}[g])\},$

where, as usual, **C** denotes the complement and the product set is the set of products:

$$S_1 \cdot S_2 = [z_1 z_2 \mid z_1 \in S_1, z_2 \in S_2].$$

The composition is obviously a commutative operation.

LEMMA 11.6.1.   **A**[f] * **A**[g] *is starlike with respect to the origin and equals*

(11.6.6) $\cap_\alpha \{v_\alpha(f) \cdot \mathbf{A}[g]\} = \cap_\beta \{v_\beta(g) \cdot \mathbf{A}[f]\},$

where $v_\alpha(f)$ *runs through the vertices of* **A**[f] *and* $v_\beta(g)$ *runs through the vertices of* **A**[g].

Proof.   If $z_1 \in \mathbf{C}(\mathbf{A}[f])$, then there is a vertex $v_\alpha(f)$ on the ray from $z = 0$ to $z = z_1$. Moreover, all the points on the ray from $v_\alpha(f)$ out to infinity belong to $\mathbf{C}(\mathbf{A}[f])$, and the latter set is the union of all such line segments. A similar statement can be made for $\mathbf{C}(\mathbf{A}[g])$. It follows that the product set $\mathbf{C}(\mathbf{A}[f]) \cdot \mathbf{C}(\mathbf{A}[g])$ consists of all points of the form

(11.6.7) $z = \rho v_\alpha(f) v_\beta(g), \quad \rho \geq 1.$

Passing over to the complementary set, we see that the product star $\mathbf{A}[f] * \mathbf{A}[g]$ is the star whose vertices are all the points of the form $v_\alpha(f) v_\beta(g)$ which are accessible by radial approach from the origin. In particular, it is clear that the product star is a star.

By symmetry it is enough to prove that the left member of (11.6.6) equals the product star. Suppose that a point z_0 belongs to $v_\alpha(f) \mathbf{A}[g]$ for all α. Then, for each α there is a point $w_\alpha \in \mathbf{A}[g]$ such that

$$z_0 = v_\alpha(f) w_\alpha.$$

This means that z_0 is never of the form (11.6.7), that is, $z_0 \in \mathbf{C}\{\mathbf{C}(\mathbf{A}[f]) \cdot \mathbf{C}(\mathbf{A}[g])\}$ or

$$\cap_\alpha \{v_\alpha(f) \cdot \mathbf{A}[g]\} \subset \mathbf{A}[f] * \mathbf{A}[g].$$

On the other hand, if z_0 belongs to the star product, then z_0 is never of the form (11.6.7). Now, for any α we can always write

$$z_0 = v_\alpha(f) w_\alpha$$

and, as we have just seen, w_α can never be of the form $\rho v_\beta(g)$ with $\rho \geq 1$. This means that $w_\alpha \in \mathbf{A}[g]$ and $z_0 \in v_\alpha(f) \cdot \mathbf{A}[g]$. This proves the opposite inclusion, so that

$$\mathbf{A}[f] * \mathbf{A}[g] = \cap_\alpha \{v_\alpha(f) \cdot \mathbf{A}[g]\}$$

as asserted.

We note again that every vertex of the product star is the product of vertices of the factor stars. We can now state and prove Hadamard's theorem.

THEOREM 11.6.1.   *We have*

(11.6.8) $\mathbf{A}[h] \supset \mathbf{A}[f] * \mathbf{A}[g].$

Proof.   We can use the argument employed in the proof of Theorem 11.4.1. If Δ is a compact subset of the right member of (11.6.8), then there exists a "scroc" C_Δ in $\mathbf{A}[f]$ which surrounds Δ and 0 while $z/t \in \mathbf{A}[g]$ for $z \in \Delta$, $t \in C_\Delta$. We can use the construction in Lemma 11.4.1 where we replace S by $\mathbf{A}[g]$. We can then form the integral

(11.6.9) $\dfrac{1}{2\pi i} \displaystyle\int_{C_\Delta} g\left(\dfrac{z}{t}\right) f(t) \dfrac{dt}{t} = h(z).$

The integral exists for $z \in \Delta$. We do not know at the outset that it represents $h(z)$ there, but at any rate it defines a holomorphic function of z in Δ since we can differentiate under the sign of integration. If $|z|$ is sufficiently small, we may contract C_Δ to a circle $|t| = \rho < R_f$, the radius of convergence of (11.6.3). It suffices that $|z| < \rho R_g$, where R_g is the radius of convergence of (11.6.2). We can then substitute these two series in the integral, multiply out, and integrate termwise. The product series (11.6.1) comes out as the result. It follows that (11.6.9) represents the analytic continuation of $h(z)$ in Δ. Hence, $h(z)$ is indeed holomorphic in the product star and (11.6.8) holds.

It should be emphasized that the inclusion in this formula may very well be proper. We saw above that $h(z) = f(z) * g(z)$ may be identically zero without either of the factors having this property. In such a case $\mathbf{A}[h]$ is the whole plane and the product star is practically arbitrary. Hadamard's theorem is often carelessly stated as asserting that $(f * g)(z)$ has no other singularities than points of the form

$$z_\alpha z_\beta,$$

where z_α is a singular point of $f(z)$ and z_β a singular point of $g(z)$. This assertion is certainly false in general. The following example illustrates this point. We take

$$f(z) = \sum_{n=0}^{\infty} (-1)^n z^{2n},$$

$$g(z) = \sum_{n=0}^{\infty} b_n z^n,$$

where

$$b_n = \begin{cases} 1 & \text{if } n = 3^k, \\ 2^{-n} & \text{if } n = 4^k, \\ 0 & \text{otherwise.} \end{cases} \qquad k = 1, 2, 3, \cdots,$$

Here

$$h(z) = \sum_{k=1}^{\infty} \left(\frac{z}{2}\right)^{4^k}.$$

Anticipating a result which will be proved in the next section, we observe that $g(z)$ and $h(z)$ have natural boundaries, namely, the circles $|z| = 1$ and $|z| = 2$ respectively, whereas $f(z)$ has the singular points $z = \pm 1$. Here $\mathbf{A}[f] * \mathbf{A}[g]$ is the disk $|z| < 1$, and $\mathbf{A}[h]$ is the disk $|z| < 2$. There is obviously no relation between the product set $\{z_\alpha z_\beta\}$ which is the circle $|z| = 1$ and the set of singularities of $h(z)$ which is the circle $|z| = 2$.

The relations between the product set $\{z_\alpha z_\beta\}$ and the singularities of $(f * g)(z)$ under various assumptions on $f(z)$ and $g(z)$ have been the object of much research, but we are far from a satisfactory understanding of the true state of affairs. We shall not pursue this question further.

There is a related composition theorem due to Adolf Hurwitz (1859–1919). Here we start out with the two series

(11.6.10)
$$a(z) = \sum_{n=0}^{\infty} a_n z^{-n-1}, \quad |z| > r_a,$$

(11.6.11)
$$b(z) = \sum_{n=0}^{\infty} b_n z^{-n-1}, \quad |z| > r_b,$$

and form the composed series

(11.6.12)
$$c(z) = \sum_{n=0}^{\infty} \left\{ \sum_{k=0}^{n} \binom{n}{k} a_k b_{n-k} \right\} z^{-n-1}$$

which converges at least for

(11.6.13)
$$|z| > r_a + r_b.$$

With these series we consider their principal stars with respect to ∞, which we denote by $\mathbf{A}_\infty[a]$, $\mathbf{A}_\infty[b]$, and $\mathbf{A}_\infty[c]$, respectively. Thus a point z_0 belongs to $\mathbf{A}_\infty[a]$ if $a(z)$ is holomorphic for every z of the form λz_0, $1 \le \lambda < \infty$. We also consider a process of addition applied to such stars. We define

(11.6.14) $$\mathbf{A}_\infty[a] \oplus \mathbf{A}_\infty[b] = \mathbf{C}\{\mathbf{C}[\mathbf{A}_\infty[a]] + \mathbf{C}[\mathbf{A}_\infty[b]]\}.$$

We recall that $\mathbf{C}[S]$ is the complement of the set S. Further, we write

$$S_1 + S_2 \equiv \{z \mid z = z_1 + z_2, z_1 \in S_1, z_2 \in S_2\}.$$

LEMMA 11.6.2. $\mathbf{A}_\infty[a] \oplus \mathbf{A}_\infty[b]$ *is a star with respect to* ∞ *which contains the exterior of the circle* $|z| = r_a + r_b$.

Proof.   We note first that $\mathbf{C}[\mathbf{A}_\infty[a]]$ and $\mathbf{C}[\mathbf{A}_\infty[b]]$ are stars with respect to the origin and so is their vector sum. Since the two summands are confined to $|z| < r_a$ and $|z| < r_b$ respectively, the sum is confined to $|z| < r_a + r_b$. From this it follows that the complement contains the exterior of the circle $|z| = r_a + r_b$ as asserted. Further, the complement of a bounded domain which is starlike with respect to the origin is a star with respect to infinity. This completes the proof.

LEMMA 11.6.3.   *Suppose that Δ is a subset of $\mathbf{A}_\infty[a] \oplus \mathbf{A}_\infty[b]$ whose distance from the boundary is positive. Then there exists a "scroc" C_Δ in $\mathbf{A}_\infty[a]$ surrounding the origin and separating 0 from Δ such that $z - t \in \mathbf{A}_\infty[b]$ if $z \in \Delta$, $t \in C_\Delta$.*

Proof.   If Δ is not in $\mathbf{C}(\mathbf{A}_\infty[a]) + \mathbf{C}(\mathbf{A}_\infty[b])$, then

$$\{\Delta - \mathbf{C}(\mathbf{A}_\infty[b])\} \cap \mathbf{C}(\mathbf{A}_\infty[a]) = \emptyset,$$

where the first set on the left is the set of all points $z - w$ such that $z \in \Delta$, $w \in \mathbf{C}(\mathbf{A}_\infty[b])$. Since the closed set $\mathbf{C}(\mathbf{A}_\infty[b])$ is bounded and contains the origin, we can find a simple closed polygon Π which contains $\mathbf{C}(\mathbf{A}_\infty[a])$ in its interior and $\Delta - \mathbf{C}(\mathbf{A}_\infty[b])$ in its exterior. If t is any point of Π, then $t \in \mathbf{A}_\infty[a]$ and $z - t \in \mathbf{A}_\infty[b]$. We can then take $C_\Delta = \Pi$ with the positive orientation so that arg t increases by 2π when t describes C_Δ.

THEOREM 11.6.2.   *We have*

(11.6.15) $$\mathbf{A}_\infty[c] \supset \mathbf{A}_\infty[a] \oplus \mathbf{A}_\infty[b].$$

Proof.   Suppose that Δ is a subset of the right member such that the distance of Δ from the boundary of the sum set is positive. We can then find a "scroc" C_Δ with the properties stated in Lemma 11.6.3. We use this to form the integral

(11.6.16) $$\frac{1}{2\pi i} \int_{C_\Delta} a(t)\, b(z - t)\, dt = c(z).$$

The integral exists since $t \in \mathbf{A}_\infty[a]$ and $z - t \in \mathbf{A}_\infty[b]$ when $z \in \Delta$. It represents a holomorphic function of z since we can differentiate with respect to z under the sign of integration. If $|z| > r_a + r_b$, we can use a circle $|t| = r > r_a$ as path of integration. If r is chosen properly, $|z - t|$ will be greater than r_b and we can substitute the two series (11.6.1) and (11.6.11) for $a(t)$ and $b(z - t)$ in the integral. A straightforward computation gives the series (11.6.12) as value of the integral. Thus, the latter does really give the analytic continuation of $h(z)$, and this function is holomorphic everywhere in $\mathbf{A}_\infty[a] \oplus \mathbf{A}_\infty[b]$. This completes the proof.

EXERCISE 11.6

1. Show that the set of convergent power series in which the nth coefficient is 0, n fixed, forms an ideal in the algebra described above in which Hadamard composition is the product operation.

2. Show that the radius of convergence of the series for $h(z)$ is not less than the product of the radii for $f(z)$ and $g(z)$.

3. Show that the existence of a "scroc" C_Δ required in formula (11.6.9) can be based directly upon the definition (11.6.5).

4. Prove that the Hadamard product of $(1 - z)^{-\alpha}$ and $(1 - z)^{-\beta}$ is the hypergeometric function

$$F(\alpha, \beta, 1; z).$$

(The interest of this example lies in the fact that the product may have a logarithmic singularity at $z = 1$ whereas the factors have algebraic branch points there if α and β are rational. Further, the continuation of the product has a logarithmic branch point at $z = 0$.)

5. Verify (11.6.3).

6. Give the details showing that the integral (11.6.16) is represented by the series (11.6.12) for large values of z.

7. Prove that differentiation under the sign of integration in (11.6.9) and (11.6.16) is legitimate by discussing in each case the difference quotient and proving that the latter converges uniformly with respect to z in Δ to the formal derivative.

11.7. Gap theorems and noncontinuable power series.   In this section we consider power series

$$(11.7.1) \qquad \sum_{n=0}^{\infty} c_n z^n \equiv f(z)$$

whose radius of convergence is $R = 1$. We know that the singularities of $f(z)$ are determined by the properties of the sequence $\{c_n\}$. In Section 5.7 it was remarked that recurrent nonperiodic irregularities in this sequence are apt to lead to noncontinuable power series. A number of such irregularities have been examined in the literature. *Lacunary series* (sometimes called "gap series") form the oldest and best-known case. They are series of the type (11.7.1) in which "a majority" of the coefficients c_n are 0. Thus, a lacunary series has the form

$$(11.7.2) \qquad \sum_{k=1}^{\infty} a_k z^{n_k}, \quad a_k \neq 0,$$

where

$$(11.7.3) \qquad \lim_{k \to \infty} \frac{k}{n_k} = 0.$$

This is sufficient to make the series noncontinuable. Another case is that in which the coefficients have only a finite number of distinct values. If these values do not ultimately form a periodic sequence, the series is noncontinuable. The same is true for series with integral coefficients when $R = 1$; either the series represents a rational function of z or it is noncontinuable. There are other results dealing with random coefficients. Thus, among the power series

$$(11.7.4) \qquad \sum_{n=0}^{\infty} \pm c_n z^n$$

there is at least one which is noncontinuable, and if $|c_n|^{\frac{1}{n}} \to 1$ the continuable power series form at most a countable subset. This fact lends some substance to the assertion that is commonly made that a power series "normally" is noncontinuable. It should be mentioned, however, that other criteria of a reasonable nature may be used which lead to the opposite conclusion.

In the following we shall consider gap series and series of the type (11.7.4). We start with Hadamard's gap theorem of 1892, for which we give a proof due to L. J. Mordell (1927).

THEOREM 11.7.1.   *The series* (11.7.2) *has its circle of convergence as a natural boundary provided there exists a fixed* $\lambda > 1$ *such that for all* k

$$(11.7.5) \qquad\qquad \frac{n_{k+1}}{n_k} \geq \lambda.$$

Proof.   We assume that $R = 1$. If the theorem were false, then there would be a point z_0 with $|z_0| = 1$ such that $f(z)$ is holomorphic in a neighborhood of $z = z_0$. We may take $z_0 = 1$ since a rotation of the z-plane does not affect the gaps. Choose a positive integer p such that

$$(11.7.6) \qquad\qquad \frac{p+1}{p} < \lambda,$$

and consider the mapping

$$(11.7.7) \qquad\qquad z = \tfrac{1}{2}(w^p + w^{p+1}).$$

The disk $|w| \leq 1$ is mapped onto a proper subset of $|z| \leq 1$, and $w = 1$ gives $z = 1$ as the only point on $|z| = 1$ under this transformation. Hence, for a given $\varepsilon > 0$ there exists an $r > 1$ such that the disk $|w| < r$ is mapped onto a domain D in the z-plane which contains $z = 1$ and which is contained in the union of the sets

$$|z| < 1, \quad |z - 1| < \varepsilon.$$

We substitute (11.7.7) for z in (11.7.2) and denote the resulting function of w by $F(w)$. Thus

$$F(w) = \sum_{k=1}^{\infty} a_k [\tfrac{1}{2}(w^p + w^{p+1})]^{n_k}.$$

Expanding the expression in brackets, we get

$$(11.7.8) \qquad F(w) = \sum_{k=1}^{\infty} a_k 2^{-n_k} \sum_{n=0}^{n_k} \binom{n_k}{n} w^{pn_k + n}.$$

Since

$$(p+1)n_k < pn_{k+1}$$

by the choice of p, we see that in this expansion every power of w occurs at most once; that is, if we sum the double series by rows, we obtain the Maclaurin series of $F(w)$. If $f(z)$ is holomorphic in $|z - 1| < \varepsilon$, we conclude from the properties of the mapping, discussed above, that the radius of convergence of (11.7.8) considered as a power series must be greater than 1.

But this leads to a contradiction right away. For it would imply that the series (11.7.8) is absolutely convergent for some $r > 1$; that is, the series

$$\sum_{k=1}^{\infty} | a_k | \, [\tfrac{1}{2}(r^p + r^{p+1})]^{n_k}$$

would have to converge. This, however, is impossible for $\tfrac{1}{2}(r^p + r^{p+1}) > 1$, and we have assumed that the series (11.7.2) has the radius of convergence 1. This completes the proof.

In 1896 Eugène Fabry (1856–1944) found the general gap theorem as a result of an investigation of necessary and sufficient conditions that a point on the circle of convergence of a power series be a singular point. It is well known that it is possible to obtain the gap theorem also with the aid of holomorphy-preserving differential operators of the type considered in Section 11.2. This device will be used below. We anticipate the relation (11.7.12), the proof of which will be given in Section 14.3.

THEOREM 11.7.2.   *The series* (11.7.2) *has its circle of convergence as a natural boundary if* (11.7.3) *holds.*

Proof.   We break up the sequence $\{n_k\}$ into two complementary sub-sequences $\{m_k\}$ and $\{p_k\}$ so that

$$f(z) = \sum_{k=1}^{\infty} h_k z^{m_k} + \sum_{k=1}^{\infty} g_k z^{p_k} \equiv h(z) + g(z)$$

with obvious notation. Here the sequence $\{m_k\}$ is chosen subject to the two following conditions:

(11.7.9) $$\lim_{k \to \infty} | h_k |^{\frac{1}{m_k}} = 1,$$

(11.7.10) $$\frac{m_{k+1}}{m_k} \geq 2.$$

The series defining $g(z)$ and $h(z)$ are both lacunary, but $h(z)$ is an Hadamard gap series having the same radius of convergence as the original series, that is, $R = 1$. We now construct a differential operator which is holomorphy preserving and which annihilates $g(z)$. We take

(11.7.11) $$G(w) = \prod_{k=1}^{\infty} \left(1 - \frac{w^2}{p_k^{\,2}} \right).$$

This entire function is of type (11.2.24). It vanishes for $w = p_k$, $k = 1, 2, 3, \cdots$, and

(11.7.12) $$\lim_{k \to \infty} \frac{1}{m_k} \log | G(m_k) | = 0,$$

as will be shown in Section 14.3. The corresponding differential operator $G(\vartheta)$ is holomorphy preserving. We have

$$(11.7.13) \qquad G(\vartheta)[f](z) = G(\vartheta)[h](z) = \sum_{k=1}^{\infty} G(m_k) h_k z^{m_k}.$$

This is also an Hadamard gap series and by (11.7.9) and (11.7.12) its radius of convergence equals 1. Since the unit circle is the natural boundary of this series, the unit circle must be the natural boundary of the original series, and the theorem is proved.

Finally, we consider power series of the type (11.7.4), for which we prove a theorem due to A. Hurwitz and G. Pólya.

THEOREM 11.7.3.   *The family of power series* (11.7.4) *contains a non-denumerable subset of power series each of which has the unit circle as natural boundary.*

Proof.   Let $f(z)$ be defined by (11.7.1). Let

$$f(z) = g(z) + h(z)$$

where

$$h(z) = \sum_{k=1}^{\infty} c_{m_k} z^{m_k}$$

and the sequence $\{m_k\}$ satisfies (11.7.10) and also

$$\lim_{k \to \infty} |c_{m_k}|^{\frac{1}{m_k}} = 1.$$

Thus $h(z)$ is an Hadamard gap series. Next we write $h(z)$ as the sum of infinitely many infinite subseries

$$h(z) = \sum_{n=1}^{\infty} h_n(z)$$

in such a manner that each term of $h(z)$ figures in one and only one of the subseries. Thus each $h_n(z)$ is an Hadamard gap series whose radius of convergence is $R = 1$, so that the unit circle is its natural boundary.

Consider the family of unending dyadic fractions

$$\alpha = \sum_{n=1}^{\infty} \frac{\delta_n}{2^n}, \quad \delta_n = 0, 1.$$

This is the set of real numbers in $(0, 1]$. We let a function $f_\alpha(z)$ correspond to α in $(0, 1]$ by

$$(11.7.14) \qquad f_\alpha(z) = g(z) + \sum_{n=1}^{\infty} (2\delta_n - 1) h_n(z).$$

Each such function admits of a power series expansion of type (11.7.4). The values of α which give continuable power series form a set which is at most countable. Suppose this is false and that the set is not countable. Each continuable power series is holomorphic on an arc of the unit circle $\theta_1 < \arg z < \theta_2$, $|z| = 1$, where θ_1 and θ_2 are rational multiples of 2π. Since the total number of such arcs is countable, there must be two distinct values α and β such that $f_\alpha(z)$ and $f_\beta(z)$ have a rational arc of holomorphy in common. Then their difference is also holomorphic on the arc in question. But this difference is

$$f_\alpha(z) - f_\beta(z) = 2 \sum_{n=1}^{\infty} (\delta_n - \eta_n) h_n(z)$$

where at least one $\delta_n - \eta_n$ is not 0. Hence, the difference is an Hadamard gap series having the unit circle as its natural boundary. This is a contradiction. It follows that the family $\{f_\alpha(z)\}$ contains at most a countable subfamily of continuable power series. This proves the theorem.

EXERCISE 11.7

1. Let a be an odd integer > 1, let $0 < b < 1$ and let $ab > 1 + \frac{3}{2}\pi$. Then

$$f(z) = \sum_{n=0}^{\infty} b^n z^{a^n}$$

is holomorphic in $|z| < 1$ and continuous in $|z| \leq 1$. Its real part is not a differentiable function of the arc on $|z| = 1$, so that $f(z)$ is not continuable. Try to prove these assertions. (This is the first example of a noncontinuable power series, published by Weierstrass in 1880. The nondifferentiability of the cosine series had been published five years earlier as the first example of a continuous nowhere-differentiable function.)

2. Ivar Fredholm (1866–1927), of integral equation fame, in 1901 gave the following example of a noncontinuable power series:

$$f(z) = \sum_{n=0}^{\infty} a^n z^{n^2}, \quad 0 < a < 1.$$

Verify that $f(z)$ is noncontinuable and show that $f(e^{i\theta})$ has continuous derivatives of all orders with respect to θ.

3. If p runs through the primes, what is the nature of the function defined by

$$\sum z^p \ ?$$

(*Hint:* The nth prime is greater than $n \log n$.)

4. The polynomial series

$$\sum_{k=0}^{\infty} [z(1-z)]^{3^k} = f(z)$$

converges inside the lemniscate

and this curve is the natural boundary of the function defined by the series. Verify! If each power is expanded in ascending powers of z and the resulting expressions are written consecutively, we obtain the power series of $f(z)$. This series has radius of convergence 1. Why? The sequence of partial sums of this series, $S_m(z)$, with $m = 2 \cdot 3^k$, $k = 0, 1, 2, 3, \cdots$, converges to $f(z)$ inside the lemniscate, that is, partly outside the circle of convergence of the power series. For this phenomenon of *overconvergence* see Section 16.7.

    5. Use Theorem 11.7.3 to show that if $h(z)$ is any power series whose radius of convergence equals 1, then there exists a power series $f(z)$ such that $|z| = 1$ is the natural boundary of $f(z)$ and $h = f * f$. This observation shows that the power series obtained by Hadamard composition have no special properties.

COLLATERAL READING

    As a general reference see

BIEBERBACH, L. *Analytische Fortsetzung.* Ergebnisse der Mathematik, New Series, No. 3. Springer-Verlag, Berlin, 1955. (This excellent monograph has an extensive and up-to-date bibliography.)

    See also

DIENES, P. *The Taylor Series*, Chaps. 9, 10, and 11. Clarendon Press, Oxford, 1931.

    In connection with Sections 11.3, 11.6, and 11.7 the following papers will provide additional background and much useful information:

PÓLYA, G. "Untersuchungen über Lücken und Singularitäten von Potenzreihen," *Mathematische Zeitschrift*, Vol. 29 (1929), pp. 549–640; Part 2, *Annals of Mathematics*, Series 2, Vol. 34 (1933), pp. 731–777.

    The reader who wants information concerning binomial series and their growth functions is referred to

NÖRLUND, N. E. *Leçons sur les Séries d'Interpolation*, Chap. 5. Gauthier-Villars, Paris, 1926.

12

ALGEBRAIC FUNCTIONS

12.1. Local properties. In this chapter we shall study the elements of the theory of algebraic functions and their integrals.

Let there be given a polynomial in two variables

$$(12.1.1) \qquad F(z, w) \equiv p_0(z)w^n + p_1(z)w^{n-1} + \cdots + p_n(z)$$

where

$$(12.1.2) \qquad p_j(z) = \sum_{k=0}^{m_j} a_{jk}z^k.$$

We assume once and for all that $F(z, w)$ is an irreducible polynomial in (z, w), that is, that $F(z, w)$ is not the product of two polynomials in (z, w). We shall also assume that the $p_j(z)$ have no common divisor except a constant.

For a fixed value of z the equation in w

$$(12.1.3) \qquad F(z, w) = 0$$

normally has n distinct finite roots

$$(12.1.4) \qquad w_1(z), w_2(z), \cdots, w_n(z).$$

These are the branches of the algebraic function which we shall study.

There are certain exceptional values of z for which the statement about the number of roots is not valid. These exceptional values form three sets. The first of these is the set of roots of

$$(12.1.5) \qquad p_0(z) = 0;$$

these we denote by

$$(12.1.6) \qquad z_{01}, z_{02}, \cdots, z_{0m_0}.$$

For such values of z, the degree in w of the equation (12.1.3) is $< n$, and as z approaches one of these values we shall see that one or more of the roots become infinite. It follows that the points (12.1.6) are singular points of the algebraic function to be considered.

Secondly, we must exclude the values of z for which (12.1.3) has multiple roots. As we shall see, these values are the roots of a certain auxiliary algebraic equation whose degree does not exceed $2m(n-1)$, where $m = \max m_j$ is the degree of $F(z, w)$ as a polynomial in z. Not necessarily all these exceptional values are singular, but, as will be shown later, the finite branch points belong to this set.

Thirdly, we must except the point at infinity, for which the equation becomes meaningless. This may or may not be a singular point of the algebraic function.

Suppose now that Δ is the closure of a bounded domain in the z-plane whose distance from the exceptional set is some positive number δ. If $z \in \Delta$, then there exists a finite R such that all the roots $w_k(z)$ of (12.1.3) are located in the fixed disk $|w| \leq R$. By formula (8.2.19) it is permitted to take

$$R = 1 + \frac{P}{P_0}, \quad \text{where} \quad P = \max |p_j(z)|, \, P_0 = \min |p_0(z)|.$$

The maximum is taken with respect to z in Δ and $j = 1, 2, 3, \cdots, n$. The minimum refers to z in Δ. For such values of z and w the function $F(z, w)$ and its partial derivatives are bounded. Thus we can find a finite M_0 such that

$$|F_{j,k}(z, w)| \leq M_0 j! \, k!,$$

where the subscripts indicate differentiation j times with respect to z and k times with respect to w.

Let us now rewrite equation (12.1.3) in a form which is suitable for the use of the implicit function theorem in a neighborhood of $z = z_0$, $w = w_0$ where w_0 is any one of the roots of

$$F(z_0, w) = 0.$$

The analogue of formula (9.4.22) becomes

(12.1.7) $w - w_0 = b_{10}(z - z_0) + \sum\sum b_{jk}(z - z_0)^j(w - w_0)^k.$

Here $2 \leq j + k$ and $j \leq m$, $k \leq n$. The coefficients are given by

$$b_{jk} = -\frac{F_{j,k}(z_0, w_0)}{j! \, k! \, F_{0,1}(z_0, w_0)}.$$

The absolute value of the factor $F_{0,1}(z_0, w_0)$ is bounded away from 0 for $z_0 \in \Delta$, $|w_0| \leq R$ and has a positive minimum in this region. It follows that we can find a finite M such that

(12.1.8) $|b_{jk}| \leq M$

for all j and k.

This means that a majorant of the right-hand side of (12.1.7) is given by

(12.1.9) $M\{(1 - z + z_0)^{-1}(1 - w + w_0)^{-1} - 1 - w + w_0\}.$

The equation

(12.1.10) $t = M[(1 - s)^{-1}(1 - t)^{-1} - 1 - t]$

is satisfied by

(12.1.11) $t = \frac{1}{2}(1 + M)^{-1}\{1 - (1 - s)^{-\frac{1}{2}}[1 - (1 + 2M)^2 s]^{\frac{1}{2}}\}.$

If the roots are given the value $+1$ for $s = 0$, this solution vanishes for $s = 0$. It is holomorphic for $|s| < r$ where

$$(12.1.12) \qquad\qquad r = (1 + 2M)^{-2}.$$

The implicit function theorem now tells us that the roots of the equation (12.1.3) have the following properties: For each $z_0 \in \Delta$ there are n functions

$$w_1(z; z_0),\, w_2(z; z_0),\, \cdots,\, w_n(z; z_0),$$

holomorphic in the disk $|z - z_0| < r_1$ such that $r_1 = \min(r, \tfrac{1}{2}\delta)$, and

$$(12.1.13) \qquad w_k(z_0; z_0) = w_k, \quad F(z, w_k(z; z_0)) \equiv 0, \quad |z - z_0| < r_1.$$

Let us now consider the question of analytic continuation of the root functions. If a function $w_j(z; z_0)$ can be continued analytically along a path in Δ, then

$$F(z, w_j(z; z_0)) \equiv 0$$

will hold everywhere along the path by the law of permanence of functional equations. The following argument shows that it is always possible to perform the analytic continuation: Let us consider a simple polygonal line Π in Δ joining two points $z = a$ and $z = b$. We intercalate points $z_0, z_1, z_2, \cdots, z_p$ such that $z_0 = a$, $z_p = b$, and z_j precedes z_k on Π if $j < k$. Further, we require that $|z_j - z_{j+1}| < r$ for each j. We start at $z = a$ with the n functions $w_j(z; a)$. These functions are holomorphic at $z = z_1$ since $|z_1 - a| < r$. We then observe that the equation

$$F(z_1, w) = 0$$

has two sets of roots, namely

$$w_1(z_1; a),\, w_2(z_1; a),\, \cdots,\, w_n(z_1; a)$$

and

$$w_1(z_1; z_1),\, w_2(z_1; z_1),\, \cdots,\, w_n(z_1; z_1).$$

Since the roots are uniquely determined, these two sets must coincide except for their order. Thus, for each j there is a unique k such that

$$w_j(z; a) \equiv w_k(z; z_1)$$

if z lies in the intersection of the two disks $|z - a| < r$ and $|z - z_1| < r$. Moreover, distinct values of j lead to distinct values of k. The same argument applies at $z = z_2$, z_3, and so on. Thus, we see that the functions $w_j(z; a)$ can be continued analytically along Π from $z = a$ to $z = b$ and the values of the continuations at $z = b$ coincide with the roots at $z = b$ taken in some order.

In particular, if Π is a simple closed polygon, we see that the functions

$$(12.1.14) \qquad\qquad w_1(z; a),\, w_2(z; a),\, \cdots,\, w_n(z; a)$$

are either left invariant or undergo a permutation when we describe the path Π. We shall see later that the functions (12.1.14) are the different branches of a single analytic function. Moreover, we can pass from any one of these branches to any other by analytic continuation along a suitably chosen closed path which does not pass through any one of the exceptional points. This is due to the fact that $F(z, w)$ is supposed to be irreducible. To prove this, however, we need information about the nature of the singularities of an algebraic function.

Our next concern will be with the points of the exceptional set. As a preliminary orientation, let us glance at the case in which the coefficients of $F(z, w)$ are real and the curve

$$F(x, y) = 0$$

has real branches. In this case the points of the exceptional set have geometric significance. The points of the first kind, where $p_0(x) = 0$, correspond to the vertical asymptotes, and those of the second kind give either vertical tangents or multiple points or both. The existence of other asymptotes and the general problem of the behavior of the curve for large values of x are geometric aspects of the discussion of the point at infinity as a possible singularity of the algebraic function.

We shall consider next the behavior of the function at two of the simplest types of exceptional points, namely, (1) the analogue of a point with a vertical tangent, and (2) the analogue of a double point with distinct nonvertical tangents. To simplify matters we place the exceptional point under consideration at the origin.

Let us first take the case in which

$$(12.1.15) \qquad \begin{aligned} &F_{0,j}(0, 0) = 0, \quad j = 0, 1, 2, \cdots, k - 1, \quad 1 < k \leqq n, \\ &F_{0,k}(0, 0) \neq 0, \quad F_{1,0}(0, 0) \neq 0. \end{aligned}$$

This is the analogue of a vertical tangent having k-point contact with the curve at the origin. Here we can rewrite the equation (12.1.3) in the form

$$(12.1.16) \qquad z = c_{20}z^2 + c_{11}zw + c_{30}z^3 + \cdots + c_{0k}w^k + \cdots,$$

where $c_{0k} \neq 0$ and no term of lower order involving w alone is present. This equation has the same form as (12.1.7) with z and w interchanged. Thus, the implicit function theorem applies and gives

$$z = c_{0k}w^k + \gamma_{k+1}w^{k+1} + \cdots.$$

It is important to realize that no terms of lower order than the kth can be present. This series can now be inverted with the aid of Theorem 9.4.3; we obtain

$$(12.1.17) \qquad w = \sum_{\nu=1}^{\infty} d_\nu z^{\nu/k}, \quad d_1 \neq 0.$$

Thus $z = 0$ is a branch point of order $k - 1$ where k branches of the algebraic function are permuted cyclically. For the other $n - k$ roots, $z = 0$ is a regular point, provided these roots are distinct.

Next we consider the case of a double point, which we place at $(0, 0)$. We can now write (12.1.3) in the form

$$(12.1.18) \qquad w^2 + a_{11}zw + a_{20}z^2 = \sum a_{jk}z^j w^k,$$

where on the right $0 \leq j \leq m$, $0 \leq k \leq n$, $3 \leq j + k$, and

$$a_{11}{}^2 \neq 4a_{20}$$

to ensure that the "tangents" be distinct. Here we substitute a power series

$$(12.1.19) \qquad w = \sum_{p=1}^{\infty} \alpha_p z^p.$$

The coefficients satisfy equations of the following form:

$$(12.1.20) \qquad \alpha_1{}^2 + a_{11}\alpha_1 + a_{20} = 0,$$

$$(12.1.21) \qquad (2\alpha_1 + a_{11})\alpha_p = P(\alpha_1, \alpha_2, \cdots, \alpha_{p-1}), \quad p > 1,$$

where $P(\cdot)$ is a polynomial in the variables indicated which is linear in the coefficients a_{jk}. Here α_1 has to be one of the two roots α and β of the equation (12.1.20). If α_1 is chosen in this manner, then $2\alpha_1 + a_{11} = \alpha - \beta \neq 0$, so that the coefficients α_p can be determined uniquely by recurrence. The convergence of the resulting series (12.1.19) may be proved by a suitable majorant argument, but it is easier to reduce the problem to an application of the implicit function theorem.

We take the root α of (12.1.20) and substitute

$$w = z(\alpha + v)$$

in (12.1.18), obtaining a result of the form

$$v(\alpha - \beta + v) - zG(z, v) = 0$$

after cancellation of a factor z^2. The left member is a polynomial in (z, v). It vanishes for $(z, v) = (0, 0)$, but its partial derivative with respect to v has the value $\alpha - \beta \neq 0$ at $(0, 0)$. The implicit function theorem applies and asserts that v can be expanded in a power series in z convergent in some neighborhood of $z = 0$. Hence w has the same property, and we are assured that the series (12.1.19) has a positive radius of convergence. In this case $z = 0$ is not a singular point of the algebraic function though the point belongs to the exceptional set.

The case in which the left member of (12.1.18) is a perfect square is totally different. Geometrically speaking we then have a *cusp* at $(0, 0)$; there are two types, depending upon the nature of the third-order terms. Suppose that

$$(12.1.22) \qquad (w - \alpha z)^2 = \sum a_{jk} z^j w^k,$$

where $j + k \geq 3$, and the third-order terms satisfy the condition

$$(12.1.23) \qquad a_{30} + a_{21}\alpha + a_{12}\alpha^2 + a_{03}\alpha^3 \equiv \gamma^2 \neq 0$$

which ensures that the origin is an *ordinary* (or *ceratoid*) *cusp*. Here we set

$$z = t^2, \quad w - \alpha z = t^3 v$$

and substitute in (12.1.22). We can cancel a factor t^6 and obtain a result of the form

$$v^2 - \gamma^2 - tG(t, v) = 0.$$

The left member is a polynomial in (t, v) which vanishes for

$$(t, v) = (0, \gamma) \quad \text{and} \quad (0, -\gamma),$$

but the partial derivative with respect to v does not vanish at either place. Hence the implicit function theorem can be used once more and shows that the two roots can be expanded in power series in t. Going back to the old variable, we obtain

$$(12.1.24) \qquad\qquad w = \alpha z + \gamma z^{\frac{3}{2}} + \sum_{p=1}^{\infty} \gamma_p z^{\frac{p+3}{2}},$$

where the square root is given its two determinations. In this case the double point gives rise to a branch point of the algebraic function, and two branches are permuted when z makes a circuit about $z = 0$.

EXERCISE 12.1

1. Given the leaf of Descartes

$$z^3 + w^3 = 3zw.$$

Here $(0, 0)$ is a double point and the two branches are tangent to $w = 0$ and $z = 0$. Find the expansions of w corresponding to these branches in integral or fractional powers of z, giving in each case the first three nonvanishing terms. (*Hint:* Set $w = z^2 u$ in the one case, $z = t^2$, $w = tv$ in the other.)

2. Given the lemniscate

$$w^2 - z^2 = (z^2 + w^2)^2.$$

It is desired to obtain the first three nonvanishing terms in the power series expansion of one of the branches through $(0, 0)$.

3. The cardioid

$$w^2 = 2z(z^2 + w^2) + (z^2 + w^2)^2$$

has an ordinary cusp at $(0, 0)$. Obtain the first three nonvanishing terms in the corresponding expansion (12.1.24).

4. The curve

$$w^2 = z^4 + w^4$$

has a so-called *tacnode* at $(0, 0)$, where the curve is tangent to itself and to $w = 0$. Prove that $z = 0$ is not a singular point of the algebraic function by finding the series expansions of the roots which vanish for $z = 0$. (*Hint:* Set $w = z^2 u$ and expand u in powers of z.)

12.2. Critical points. We now take up the discussion of the exceptional points in the general case. We shall first obtain the nature of the representation of the roots in a neighborhood of an exceptional point. We shall get more precise information later, but general methods of complex function theory suffice for the characterization of algebraic functions.

We start with an exceptional point of the second kind, which we place at the origin. We suppose that the equation

$$(12.2.1) \qquad F(0, w) = 0$$

has k equal roots, where $1 < k \leq n$, and that $w = 0$ is this k-fold root. Since the roots are continuous functions of z, we conclude that the equation

$$(12.2.2) \qquad F(z, w) = 0$$

has k roots

$$w_1(z), w_2(z), \cdots, w_k(z)$$

which tend to zero with k. These functions of z are defined in a punctured disk $0 < |z| < r$, where they are locally holomorphic but normally not single-valued. Let C be a circle $|z| = \rho < r$ to which we give the positive orientation such that arg z increases by 2π when we describe C once. Let us start at $z = \rho$ with the function $w_1(z)$ and describe C once. If $w_1(z)$ returns to its original value, then $w_1(z)$ is holomorphic in $|z| < r$ since it has the value 0 at $z = 0$ and is holomorphic in the punctured disk. In this case $w_1(z)$ may be represented by a convergent power series in z in the disk.

If $w_1(z)$ is not single-valued in $0 < |z| < r$, then the circuit C will carry $w_1(z)$ into another of the roots, say $w_2(z)$. Describing C once more, we see that $w_2(z)$ cannot be carried into itself, but may possibly be carried back into $w_1(z)$. In the latter case, the two roots $w_1(z)$ and $w_2(z)$ form a *cycle* and are simply permuted by C. It follows that

$$w_1(t^2)$$

is single-valued in the punctured disk $0 < |t| < r^{\frac{1}{2}}$. Since it has the value 0 at the center, it is holomorphic in $|t| < r^{\frac{1}{2}}$ and can be expanded in a power series in t. This means an expansion of $w_1(z)$ in ascending powers of $z^{\frac{1}{2}}$ which also represents $w_2(z)$ if the square root is unrestricted.

In general, there exists an integer κ, $1 \leq \kappa \leq k$, such that κ of the roots form a cycle whose elements are permuted cyclically as z describes C. This means that any one of these functions is carried back into itself when we describe C κ times. Thus, $w_1(t^\kappa)$ is single-valued and consequently holomorphic in a neighborhood of $t = 0$. This in turn implies that we have a series

$$(12.2.3) \qquad \sum_{\mu=1}^{\infty} c_\mu z^{\mu/\kappa} ;$$

for each determination of the root $z^{1/\kappa}$, (12.2.3) represents one of the functions of the cycle.

If $\kappa = k$, our cycle accounts for all the roots which tend to zero with z. If $\kappa < k$, then we examine the remaining roots in the same manner. We find a set of positive integers

$$(12.2.4) \qquad\qquad \kappa_1, \kappa_2, \cdots, \kappa_\alpha,$$

with

$$(12.2.5) \qquad\qquad \kappa_1 + \kappa_2 + \cdots + \kappa_\alpha = k.$$

To each of these integers $\kappa_\beta (1 \leq \beta \leq \alpha)$ corresponds a cycle of roots represented by a series of type (12.2.3), and for each determination of z^{1/κ_β} we get one of the branches of the cycle. These series, α in number, represent all the roots of (12.2.2) which tend to zero with z. The α elements of the function $w(z)$ are normally algebraic, but if one of the κ's equals 1, then the corresponding element is regular, and if $\alpha = k$, then all elements are regular. In addition to these elements which vanish for $z = 0$, there are others corresponding to the roots of (12.2.1) which are different from 0. If some of the latter roots are also multiple roots, then there will be other algebraic elements. If they are all simple, then we have to add $n - k$ regular elements to give the complete description of $w(z)$ at the point $z = 0$. If there is at least one algebraic element in the set, we say that the exceptional point $z = 0$ is a *critical point* of the algebraic function $f(z)$. This takes care of the exceptional points of the second kind, except for the actual determination of the numbers κ and of the leading coefficient of the series (12.2.3). This will be done in the following section.

The discussion of exceptional points of the first kind and of the point at infinity is easily reduced to the case which we have just studied. Suppose then that $z = a$ is a root of

$$p_0(z) = 0.$$

We may assume $a = 0$. Then the equation

$$F(0, w) = 0$$

has $\nu < n$ roots. To each of the latter correspond regular or algebraic elements which we can discuss by the method developed above. In order to handle the infinite roots which now present themselves, we set

$$w = \frac{1}{u}$$

and form

$$(12.2.6) \qquad G(z, u) = u^n F\left(z, \frac{1}{u}\right) = p_n(z)u^n + \cdots + p_0(z).$$

Here

$$(12.2.7) \qquad\qquad G(0, u) = 0$$

has a root of multiplicity $n - \nu$ at $u = 0$. Using the results obtained above, we see that the equation

$$(12.2.8) \qquad\qquad G(z, u) = 0$$

has $n - \nu$ roots $u_\kappa(z)$ which tend to zero with z. These roots arrange themselves into a certain number of cycles and each of them is represented by a series of the form

$$(12.2.9) \qquad u(z) = \sum_{\mu=1}^{\infty} b_\mu z^{\mu/\kappa},$$

where the κ different determinations of $z^{1/\kappa}$ give the κ different members of the cycle in question. If now

$$b_1 = b_2 = \cdots = b_{\lambda-1} = 0, \quad b_\lambda \neq 0,$$

we see that

$$(12.2.10) \qquad w(z) = z^{-\lambda/\kappa} \sum_{\mu=0}^{\infty} c_\mu z^{\mu/\kappa}, \quad c_0 \neq 0.$$

Thus, an exceptional point of the first kind is always singular. It is normally critical, but the algebraic elements have poles instead of zeros at the point in question. Even if all branches are single-valued, there are necessarily some polar elements present. For the terminology, see Section 10.6.

Finally, we come to the infinitary elements. We set

$$z = \frac{1}{s}.$$

Then

$$(12.2.11) \qquad H(s, w) \equiv s^m F\left(\frac{1}{s}, w\right) \equiv \sum_{k=0}^{n} q_k(s)\, w^{n-k}$$

is a polynomial in (s, w), and the exceptional point $z = \infty$ for $F(z, w)$ is mapped into the point $s = 0$. The latter is not necessarily an exceptional point with respect to $H(s, w)$. If not, then all branches of $w(z)$ are holomorphic at infinity, and we have representations of the form

$$(12.2.12) \qquad w_j(z) = \alpha_j + \sum_{k=1}^{\infty} b_{jk} z^{-k}$$

for the different branches. If $s = 0$ is exceptional, however, then the equation

$$H(0, w) = 0$$

either has fewer than n (finite) roots, and/or some of the roots are multiple. In either case, we can reduce the discussion to cases previously considered. As a result we see that the infinitary elements are of the form

$$(12.2.13) \qquad \sum_{k=0}^{\infty} b_k\, z^{-(\nu+k)/\kappa},$$

where κ and ν are integers, $1 \leq \kappa \leq n$, and ν may be positive, zero, or negative. Normally, $z = \infty$ is a singular point which is also critical, but, as remarked above, it can happen that all elements are regular.

This completes the discussion of the singularities of the algebraic function. It still remains to prove, however, that the n root functions

$$w_1(z), w_2(z), \cdots, w_n(z)$$

are really branches of the same analytic function, so that one can obtain all these functions by analytic continuation of a single one of them along suitably chosen closed paths. Offhand there is the possibility that analytic continuation of $w_1(z)$, for instance, yields only a proper subset of the roots, say

$$w_1(z), w_2(z), \cdots, w_k(z),$$

while the remaining roots form another invariant subset. To eliminate this possibility, we form the elementary symmetric functions

$$E_1(z) = \sum_1^k w_j(z),$$
$$E_2(z) = \sum w_{j_1}(z) w_{j_2}(z),$$
$$\cdots \cdots \cdots \cdots \cdots \cdots$$
$$E_k(z) = w_1(z) w_2(z) \cdots w_k(z).$$

These expressions represent single-valued analytic functions since analytic continuation merely permutes the k roots involved and leaves their symmetric functions unchanged. Further, the functions $E_j(z)$ have only a finite number of singular points, and these must be located at singular points of the roots. The singularities of the single-valued functions $E_j(z)$ are isolated and are, consequently, either poles or essential singular points. The latter possibility is easily excluded. If $z = a$ is a zero of $p_0(z)$ and hence a singular point of some of the roots, then such roots become infinite as (fractional) powers of $|z - a|^{-1}$. Thus this implies that $E_j(z)$ becomes infinite as a power of $|z - a|^{-1}$ as $z \to a$, so that $z = a$ is a pole of $E_j(z)$ or possibly a regular point. The latter case will arise if none of the roots $w_j(z)$ with $1 \leq j \leq k$ becomes infinite at $z = a$. If, on the other hand, $z = a$ is a critical point corresponding to an exceptional point of the second kind, then each $w_j(z)$ tends to a finite limit and $z = a$ is a removable singularity of $E_j(z)$, that is, each $E_j(z)$ can be defined so as to be holomorphic in some neighborhood of $z = a$. Similarly, we see that $z = \infty$ is at most a pole of $E_j(z)$. Thus the functions $E_j(z)$ are holomorphic in the extended plane save for poles, that is, each $E_j(z)$ is a rational function of z and its finite poles are to be found among the zeros of $p_0(z)$. Let us now form

(12.2.14) $R(z, w) \equiv w^k - E_1(z)w^{k-1} + E_2(z)w^{k-2} - \cdots + (-1)^k E_k(z).$

Then

$$R(z, w) = 0$$

has the roots

$$w_1(z), w_2(z), \cdots, w_k(z),$$

which are also roots of

$$F(z, w) = 0.$$

It follows that $F(z, w)$ as a polynomial in w must be divisible by $R(z, w)$. The division algorithm then gives

$$F(z, w) = R(z, w)Q(z, w),$$

where $Q(z, w)$ is a polynomial of degree $n - k$ in w whose coefficients are rational functions of z. This contradicts our assumption that $F(z, w)$ is irreducible in the field of rational functions. From this we conclude that it is possible to pass from one root function $w_j(z)$ to any other function $w_k(z)$ by analytic continuation, and the various roots are simply the n different branches of the same analytic function $w(z)$. We call the latter an *algebraic function*.

The results obtained so far may be summarized as

THEOREM 12.2.1. *Suppose that*

$$F(z, w) = \sum_{j=0}^{n} p_j(z)\, w^{n-j},$$

as a polynomial in w, is irreducible in the field of rational functions and that the $p_j(z)$ are polynomials in z. Then the equation

$$F(z, w) = 0$$

defines an analytic function $w(z)$ for all z. The singularities of $w(z)$ come from three different sources: (i) *The zeros of $p_0(z)$ are infinitudes of one or more of the branches of $w(z)$.* (ii) *The roots of the equation in z obtained by elimination of w from the simultaneous equations*

(12.2.15) $$F(z, w) = 0, \quad F_{0,1}(z, w) = 0$$

may be critical points of some or all branches, but all branches tend to finite limits as z approaches such a point. (iii) *The point at infinity may be a pole or a critical point for some of the branches. At any other point, $w(z)$ has n regular, distinct elements. At the singular points the branches form cycles and there are one or more algebraic elements besides regular and polar elements.*

The elimination of w between the two equations (12.2.15) leads to the *discriminant* $D(z)$ of $F(z, w)$ regarded as a polynomial in w. For the following concepts and formulas the reader is referred to L. E. Dickson, *First Course in the Theory of Equations* (John Wiley and Sons, Inc., New York, 1921), or similar texts. We have

(12.2.16)
$$D(z) = [p_0(z)]^{2n-2} \prod_{1 \leq j < k \leq n} [w_j(z) - w_k(z)]^2$$
$$= (-1)^{\frac{1}{2}n(n-1)}[p_0(z)]^{-1} R[F, F_w],$$

where $R[F, F_w]$ is the *resultant* of F and F_w as polynomials in w. This resultant can be written as a determinant $(R_{jk}(z))$ of $2n - 1$ rows and columns where, with $p_\alpha(z) \equiv 0$ if $\alpha < 0$ or $> n$, we have

(12.2.17) $$R_{j,k}(z) = \begin{cases} p_{k-j}(z), & 1 \leq j \leq n - 1, \\ (j - k + 1)p_{n+k-j-1}(z), & n \leq j \leq 2n - 1. \end{cases}$$

$D(z)$ is obviously a symmetric function of the roots $w_j(z)$. As such, $D(z)$ is a rational function of the coefficients $p_j(z)$ of the polynomial $F(z, w)$. As a matter of fact, it is actually a homogeneous form of degree $2(n - 1)$ in these coefficients.

Assuming that the $p_j(z)$ are polynomials of degree m at most, we see that $D(z)$ is a polynomial in z whose degree does not exceed $2m(n-1)$. A necessary and sufficient condition that the equation

$$F(z_0, w) = 0$$

have multiple roots is that $z = z_0$ be a root of the equation

(12.2.18) $$D(z) = 0.$$

These roots thus give the set of exceptional points of the second kind, introduced in Section 12.1, of which the finite branch points (z finite, w finite) form a subset.

EXERCISE 12.2

1. Compute $D(z)$ and find its roots if

$$w^2 + z^3 w - z^3 = 0.$$

2. Determine the singular points of the algebraic function defined by the equation in Problem 1, Exercise 12.1. Let ω be a primitive sixth root of unity ($\omega^3 = -1$), substitute

$$w = \omega z(1 + u)$$

in the equation, and discuss the properties of u for large values of z. Use the result to show that the point at infinity is a simple pole for each of the branches of $w(z)$.

3. Determine the location and the nature of the singularities if $w(z)$ is defined by the equation of the lemniscate of Problem 2, Exercise 12.1.

4. Show that the reciprocal of an algebraic function, not identically zero, is algebraic.

5. Show that the sum of two algebraic functions is algebraic. (*Hint:* If $u_1, u_2, \cdots, u_m$ and $v_1, v_2, \cdots, v_n$ are the determinations of the two functions, form the algebraic equation satisfied by $u_j + v_k$, $j = 1, \cdots, m$, $k = 1, \cdots, n$ and show that its coefficients are rational functions of z.)

6. Show that the product of two algebraic functions is algebraic. (*Hint:* Discuss $u_j v_k$.)

7. Obtain explicit equations in Problems 5 and 6 for the case in which $m = n = 2$.

8. Prove that algebraic functions form a field.

9. Let $w(z)$ be the algebraic function defined by (12.1.3) and let m be an integer, $m \neq 0$. Prove the existence of a unique set of rational functions $R_k(z)$ such that

$$[w(z)]^m = \sum_{k=0}^{n-1} R_k(z)[w(z)]^k.$$

12.3. Newton's diagram. We still have the problem of determining the actual expansions at the singular points. This problem can be handled with the aid of a device known variously as *Newton's diagram* or *polygon*, as the *algebraic* or the *analytical triangle*, or as *the method of Puiseux.* Sir Isaac Newton (1642–1727) introduced the device as an aid in curve tracing. The triangle names arose in connection with the use of homogeneous coordinates in algebraic geometry. In 1850 the device was adapted to the discussion of algebraic functions by Victor Alexandre Puiseux (1820–1883). It is very convenient for obtaining first approximations, but it may also be developed as a method of successive approximations. We shall combine it with the implicit function theorem to obtain the desired information.

Let us consider the equation

(12.3.1) $$F(z, w) \equiv \sum a_{jk} z^j w^k = 0, \quad a_{00} = 0.$$

We want to find the nature of the various branches of $w(z)$ which tend to zero with z. We know a priori the existence of a positive rational number α and of a complex number a for each such branch, such that

(12.3.2) $$\lim_{z \to 0} z^{-\alpha} w(z) = a.$$

The first problem is to determine α and a. This gives a first approximation

$$w(z) \sim a z^\alpha,$$

but naturally we want to find further terms in the expansion or, at least, what fractional power of z will be present.

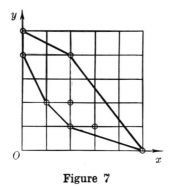

Figure 7

For this purpose we construct the Newton diagram of $F(z, w)$. In an (x, y)-plane we plot the points (j, k) such that $a_{jk} \neq 0$ in (12.3.1). We then form the least convex polygon which contains these points. See Figure 7, which represents the diagram for the sextic

(12.3.3) $$z^2 w^4 + w^5 + 2w^4 + 2z^2 w^2 - zw^2 + z^2 w + z^3 w - z^5 = 0.$$

The polygon always has at least one point on each of the axes since $F(z, w)$ is irreducible. Let $Z = (\rho, 0)$ and $W = (0, \sigma)$ be the points on the axes nearest to the origin. Thus σ is the number of roots which tend to zero with z. The polygon has a broken line joining W with Z, and this line consists of at most σ segments. Each of these segments has the property that all points (j, k) of the diagram lie either on the segment or above the line produced. It is only this part of the polygon which is of interest for the discussion at $z = 0$, but the line segments "facing" infinity play a similar role for the discussion of the infinitary elements.

We start with the segment of $[W, \cdots, Z]$ whose left endpoint is W. Suppose the equation of the line on this segment is

$$px + qy = r.$$

Here p, q, r are positive integers and we may assume that p and q are relatively prime. Further, $p \leqq \sigma, q \leqq \rho$. The slope of this line is $-p/q$. Suppose that the lower endpoint of the line segment is $(\sigma - \mu p, \mu q)$ where μ is an integer, $1 \leqq \mu$. Since p and q are relatively prime, μ is one unit less than the number of lattice points (= points with integral coordinates) on the segment. We shall see that there are μp branches of $w(z)$ associated with this segment, and for each of them the number α of (12.3.2) equals q/p, that is, the negative reciprocal of the slope of the segment.

To verify this, we substitute

(12.3.4) $z = t^p, \quad w = t^q u$

in the equation and obtain

(12.3.5) $\sum a_{jk} t^{jp+kq} u^k = 0.$

For every point (j, k) which is on the line we have

$$jp + kq = r;$$

for all other points

$$jp + kq > r.$$

Thus (12.3.5) is divisible by t^r, and after division we have an expression of the form

(12.3.6) $G(t, u) \equiv \sum' a_{jk} u^k + \sum'' a_{jk} t^{jp+kq-r} u^k = 0,$

where the first summation extends over the points on the line segment and the second takes in the remaining points.

The first sum contains at least the two terms corresponding to the endpoints of the line segment. In any case this sum gives an expression of the form

$$u^{\sigma - \mu p} P_\mu(u^p)$$

where $P_\mu(v)$ is a polynomial of degree μ, $1 \leqq \mu < \dfrac{\sigma}{p}$, such that $P_\mu(0) \neq 0$.

Suppose that

$$(12.3.7) \qquad\qquad P_\mu(v) = 0$$

has the roots

$$(12.3.8) \qquad\qquad \alpha_1{}^p, \alpha_2{}^p, \cdots, \alpha_\mu{}^p$$

which, for the time being, we assume to be distinct. The occurrence of multiple roots is the plague of algebraic function theory and of algebraic geometry. We shall return to the excluded case later.

We note that the second summation in (12.3.6) cannot be void since this would imply that $F(z, w)$ is reducible. We can rewrite (12.3.6) as follows:

$$(12.3.9) \qquad G(t, u) = u^{\sigma - \mu p} P_\mu(u^p) + t\, G_1(t, u) = 0,$$

where $G_1(t, u)$ is a polynomial in (t, u). Then $G(t, u)$ vanishes for $t = 0$, $u = \alpha_\lambda$, $\lambda = 1, 2, \cdots, \mu$, but $G_{01}(t, u) \neq 0$ for any such combination (t, u). Incidentally, α_λ is some pth root of $\alpha_\lambda{}^p$; it is immaterial which one. The implicit function theorem then applies and shows that equation (12.3.9) is satisfied by μ distinct power series in t,

$$u_\lambda(t) = \alpha_\lambda + \sum_{k=1}^{\infty} c_{\lambda k} t^k.$$

Hence, returning to $w(z)$ and taking into account the substitution (12.3.4), we obtain μ cycles with p branches each, given by

$$(12.3.10) \qquad w_\lambda(z) = z^{q/p}\left\{ \alpha_\lambda + \sum_{k=1}^{\infty} c_{\lambda k} z^{k/p} \right\}, \quad \lambda = 1, 2, \cdots, \mu.$$

These are the μp branches corresponding to the first segment of the Newton polygon.

We can handle the other sides in the same manner. To keep better track of the branches, let us set $p_1 = p$, $\mu_1 = \mu$. Suppose that the difference of the x-coordinates of the endpoints of the second line segment is $\mu_2 q_2$ and that of the y-coordinates is $\mu_2 p_2$, so that the slope of the line is $-p_2/q_2$. Here the slope is necessarily different from the slope of the first line. We consequently get a new exponent α, namely q_2/p_2. Corresponding to this side we now have μ_2 cycles with p_2 branches in each cycle and representations of the form (12.3.10), in which p, q must be replaced by p_2, q_2. In this manner we proceed along the different sides. From the geometry of the polygon we see that if there are β sides, then

$$(12.3.11) \qquad \sum_1^\beta \mu_i p_i = \sigma, \quad \sum_1^\beta \mu_i q_i = \rho.$$

Only the first of these relations is of direct interest to us. It asserts that the total number of branches of $w(z)$ obtainable in this manner is σ, that is, all branches which tend to zero with z are accounted for.

We assumed above that the roots of (12.3.7) are distinct. We shall indicate what modifications are necessary if this assumption does not hold. A complete discussion involves consideration of a number of possibilities and is quite laborious. We shall settle one case and call attention to the other possible situations. Suppose then that α^p is a ϖ-fold root of (12.3.7) where $\varpi \leq \mu$. We set

$$u = \alpha + u_1$$

and substitute in (12.3.9). The result is of the form

(12.3.12) $$G_2(t, u) \equiv u_1{}^\varpi Q(u_1) + t\, G_1(t, \alpha + u_1) = 0,$$

where $Q(0) \equiv a \neq 0$. Suppose for the sake of simplicity that $G_1(t, \alpha)$ does not vanish for $t = 0$ and set $G_1(0, \alpha) = b$. The Newton diagram for $G_1(t, u_1)$ then has a single side facing the origin, and this side joins $(0, \varpi)$ with $(1, 0)$. We can then use the analogue of the substitution (12.3.4), which in the present case becomes

(12.3.13) $$t = s^\varpi, \quad u_1 = sv.$$

After cancellation of a factor s^ϖ, we are left with

(12.3.14) $$G_3(s, v) \equiv v^\varpi Q(sv) + G_1(s^\varpi, \alpha + sv) = 0.$$

For $s = 0$ this gives

$$av^\varpi = -b.$$

We conclude, again with the aid of the implicit function theorem, that a holomorphic function $v(s)$ exists such that in some neighborhood of $s = 0$ we have

$$G_3(s, v(s)) \equiv 0 \quad \text{while} \quad v(0) = \left(-\frac{b}{a}\right)^{1/\varpi}.$$

Going back to the original variables, we find an expansion of the form

(12.3.15) $$w(z) = z^{q/p}\left\{\alpha + \sum_{k=1}^{\infty} c_k z^{k/p\varpi}\right\}.$$

Thus ϖ of the cycles (12.3.10) join in forming a single cycle with ϖp branches.

If $G_1(t, \alpha)$ does vanish for $t = 0$, the Newton polygon of $G_3(t, u_1)$ may conceivably have several sides facing the origin. If so, we have to consider each side separately. For each side we get a cycle of type (12.3.15), but with ϖ replaced by a smaller integer ϖ_i. Now $\Sigma\, \varpi_i = \varpi$. This presupposes that in the process we do not run into multiple roots of the corresponding equation (12.3.7). It should be noticed, however, that the new equations (12.3.7) are definitely of lower degree than the original one. This reduces the chances of multiple roots and it also ensures that if the reduction process is repeated sufficiently often, we will reach equations with simple roots after a finite number of steps. We shall desist from further details.

For the discussion of the infinitary elements, either we can make a preliminary transformation

$$z = \frac{1}{t}$$

and discuss the transformed equation with the aid of Newton's diagram, or we can use the diagram of the original equation, where we now consider the sides facing infinity instead of zero. We leave the details to the reader.

EXERCISE 12.3

1. Use the Newton diagram to discuss the branches of $w(z)$ which vanish with z if $w(z)$ is defined by (12.3.3). Here

$$p_1 = 2, p_2 = p_3 = 1; \quad q_1 = q_2 = 1, q_3 = 3; \quad \mu_1 = \mu_2 = \mu_3 = 1.$$

2. Discuss the same equation at infinity. There is one element which is single-valued and has a pole; the other four branches form a cycle.

3. Find the first three terms in the expansion of $w(z)$ at $z = 0$ if

$$w^2 + z^3 w - z^3 = 0,$$

using the Newton diagram repeatedly. Check your result by solving the quadratic and expanding the roots in fractional powers of z.

4. The function defined by

$$w(w^2 - z^2) = z^4$$

has three regular elements at the origin and a single cycle of three branches at infinity. Verify, and get the leading terms of the expansions. Find the finite critical points and show that in each case there is one regular element and an algebraic element with two branches.

12.4. Riemann surfaces; some concepts of algebraic geometry. With each irreducible polynomial $F(z, w)$ of degrees m in z and n in w there is associated a Riemann surface $\mathfrak{G}[w]$, the graph of $w(z)$ in the sense of Section 10.6. The places on $\mathfrak{G}$ are the pairs $(z_0, w_j(z))$ where $w_j(z)$ is an element of $w(z)$, defined at $z = z_0$. The element can be regular, polar, algebraic, or infinitary. Theorem 12.2.1 asserts that if z_0 is a regular point of $w(z)$, then there are n distinct places $(z_0, w_j(z))$ with the trace z_0. We say that $\mathfrak{G}$ has n *sheets*. At a singular point there are usually fewer places, but every z_0 of the extended plane is the carrier of at least one place.

The neighborhoods of a given place are given by Definition 10.6.1. If z_0 is a regular point, then we take a circular disk $| z - z_0 | < \delta$, where δ is less than the distance from z_0 to the set of singular points of $w(z)$. A neighborhood of the place $\mathfrak{p}_0 = (z_0, w_j(z))$ is the set of places $\mathfrak{p} = (z_1, w_j(z; z_1))$ where $| z_1 - z_0 | < \delta$

and $w_j(z; z_1)$ is the rearrangement at $z = z_1$ of the power series in $z - z_0$ which defines $w_j(z)$. Suppose now that z_0 is the carrier of an algebraic element $w_0(z)$ with r branches which permute cyclically about z_0. A neighborhood of $\mathfrak{p}_0 = (z_0, w_0(z))$ is then the set of all places $\mathfrak{p} = (z_1, w_{0j}(z))$ where $|z_1 - z_0| < \delta$, the distance from z_0 to the remaining singular points, and $w_{0j}(z)$ is any one of the r power series obtainable from $w_0(z)$ by direct rearrangement at $z = z_1$. Compare formulas (10.6.5) and (10.6.6). Here we note that every such $z_1 \neq z_0$ is the carrier of r regular elements belonging to the neighborhood of $\mathfrak{p}_0$. If $r < n$, then z_0 is the carrier of other elements besides $w_0(z)$ and, hence, of other places besides $\mathfrak{p}_0$. Corresponding to each such place there is an associated system of neighborhoods. The same considerations apply for polar elements and, with suitable modifications, also at $z = \infty$.

If $n > 1$, as we may assume, there are always algebraic elements, since $F(z, w)$ is supposed to be irreducible. There are at least two algebraic elements having distinct centers. If there is only one finite critical point $z = a$, then $z = \infty$ is necessarily critical. The nth root of z is an example of an algebraic function having only one finite critical point. The number of critical points is finite, and an upper bound for this number may be given in terms of the degrees m and n.

In any case, the number of critical points is not so important as the degree of ramification of the surface induced by these points. To measure the ramification, we use the number r which gives the number of branches in a cycle. We call $(r - 1)$ the *ramification* of the place in question. It is zero for a place whose second coordinate is a regular element and positive for algebraic elements. We sum $(r - 1)$ over the surface, that is, over the places on the surface, and set

(12.4.1) $$\rho = \sum (r - 1),$$

which we call the *total ramification of the surface*. It turns out to be an even integer ≥ 2. The ratio

(12.4.2) $$\vartheta = \frac{\rho}{n}$$

is the *ramification index* of the surface.

In the remaining portion of the section we shall mention briefly a number of concepts of algebraic geometry which have an obvious bearing on the theory of algebraic functions. Some of these concepts, such as tangents, double points, and cusps, have figured already in Section 12.1 and will be encountered again in the following. For proofs and further details we refer the reader to the treatises quoted in the Collateral Reading at the end of this chapter.

The basic concept for us is that of the *genus* of a Riemann surface of an algebraic function or, equivalently, of an algebraic curve. This notion can be defined in terms of the total ramification of the surface and the number of its sheets by the formula

(12.4.3) $$p = \tfrac{1}{2}\rho - n + 1.$$

The number p turns out to be a non-negative integer.

As an example we consider the special case

$$(12.4.4) \qquad w^2 = P(z),$$

which will figure prominently in the following. Here $P(z)$ is assumed to be a polynomial of degree $2p + 1$ or $2p + 2$ whose zeros are distinct. Here $n = 2$. There are $2p + 2$ branch points, namely, the zeros of $P(z)$ plus the point at infinity when the degree of $P(z)$ is odd. For each of these critical points we have $r - 1 = 1$ and

$$\tfrac{1}{2}(2p + 2) - 2 + 1 = p,$$

so that the genus of the corresponding Riemann surface or algebraic curve is p.

The genus enters prominently into many other connections, of which we shall list a few.

The *order* of the algebraic curve

$$(12.4.5) \qquad C: \quad F(z, w) = 0$$

is $N = \max(m, n)$ in our previous notation. Such a curve may have *nodes*, that is, double points with distinct tangents. The maximum number of such points is found to be

$$\tfrac{1}{2}(N - 1)(N - 2).$$

Thus a cubic cannot have more than one node or cusp if $F(z, w)$ is irreducible. For if there were two nodes, for instance, then a straight line through these two points would intersect the cubic in at least four points; this is impossible unless the cubic factors into a straight line and a conic, that is, unless $F(z, w)$ is reducible. Similarly a quartic can have at most three nodes, for if there were four, then we could pass a conic through the four nodes and a fifth arbitrarily chosen point on the quartic. The conic and the quartic would then have at least nine intersections, whereas the maximum number of intersections of a nondegenerate quartic with a conic is $4 \cdot 2 = 8$. Thus the quartic would have to factor into two conics, and $F(z, w)$ would again be reducible.

The maximum number of nodes listed above is not necessarily reached. Suppose we have actually δ nodes and, in addition, κ ordinary cusps. Then

$$(12.4.6) \qquad \tfrac{1}{2}(N - 1)(N - 2) - \delta - \kappa = p,$$

the genus of the curve. In this connection algebraic geometers often use the term *deficiency* instead of "genus."

Besides the order N of a curve, its *class* K is also important. This concept is obtained as follows. The condition that the straight line

$$uz + vw + 1 = 0$$

be a tangent to the curve C of (12.4.5) is that u and v satisfy a certain equation

$$(12.4.7) \qquad G(u, v) = 0,$$

known as the *tangential equation* of C. The degree K of this equation is the *class* of C. The maximum value of K is $N(N - 1)$, but if point singularities are present this number is reduced. Thus, if there are δ nodes and κ cusps we have

(12.4.8) $$K = N(N - 1) - 2\delta - 3\kappa.$$

This is one of Plücker's formulas, named after Julius Plücker (1801–1868), who laid the foundations of algebraic geometry in a series of monographs during the period 1828–1839. Plücker introduced the very fruitful *principle of duality* in geometry: *Any proposition of plane projective geometry* (deducible from certain postulates) *remains valid when the words "points" and "lines" are interchanged.* This dualism extends to the singularities of algebraic curves. To a node corresponds a *bitangent,* that is, a line having contact with the curve at two distinct points, and to a cusp corresponds an *inflection tangent.* If the number of bitangents of C is τ and the number of inflections ι, then conversely

(12.4.9) $$N = K(K - 1) - 2\tau - 3\iota,$$

and we also have

(12.4.10) $$p = \tfrac{1}{2}(K - 1)(K - 2) - \tau - \iota.$$

These formulas presuppose that the curve has only ordinary double points and cusps (ordinary bitangents and inflection tangents). If higher singularities are present, it is necessary to "resolve" the singularities into simpler ones. Thus, a k-tuple point with distinct tangents counts as

$$\tfrac{1}{2}k(k - 1)$$

nodes. The reader will note that the last quantity equals the number of intersections of k straight lines in general position.

The actual resolution can be made by means of *birational transformations,* more precisely, by repeated use of *quadratic transformations.* The reader is already familiar with a special case of a quadratic transformation, namely, inversion in a circle, introduced in Section 2.1. In the general case, we simply replace the circle by a general conic Γ. One way of defining a quadratic transformation is to take a point O not on Γ and let the point X have the image Y under the transformation if Y is the intersection of the line OX produced with the polar of X with respect to Γ. The latter is defined as the line joining the points of contact of the two tangents which can be drawn to Γ from X. An alternate definition is based on the use of homogeneous coordinates in the (x, y)-plane. Three straight lines which intersect in the points O, P, Q serve as lines $x_1 = 0$, $x_2 = 0$, $x_3 = 0$. Each point (x, y) corresponds to a triple (x_1, x_2, x_3) where, for instance, x_1, x_2, x_3 are proportional to the distances of the point from the three lines. The image $Y = (y_1, y_2, y_3)$ of the point $X = (x_1, x_2, x_3)$ is then defined by

$$\rho y_1 = x_2 x_3, \quad \rho y_2 = x_3 x_1, \quad \rho y_3 = x_1 x_2.$$

Here obviously

$$y_1 : y_2 : y_3 = \frac{1}{x_1} : \frac{1}{x_2} : \frac{1}{x_3}.$$

The transformation is one to one except on the base lines. A very simple special case of a quadratic transformation is defined in nonhomogeneous coordinates by

$$x = s, \quad y = st.$$

A quadratic transformation takes an algebraic curve into an algebraic curve of the same genus.

This important property can be utilized in a number of ways. One important use is for the resolution of multiple singularities. By applying a suitable sequence of quadratic transformations to the given curve, one can obtain an equivalent curve having only nodes. If the image curve is

(12.4.11) $$C_1: \quad F_1(z_1, w_1) = 0,$$

then the correspondence between the two curves is given by a birational transformation

(12.4.12) $$\begin{cases} z = R_1(z_1, w_1), \\ w = R_2(z_1, w_1), \end{cases} \begin{cases} z_1 = S_1(z, w), \\ w_1 = S_2(z, w), \end{cases}$$

where R_1, R_2, S_1, S_2 are rational functions of their two arguments. The genus of C_1 is the same as that of C. Further, the number of nodes corresponding to a multiple point is invariant under such birational transformations.

The same device may be used to reduce the curve C to a suitable normal form. Thus, there exists a polynomial $P(s)$ of degree $2p + 1$ or $2p + 2$ whose roots are distinct such that C is given parametrically by rational functions of s and t:

(12.4.13) $$\begin{cases} z = R_1(s, t), \\ w = R_2(s, t), \end{cases} \quad \text{where} \quad t^2 = P(s).$$

We may regard the following as the normal form of C

(12.4.14) $$t^2 = P(s).$$

If $p = 0$, we may go one step farther, for then

$$s = t^2 + \alpha,$$

so that actually

(12.4.15) $$z = R_1(t), \quad w = R_2(t).$$

Conversely, if C admits of such a representation by rational functions of a parameter, then C is of genus zero. Such a curve is known as a *rational* or *unicursal* curve, where the second appellation refers to the fact that the curve can be drawn by a pen which never leaves the plane of the paper except to pass from one end of an asymptote to the other or to insert isolated points.

In the case $p = 1$ we can take the normal form to be

(12.4.16) $$\begin{aligned} t^2 &= 4(s - e_1)(s - e_2)(s - e_3) \\ &= 4s^3 - g_2 s - g_3 \end{aligned}$$

if

$$e_1 + e_2 + e_3 = 0$$

as we may assume. This is the *elliptic* case, and we shall see in the next chapter that the coordinates of (12.4.16) may be represented in terms of the meromorphic *elliptic* $\wp$-function of Weierstrass by

$$(12.4.17) \qquad\qquad s = \wp(u), \quad t = \wp'(u).$$

This means that any algebraic curve of genus one can be represented by rational functions of $\wp(u)$ and $\wp'(u)$. Conversely, every curve which admits of such a representation is of genus one.

The case $p > 1$ is known as the *hyperelliptic* case. In this case it is not possible to represent the coordinates by functions of a parameter u which are meromorphic in the finite plane (theorem of Picard—see Theorem 14.7.2), but there are representations by means of *automorphic functions* which are meromorphic in the unit circle and have the latter as a natural boundary.

EXERCISE 12.4

1. What is the genus of the surface if $F(z, w)$ is linear in z?

2. Show that the ramification index ϑ is < 2, $= 2$, or > 2 according as $p = 0$, $= 1$, or > 1 respectively.

3. Consider the surface of

$$w(w^2 - z^2) = z^4.$$

Determine the critical points, the total ramification, the ramification index, and the genus.

4. Replace z by x and w by y in the preceding problem and plot the resulting curve in the real (x, y)-plane. It has a triple point with distinct tangents at the origin. Show that a straight line $y = tx$ through the origin intersects the curve in a single point besides the origin and get the parametric representation of the curve.

5. The triple point in the preceding problem can be resolved by the special quadratic transformation

$$x = s, \quad y = st.$$

The image curve has no singular points in the finite plane. Verify that its genus is zero.

6. Formula (12.4.9) shows that the semicubical parabola

$$y^2 = x^3$$

is of class three since it has a cusp at $(0, 0)$. Verify that the tangential equation is

$$27v^2 + 4u^3 = 0.$$

The curve is rational. Find a parametric representation.

7. The leaf of Descartes (Problem 1, Exercise 12.1) is a rational curve. Find a parametric representation.

8. Discuss the corresponding Riemann surface. Find the critical points, ρ, and ϑ.

9. Verify that the curve is of class four and has the tangential equation

$$(1 - a^2 uv)^2 = 4a^2(u - av^2)(v - au^2).$$

10. Verify that the surface defined by

$$w^6 = (z - a)^3(z - b)^4$$

is of genus one if $a \neq b$. Obtain a representation of type (12.4.13) by setting $z = b + s^3$ and solving for w.

11. Same question for

$$w^4 = (z - a)(z - b)^2,$$

where a similar substitution leads to the desired representation.

12.5. Rational functions on the surface and Abelian integrals.

Let us consider a Riemann surface $\mathfrak{G}$ defined by an irreducible polynomial of degree $n > 1$ in w

$$(12.5.1) \qquad\qquad F(z, w) = 0.$$

Let $R(z, w)$ be a rational function of the two variables z and w. We shall study $R(z, w)$ on the surface $\mathfrak{G}$, that is, for values z, w which satisfy the equation (12.5.1). We shall suppose that $R(z, w)$ is neither identically zero nor identically infinity on $\mathfrak{G}$. This means that in the representation

$$(12.5.2) \qquad\qquad R(z, w) = \frac{P(z, w)}{Q(z, w)}$$

neither of the polynomials $P(z, w)$ and $Q(z, w)$ is divisible by $F(z, w)$.

This being the case, we can use the Euclidean algorithm to determine two polynomials $A(z, w)$ and $B(z, w)$ such that

$$(12.5.3) \qquad Q(z, w)A(z, w) - F(z, w)B(z, w) \equiv C(z)$$

is a polynomial in z alone. On $\mathfrak{G}$ we have $F(z, w) = 0$ and

$$Q(z, w)A(z, w) = C(z),$$

so that

$$R(z, w) = \frac{A(z, w)}{C(z)} P(z, w) \equiv \frac{P_1(z, w)}{C(z)},$$

where $P_1(z, w)$ is also a polynomial in z, w. But on $\mathfrak{G}$ we have

$$(12.5.4) \qquad w^n = R_{n,0}(z) + R_{n,1}(z) w + \cdots + R_{n,n-1}(z) w^{n-1},$$

where the R's are rational functions of z. It follows that for any $m \geqq n$ we can find rational functions $R_{m,j}(z)$ such that

$$w^m = R_{m,0}(z) + R_{m,1}(z)\, w + \cdots + R_{m,n-1}(z)\, w^{n-1}.$$

If these expressions are substituted in $P_1(z, w)$, we obtain

(12.5.5) $$\qquad R(z, w) = R_0(z) + R_1(z)\, w + \cdots + R_{n-1}(z)\, w^{n-1}$$

as a normal form of $R(z, w)$ on $\mathfrak{G}$. If we divide this expression by w^n and multiply by the right member of (12.5.4), we obtain an expression which may be reduced to the form

(12.5.6) $$\qquad R(z, w) = S_0(z) + S_1(z)\, w^{-1} + \cdots + S_{n-1}(z)\, w^{-n+1}.$$

This is the second normal form of $R(z, w)$ on $\mathfrak{G}$.

The function $R(z, w)$ is single-valued on $\mathfrak{G}$, for at the trace of each place $\mathfrak{p}_0$ we have a unique value of w as well as of z. At every point (z_0, w_0) we have local representations of $R(z, w)$ by means of series of the form

(12.5.7) $$\qquad R(z, w) = t^\mu \sum_{\nu=0}^{\infty} \alpha_\nu t^\nu, \quad \alpha_0 \neq 0,$$

where μ is an integer. Suppose that $\mathfrak{p}_0 = (z_0, w_0(z))$ and $w_0 = w_0(z_0)$. If $w_0(z)$ is a regular element, we can take

$$t = z - z_0.$$

For $w_0(z)$ we have a Taylor series in $(z - z_0)$ and similarly for the powers of $w_0(z)$. If the rational functions $R_j(z)$ of (12.5.5) are all holomorphic at $z = z_0$, we obtain a Taylor series for $R(z, w)$ in $(z - z_0)$, and the exponent μ is $\geqq 0$. We can obtain a negative value of μ only if one of the functions $R_j(z)$ has a pole at $z = z_0$ and if this pole is not canceled by a zero of $w_0(z)$.

If $w_0(z)$ is an algebraic element with r branches in the cycle, then we have

(12.5.8) $$\qquad z = z_0 + t^r, \quad w_0(z) = t^\lambda \sum_{n=0}^{\infty} a_n t^n,$$

where λ is an integer. We have similar expansions for the functions $R_j(z)$ and for the powers of $w_0(z)$. If these various series are substituted in (12.5.5), we obtain the series (12.5.7). This discussion is also valid for polar elements. Finally, at infinity we have

(12.5.9) $$\qquad z = t^{-r}, \quad w_0(z) = t^\lambda \sum_{\nu=0}^{\infty} b_\nu t^\nu,$$

and again a series expansion of type (12.5.7) is obtained for $R(z, w)$.

Thus for every place $\mathfrak{p}$ on the surface we have a corresponding series (12.5.7). We note that in general the integer $\mu = 0$. There can be at most a finite number of places on $\mathfrak{G}$ where $\mu \neq 0$. If at such a place $\mu < 0$, we say that (z_0, w_0) is a *pole* of $R(z, w)$ of multiplicity $-\mu$. If $\mu > 0$, we speak of a *zero* of order μ instead.

Let us now choose a place $p_0 = (z_0, w_0(z))$ with $w_0(z_0) = w_0$ such that $w_0(z)$ is a regular element of $w(z)$ and (z_0, w_0) is not a pole of $R(z, w)$. Let us continue $w_0(z)$ along a rectifiable path C which does not pass through a singularity of $w_0(z)$ or a pole of $R(z, w)$ when the latter function is continued along the same path C. For such a path the so-called *Abelian integral*

$$(12.5.10) \qquad A(z, z_0) \equiv \int_{(z_0, w_0)}^{(z, w)} R(Z, W)\, dZ$$

is well defined, since $R(Z, W)$ is a holomorphic function of Z in a neighborhood of every point on C. $A(z, z_0)$ is clearly a continuous function of z on C. More can be said, however. If $z_1 \in C$, then there exists a neighborhood of z_1 in which $A(z, z_0)$ defined by

$$\left\{ \int_{(z_0, w_0)}^{(z_1, w_1)} + \int_{(z_1, w_1)}^{(z, w)} \right\} R(Z, W)\, dZ$$

exists and is a holomorphic function of z with

$$A'(z, z_0) = R(z, w_0(z)).$$

This implies that $A(z, z_0)$ is locally holomorphic on $\mathfrak{G}$ if we omit the singular points of $w(z)$ and the poles of $R(z, w)$. Normally this function is not single-valued on $\mathfrak{G}$, much less single-valued as a function of z.

It is a fairly simple matter to study the behavior of $A(z, z_0)$ as z approaches one of the possible singular points. We simply use the information that can be read off from formulas (12.5.7) and (12.5.8). If z_1 is such a point, then these formulas with z_0 replaced by z_1 show that

$$(12.5.11) \qquad A(z, z_0) = A - r \int_t^a s^{\mu + r - 1} \left\{ \sum_{\nu=0}^{\infty} \alpha_\nu s^\nu \right\} ds,$$

where a and A are suitably chosen constants and the infinite series converges uniformly for the values of t under consideration. We can then integrate term by term. There are two cases.

(i) If $\mu + r > 0$, then we get a convergent power series in t and we conclude that $A(z, z_0)$ is a holomorphic function of

$$t = (z - z_1)^{1/r}$$

in some neighborhood of $z = z_1$. We note that $r = 1$ unless $z = z_1$ is a critical point of $w_0(z)$. In either case $A(z, z_0)$ is a holomorphic function on the surface, that is, relative to the local parameter t.

(ii) If $\mu + r \leq 0$, the integrated series will involve negative powers of t and possibly also a term in $\log t$. In the first case $A(z, z_0)$ is single-valued when restricted to a neighborhood of z_1 on the r sheets of $\mathfrak{G}$ which participate in the cycle, and $A(z, z_0)$ has a pole at $z = z_1$. If a term in $\log t$ is present, however, $A(z, z_0)$ is obviously infinitely many-valued.

Similar considerations hold at infinity. Here we have to replace (12.5.8) by

(12.5.9). $A(z, z_0)$ is holomorphic at infinity if and only if $\mu - r > 0$. For $\mu - r \leqq 0$ we have a pole and/or a logarithmic point.

Let us now turn to special cases. We shall consider the cases when $p = 0$ or 1 and indicate briefly what happens for $p > 1$. If $p = 0$ we have a parametric representation of z and w by rational functions of a parameter:

$$z = R_1(s), \quad w = R_2(s),$$

while conversely

$$s = R_3(z, w).$$

We have then

$$A(z, z_0) = \int_{s_0}^{s} R[R_1(\sigma), R_2(\sigma)]R_1'(\sigma)\, d\sigma$$
$$\equiv \int_{s_0}^{s} R_0(\sigma)\, d\sigma,$$

where R_0 is also a rational function. It follows by elementary calculus that

(12.5.12)
$$A(z, z_0) = R_4(s) + \log R_5(s)$$
$$= R_6(z, w) + \log R_7(z, w),$$

where all the R's denote rational functions. Thus, if $p = 0$ the transcendental part of $A(z, z_0)$ reduces to the logarithm of an algebraic function.

The case $p = 1$ is more interesting, for here we shall encounter three new classes of transcendental functions. In view of (12.4.13) we can restrict ourselves to the binomial case with $n = 2$, that is, $\mathfrak{S}$ is simply the surface of the cubic

(12.5.13)
$$w^2 = 4(z - e_1)(z - e_2)(z - e_3)$$
$$= 4z^3 - g_2 z - g_3$$

in the normalization of Weierstrass. By (12.5.6) we have then

$$R(z, w) = \frac{P_0(z)}{Q(z)} + \frac{P_1(z)}{Q(z)}\frac{1}{w(z)},$$

where P_0, P_1, and Q are polynomials in z. The first of these fractions evidently will give a contribution to $A(z, z_0)$ which is a rational function of z plus the logarithm of such a function. The second term on the right is more promising. Here we start by expanding $P_1(z)/Q(z)$ in partial fractions

$$\frac{P_1(z)}{Q(z)} = P_2(z) + \sum \sum \frac{a_{jk}}{(z - a_j)^k},$$

where $P_2(z)$ is a polynomial. We are thus led to a set of integrals of the form

(12.5.14)
$$W_k(a) \equiv \int (z - a)^k [w(z)]^{-1}\, dz,$$

where k is an integer. Our object will be to show that every such integral is a linear combination with constant coefficients of the three integrals

(12.5.15)
$$W_1(0), \quad W_0(0), \quad W_{-1}(a)$$

plus a rational function of z and $w(z)$. These three integrals are known as the *elliptic normal integrals of the second, first, and third kind* respectively. To prove the result we shall derive a recurrence relation.

Let us expand $P(z)$ in powers of $z - a$ so that

$$P(z) = A(z - a)^3 + 3B(z - a)^2 + 3C(z - a) + D.$$

We then form

$$(12.5.16) \quad \frac{d}{dz}[(z - a)^k \sqrt{P(z)}] = \frac{1}{\sqrt{P(z)}}\{k(z - a)^{k-1}P(z) + \tfrac{1}{2}(z - a)^k P'(z)\}$$

and note that the quantity between the braces may be written

$$(k + \tfrac{3}{2})A(z - a)^{k+2} + 3(k + 1)B(z - a)^{k+1}$$
$$+ 3(k + \tfrac{1}{2})C(z - a)^k + kD(z - a)^{k-1}.$$

Upon integrating (12.5.16) we consequently get

$$(12.5.17) \quad \begin{aligned}(z - a)^k \sqrt{(Pz)} &= (k + \tfrac{3}{2})A W_{k+2}(a) + 3(k + 1)B W_{k+1}(a) \\ &\quad + 3(k + \tfrac{1}{2})C W_k(a) + kD W_{k-1}(a).\end{aligned}$$

This is the desired relation.

Since $A = 4$ we can always solve for $W_{k+2}(a)$. Thus, if $k \geq 2$ we can, by repeated use of the recurrence relation, express $W_k(a)$ as an elementary function plus a linear combination of

$$W_1(a) = W_1(0) - a W_0(0) \quad \text{and} \quad W_0(a) = W_0(0).$$

The result is of the form

$$(12.5.18) \quad W_k(a) = M_{k-2}(z)\sqrt{P(z)} + \alpha W_1(0) + \beta W_0(0),$$

where $M_{k-2}(z)$ is a polynomial in z of degree $k - 2$.

If $k < -1$ we want to solve for the W_j with the lowest subscript in (12.5.17). Now $D = P(a)$, which is 0 if and only if a equals e_1, e_2, or e_3. Since $C = \tfrac{1}{3}P'(a)$, either C or D is different from 0. If follows again that we can reduce any integral $W_k(a)$ with $k < -1$ to an elementary function plus a linear combination of $W_{-1}(a)$ and $W_0(0)$ to which should be added a multiple of $W_1(0)$ unless $P(a) = 0$. The result is of the form

$$W_{-m}(a) = N(z)\sqrt{P(z)} + \alpha W_{-1}(a) + \beta W_0(0) + \gamma W_1(0).$$

Here $N(z)$ is a polynomial in $(z - a)^{-1}$ of degree m and with a double zero at ∞ if $P(a) = 0$, but of degree $m - 1$ and with a simple zero at ∞ if $P(a) \neq 0$.

Summing up we see that the general elliptic integral $A(z, z_0)$ is of the form

$$(12.5.19) \quad \begin{aligned}A(z, z_0) &= R_1(z, w(z)) + \log R_2(z, w(z)) + \alpha W_1(0) + \beta W_0(0) \\ &\quad + \gamma_1 W_{-1}(a_1) + \gamma_2 W_{-1}(a_2) + \cdots + \gamma_\nu W_{-1}(a_\nu),\end{aligned}$$

where the a's are the zeros of $Q(z)$.

The three types of normal integrals (12.5.15) are entirely different in their analytical behavior, which may be determined in the way indicated above for

$A(z, z_0)$. They are all three infinitely many-valued as functions of z. The integral of the first kind is locally holomorphic everywhere on the surface $\mathfrak{S}$. This is obvious in the neighborhood of ordinary points, but it holds also at the branch points. At $z = e_j$ we have $\mu = -1$, $r = 2$, so that $\mu + r > 0$, and at $z = \infty$ we have $\mu = 3$, $r = 2$, so that $\mu - r > 0$, and it follows that $W_0(0)$ is holomorphic as a function of the local parameter t. The integral of the second kind is also holomorphic in t at the finite branch points, but has a simple pole at infinity since now $\mu - r = -1$. Finally, an integral of the third kind normally has a logarithmic branch point at $z = a$ and is holomorphic in t at the branch points of $w(z)$. The picture changes if $a = e_j$; in this case $z = a$ is a simple pole.

We observed above that the integral of the first kind is infinitely many-valued. In this respect its behavior is similar to that of $\log z$, whose various determinations differ by multiples of $2\pi i$. Like the logarithm, the function $W_0(0)$ also has *additive periods*, but these periods have as their basis *two* fundamental periods $2\omega_1$ and $2\omega_2$ which are linearly independent over the real field. Every determination of $W_0(0)$ is of the form

$$(12.5.20) \qquad \pm W_0^*(0) + 2m\omega_1 + 2n\omega_3,$$

where $W_0^*(0)$ is a suitably chosen principal determination and m and n are arbitrary integers. The quantities ω_1 and ω_3 are integrals of $[P(z)]^{-\frac{1}{2}}$ taken between branch points of $w(z)$.

Let us normalize the integral of the first kind by setting

$$(12.5.21) \qquad W_0(z) = -\int_z^\infty [4(s - e_1)(s - e_2)(s - e_3)]^{-\frac{1}{2}}\, ds.$$

To treat a concrete case, suppose that the e's are real and $e_3 < e_2 < e_1$. To make the square root single-valued, we cut the s-plane along the real axis from $-\infty$ to e_1 and choose that determination which is positive for s real, $s > e_1$. The integral taken along a horizontal line in the s-plane from $s = z$ to ∞ is then uniquely defined and is, by definition, the principal value of $W_0(z)$ in the cut plane.

This function maps the upper half-plane $\Im(z) > 0$ on the interior of a rectangle R_1 in the W-plane with vertices at the points

$$0, \ -\omega_1, \ -\omega_1 + \omega_3, \ \omega_3,$$

where

$$(12.5.22) \qquad \begin{aligned} \omega_1 &= \int_{e_1}^\infty [4(s - e_1)(s - e_2)(s - e_3)]^{-\frac{1}{2}}\, ds, \\ \omega_3 &= \int_{-\infty}^{e_3} [4(s - e_1)(s - e_2)(s - e_3)]^{-\frac{1}{2}}\, ds. \end{aligned}$$

In the first integral the integrand is real positive, in the second purely imaginary with a positive imaginary part. See Figures 8 and 9.

It is a simple matter to verify these properties of the mapping. Since, however, $W_0(z)$ is a special case of the Schwarz-Christoffel functions which map half-

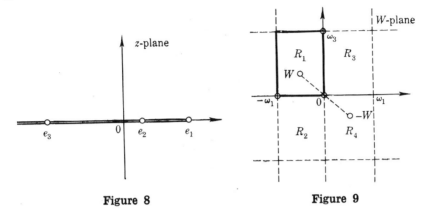

Figure 8   Figure 9

planes onto polygonal domains, we refer the reader to the discussion of these mapping functions in Section 17.6.

At this juncture we merely call attention to the fact that the discussion of the various determinations of $W_0(z)$ can be based on the principle of symmetry of Schwarz (see Section 7.7) since we have assumed the e's to be real. First, it is clear that the lower half-plane is mapped on the rectangle R_2 of Figure 9 where R_2 is symmetric to R_1 with respect to the interval $(-\omega_1, 0)$, which corresponds to the interval (e_1, ∞) in the z-plane. The open rectangles R_1 and R_2 plus the interval $(-\omega_1, 0)$ form the range of the principal determination of $W_0(z)$. Next, we can reflect each of the half-planes in any one of the intervals $(-\infty, e_3)$, (e_3, e_2), and (e_2, e_1). Using the first of these intervals we obtain the two rectangles R_3 and R_4 in the figure. Here the two points W and $-W$ correspond to the same point z in the upper half-plane. Proceeding in this manner with repeated reflections we obtain the W-plane covered by a rectangular net such that each rectangle corresponds to a copy of the upper or the lower z-half-plane. In particular, we see that the same point z corresponds to an infinite set of points

$$\pm W + 2m\omega_1 + 2n\omega_3,$$

which is the set of values of $W_0(z)$. This confirms (12.5.20).

One very important feature should be observed: although $W_0(z)$ is infinitely many-valued, the inverse function $z = \wp(W)$ is single-valued. This fact is basic for the theory of elliptic functions, to which the next chapter is devoted.

We have a similar situation in the general case where the e's are not real, but instead of a net of rectangles symmetric with respect to the axes in the W-plane, we obtain a net of parallelograms usually without any properties of symmetry with respect to the axes.

We add a last remark concerning elliptic integrals. The integral of the second kind also has a system of two additive periods $2\eta_1$ and $2\eta_3$, and the various determinations of $W_1(z)$ are of the form

$$(12.5.23) \qquad W_1^*(z) + 2m\eta_1 + 2n\eta_3.$$

The set of determinations of an integral of the third kind has a complicated structure and will not be given here.

We shall be quite brief in our description of the case $p > 1$. Here we take

$$(12.5.24) \qquad w^2 = (z - a_1)(z - a_2) \cdots (z - a_{2p+1}) \equiv P(z)$$

and

$$R(z,w) = \frac{P_0(z)}{Q(z)} + \frac{P_1(z)}{Q(z)} \frac{1}{\sqrt{P(z)}}.$$

As in the case $p = 1$ we are led to integrals $W_k(a)$ of the type (12.5.14). The recurrence relation corresponding to (12.5.17) now contains $2p + 2$ integrals $W_j(a)$ and may be used to express any integral with $k \geq 2p$ as an elementary function plus a linear combination of the p integrals of the second kind

$$(12.5.25) \qquad A_p(0), \cdots, A_{2p-1}(0)$$

and of the p integrals of the first kind

$$(12.5.26) \qquad A_0(0), \cdots, A_{p-1}(0).$$

If $k \leq -1$ we need in addition the integral of the third kind $A_{-1}(a)$.

The integrals of the first kind are holomorphic everywhere on the surface as functions of the local parameter t. Each of them has a system of additive periods, $2p$ in number, so there is a period matrix of p rows and $2p$ columns. None of these functions has a single-valued inverse, but there exist single-valued meromorphic functions of p variables $W_0, W_1, \cdots, W_{p-1}$ which are the values of the integrals of the first kind for a given value of z. Such Abelian functions fall outside the scope of this treatise.

EXERCISE 12.5

1. Find the analogue of the recurrence relation (12.5.17) for the case of a polynomial of degree four.

2. If $k \neq \pm 1$ find the genus of the surface defined by

$$w^2 = (1 - z^2)(1 - k^2 z^2).$$

Find integrals of the first and second kind.

3. If $w^2 = R(z)$, a rational function of z, verify directly that on the surface

$$R(z, w) = R_0(z) + R_1(z)w(z).$$

4. Show that in the formulas (12.5.22) the limits of integration may be changed from (e_1, ∞) to (e_3, e_2) and from $(-\infty, e_3)$ to (e_2, e_1) without change of the values of the integrals provided the square root is chosen properly.

5. The surface defined by

$$w^3 = (z - a_1)(z - a_2)(z - a_3),$$

where the a's are distinct, is of genus one. Find an integral of the first kind by inspection.

COLLATERAL READING

As general references see

APPELL, P., and GOURSAT, É. *Théorie des Fonctions Algébriques et de leurs Intégrales*. Gauthier-Villars, Paris, 1895.

BLISS, G. A. *Algebraic Functions*. Colloquium Publications, Vol. 16. American Mathematical Society, New York, 1933.

For Section 12.4 consult

SPRINGER, G. *Introduction to Riemann Surfaces*. Addison-Wesley Publishing Company, Inc., Reading, Massachusetts, 1957.

WALKER, R. J. *Algebraic Curves*. Princeton University Press, Princeton, New Jersey, 1950.

13

ELLIPTIC FUNCTIONS

13.1. Doubly-periodic functions. We have already encountered a number of analytic functions which are simply-periodic functions, that is, they are such that there exists a number $\omega \neq 0$ with the property that

(13.1.1) $$f(z + \omega) = f(z)$$

for all z in the domain of existence of $f(z)$. As examples we may list

$$\sin z, \quad \tan z, \quad e^z$$

with the periods

$$2\pi, \quad \pi, \quad 2\pi i$$

respectively. We say that ω is a *primitive period* of $f(z)$ if ω is a period and no submultiple of ω has this property.

The existence of doubly-periodic analytic functions was established in 1827 by N. H. Abel and C. G. J. Jacobi, independently of each other.

DEFINITION 13.1.1. *An analytic function $f(z)$ is doubly-periodic with periods ω and ϖ if*

(13.1.2) $$\Im\left(\frac{\varpi}{\omega}\right) > 0$$

and

(13.1.3) $$f(z + \omega) = f(z), \quad f(z + \varpi) = f(z)$$

for all z in the domain of definition of $f(z)$.

DEFINITION 13.1.2. *An analytic function $f(z)$ is said to be elliptic if (i) $f(z)$ is doubly-periodic and (ii) $f(z)$ is holomorphic save for poles in the finite plane.*

As an explicit example of an elliptic function we quote

(13.1.4) $$F(z) = \sum_{m=-\infty}^{\infty} \sum_{n=-\infty}^{\infty} (z - m - ni)^{-3}.$$

This function has the periods 1 and i and has triple poles at the Gaussian integers. An example of a doubly-periodic function which is not elliptic is furnished by $\exp[F(z)]$.

The numbers

(13.1.5) $$m\omega + n\varpi, \quad m, n = 0, \pm 1, \pm 2, \cdots$$

are clearly periods of $f(z)$ if $f(z)$ is doubly-periodic with the periods ω and ϖ.

124

We say that ω and ϖ are *primitive periods* of $f(z)$ if all the periods of $f(z)$ are of the form (13.1.5). Let ω and ϖ be a pair of primitive periods of $f(z)$. Let a, b, c, d be integers such that

(13.1.6) $$ad - bc = 1,$$

and set

(13.1.7) $$\omega^* = a\omega + b\varpi, \quad \varpi^* = c\omega + d\varpi.$$

Then conversely

$$\omega = d\omega^* - b\varpi^*, \quad \varpi = -c\omega^* + a\varpi^*,$$

so that ω^* and ϖ^* also form a pair of primitive periods.

The set of periods (13.1.5) forms a "lattice" in the z-plane and determines a set of congruent parallelograms, known as *period parallelograms* of $f(z)$. See Figure 10. It is clear that the lattice points do not determine the net of parallelograms uniquely, but each pair of primitive periods determines such a net in an obvious manner.

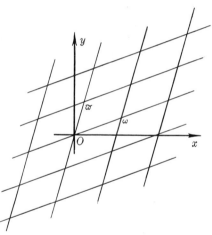

Figure 10

The range of a doubly-periodic function coincides with the set of values which it assumes in a period parallelogram.

An important consequence of this fact is

THEOREM 13.1.1. *If a doubly-periodic function is an entire function, then it is a constant.*

For an entire function must be bounded in a period parallelogram and hence bounded in the entire plane. It is then a constant by the theorem of Liouville.

Before going any farther in our study of the properties of doubly-periodic functions, let us note the following result due to Jacobi.

THEOREM 13.1.2. *The only triply-periodic analytic functions are constants.*

Proof. Suppose that $f(z)$ has three periods ω_1, ω_2, ω_3 such that

$$\omega_{kmn} \equiv k\omega_1 + m\omega_2 + n\omega_3$$

is zero if and only if $k = m = n = 0$, where k, m, n are arbitrary integers. This implies that no two elements of the set $\{\omega_{kmn}\}$ can coincide. It also implies that the set is dense in the plane. For our purposes it is enough to prove that the origin is a limit point of the set. In order to prove this we shall use the famous *box principle* of Dirichlet:

If k objects are placed in j boxes where $j < k$, then there is at least one box which contains more than one object.

We apply this principle as follows: Let N be a positive integer and let P_N be the subset of periods for which $|k|$, $|m|$, $|n| \leq N$. Then P_N contains $(2N + 1)^3$ points and is located in the square Q_N with center at the origin and sides of length

$$2N(|\omega_1| + |\omega_2| + |\omega_3|) \equiv 2NS$$

parallel to the coordinate axes. Now in the set of positive integers there is always a square between any two consecutive cubes. (Why?) Hence we can find an integer M such that

$$(2N)^3 \leq M^2 < (2N + 1)^3.$$

We now subdivide Q_N into M^2 congruent subsquares. By the box principle there is one of these subsquares which contains two or more elements of P_N. Suppose that $\omega_{\alpha\beta\gamma}$ and ω_{kmn} are located in the same subsquare. Their distance apart is then at most equal to the length of the diagonal of the subsquare, that is,

$$|\omega_{\alpha\beta\gamma} - \omega_{kmn}| \leq SN^{-\frac{1}{2}}.$$

But the difference $\omega_{\alpha\beta\gamma} - \omega_{kmn}$ is also a member of the period set and has a distance from the origin which is less than $SN^{-\frac{1}{2}}$. Here N is arbitrary, so the origin must be a limit point of the set.

But this implies that if $z = a$ belongs to the domain of definition of $f(z)$, then $f(z)$ takes on the value $f(a)$ in any punctured neighborhood of $z = a$. This is impossible unless $f(z)$ is a constant.

In this connection we should also consider the meaning of condition (13.1.2). Here the important fact is that $\Im(\varpi/\omega) \neq 0$; if this holds, then we can always assume $\Im(\varpi/\omega) > 0$. Suppose contrariwise that the quotient ϖ/ω is real. Without restricting the generality, we may then assume that ω and ϖ are real and positive. There are two alternatives.

(1) $\varpi/\omega = p/q$ where p and q are relatively prime positive integers. In this case we can always find an ω_0 such that

$$\varpi = p\omega_0, \quad \omega = q\omega_0$$

and two integers k and l such that $kp - lq = 1$. We have then

$$k\varpi - l\omega = kp\omega_0 - ql\omega_0 = \omega_0,$$

so that ω_0 is also a period. In this case, neither ω nor ϖ is a primitive period, but either ω_0 or one of its submultiples is primitive. Thus $f(z)$ is really simply-periodic.

(2) ϖ/ω is irrational. Here we observe that the set

$$\omega_{jk} \equiv j\omega - k\varpi$$

is dense on the real axis and, in particular, the origin is a limit point of the set. The proof can be based on the box principle. If N is a given positive integer, one can show the existence of an ω_{jk} with $|j| \leq 2N$, $|k| \leq 2N$ such that

(13.1.8) $|\omega_{jk}| \leq SN^{-1}, \quad S = |\omega| + |\varpi|,$

whence it follows that the origin is a limit point of the set of periods. Thus there are infinitesimal periods and we conclude as above that $f(z)$ must be a constant.

We see from this discussion that the condition (13.1.2) is necessary for double periodicity.

It was shown above that a non-constant doubly-periodic analytic function could not be entire. It follows that every elliptic function must have a certain number of poles. The next theorem gives further information.

THEOREM 13.1.3.   *The sum of the residues of the poles in the period parallelogram of a non-constant elliptic function equals zero.*

Proof.   Consider the period parallelogram Π with vertices at

$$0, \omega, \omega + \varpi, \varpi.$$

We may suppose that $f(z)$ has no poles on the sides of this figure. If this is not true, we replace $f(z)$ by $f(z - a)$ where a is suitably chosen. Let us integrate $f(z)$ along the sides of the period parallelogram. This gives

$$\int_\Pi f(z)\, dz = \omega \int_0^1 f(\omega t)\, dt + \varpi \int_0^1 f(\omega + \varpi t)\, dt$$
$$+ \omega \int_0^1 f(\omega + \varpi - \omega t)\, dt + \varpi \int_0^1 f(\varpi - \varpi t)\, dt.$$

Here the first and the third integrals in the right member cancel since

$$\int_0^1 f(\omega + \varpi - \omega t)\, dt = \int_0^1 f(-\omega t)\, dt = -\int_0^1 f(\omega s)\, ds,$$

and so do the second and fourth integrals since

$$f(\omega + \varpi t) = f(\varpi t), \quad f(\varpi - \varpi t) = f(-\varpi t).$$

Thus the value of the integral is zero. On the other hand, the integral equals $2\pi i$ times the sum of the residues inside the contour of integration. This proves the theorem.

DEFINITION 13.1.3.   *The sum of the orders of the poles of an elliptic function in the period parallelogram is called the order of the function.*

THEOREM 13.1.4.   *The order of an elliptic function is ≥ 2.*

Proof.   This is implied by the preceding theorem. For if $f(z)$ has a single pole, the principal part cannot involve a term in $(z - a)^{-1}$ since the residue is zero. The order of this pole must then be at least two.

The order of an elliptic function governs its behavior.

THEOREM 13.1.5.   *An elliptic function of order $m \geq 2$ assumes every value m times in the period parallelogram if multiple values are counted with proper multiplicities.*

Proof.   We use Theorem 13.1.3 once more. If $f(z)$ is an elliptic function, so is its derivative, and for any choice of a the function

$$\frac{f'(z)}{f(z) - a}$$

will have the same property. But by Theorem 9.2.1 the sum of the residues of this function equals the number of zeros of the denominator minus the number of poles. The latter is m by definition, and the sum of the residues is 0. It follows that the value a is assumed m times in the period parallelogram if multiple roots are counted with their proper multiplicities.

Theorem 9.2.4 also has an interesting application to the theory of elliptic functions.

THEOREM 13.1.6.   *Suppose that the poles and the a-values of an elliptic function $f(z)$ are located at the points*

$$\pi_1, \pi_2, \cdots, \pi_m \quad and \quad \alpha_1, \alpha_2, \cdots, \alpha_m$$

respectively, where multiple values are repeated as often as indicated by their multiplicity. Then the difference

$$\pi_1 + \pi_2 + \cdots + \pi_m - \alpha_1 - \alpha_2 - \cdots - \alpha_m$$

is a period of $f(z)$.

Proof.   We apply Theorem 9.2.4 with $g(z) = z$ to the function $f(z) - a$ and integrate around a period parallelogram in the positive sense. If this causes difficulties owing to the location of the points α_j and π_k, we can always find a number b such that none of these points are located on the parallelogram Π with vertices at

$$b, b + \omega, b + \omega + \varpi, b + \varpi.$$

Then on the one hand we have

$$\frac{1}{2\pi i} \int_\Pi z \frac{f'(z)}{f(z) - a} \, dz = \sum_{j=1}^{m} \alpha_j - \sum_{j=1}^{m} \pi_j,$$

and on the other, the integral in the left member is the sum of the four integrals along the sides of Π. If we group together the integrals taken along parallel sides and use the double periodicity of $f(z)$, we get

$$\frac{1}{2\pi i} \int_b^{b+\omega} [z - (z + \varpi)] \frac{f'(z)}{f(z) - a} \, dz - \frac{1}{2\pi i} \int_b^{b+\varpi} [z - (z + \omega)] \frac{f'(z)}{f(z) - a} \, dz$$

$$= -\frac{\varpi}{2\pi i} \int_b^{b+\omega} \frac{f'(z)}{f(z) - a} \, dz + \frac{\omega}{2\pi i} \int_b^{b+\varpi} \frac{f'(z)}{f(z) - a} \, dz$$

$$= -\frac{\varpi}{2\pi i} \{\log [f(z) - a]\}_{z=b}^{z=b+\omega} + \frac{\omega}{2\pi i} \{\log [f(z) - a]\}_{z=b}^{z=b+\varpi}.$$

Since $f(z)$ takes on the same values at the three points b, $b + \omega$, and $b + \varpi$, the logarithms can differ only by multiples of $2\pi i$. We have thus

$$\sum_{j=1}^{m} \alpha_j - \sum_{j=1}^{m} \pi_j = m\omega + n\varpi,$$

as asserted. This relation is due to Liouville. It is often written

$$(13.1.9) \qquad \sum_{j=1}^{m} \alpha_j - \sum_{j=1}^{m} \pi_j \equiv 0 \quad (\text{modd. } \omega, \varpi),$$

where the right member is read "congruent to zero modulis ω, ϖ."

In this section we have used the phrase "in the period parallelogram" somewhat loosely. Because of the double periodicity one has to be careful how much of the boundary is considered as belonging to the period parallelogram. It is clear that only one of the four vertices can be claimed and only two of the sides. We shall understand that the lower left vertex and the two open sides adjacent to this vertex are included in the two-dimensional point set referred to as the period parallelogram.

EXERCISE 13.1

1. Prove that there is at least one square between any two consecutive cubes. Estimate the number of squares between $(2N)^3$ and $(2N + 1)^3$ when N is large.

2. Prove (13.1.8).

3. Prove that if α is irrational then there exist infinitely many pairs of integers j, k and an absolute constant C such that

$$\left| \alpha - \frac{j}{k} \right| < \frac{C}{k^2}.$$

(*Hint:* Reduce the problem to the case $0 < \alpha < \frac{1}{2}$ and use the box principle on the numbers $k\alpha - j$, $|j| \leq N$, $|k| \leq N$. This method can be made to give $C \leq \frac{3}{2}$. The best value of C is $5^{-\frac{1}{2}}$.)

4. If $f(z)$ is an elliptic function of order m, show that $f'(z)$ is also elliptic and that its order n satisfies $m + 1 \leq n \leq 2m$.

5. If $f(z)$ is an elliptic function of order m and if $P(w)$ is a polynomial of degree n, show that $P[f(z)]$ is an elliptic function of order mn.

6. The elliptic function $F(z)$ defined by (13.1.4) is of order three. Show that it has three zeros in the period parallelogram. Show that $F(z)$ is an odd function and, hence, that the zeros are located at $z = \frac{1}{2}, \frac{1}{2}i$, and $\frac{1}{2}(1 + i)$.

7. Show that if a doubly-periodic function with periods $2\omega_1$ and $2\omega_3$ is even, then its derivative vanishes at $z = \omega_1, \omega_3$, and $\omega_1 + \omega_3$.

8. Show that the area of the period parallelogram is independent of the choice of the primitive periods.

13.2. The functions of Weierstrass. Let ω_1 and ω_3 be two complex numbers such that

$$\Im\left(\frac{\omega_3}{\omega_1}\right) > 0.$$

We shall construct a meromorphic function having the primitive periods $2\omega_1$ and $2\omega_3$. We set

$$\omega_2 = -\omega_1 - \omega_3,$$

and

(13.2.1) $\qquad \omega_{m,n} = 2m\omega_1 + 2n\omega_3, \quad m, n = 0, \pm 1, \pm 2, \cdots.$

LEMMA 13.2.1. *The series*

(13.2.2) $$\sum_{-\infty}^{\infty} \sum_{-\infty}^{\infty}{}' |\omega_{m,n}|^{-\alpha}$$

converges for $\alpha > 2$ and diverges for $\alpha \leq 2$.

Here and in the following a prime after a summation sign indicates that the combination $m = 0, n = 0$ is to be omitted.

Proof. Since we are dealing with a series whose terms are positive, we may group the terms in any way we please without affecting the convergence. Let S_k denote the partial sum of all terms with $|m| \leq k, |n| \leq k$, and set

$$U_k = S_k - S_{k-1}, \quad k = 1, 2, 3, \cdots.$$

Our series is convergent if and only if the series

$$\sum_1^{\infty} U_k$$

converges. Here U_k contains $8k$ terms of the series (13.2.2), and these terms are of the form

$$|2k\omega_1 + 2n\omega_3|^{-\alpha} \quad \text{or} \quad |2m\omega_1 + 2k\omega_3|^{-\alpha}$$

where $\mid m \mid\, \leq\, k,\, \mid n \mid\, \leq\, k$. The corresponding lattice points lie on the sides of a parallelogram which is the image of the basic period parallelogram under the magnification

$$w = 2k(z + \omega_2).$$

It follows that there exist two positive numbers a and b, independent of k, such that

$$ak < \mid z \mid < bk,$$

if z is a lattice point corresponding to one of the terms in U_k. This gives

(13.2.3) $$8b^{-\alpha}k^{1-\alpha} < U_k < 8a^{-\alpha}k^{1-\alpha},$$

so that the series (13.2.2) converges if and only if $\alpha > 2$.

We can now define the Weierstrass $\wp$-function by the series

(13.2.4) $$\frac{1}{z^2} + \sum_{-\infty}^{\infty} \sum_{-\infty}^{\infty}{}' \left\{ \frac{1}{(z - \omega_{m,n})^2} - \frac{1}{\omega_{m,n}{}^2} \right\} \equiv \wp(z; \omega_1, \omega_3),$$

which we shall denote simply by $\wp(z)$ if there is no danger of ambiguity. It follows from the discussion in Section 8.5 that this double series is uniformly convergent in any bounded domain which has a positive distance from the set $\{\omega_{m,n}\}$. The latter set gives the poles of $\wp(z)$. They are all double poles. Further

(13.2.5)
$$\wp(z + 2\omega_1; \omega_1, \omega_3) = \wp(z; \omega_1, \omega_3),$$
$$\wp(z + 2\omega_3; \omega_1, \omega_3) = \wp(z; \omega_1, \omega_3),$$

so that $\wp(z)$ is an elliptic function of order two.

To prove the periodicity we consider partial sums of the series. Let R be a closed subregion of a period parallelogram such that the distance of R from the poles is positive, and set

$$\wp_N(z) = z^{-2} + \sum_{-N}^{N} \sum_{-N}^{N}{}' \left\{ (z - \omega_{m,n})^{-2} - (\omega_{m,n})^{-2} \right\}.$$

Then

$$\lim_{N \to \infty} \wp_N(z) = \wp(z)$$

uniformly with respect to z in R. We have

$$\wp_N(z - 2\omega_1) - \wp_N(z) = \sum_{n=-N}^{N} (z - \omega_{N+1,n})^{-2}$$
$$- \sum_{n=-N}^{N} (z - \omega_{-N,n})^{-2}.$$

Proceeding as in the proof of (13.2.3), we see that

$$\sum_{n=-N}^{N} \mid z - \omega_{N+1,n} \mid^{-2} < (2N + 1)a^{-2}N^{-2} \to 0$$

as $N \to \infty$, and a similar estimate holds for the other sum. It follows that

$$\wp_N(z - 2\omega_1) - \wp_N(z) \to 0$$

as $N \to \infty$ uniformly with respect to z in R. This shows that $2\omega_1$ is a period. The same method works for $2\omega_3$.

The function $\wp(z; \omega_1, \omega_3)$ is homogeneous of degree -2 in its three arguments, that is,

$$(13.2.6) \qquad \wp(\lambda z; \lambda\omega_1, \lambda\omega_3) = \lambda^{-2}\wp(z; \omega_1, \omega_3).$$

In particular, $\wp(z)$ is an even function of z,

$$(13.2.7) \qquad \wp(-z; \omega_1, \omega_3) = \wp(z; \omega_1, \omega_3).$$

The series (13.2.4) can obviously be differentiated term by term so that

$$(13.2.8) \qquad \wp'(z; \omega_1, \omega_3) = -2 \sum_{-\infty}^{\infty} \sum_{-\infty}^{\infty} (z - \omega_{m,n})^{-3}.$$

This is an elliptic function of order three. Incidentally, the function $F(z)$ of formula (13.1.4) turns out to be $-\frac{1}{2}\wp'(z; \frac{1}{2}, \frac{1}{2}i)$.

Using the result of Problem 7 in the preceding Exercise we see that

$$(13.2.9) \qquad \wp'(\omega_1) = \wp'(\omega_2) = \wp'(\omega_3) = 0.$$

Since $\wp'(z)$ is of order three and the points ω_1, $-\omega_2$, and ω_3 belong to the same period parallelogram, we have found all the zeros of $\wp'(z)$ in this parallelogram. We set

$$(13.2.10) \qquad \wp(\omega_k; \omega_1, \omega_3) \equiv e_k, \quad k = 1, 2, 3,$$

and note that these numbers are distinct. For the elliptic function $\wp(z) - e_k$ is of order two and has a double zero at $z = \omega_k$. No other zero can occur in the period parallelogram and in particular, $e_j - e_k \neq 0$ if $j \neq k$. We shall prove later that

$$(13.2.11) \qquad e_1 + e_2 + e_3 = 0.$$

In addition to the $\wp$-function, Weierstrass introduced two related functions defined by

$$(13.2.12) \quad \zeta(z; \omega_1, \omega_3) = \frac{1}{z} + \sum_{-\infty}^{\infty} \sum_{-\infty}^{\infty}{}' \left\{ \frac{1}{z - \omega_{m,n}} + \frac{1}{\omega_{m,n}} + \frac{z}{(\omega_{m,n})^2} \right\},$$

$$(13.2.13) \quad \sigma(z; \omega_1, \omega_3) = z \prod_{-\infty}^{\infty} \prod_{-\infty}^{\infty}{}' \left\{ \left(1 - \frac{z}{\omega_{m,n}}\right) \exp\left[\frac{z}{\omega_{m,n}} + \frac{1}{2}\left(\frac{z}{\omega_{m,n}}\right)^2 \right] \right\}.$$

The convergence of these expansions is a consequence of the general theory of such series and products presented in Chapter 8 together with the convergence of the series (13.2.2) for $\alpha = 3$.

Both functions are odd functions of z:

$$(13.2.14) \qquad \zeta(-z) = -\zeta(z), \quad \sigma(-z) = -\sigma(z).$$

They are related to the $\wp$-function as shown by

(13.2.15)
$$\frac{d}{dz}\zeta(z; \omega_1, \omega_3) = -\wp(z; \omega_1, \omega_3),$$

(13.2.16)
$$\frac{d}{dz}\log \sigma(z; \omega_1, \omega_3) = \zeta(z; \omega_1, \omega_3).$$

Neither function can be doubly-periodic, but replacing z by $z + \omega_k$ gives rise to a simple transformation. Thus

(13.2.17)
$$\zeta(z + 2\omega_k) - \zeta(z) = 2\eta_k, \quad k = 1, 2, 3,$$

where the value of the constant η_k is found by substituting $-\omega_k$ for z and using (13.2.14). This gives

(13.2.18)
$$\eta_k = \zeta(\omega_k).$$

Here (13.2.17) follows from the relation

$$\frac{d}{dz}[\zeta(z + 2\omega_k) - \zeta(z)] = 0,$$

which is read off from (13.2.15).

The corresponding formulas for the sigma function read

(13.2.19)
$$\sigma(z + 2\omega_k) = -e^{2\eta_k(z + \omega_k)}\sigma(z).$$

We come now to the less obvious properties of the $\wp$-function such as the differential equation and the addition theorem. The reader who remembers the discussion in Section 12.5 will be prepared for the former at least. Actually, the common feature of both is an algebraic relation between two elliptic functions which have the same primitive periods. We shall take up this aspect of the problem in the next section. Here we give a direct derivation of the desired relations.

We start by studying the series expansion of $\wp(z)$ at the origin. Since $\wp(z)$ is an even function of z, we must have

$$\wp(z) = \frac{1}{z^2} + A_0 + A_1 z^2 + A_2 z^4 + \cdots.$$

From the series

$$\frac{1}{(z - \omega)^2} - \frac{1}{\omega^2} = \frac{2z}{\omega^3} + \frac{3z^2}{\omega^4} + \frac{4z^3}{\omega^5} + \frac{5z^4}{\omega^6} + \cdots$$

we read off that

$$A_0 = 0, \quad A_1 = 3\sum\sum{}' (\omega_{m,n})^{-4}, \quad A_2 = 5\sum\sum{}' (\omega_{m,n})^{-6}, \cdots.$$

With a view to simplifying the later formulas, we define

$$(13.2.20) \qquad g_2 = 60 \sum_{-\infty}^{\infty} {\sum_{-\infty}^{\infty}}' (\omega_{m,n})^{-4}, \quad g_3 = 140 \sum_{-\infty}^{\infty} {\sum_{-\infty}^{\infty}}' (\omega_{m,n})^{-6},$$

so that

$$(13.2.21) \qquad \wp(z) = \frac{1}{z^2} + \frac{1}{20} g_2 z^2 + \frac{1}{28} g_3 z^4 + \cdots$$

and

$$\wp'(z) = -\frac{2}{z^3} + \frac{1}{10} g_2 z + \frac{1}{7} g_3 z^3 + \cdots.$$

We now form

$$[\wp'(z)]^2 = \frac{4}{z^6} - \frac{2}{5} g_2 \frac{1}{z^2} - \frac{4}{7} g_3 + O(z^2),$$

$$4[\wp(z)]^3 = \frac{4}{z^6} + \frac{3}{5} g_2 \frac{1}{z^2} + \frac{3}{7} g_3 + O(z^2),$$

the difference of which is

$$-g_2 \frac{1}{z^2} - g_3 + O(z^2) = -g_2 \wp(z) - g_3 + O(z^2).$$

Let us now form the function

$$[\wp'(z)]^2 - 4[\wp(z)]^3 + g_2 \wp(z) + g_3.$$

This is an elliptic function whose order is at most six. Its only possible pole in the period parallelogram is at $z = 0$, but there it has a zero whose order is at least two. By Theorem 13.1.1 this function must be a constant, and since it vanishes at the origin, it must vanish identically. Hence we have

$$(13.2.22) \qquad \begin{aligned} [\wp'(z)]^2 &= 4[\wp(z)]^3 - g_2 \wp(z) - g_3 \\ &= 4[\wp(z) - e_1][\wp(z) - e_2][\wp(z) - e_3], \end{aligned}$$

where the second expression follows from (13.2.9) and (13.2.10). Incidentally, we have also proved (13.2.11).

Thus, $\wp(z)$ is a solution of the differential equation

$$(13.2.23) \qquad \left(\frac{dw}{dz}\right)^2 = 4(w - e_1)(w - e_2)(w - e_3),$$

and the general solution of this equation is given by

$$w = \wp(z + a),$$

where a is an arbitrary constant. We can of course also solve the differential equation by setting

$$(13.2.24) \qquad z + a = -\int_w^{\infty} [4(s - e_1)(s - e_2)(s - e_3)]^{-\frac{1}{2}} \, ds;$$

that is, z as a function of w is an elliptic integral of the first kind. We recall that z is an infinitely many-valued function of w, whereas w has been found to be a single-valued function of z.

We can throw some further light on the discussion in the preceding chapter. We recall formula (12.4.16),

$$(13.2.25) \qquad t^2 = 4(s - e_1)(s - e_2)(s - e_3),$$

which is the normal form of a curve of genus one. We see now, indeed, that we can obtain a parametric representation of the cubic by

$$(13.2.26) \qquad s = \wp(z; \omega_1, \omega_3), \quad t = \wp'(z; \omega_1, \omega_3)$$

provided we can determine the periods $2\omega_1$, $2\omega_3$ when the constants e_1, e_2, e_3 are known and satisfy (13.2.11). This is a transcendental problem of considerable difficulty. We have already obtained a solution, however, for the case in which the e's are real and $e_3 < e_2 < e_1$. By formula (12.5.22) we then have

$$(13.2.27) \qquad \begin{aligned} \omega_1 &= \int_{e_1}^{\infty} [4(s - e_1)(s - e_2)(s - e_3)]^{-\frac{1}{2}} \, ds, \\ \omega_3 &= \int_{-\infty}^{e_3} [4(s - e_1)(s - e_2)(s - e_3)]^{-\frac{1}{2}} \, ds, \end{aligned}$$

where the argument of the first integrand is 0 and that of the second is $\frac{1}{2}\pi$. Here e_1 and e_3 can be taken as independent variables with $e_2 = -e_1 - e_3$. By analytic continuation we can use these formulas to define $\omega_1(e_1, e_3)$ and $\omega_3(e_1, e_3)$ for all real and complex values of the variables. This is a possible mode of attack, but keeping track of the infinitely many determinations of the integrals as e_1 and e_3 vary is no easy task.

Another avenue of approach is furnished by formulas (13.2.20), which determine ω_1 and ω_3 implicitly in terms of g_2 and g_3. On the face of it, this looks like a more difficult task; the resulting information concerning the so-called *modular function* is highly rewarding, however. We shall return to this problem in Section 13.6.

We append one further remark concerning elliptic integrals of the second kind. Let us define

$$(13.2.28) \qquad J(w) = \zeta(z; \omega_1, \omega_3) \quad \text{with} \quad z = \int_w^{\infty} [P(s)]^{-\frac{1}{2}} \, ds,$$

where $P(s)$ is the right member of (13.2.25). Then

$$\frac{dJ}{dw} = \frac{d\zeta}{dz} \cdot \frac{dz}{dw} = \frac{\wp(z)}{[P(w)]^{1/2}} = \frac{w}{[P(w)]^{1/2}}.$$

It follows that $J(w)$ is an integral of the second kind. Moreover, we see that as z increases by $2m\omega_1 + 2n\omega_3$, then $J(w)$ increases by $2m\eta_1 + 2n\eta_3$ in agreement with (12.5.23).

Finally we shall take up the *addition theorem*. We can obtain this theorem by the method used in deriving the differential equation, but the following proof is shorter and more elegant. In principle it is an application of a famous theorem discovered by Abel in 1826.

THEOREM 13.2.1.   *If $Z(s, t)$ is an integral of the first kind attached to a curve C, if C_α is a family of curves of fixed degree n, then the sum of the values of $Z(s, t)$ for $(s, t) \in C \cap C_\alpha$ is independent of α up to periods of $Z(s, t)$.*

Proof of Addition Theorem.   In the following, C will be the cubic (13.2.25), and $Z(s, t)$ will be the parameter z which enters in the representation (13.2.26). Finally C_α will be a family of straight lines

$$t = as + b.$$

Each such line intersects the cubic in three points,

$$(s_1, t_1), \quad (s_2, t_2), \quad \text{and} \quad (s_3, t_3),$$

and elimination of t between the two equations shows that

(13.2.29) $$s_1 + s_2 + s_3 = \tfrac{1}{4}a^2.$$

Let the corresponding parameter values be z_1, z_2, z_3. These numbers are zeros of the elliptic function

$$\wp'(z) - a\wp(z) - b \equiv L(z),$$

and they are not congruent (modd. $2\omega_1, 2\omega_3$). $L(z)$ is of order three and has a triple pole at the origin. By (13.1.9) we then have

$$z_1 + z_2 + z_3 \equiv 0 \quad (\text{modd. } 2\omega_1, 2\omega_3)$$

in agreement with Abel's theorem. Here we may assume that the z's are chosen so that the sum actually equals 0. We now have

(13.2.30) $$\wp'(z_k) - a\wp(z_k) - b = 0, \quad k = 1, 2, 3.$$

From the first two of these equations we obtain

$$a = \frac{\wp'(z_1) - \wp'(z_2)}{\wp(z_1) - \wp(z_2)}.$$

This expression we substitute in (13.2.29). Since $s_k = \wp(z_k)$ we get

$$\wp(z_1) + \wp(z_2) + \wp(z_3) = \frac{1}{4}\left\{\frac{\wp'(z_1) - \wp'(z_2)}{\wp(z_1) - \wp(z_2)}\right\}^2,$$

or

(13.2.31) $$\wp(u + v) = -\wp(u) - \wp(v) + \frac{1}{4}\left\{\frac{\wp'(u) - \wp'(v)}{\wp(u) - \wp(v)}\right\}^2$$

for any choice of u and v. This is one form of the addition theorem.

We get another form by observing that the determinant of the system (13.2.30) must be zero, so that

(13.2.32)
$$\begin{vmatrix} \wp'(z_1) & \wp(z_1) & 1 \\ \wp'(z_2) & \wp(z_2) & 1 \\ \wp'(z_3) & \wp(z_3) & 1 \end{vmatrix} = 0 \quad \text{if } z_1 + z_2 + z_3 = 0,$$

or, expanded,

(13.2.33)
$$\wp'(z_1)[\wp(z_2) - \wp(z_3)] + \wp'(z_2)[\wp(z_3) - \wp(z_1)] \\ + \wp'(z_3)[\wp(z_1) - \wp(z_2)] = 0.$$

The addition theorem asserts that $\wp(u + v)$ is a rational function of $\wp(u)$, $\wp'(u)$, $\wp(v)$, $\wp'(v)$ and, in view of (13.2.22), an algebraic function of $\wp(u)$ and $\wp(v)$.

EXERCISE 13.2

1. Show that an elliptic function of order two with simple poles at $z = a$ and $z = b$ is defined by $\zeta(z - a) - \zeta(z - b)$.

2. Same question for the sigma quotient

$$\frac{\sigma(z - c)\sigma(z - d)}{\sigma(z - a)\sigma(z - b)}, \quad c + d = a + b.$$

3. Express $\wp(z) - e_k$ as a quotient of sigma functions.

4. Integrate $\zeta(z)$ along a suitably chosen parallelogram to obtain

$$\omega_3 \eta_1 - \omega_1 \eta_3 = \tfrac{1}{2}\pi i.$$

5. Verify that

$$\wp''(z) = 6[\wp(z)]^2 - \tfrac{1}{2}g_2.$$

6. Find $\wp'''(z)$ in terms of $\wp(z)$ and $\wp'(z)$.

7. $\wp(2z)$ is a rational function of $\wp(z)$. Use (13.2.31) to find this function.

8. Prove the addition theorem by discussing the poles of $\wp(u + a) + \wp(u)$ where a is not congruent to 0. The idea is to construct another elliptic function with the same poles and principal parts so that the difference must be a constant.

9. Show that the $\wp$-function is multiplied by λ^{-2} if z, g_2, g_3 are replaced by $\lambda z, \lambda^{-4}g_2, \lambda^{-6}g_3$ respectively.

10. The coefficients of the higher powers of z in (13.2.21) are polynomials in g_2 and g_3. More precisely, the coefficient of z^{2k-2} is of the form

$$\sum c_{m,n} g_2{}^m g_3{}^n, \quad 2m + 3n = k,$$

where the c's are positive rational numbers. Verify, and find the coefficient of z^6.

13.3. Some further properties of elliptic functions. Let $\mathbf{E} = \mathbf{E}(\omega_1, \omega_3)$ denote the class of all elliptic functions with $2\omega_1$ and $2\omega_3$ as periods.

THEOREM 13.3.1. *If $f(z)$ and $g(z)$ belong to $\mathbf{E}$, then there exists a polynomial $F(Z_1, Z_2)$ such that*

$$(13.3.1) \qquad\qquad F(f(z), g(z)) \equiv 0.$$

Proof. Let $F(Z_1, Z_2)$ be an arbitrary polynomial of total degree n. Such a polynomial involves $\frac{1}{2}n(n + 3)$ essential constants, not counting the constant term. Let the poles of $f(z)$ and of $g(z)$ in the basic period parallelogram be located at the points $a_1, a_2, \cdots, a_m$ and let μ_j be the larger of the multiplicities of the poles at $z = a_j$ (one multiplicity may be zero). Set

$$M = \mu_1 + \mu_2 + \cdots + \mu_m.$$

We now form

$$F[f(z), g(z)].$$

This is an elliptic function of order $\leq Mn$, and its poles, if any, are located at the points a_j. The point $z = a_j$ is normally a pole of order $n\mu_j$, but the corresponding principal part will be identically zero if the coefficients of $F(Z_1, Z_2)$ satisfy a certain set of $n\mu_j$ linear and homogeneous equations. Thus, we can annihilate all principal parts corresponding to presumptive poles in the period parallelogram by imposing a total of nM linear and homogeneous conditions on the coefficients. These equations do not involve the constant term of $F(Z_1, Z_2)$. If now

$$n + 3 > 2M,$$

we have more coefficients than equations, and it is always possible to find a non-trivial solution of the linear system. Let $F(Z_1, Z_2)$ be the polynomial of lowest degree obtainable in such a manner. Then $F[f(z), g(z)]$ is an elliptic function without poles and hence a constant. We can then choose the so far arbitrary constant term of the polynomial so that this constant is zero. This completes the proof.

COROLLARY. *The genus of the curve*

$$(13.3.2) \qquad\qquad F(Z_1, Z_2) = 0$$

is zero or one.

It is obviously zero if $g(z)$ is a rational function of $f(z)$ or vice versa. That it cannot exceed one follows from a theorem due to Picard according to which a relation of type (13.3.1) cannot be satisfied by functions meromorphic in the finite plane when the genus exceeds one.

This theorem has a number of consequences. In particular, it implies that every elliptic function satisfies a first-order differential equation and admits of an addition theorem.

THEOREM 13.3.2.   *An elliptic function of order m satisfies a differential equation of the form*

(13.3.3)
$$\left(\frac{dw}{dz}\right)^m + \sum_{k=1}^{m} P_k(w)\left(\frac{dw}{dz}\right)^{m-k} = 0,$$

where $P_k(w)$ is a polynomial in w with constant coefficients of degree not exceeding $2k$. In particular, $P_{m-1}(w) \equiv 0$.

Proof.   Since $f(z)$ and $f'(z)$ have the same primitive periods, the preceding theorem shows the existence of a polynomial $F(Z_1, Z_2)$ such that

$$F[f(z), f'(z)] \equiv 0.$$

Suppose that

$$F(Z_1, Z_2) \equiv \sum_{k=0}^{n} P_k(Z_1) Z_2^{n-k}.$$

It remains to prove that the polynomial has the special structure stated in the theorem.

First we observe that $P_0(Z_1)$ must be a constant. For otherwise there would be a value of Z_1 such that $P_0(Z_1) = 0$. The discussion in Section 12.2 shows that in such a case there are some determinations of the algebraic function $Z_2(Z_1)$ defined by the corresponding equation (13.3.2) which become infinite as Z_1 tends to the critical value under consideration. But this is out of the question here for $Z_1 = w(z)$ and $Z_2 = w'(z)$ and, hence, Z_1 and Z_2 are cofinite. This requires that $P_0(Z_1)$ is a constant. We take $P_0(Z_1) \equiv 1$.

Next we observe that since $w(z)$ has no singularities other than poles in the period parallelogram, all the branches of $Z_2(Z_1)$ must become infinite with Z_1. If $z = a$ is a pole of $w(z)$ of order μ, then it is also a pole of $w'(z)$ of order $\mu + 1$, that is, there must be a branch of the form

(13.3.4)
$$Z_2(Z_1) = CZ_1^{\frac{\mu+1}{\mu}}[1 + o(1)]$$

as $Z_1 \to \infty$. Let us now consider the Newton diagram of $F(Z_1, Z_2)$. The part of the diagram which is of interest here is the portion which faces the infinite region. This consists of a polygonal line joining $(0, n)$ on the Z_2-axis with some point $(p, 0)$ on the Z_1-axis. This line is part of a convex polygon, and formula (13.3.4) shows that all the line segments which make up this part of the polygon have negative slopes of the form $-\dfrac{\mu}{\mu+1}$, where μ is an integer ≥ 1. In particular, the whole diagram lies on or below the line

(13.3.5)
$$Z_2 - n = -\frac{\mu_0}{\mu_0 + 1} Z_1,$$

where μ_0 is the greatest lower bound of the multiplicities of the poles of $f(z)$.

This asserts, in particular, that

$$p \leqq \frac{\mu_0 + 1}{\mu_0} n \leqq 2n.$$

This means that every $P_k(w)$ is of degree $\leqq 2n$. More precisely, the polynomial $P_k(w)$ gives rise to the lattice points of ordinate $n - k$ in the diagram. Since these points lie on or below the line (13.3.5), we see that the degree of $P_k(w)$ cannot exceed

$$\frac{\mu_0 + 1}{\mu_0} k \leqq 2k.$$

The analogue of formula (12.3.11) tells us that

$$\sum \mu_j = n.$$

Thus the geometry of the diagram shows that the sum of the multiplicities of the poles in the period parallelogram equals the degree of $F(Z_1, Z_2)$ in Z_2. But $\sum \mu_j = m$, the order of $f(z)$. It follows that $n = m$.

It remains only to prove that $P_{m-1}(w) \equiv 0$. This is a subtle point. The theory of symmetric functions of the roots of an algebraic equation shows that, aside from the sign, $P_{m-1}(w)/P_m(w)$ is the sum of the reciprocals of the roots of (13.3.2) as an equation in Z_2, that is,

$$\pm \frac{P_{m-1}(w)}{P_m(w)} = \sum_{j=1}^{m} \frac{1}{w'(z_j)} = \sum_{j=1}^{m} \frac{dz_j}{dw},$$

where the summation extends over those values of z for which $w(z)$ takes on the value w. But by Theorem 13.1.6 the sum of these values is independent of w, so its derivative with respect to w must be identically 0, as asserted. This completes the proof.

The structure of the set $\mathbf{E}(\omega_1, \omega_3)$ is rather interesting. It is obviously a field over the complex field. By Theorem 13.3.1 it contains the roots of infinitely many algebraic equations whose coefficients are in $\mathbf{E}$. The field $\mathbf{E}$ is not algebraically closed, however, for, in general, the roots of an algebraic equation with coefficients in $\mathbf{E}$ are not single-valued functions of z. They are doubly-periodic but normally not elliptic functions.

An important property of $\mathbf{E}$ is that it is *finitely generated*. More precisely, we can find two elements of $\mathbf{E}$ such that every element of $\mathbf{E}$ is a rational function of these two generators. One of the generators may be chosen arbitrarily; the second one must then be found outside the subfield generated by the first generator.

We shall show that $\wp(z; \omega_1, \omega_3)$ and $\wp'(z; \omega_1, \omega_3)$ may be used as generators of $\mathbf{E}(\omega_1, \omega_3)$. From the differential equation (13.2.22) it follows that any polynomial in $\wp(z)$ and $\wp'(z)$ may be reduced to the form

$$P[\wp(z)] + Q[\wp(z)]\wp'(z),$$

where $P(w)$ and $Q(w)$ are polynomials in w with constant coefficients. Similarly every rational function of $\wp(z)$ and $\wp'(z)$ is of the form

$$(13.3.6) \qquad R[\wp(z)] + S[\wp(z)]\wp'(z),$$

where $R(w)$ and $S(w)$ are rational functions of w with constant coefficients.

THEOREM 13.3.3.   *Every element of* $\mathbf{E}$ *is of the form* (13.3.6).

Proof.   It is clear that every such function belongs to $\mathbf{E}$. It is an even function of z if and only if $S(w) \equiv 0$, an odd function if and only if $R(w) \equiv 0$.

Suppose now that $f(z) \in \mathbf{E}$ and $f(-z) = f(z)$. For the moment we suppose that $f(z) \neq 0, \infty$ when $z \equiv 0 \pmod{\omega_1, \omega_3}$. Since $f(z)$ is an even function of z, its order is $2k$, an even integer. We can then find k numbers $a_1, a_2, \cdots, a_k$ such that every zero of $f(z)$ is of the form $\pm a_j + \omega_{m,n}$. Further, we can find k numbers $b_1, b_2, \cdots, b_k$ such that every pole of $f(z)$ is of the form $\pm b_j + \omega_{m,n}$. It is understood that multiple zeros are repeated in the sequence $a_1, a_2, \cdots, a_k$ as often as indicated by the multiplicity, and similarly for the poles. Then

$$(13.3.7) \qquad f(z) = f(0) \prod_{j=1}^{k} \frac{\wp(z) - \wp(a_j)}{\wp(z) - \wp(b_j)}.$$

For the right-hand side is an elliptic function having the same zeros and poles as $f(z)$, so that the quotient of the two sides is an elliptic function without zeros and poles, that is, a constant whose value can be found by letting z approach 0.

This procedure must be slightly modified if $f(z)$ has a zero or a pole at the origin or at one of the half-periods. Such a zero or pole is always of even order. We can then find an integral power of $\wp(z)$ which behaves like $f(z)$ at $z = 0$ and an integral power of $\wp(z) - e_\alpha$ which behaves like $f(z)$ at $z = \omega_\alpha$ where $\alpha = 1$, 2, or 3. The result is

$$(13.3.8) \quad f(z) = A[\wp(z)]^\mu \prod_{\alpha=1}^{3} [\wp(z) - e_\alpha]^{\nu_\alpha} \prod_{j=1}^{m} [\wp(z) - \wp(a_j)] \prod_{j=1}^{n} [\wp(z) - \wp(b_j)]^{-1},$$

where

$$2(\mu + \nu_1 + \nu_2 + \nu_3) + m - n = 0.$$

Thus, if $f(z) \in \mathbf{E}$ and if $f(-z) = f(z)$, then $f(z)$ is a rational function of $\wp(z)$.

Now if $f(z)$ is odd, then $f(z)/\wp'(z)$ is even. By the previous result we then have that $f(z)$ equals $\wp'(z)$ times a rational function of $\wp(z)$. The general case is reduced to these special cases since

$$f(z) = \tfrac{1}{2}[f(z) + f(-z)] + \tfrac{1}{2}[f(z) - f(-z)]$$

is the sum of an even function and an odd function both of which belong to $\mathbf{E}$. This completes the proof.

Thus $\mathbf{E}$ is generated by $\wp(z)$ and $\wp'(z)$. This suffices for our needs, but it should be mentioned that much more general results hold. Two functions $g(z)$

and $h(z)$ in **E** can serve as generators of **E** if they separate points in a sense which we shall not take time to explain.

Another consequence of the particular structure of **E** is the fact that the elements of **E** possess addition theorems just as $\wp(z)$ does.

THEOREM 13.3.4. *If $f(z) \in$ **E**, then $f(u + v)$ is an algebraic function of $f(u)$ and $f(v)$.*

REMARK. It is possible to make this statement a little stronger: $f(u + v)$ *is a rational function of $f(u)$, $f'(u)$, $f(v)$, and $f'(v)$.* Since $f'(u)$ is an algebraic function of $f(u)$ by Theorem 13.3.2, we see that Theorem 13.3.4 is a consequence of this observation. We shall not prove the stronger statement.

Proof. To simplify the notation, let us write

$$\wp(u) = p, \quad \wp(v) = q, \quad \wp'(u) = r, \quad \wp'(v) = s.$$

Further let P, Q, R, and S denote rational functions of p and q, symmetric in their arguments, and let T, U, V, and W be rational functions of a single variable. All these functions have constant coefficients. Now the functions $f(u + v)$, $f(u)$, and $f(v)$ are elements of **E** considered as functions of u or of v. As such they are representable by formula (13.3.6) in terms of the corresponding $\wp$-function and its derivative. This leads to the following five formulas

$$(13.3.9) \quad \begin{cases} f(u + v) = P(p, q) + Q(p, q)r + R(p, q)s + S(p, q)rs, \\ f(u) = T(p) + U(p)r, \\ f(v) = V(q) + W(q)s, \\ r^2 = 4p^3 - g_2 p - g_3, \\ s^2 = 4q^3 - g_2 q - g_3. \end{cases}$$

The first of these formulas calls for some comments. Here $f(u + v)$ is a symmetric function of u and v. If we write the representation of $f(u + v)$ first in terms of p and r and then in terms of q and s, we see that these representations can be unified into a single one which must be symmetric in the variables p and q as well as in r and s and linear in each of r and s. The first formula is the result of this consideration.

Using these five equations, we can now eliminate the four auxiliary variables p, q, r, and s. The result is of the form

$$(13.3.10) \quad F[f(u + v), f(u), f(v)] = 0$$

where F is a polynomial in its three arguments since p, q, r, and s enter rationally in the eliminants. This is the required addition theorem.

Weierstrass proved in his lectures that a function $f(z)$ which possesses an algebraic addition theorem is either an algebraic function of z, or an algebraic function of $\exp(2\pi i z/\omega)$ for a suitably chosen constant ω, or an algebraic function of $\wp(z; \omega_1, \omega_3)$ where ω_1 and ω_3 are suitably chosen.

EXERCISE 13.3

1. Prove that $\wp^{(2k)}(z)$ is a polynomial in $\wp(z)$ of degree $k+1$ whose coefficients are polynomials in g_2 and g_3. What is the corresponding result for $\wp^{(2k+1)}(z)$?

2. Show that $[\wp(z) - e_1]^{-1}$ is an even function of $z - \omega_1$ and find its principal part at $z = \omega_1$.

3. Show that $[\wp(z) - e_1]^{\frac{1}{2}}$ is a single-valued function of z, but not an element of **E** because $2\omega_1$, $4\omega_3$ is a pair of primitive periods.

4. By Theorem 13.3.2, $\wp'(z)$ satisfies a differential equation of the form

$$(w')^3 + P_1(w)(w')^2 + P_3(w) = 0,$$

where $P_1(w)$ and $P_3(w)$ are polynomials of degree ≤ 2 and ≤ 6 respectively. Use the fact that $\wp'(z)$ has triple poles to lower these limits. Use this and parity considerations to prove that $P_1(w)$ is actually a constant and $P_3(w)$ is an even polynomial of degree four. Use (13.2.21) and the point $z = \omega_1$ to obtain

$$P_1(w) = -\tfrac{3}{2}g_2,$$
$$P_3(w) = -\tfrac{27}{2}w^4 - 27g_3w^2 - 2(3e_1^2 - g_2)(6e_1^2 - \tfrac{1}{2}g_2)^2.$$

The constant term of $P_3(w)$ must be a symmetric function of the roots e_1, e_2, e_3. Its value turns out to be

$$\tfrac{1}{2}(g_2^3 - 27g_3^2) = 2^3 (e_1 - e_2)^2 (e_2 - e_3)^2 (e_3 - e_1)^2.$$

5. Verify that the function $f(u) = \dfrac{1 - ku}{1 + ku}$ has the addition theorem

$$[3 + f(u) + f(v) - f(u)f(v)]f(u+v) = -1 + f(u) + f(v) + 3f(u)f(v)$$

for any value of k including $k = 0$. Show that $f(u) \equiv -1$ is also a solution.

What functions admit of the following addition theorems?

6. $f(u+v) = f(u)f'(v) + f(v)f'(u)$.

7. $f(u+v) = \dfrac{f(u) + f(v)}{1 - f(u)f(v)}$.

8. Show that there are three distinct constant solutions of Problem 7.

9. Setting $v = u$ in Problem 7, we get

$$f(2u) = \frac{2f(u)}{1 - [f(u)]^2}.$$

Suppose that we know that this equation is satisfied by a holomorphic function $f(u)$ for $|u| < \rho$ and that $f(u) \neq \pm 1$ for such values. Show that $f(u)$ can be continued as a meromorphic function of u in the whole (finite) plane.

13.4. On the functions of Jacobi. So far the functions of Weierstrass have been kept in the foreground, but neither are they the oldest elliptic functions discovered nor are they the ones best suited for the applications to arithmetic, geometry, and mechanics. The functions introduced by Jacobi antedate those of Weierstrass by some thirty years, and the student cannot afford to ignore them.

In Section 12.4 we mentioned that the curves of genus one have two equivalent normal forms $y^2 = P(x)$, where $P(x)$ is a polynomial of degree three or four with distinct roots. The functions of Weierstrass are attached to the cubic case and are based on the normal form

$$y^2 = 4(x - e_1)(x - e_2)(x - e_3), \quad e_1 + e_2 + e_3 = 0.$$

The functions of Jacobi on the other hand belong to the quartic case. The normal form is usually that of A. M. Legendre (1752–1833),

$$(13.4.1) \qquad y^2 = (1 - x^2)(1 - k^2 x^2),$$

where $k \neq 0, +1, -1$. The basic differential equation is now

$$(13.4.2) \qquad \left(\frac{dw}{dz} \right)^2 = (1 - w^2)(1 - k^2 w^2).$$

With a view toward studying the properties of $w(z)$ without the use of explicit representations, we shall develop briefly some methods used in the theory of differential equations.

We shall need the *basic existence theorem*.

THEOREM 13.4.1. *Suppose that $f(z, w)$ is holomorphic for $|z - z_0| < a$, $|w - w_0| < b$ and that*

$$(13.4.3) \qquad |f(z, w)| \leqq M, \quad |f(z, w_1) - f(z, w_2)| \leqq K |w_1 - w_2|$$

for such values. Let

$$(13.4.4) \qquad r = \min \left[a, \frac{1}{K} \log \left(1 + \frac{bK}{M} \right) \right].$$

Then the differential equation

$$(13.4.5) \qquad \frac{dw}{dz} = f(z, w)$$

has a unique solution, $w = w(z; z_0, w_0)$, which is holomorphic for $|z - z_0| < r$ and tends to w_0 as $z \to z_0$.

Proof. We use the method of successive approximations. The differential equation, together with the initial conditions, is equivalent to the integral equation

$$(13.4.6) \qquad w(z) = w_0 + \int_{z_0}^{z} f(s, w(s)) \, ds,$$

where the integral is taken along the line segment joining $s = z_0$ with $s = z$. We define

$$w_0(z) \equiv w_0,$$

(13.4.7)

$$w_n(z) = w_0 + \int_{z_0}^{z} f(s, w_{n-1}(s))\, ds, \quad n > 0.$$

We suppose that $|z - z_0| < r$ and proceed to prove that each $w_n(z)$ is well defined and that its values satisfy $|w_n(z) - w_0| < b$. Now

$$w_1(z) = w_0 + \int_{z_0}^{z} f(s, w_0)\, ds$$

is certainly well defined as a holomorphic function of z for $|z - z_0| < r \le a$. Further,

$$|w_1(z) - w_0| \le M\,|z - z_0| < Mr \le \frac{M}{K} \log\left(1 + \frac{bK}{M}\right) < b$$

since $\log(1 + u) < u$ for $0 < u$. Suppose that we have verified existence and holomorphy of $w_n(z)$ for $|z - z_0| < r$, $n \le k$, together with the inequalities

(13.4.8) $|w_n(z) - w_{n-1}(z)| \le M\, \dfrac{K^{n-1}}{n!}\,|z - z_0|^n, \quad n = 1, 2, \cdots, k.$

Adding the inequalities we get

$$|w_k(z) - w_0| \le \sum_{n=1}^{k} |w_n(z) - w_{n-1}(z)| \le M \sum_{n=1}^{k} \frac{K^{n-1}}{n!}|z - z_0|^n$$

(13.4.9)

$$< \frac{M}{K}[e^{K|z-z_0|} - 1] < \frac{M}{K}[e^{Kr} - 1] \le b$$

by the definition of r. It follows that $f(s, w_k(s))$ is well defined as a holomorphic function of s on the path of integration provided $|z - z_0| < r$. This implies that $w_{k+1}(z)$ is also well defined as a holomorphic function of z. Further, from

$$w_{k+1}(z) - w_k(z) = \int_{z_0}^{z} [f(s, w_k(s)) - f(s, w_{k-1}(s))]\, ds$$

and the Lipschitz condition we get

$$|w_{k+1}(z) - w_k(z)| \le K \int_{z_0}^{z} |w_k(s) - w_{k-1}(s)|\,|ds|$$

$$\le M \frac{K^k}{k!} \int_{z_0}^{z} |s - z_0|^k\,|ds| = M \frac{K^k}{(k+1)!}|z - z_0|^{k+1},$$

so that (13.4.8) holds for all k. The same is then true for (13.4.9).

Thus we see that the sequence $\{w_n(z)\}$ is well defined for $|z - z_0| < r$. Further, (13.4.8) implies that

$$w(z) \equiv \lim_{n \to \infty} w_n(z)$$

exists uniformly with respect to z in $|z - z_0| < r$. Thus $w(z)$ is holomorphic in this disk. Moreover, from (13.4.7) and the uniform convergence it follows that $w(z)$ satisfies (13.4.6) and the initial condition. It is consequently the desired solution of the differential equation.

It remains to show uniqueness. Suppose that $W(z)$ is a solution of (13.4.6). We then have in some disk $|z - z_0| < r_1 \leqq r$

$$W(z) - w_n(z) = \int_{z_0}^{z} [f(s, W(s)) - f(s, w_{n-1}(s))]\, ds,$$

and by the Lipschitz condition

$$|W(z) - w_n(z)| \leqq K \int_{z_0}^{z} |W(s) - w_{n-1}(s)|\, |ds|$$

$$\leqq K^2 \int_{z_0}^{z} \int_{z_0}^{t} |W(s) - w_{n-2}(s)|\, |ds|\, |dt|$$

$$= K^2 \int_{z_0}^{z} |s - z_0|\, |W(s) - w_{n-2}(s)|\, |ds|,$$

where we have used a familiar device for the evaluation of an iterated integral. Repeating the process we obtain

$$|W(z) - w_n(z)| \leqq \frac{K^n}{(n-1)!} \int_{z_0}^{z} |s - z_0|^{n-1} |W(s) - w_0|\, |ds|$$

$$< A \frac{K^n}{(n-1)!} \int_{z_0}^{z} |s - z_0|^{n-1} |ds| = A \frac{K^n}{n!} |z - z_0|^n,$$

if A is an upper bound for $|W(s) - w_0|$ in the disk $|s - z_0| < r_1$. It follows that

$$\lim_{n \to \infty} |W(z) - w_n(z)| = 0,$$

that is, the solution $W(z)$ must coincide with the solution $w(z) = \lim_{n \to \infty} w_n(z)$ which we have already constructed. Thus the solution is unique and the existence theorem is proved.

This result will now be applied to equation (13.4.2).

THEOREM 13.4.2. *Every solution of (13.4.2) is a single-valued function of z whose only singularities are simple poles.*

Proof. The equation does not contain z explicitly, so that if $w(z)$ is a solution so is $w(z - c)$ for any c. We can take z_0 arbitrarily and the quantity a of Theorem 13.4.2 is ∞ for *any* value of z_0. We have

$$f(z, w) \equiv (1 - w^2)^{\frac{1}{2}} (1 - k^2 w^2)^{\frac{1}{2}},$$

which is a two-valued function of w having simple branch points at $w = \pm 1$, $\pm 1/k$. This implies that if $w(z)$ is a solution of (13.4.2) so is $-w(z)$. It will appear from the discussion below that these solutions are actually distinct and cannot be carried into each other by analytic continuation. We can take w_0 anywhere in the finite plane except at the branch points, which require special discussion. With this point as center we draw a circle $| w - w_0 | = b$ where b is less than the distance from w_0 to the branch points. In the disk bounded by this circle, the square root is uniquely determined, once the value at the center is given. Further, $f(z, w)$ is bounded and satisfies a Lipschitz condition in the disk.

It follows that the equation $w' = f(z, w)$ has a unique solution $w(z; z_0, w_0)$ which is holomorphic in a disk $| z - z_0 | < r$. This is also a solution of (13.4.2), and so is $-w(z; z_0, w_0)$. By the law of permanence of functional equations, Theorem 10.7.1, any analytic continuation of $w(z; z_0, w_0)$ will remain a solution of $w' = f(z, w)$ if the right-hand member is continued along the same path.

The existence theorem as stated gives us no information about what happens at the branch points of the square root. Suppose that we take $w_0 = 1$. The square root does not satisfy a Lipschitz condition in any domain of the w-plane which contains $w = 1$. Actually uniqueness is lost. We see that

$$w(z) \equiv 1$$

is a solution, but this does not exhaust the possibilities. Set

$$w = 1 - u^2$$

and assume that $u(z) \not\equiv 0$. Substitution in (13.4.2) gives, after simplification and extraction of the square root,

$$(13.4.10) \qquad \frac{du}{dz} = \tfrac{1}{2}(2 - u^2)^{\frac{1}{2}}[1 - k^2(1 - u^2)^2]^{\frac{1}{2}},$$

where the right-hand side is holomorphic and satisfies a Lipschitz condition in some neighborhood of $u = 0$. It is immaterial how we choose the roots in (13.4.10) since $u(z)$ and $-u(z)$ give the same value for $w(z)$. The existence theorem applies to (13.4.10). For a given choice of the square roots there is a unique solution, $u = u(z; z_0, 0)$, which is holomorphic in some neighborhood of $z = z_0$. It follows that (13.4.2) has a solution

$$(13.4.11) \qquad \begin{aligned} w(z; z_0, 1) &\equiv 1 - [u(z; z_0, 0)]^2 \\ &= 1 - \tfrac{1}{2}(1 - k^2)(z - z_0)^2 + \cdots. \end{aligned}$$

Moreover, $f(z, w(z; z_0, 1))$ is a holomorphic function of z in some neighborhood of $z = z_0$ though $f(z, w)$ has a branch point at $w = 1$.

Similar results hold at the other branch points of $f(z, w)$. There is the constant solution, $w(z) \equiv w_0$, and also a solution $w(z; z_0, w_0)$ which is holomorphic together with $f(z, w(z; z_0, w_0))$ in some neighborhood of $z = z_0$; this solution is not identically equal to w_0.

The last exceptional initial value is $w_0 = \infty$. In this case we want to study a solution which becomes infinite as z approaches a given point $z = z_0$. We resort to the usual device of introducing the reciprocal function. We set

$$w = \frac{1}{kv}$$

and find that v also satisfies equation (13.4.2). From this we conclude that $v(z)$ must be either $w(z; z_0, 0)$ or its negative, that is,

$$(13.4.12) \qquad w(z; z_0, \infty) = \pm \frac{1}{kw(z; z_0, 0)}.$$

This shows that a solution of (13.4.2) which becomes infinite as $z \to z_0$ must have a simple pole at $z = z_0$ and the residue is either $1/k$ or $-1/k$.

In this manner we can construct solutions $w(z; z_0, w_0)$ of (13.4.2) where w_0 is either an ordinary initial value or one of the exceptional values ± 1, $\pm 1/k$, or ∞.

There is still one possibility which has not been covered in our discussion. It could possibly happen that a solution $w(z)$ is defined in some partial neighborhood of $z = z_0$ but that it does not tend to a definite limit, finite or infinite, as $z \to z_0$ in the domain of definition D of $w(z)$. We have to show that this possibility cannot arise for the equation under consideration. Let us denote by $S(\rho)$ the closure of the set of values taken on by $w(z)$ in $D \cap D_\rho$ where $D_\rho = [z \mid |z - z_0| < \rho]$. Set $S = \cap_\rho S(\rho)$. Then by assumption S cannot reduce to a single point. Suppose that S contains an ordinary point w_0, that is, w_0 is finite and not a branch point of the square root. Then we can find a sequence $\{z_n\} \subset D$ such that $\lim z_n = z_0$ and $\lim w(z_n) \equiv \lim w_n = w_0$. Further, for n large we can find a disk $|w - w_n| < b$ which contains $w = w_0$ and in which the square root is bounded and satisfies a fixed Lipschitz condition. Then there exists a unique solution $w(z; z_n, w_n)$ which is holomorphic in a disk $|z - z_n| < r$, where r is independent of n. If n is sufficiently large, this disk will contain $z = z_0$ so that the solution is holomorphic at $z = z_0$. On the other hand, $w(z)$ and $w(z; z_n, w_n)$ take on the same value at $z = z_n$, namely $w = w_n$, and by the uniqueness theorem these two solutions must coincide in their common domain of definition. This means that $w(z; z_n, w_n)$ is an analytic continuation of $w(z)$, and we must have $S = \{w_0\}$.

The remaining possibility is that S reduces to $\{+1, -1, +1/k, -1/k, \infty\}$ or a subset thereof containing at least two points. Suppose, for instance, that $+1 \in S$. We know that $w(z) \not\equiv 1$. We now use the transformation $w = 1 - u^2$ and obtain (13.4.10) as the differential equation for $u(z)$. This equation now has a solution $u(z)$, defined in some partial neighborhood of $z = z_0$, which does not tend to a definite limit as $z \to z_0$. There is a limiting set S_1, corresponding to the set S, and S_1 contains 0 among other values. We can then find a sequence $\{z_n\} \subset D$ such that $z_n \to z_0$ and $u_n \equiv u(z_n) \to 0$. The right member of (13.4.10) is holomorphic in some neighborhood of $u = 0$ which contains all u_n with $n > n_0$, and in this neighborhood the right member satisfies a fixed Lipschitz condition. Consequently there exists a solution $u(z; z_n, u_n)$ of (13.4.10) which is holomorphic in a disk $\mid z - z_n \mid < r$, where r is independent of n for $n > n_0$. For large n this disk contains $z = z_0$ so that $u(z; z_n, u_n)$ is holomorphic at $z = z_0$. But

$$w(z) = 1 - [u(z; z_n, u_n)]^2$$

in their common domain of definition. We conclude that $w(z)$ is actually holomorphic at $z = z_0$ and tends to the limit 1 as $z \to z_0$. Again S reduces to a single point, this time $S = \{1\}$. We handle the other possibilities in the same manner.

　　We are now ready to evaluate our results. Suppose that we start at $z = z_0$ with a solution $w(z; z_0, w_0)$ and continue analytically along a path C from z_0. Suppose the other endpoint of C is $z = z_1$. Then as $z \to z_1$ along C we have seen that $w(z; z_0, w_0)$ must tend to a definite limit. If this limit is finite, $w = w_1$, then the continuation is holomorphic at $z = z_1$ and coincides with $w(z; z_1, w_1)$ in some neighborhood of $z = z_1$. If, on the other hand, the limit is infinite, then the continuation has a simple pole at $z = z_1$ and coincides with $w(z; z_1, \infty)$ in some neighborhood of $z = z_1$. We see that $w(z; z_0, w_0)$ can be continued analytically anywhere in the finite plane without ever encountering other singularities than simple poles. By the extended theorem of monodromy (Problem 1 of Exercise 10.3) we conclude that the solution is a meromorphic function. This completes the proof.

　　We shall be concerned mainly with the solution $w(z; 0, 0)$ for which we write

(13.4.13)　　　　　　　　　　$\operatorname{sn} z$　　or　　$\operatorname{sn}(z, k)$.

Here k is known as the *modulus* of the function. This is the notation of C. Gudermann (1838). Jacobi (1829) wrote "sin am z," for which read "sine amplitude z." It is the sine of a function am z which is defined implicitly by

$$(13.4.14) \qquad \int_0^v (1 - k^2 \sin^2 \theta)^{-\frac{1}{2}}\, d\theta = z, \quad v = \operatorname{am} z.$$

　　It is desirable also to consider two other functions

(13.4.15)　　　　　　　$\operatorname{cn} z = \cos \operatorname{am} z, \quad \operatorname{dn} z = \Delta \operatorname{am} z,$

which are defined by

(13.4.16)　　　　$\operatorname{cn} z = (1 - \operatorname{sn}^2 z)^{\frac{1}{2}}, \quad \operatorname{dn} z = (1 - k^2 \operatorname{sn}^2 z)^{\frac{1}{2}},$

where the square roots take the value $+1$ for $z = 0$. The differential equation then gives

$$(13.4.17) \qquad \frac{d}{dz} \operatorname{sn} z = \operatorname{cn} z \, \operatorname{dn} z.$$

We shall prove later the parity relations

(13.4.18)　　　$\operatorname{sn}(-z) = -\operatorname{sn} z, \quad \operatorname{cn}(-z) = \operatorname{cn} z, \quad \operatorname{dn}(-z) = \operatorname{dn} z,$

of which the latter two are obvious consequences of the first.

THEOREM 13.4.3.   *The function* sn z *is doubly-periodic with the primitive periods* $4K$ *and* $2iK'$ *where*

(13.4.19) $$K = \int_0^1 (1 - s^2)^{-\frac{1}{2}}(1 - k^2 s^2)^{-\frac{1}{2}}\, ds,$$

(13.4.20) $$iK' = \int_1^{1/k} (1 - s^2)^{-\frac{1}{2}}(1 - k^2 s^2)^{-\frac{1}{2}}\, ds.$$

It is an odd function of z *and we have also*

(13.4.21) $$\operatorname{sn}(2K - z) = \operatorname{sn} z.$$

There are simple zeros at the points $z \equiv 0$ (modd. $2K$, $2iK'$) *and simple poles at* $z \equiv iK'$ (modd. $2K$, $2iK'$).

REMARK.   If $0 < k < 1$ we have $K' > 0$ by our choice of the radicals. It may be shown that for any k ($\neq 0$, ± 1, ∞) we have

(13.4.22) $$\Im\left(\frac{iK'}{K}\right) > 0.$$

Proof.   We obtain these results by a study of z as a function of w determined by equation (13.4.2). In other words, we consider the normal elliptic integral of the first kind

(13.4.23) $$I(w) = \int_0^w (1 - s^2)^{-\frac{1}{2}}(1 - k^2 s^2)^{-\frac{1}{2}}\, ds.$$

To fix the ideas, we suppose that k lies in the first quadrant, including the boundary, but we exclude the case where k is real and > 1. Then $1/k$ lies in the fourth quadrant, including the boundary, but $1/k$ does not take on real values belonging to the interval $(0, 1)$.

We introduce a system of cuts in the plane from $w = 1$ to $1/k$ to $1/k + \infty$ and from -1 to $-1/k$ to $-1/k - \infty$. See Figure 11. We choose the square roots

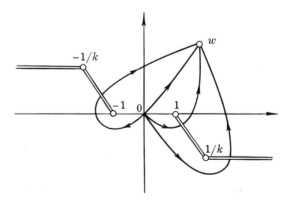

Figure 11

to be $+1$ for $s = 0$. The resulting value of $I(w)$ is denoted by $I_0(w)$, and it is, by definition, the principal determination of $I(w)$. Since the integrand is an even function of s we see that $I_0(w)$ is odd:

$$(13.4.24) \qquad I_0(-w) = -I_0(w).$$

We get other determinations of $I(w)$ if w is allowed to cross the cuts. These new determinations can all be expressed in terms of $I_0(w)$ and the two integrals K and iK' defined by (13.4.19) and (13.4.20). The figure shows three paths joining $s = 0$ and $s = w$ each of which crosses one of the cuts. The path which intersects $[1, 1/k]$ may be deformed into a loop around $[0, 1]$ plus the path in the cut plane, where, however, the integrand must be taken with opposite sign since the first square root becomes -1 after the loop is described. The integral along the loop reduces to

$$2I_0(1) = 2K,$$

so that the new determination is

$$2K - I_0(w).$$

We get a determination

$$I_0(w) + 4K$$

if after crossing the cut $[1, 1/k]$ we also cross $[-1, -1/k]$ before we proceed to w. See Figure 11. For the integrals along the loops give $2K$ each, and the square roots are back to their original determinations after the two loops have been described.

The path which intersects $(1/k, 1/k + \infty)$ is found to give the determination

$$I_0(w) + 2iK'.$$

Proceeding in this manner we see that all determinations of $I(w)$ are given by

$$(13.4.25) \qquad \begin{aligned} &I_0(w) + 4mK + 2niK', \\ &2K - I_0(w) + 4mK + 2niK', \end{aligned}$$

where m and n are arbitrary integers.

Now $I(w) = z$ is the inverse function of $w = \operatorname{sn} z$. Hence (13.4.25) implies that

$$(13.4.26) \qquad \begin{aligned} &\operatorname{sn}(z + 4mK + 2niK') = \operatorname{sn} z, \\ &\operatorname{sn}(2K - z + 4mK + 2niK') = \operatorname{sn} z. \end{aligned}$$

Further, from (13.4.24) we get that $\operatorname{sn}(-z) = -\operatorname{sn} z$.

It remains to prove the assertions concerning zeros and poles. Those concerning the zeros follow directly from (13.4.25) and (13.4.26) since $\operatorname{sn} 0 = 0$. To study the poles we first observe that

$$(13.4.27) \qquad \int_{1/k}^{1/k + \infty} (1 - s^2)^{-\frac{1}{2}}(1 - k^2 s^2)^{-\frac{1}{2}} \, ds = -\int_0^1 (1 - t^2)^{-\frac{1}{2}}(1 - k^2 t^2)^{-\frac{1}{2}} \, dt$$

$$= -K.$$

This is obtained by putting $s = (kt)^{-1}$, but this substitution leaves the sign in doubt. We note that if k is real, $0 < k < 1$, then the left member is real negative and so is the right. Since both sides are analytic functions of k, we conclude that for the values of k under consideration we have to take the minus sign in the right member.

We now let $w \to \infty$ in the right half-plane above the cuts. The integral $I_0(w)$ obviously tends to a finite limit, and we can deform the path of integration so that it goes from 0 to 1, from 1 to $1/k$, and from $1/k$ to $1/k + \infty$. The integrals along the three line segments have the values K, iK', and $-K$ respectively, so we obtain

$$I_0(\infty) = iK'.$$

It follows that

$$\operatorname{sn}(iK') = \infty.$$

We have then also by (13.4.18) and (13.4.21)

$$\operatorname{sn}(2K + iK') = \operatorname{sn}(-iK') = -\operatorname{sn}(iK') = \infty.$$

Now $\operatorname{sn} z$ has two zeros in the period parallelogram determined by the periods $4K$ and $2iK'$, and these zeros are simple by (13.4.17) since $\operatorname{cn} 0 = \operatorname{dn} 0 = 1$. It follows that $\operatorname{sn} z$ is a doubly-periodic elliptic function of order two. In the period parallelogram it has poles at $z = iK'$ and $2K + iK'$. We know that these poles are simple and that their residues are either $1/k$ or $-1/k$ as shown in the proof of Theorem 13.4.2. Actually the residue is $1/k$ at $z = iK'$ and congruent poles (modd. $4K, 2iK'$) and $-1/k$ at $z = iK' + 2K$ and congruent poles.

To prove the last assertion, we start by observing that the residues are analytic functions of k. It is then sufficient to determine the sign when k is real and $0 < k < 1$. In this case $K' > 0$ and $\operatorname{sn}(iy, k)$ equals i times a positive number as long as $0 < y < K'$. Assuming this for a moment, we see that

$$(iy - iK') \operatorname{sn}(iy, k) > 0, \quad 0 < y < K'.$$

The limit of this expression as $y \to K'$, which is the desired residue, must then be positive; that is, the residue is $1/k$ and not $-1/k$. Similarly, if $z = 2K + iy$, $0 < y < K'$, then

$$(z - 2K - iK') \operatorname{sn}(z, k) = i(y - K') \operatorname{sn}(iy + 2K, k)$$
$$= i(y - K') \operatorname{sn}(-iy, k) = -i(y - K') \operatorname{sn}(iy, k),$$

and this expression is negative, so its limit must be $-1/k$.

It remains to prove the assertion on $\operatorname{sn}(iy, k)$. For this purpose it suffices to study $I_0(w)$ on the positive imaginary axis. We have for real values of k

$$I_0(iv) = i \int_0^v (1 + t^2)^{-\frac{1}{2}} (1 + k^2 t^2)^{-\frac{1}{2}} \, dt,$$

where the integral is positive and tends to K' as $v \to +\infty$, since $I_0(w)$ tends to

iK' as $w \to \infty$ above the cuts. From sn $(I_0(iv), k) = iv$ we see that sn (iy, k) has the stated property.

In view of (13.4.12) we have

$$(13.4.28) \qquad \text{sn} \, (z - iK') = [k \, \text{sn} \, z]^{-1}.$$

We can also verify (13.4.22). The function sn z is obviously doubly-periodic and not a constant. Further, $4K$ and $2iK'$ are primitive periods. It follows that $\Im(iK'/K) \neq 0$, and as a continuous function of k the expression must keep a constant sign. The sign is plus for k real, $0 < k < 1$, consequently always plus. This argument presupposes that K and K', as functions of k, do not vanish for any k. We shall return to this question in Problem 11 of Exercise 13.5. This completes the proof of Theorem 13.4.3.

Let us look briefly at the properties of the functions cn z and dn z. From

$$I_0(1) = K, \quad I_0(-1) = -K$$

we get

$$\text{sn} \, K = 1, \quad \text{sn} \, (-K) = -1,$$
$$\text{cn} \, K = 0, \quad \text{cn} \, (-K) = 0.$$

It follows from the proof of Theorem 13.4.2 that

$$w(z; K, 1) = w(z; 0, 0) = w(z; -K, -1)$$

since all three functions are solutions of (13.4.2) and the first and the second agree for $z = K$ while the second and the third agree for $z = -K$. We also called attention to the fact that for solutions with branch points as initial values, an expression like

$$\{1 - [w(z; K, 1)]^2\}^{\frac{1}{2}}$$

is holomorphic in a neighborhood of $z = K$. Similarly

$$\{1 - [w(z; -K, -1)]^2\}^{\frac{1}{2}}$$

is holomorphic in a neighborhood of $z = -K$. From this we conclude that $(1 - \text{sn}^2 z)^{\frac{1}{2}} = \text{cn} \, z$ is a single-valued doubly-periodic function. It has simple zeros at $z = K$ (modd. $2K, 2iK'$), and its poles coincide with those of sn z. The residues, however, are $-i/k$ at $z = iK'$ and $+i/k$ at $z = iK' + 2K$. The periods are also different from those of sn z. Although $4K$ is a primitive period, $2iK'$ is not, since cn $(2iK') = -1$ and not $+1$. Here $2(K + iK')$ can be taken as the other primitive period. We have

$$(13.4.29) \qquad \frac{d}{dz} \, \text{cn} \, z = -\text{sn} \, z \, \text{dn} \, z,$$

and the analogue of (13.4.28) is

$$(13.4.30) \qquad \text{cn} \, (z - iK) = \frac{i \, \text{dn} \, z}{k \, \text{sn} \, z}.$$

Similarly one shows that dn z is an elliptic function. It has $2K$ and $4iK'$ as primitive periods. There are simple zeros at $z \equiv K + iK'$ (modd. $2K, 2iK'$). The poles are the same as those of sn z, but the residue at $z = iK'$ is $-i$ and at $z = iK' + 2K$ it is $+i$. Further,

(13.4.31)
$$\frac{d}{dz} \operatorname{dn} z = -k^2 \operatorname{sn} z \operatorname{cn} z,$$

(13.4.32)
$$\operatorname{dn}(z - iK') = i \frac{\operatorname{cn} z}{\operatorname{sn} z}.$$

We have still to consider the addition theorems. Each of the three addition theorems involves all three functions, so it is convenient to consider them simultaneously. Let α be a complex number not congruent to iK' (modd. $2K, 2iK'$) and form the products

$$\operatorname{sn} z \operatorname{sn}(z - \alpha), \quad \operatorname{cn} z \operatorname{cn}(z - \alpha), \quad \operatorname{dn} z \operatorname{dn}(z - \alpha).$$

These are elliptic functions of order two with primitive periods $2K$ and $2iK'$, as is easily verified. Each of them has a simple pole at the points iK' and $iK' + \alpha$ (modd. $2K, 2iK'$). Using what has been said above concerning the residues of sn z, cn z, and dn z at $z = iK'$, as well as formulas (13.4.28), (13.4.30), and (13.4.32), we find that the residues of the three products at $z = iK'$ are

$$-\frac{1}{k^2 \operatorname{sn} \alpha}, \quad \frac{\operatorname{dn} \alpha}{k^2 \operatorname{sn} \alpha}, \quad \frac{\operatorname{cn} \alpha}{\operatorname{sn} \alpha}$$

respectively. It follows that each of the functions

$$-k^2 \operatorname{sn} \alpha \operatorname{sn} z \operatorname{sn}(z - \alpha), \quad k^2 \frac{\operatorname{sn} \alpha}{\operatorname{dn} \alpha} \operatorname{cn} z \operatorname{cn}(z - \alpha), \quad \frac{\operatorname{sn} \alpha}{\operatorname{cn} \alpha} \operatorname{dn} z \operatorname{dn}(z - \alpha)$$

has a simple pole of residue 1 at $z = iK'$ and hence a simple pole of residue -1 at $z = iK' + \alpha$. The difference between any two of these functions is then an elliptic function without poles, that is, a constant. The value of this constant can be found by substituting 0 for z. This leads to the following identities:

(13.4.33) $\operatorname{dn} \alpha \operatorname{sn} z \operatorname{sn}(z - \alpha) + \operatorname{cn} z \operatorname{cn}(z - \alpha) = \operatorname{cn} \alpha$,

(13.4.34) $k^2 \operatorname{cn} \alpha \operatorname{sn} z \operatorname{sn}(z - \alpha) + \operatorname{dn} z \operatorname{dn}(z - \alpha) = \operatorname{dn} \alpha$.

Here we set $\alpha = u + v$, $z = u$ and obtain the system of equations

$$\operatorname{cn} u \operatorname{cn} v - \operatorname{sn} u \operatorname{sn} v \operatorname{dn}(u + v) = \operatorname{cn}(u + v),$$

$$\operatorname{dn} u \operatorname{dn} v - k^2 \operatorname{sn} u \operatorname{sn} v \operatorname{cn}(u + v) = \operatorname{dn}(u + v),$$

from which we obtain cn $(u + v)$ and dn $(u + v)$. Next, we set $\alpha = -v$, $z = u$ in (13.4.33) and obtain

$$\operatorname{sn} u \operatorname{dn} v \operatorname{sn}(u + v) = \operatorname{cn} v - \operatorname{cn} u \operatorname{cn}(u + v).$$

The result it the following set of expressions:

$$(13.4.35) \quad \begin{cases} \operatorname{sn}(u+v) = \dfrac{\operatorname{sn} u \operatorname{cn} v \operatorname{dn} v + \operatorname{sn} v \operatorname{cn} u \operatorname{dn} u}{1 - k^2 \operatorname{sn}^2 u \operatorname{sn}^2 v}, \\[2ex] \operatorname{cn}(u+v) = \dfrac{\operatorname{cn} u \operatorname{cn} v - \operatorname{dn} u \operatorname{dn} v \operatorname{sn} u \operatorname{sn} v}{1 - k^2 \operatorname{sn}^2 u \operatorname{sn}^2 v}, \\[2ex] \operatorname{dn}(u+v) = \dfrac{\operatorname{dn} u \operatorname{dn} v - k^2 \operatorname{sn} u \operatorname{sn} v \operatorname{cn} u \operatorname{cn} v}{1 - k^2 \operatorname{sn}^2 u \operatorname{sn}^2 v}. \end{cases}$$

We recall that each of the three functions $\operatorname{sn} z$, $\operatorname{cn} z$, $\operatorname{dn} z$ is an algebraic function of any one of the other two functions. It follows, for instance, that $\operatorname{sn}(u+v)$ is an algebraic function of $\operatorname{sn} u$ and $\operatorname{sn} v$. Further, since $\operatorname{cn} u \operatorname{dn} u$ is the derivative of $\operatorname{sn} u$, we see that $\operatorname{sn}(u+v)$ is a rational function of $\operatorname{sn} u$, $\operatorname{sn} v$ and their derivatives. Similar results hold for the other functions. We have thus obtained the addition theorems of the three Jacobi functions.

EXERCISE 13.4

1. If $|z| < |K'|$, the function $\operatorname{sn}(z, k)$ can be expanded in a power series of the form

$$\operatorname{sn}(z, k) = \sum_{n=0}^{\infty} \frac{(-1)^n}{(2n+1)!} P_n(k^2) z^{2n+1},$$

where $P_n(s)$ is a polynomial in s of degree n with positive integral coefficients. Compute the first three polynomials and try to prove the general statement by induction or otherwise.

2. What is the analogue of Theorem 13.4.2 for the equation

$$\left(\frac{dw}{dz}\right)^2 = 4(w - e_1)(w - e_2)(w - e_3), \quad e_1 + e_2 + e_3 = 0?$$

This is the equation satisfied by the Weierstrass $\wp$-function, so the results must reflect properties of the latter which are already known to the reader.

3. Find a path of integration in (13.4.23) which gives the integral the value $2K - I_0(w) - 2iK'$.

4. Prove that $K' = K$ if $k^2 = -1$ (lemniscatic case).

5. Prove formulas (13.4.29) and (13.4.31).

6. These formulas imply that the functions $\operatorname{cn} z$ and $\operatorname{dn} z$ satisfy differential equations of the form

$$\left(\frac{dw}{dz}\right)^2 = A + Bw^2 + Cw^4.$$

Find the coefficients for the two cases.

7. Verify the assertions made in the text concerning the residues of cn z and dn z.

8. Prove formulas (13.4.30) and (13.4.32).

9. Verify the assertions concerning the periods of cn z and dn z.

10. Prove that

$$\wp(u; g_2, g_3) = e_3 + (e_1 - e_3)\left\{\operatorname{sn}\left[u(e_1 - e_3)^{\frac{1}{2}}, \left(\frac{e_2 - e_3}{e_1 - e_3}\right)^{\frac{1}{2}}\right]\right\}^{-2}.$$

13.5. The theta functions. In the theory of the functions of Jacobi there is a class of functions, known as the *theta functions*, which play a role analogous to that of the sigma functions of Weierstrass. They are entire functions of the variable z. They also depend upon a parameter, denoted by either q or τ, where

(13.5.1) $$q = e^{\pi i \tau}$$

and $\Im(\tau) > 0$ so that $|q| < 1$. We write either $\vartheta_j(z, q)$ or $\vartheta_j(z \mid \tau)$, where the subscript takes on the values $1, 2, 3$, and 4. If τ or q is fixed in the discussion and there is no ambiguity, we omit the parameter and write simply $\vartheta_j(z)$. The notation ϑ_j is reserved for the values of these functions for $z = 0$:

(13.5.2) $$\vartheta_j = \vartheta_j(0, q), \quad j = 2, 3, 4.$$

Similarly one writes

(13.5.3) $$\vartheta_j' = \vartheta_j'(0, q), \quad j = 1.$$

As we shall see in a moment, the other values of j give functions of q which are identically 0.

We define the theta functions by the series

$$\vartheta_1(z, q) = 2 \sum_{n=0}^{\infty} (-1)^n q^{\frac{1}{4}(2n+1)^2} \sin (2n + 1)z,$$

$$\vartheta_2(z, q) = 2 \sum_{n=0}^{\infty} q^{\frac{1}{4}(2n+1)^2} \cos (2n + 1)z,$$

(13.5.4)

$$\vartheta_3(z, q) = 1 + 2 \sum_{n=1}^{\infty} q^{n^2} \cos 2nz,$$

$$\vartheta_4(z, q) = 1 + 2 \sum_{n=1}^{\infty} (-1)^n q^{n^2} \cos 2nz.$$

These functions are clearly related. Actually we have the following relations, which are easily verified:

(13.5.5)
$$\vartheta_1(z, q) = -ie^{iz + \frac{1}{4}\pi i \tau} \vartheta_4(z + \tfrac{1}{2}\pi\tau, q),$$
$$\vartheta_2(z, q) = \vartheta_1(z + \tfrac{1}{2}\pi, q),$$
$$\vartheta_3(z, q) = \vartheta_4(z + \tfrac{1}{2}\pi, q).$$

The series (13.5.4) converge very rapidly. Thus, the nth term of $\vartheta_3(z, q)$ does not exceed

$$| q |^{n^2} e^{n | z |}$$

in absolute value, the nth root of which tends to zero as $n \to \infty$. Similar estimates hold for the other series. They consequently represent entire functions of z for fixed q. These entire functions have simple periodicity properties. It is obvious that

$$(13.5.6) \qquad \vartheta_j(z + \pi) = (-1)^{[(1+j)/2]} \vartheta_j(z)$$

where $[s]$ is the largest integer $\leqq s$. There are also simple transformations of the functions under the shift $z \to z + \pi\tau$. We have

$$(13.5.7) \qquad \vartheta_j(z + \pi\tau) = (-1)^{1 + [j/2]} q^{-1} e^{-2iz} \vartheta_j(z).$$

From these relations we get

$$(13.5.8) \qquad \frac{\vartheta_j{}'(z + \pi)}{\vartheta_j(z + \pi)} = \frac{\vartheta_j{}'(z)}{\vartheta_j(z)}, \quad \frac{\vartheta_j{}'(z + \pi\tau)}{\vartheta_j(z + \pi\tau)} = -2i + \frac{\vartheta_j{}'(z)}{\vartheta_j(z)}$$

for each value of j.

THEOREM 13.5.1.   *The zeros of $\vartheta_j(z \mid \tau)$ are congruent* (modd. $\pi, \pi\tau$) *to* 0, $\frac{1}{2}\pi, \frac{1}{2}\pi + \frac{1}{2}\pi\tau, \frac{1}{2}\pi\tau$ *for $j = 1, 2, 3, 4$ respectively.*

Proof.   Since $\vartheta_1(0 \mid \tau) = 0$, we conclude with the aid of (13.5.5)–(13.5.7) that $\vartheta_j(z \mid \tau)$ has zeros at the points stated in the theorem. All that remains to be shown, then, is that there are no other zeros. This follows from (13.5.8). Let Π be a parallelogram with vertices at the points

$$a, a + \pi, a + \pi + \pi\tau, a + \pi\tau.$$

We may suppose that a is so chosen that $\vartheta_j(z)$ has no zeros on Π. The number of zeros of $\vartheta_j(z)$ inside Π is then given by

$$\frac{1}{2\pi i} \int_\Pi \frac{\vartheta_j{}'(z)}{\vartheta_j(z)} \, dz.$$

The integral along $[a + \pi\tau, a]$ cancels that along $[a + \pi, a + \pi + \pi\tau]$ by (13.5.8) while the integrals along the other two sides give the value 1. Thus there is one and only one zero of $\vartheta_j(z)$ inside Π, that is, there are no zeros in addition to those already found. This completes the proof.

The proof also shows that each zero is simple.

COROLLARY. $\vartheta_1{}'(0 \mid \tau) \neq 0$ *for* $\Im(\tau) > 0$.

It is obvious that the quotient of any two distinct theta functions is an elliptic function. Consider, for instance,

$$\frac{\vartheta_1(z, q)}{\vartheta_4(z, q)}.$$

It has the periods 2π and $\pi\tau$ and it has simple poles at $z \equiv \frac{1}{2}\pi\tau$ (modd. π, $\pi\tau$). It has simple zeros at $z \equiv 0$ (modd. π, $\pi\tau$). These are essentially the properties of a sine amplitude for a suitable choice of variable and modulus. In fact, we have

$$(13.5.9) \qquad \operatorname{sn}\left(\frac{2Kz}{\pi}, k\right) = \frac{\vartheta_3}{\vartheta_2} \cdot \frac{\vartheta_1(z, q)}{\vartheta_4(z\, q)},$$

provided k is so chosen that

$$(13.5.10) \qquad \tau = i\,\frac{K'}{K}.$$

For the two sides of (13.5.9) have the same poles and the same periods, so that their quotient is a constant whose value is found by setting $z = \frac{1}{2}\pi$ and using the values

$$\operatorname{sn} K = 1, \quad \vartheta_1(\tfrac{1}{2}\pi) = \vartheta_3(0) = \vartheta_3, \quad \vartheta_4(\tfrac{1}{2}\pi) = \vartheta_2(0) = \vartheta_2.$$

Similarly we find that

$$(13.5.11) \qquad \operatorname{cn}\left(\frac{2Kz}{\pi}, k\right) = \frac{\vartheta_4}{\vartheta_2} \cdot \frac{\vartheta_2(z, q)}{\vartheta_4(z, q)},$$

$$(13.5.12) \qquad \operatorname{dn}\left(\frac{2Kz}{\pi}, k\right) = \frac{\vartheta_4}{\vartheta_3} \cdot \frac{\vartheta_3(z, q)}{\vartheta_4(z, q)}.$$

If τ or, equivalently, q is given, then it is a fairly simple problem to compute k, K, and K' as functions of q. This is a problem of considerable practical importance for the applications, so we shall give the solution. Differentiating both sides of (13.5.9) with respect to z and setting $z = 0$ in the result, we get the relation

$$\frac{2}{\pi} K = \frac{\vartheta_3 \vartheta_1{}'}{\vartheta_2 \vartheta_4},$$

from which we can compute K. Since actually

$$(13.5.13) \qquad \vartheta_1{}' = \vartheta_2 \vartheta_3 \vartheta_4,$$

the formula reduces to

$$(13.5.14) \qquad K = \tfrac{1}{2}\pi\vartheta_3{}^2 = \tfrac{1}{2}\pi\left\{1 + 2\sum_{n=1}^{\infty} q^{n^2}\right\}^2.$$

The series converges rapidly unless $|q|$ is close to 1.

Similarly, equating the residues at $z = \frac{1}{2}\pi\tau$ of the two members of (13.5.9), we get the identity

$$\frac{\pi}{2kK} = \frac{\vartheta_3\vartheta_1(\tfrac{1}{2}\pi\tau)}{\vartheta_2\vartheta_4{}'(\tfrac{1}{2}\pi\tau)} = \frac{\vartheta_3\vartheta_4}{\vartheta_2\vartheta_1{}'}.$$

Here we have used the first formula under (13.5.5) to simplify the middle member. If we multiply the two formulas which contain K and ϑ_1', these quantities cancel and we are left with

$$(13.5.15) \qquad k = \frac{\vartheta_2^2}{\vartheta_3^2},$$

which is also suitable for numerical computation. Finally we have

$$K' = -iK\tau.$$

It is possible to obtain series expansions which converge faster than those which we have already listed. The method is based upon the interesting identity

$$(13.5.16) \qquad \vartheta_2^4 + \vartheta_4^4 = \vartheta_3^4$$

or

$$16q(1 + q^{1\cdot2} + q^{2\cdot3} + q^{3\cdot4} + \cdots)^4 + (1 - 2q + 2q^4 - 2q^9 + \cdots)^4$$
$$= (1 + 2q + 2q^4 + 2q^9 + \cdots)^4,$$

which we now proceed to prove. For this purpose we consider the function

$$\frac{a\vartheta_3^2(z) + b\vartheta_2^2(z)}{\vartheta_4^2(z)}$$

which is doubly-periodic with periods π and $\pi\tau$ for any choice of the constants a and b. Normally it has a double pole at $z = \frac{1}{2}\pi\tau$, where the value of the numerator is

$$q^{-\frac{1}{2}}(a\vartheta_2^2 + b\vartheta_3^2).$$

If we choose

$$a = \vartheta_3^2, \quad b = -\vartheta_3^2,$$

the resulting elliptic function will have at most a simple pole in the period parallelogram and must therefore be a constant whose value we can find by setting $z = \frac{1}{2}\pi$. The result is the identity

$$(13.5.17) \qquad \vartheta_3^2\vartheta_3^2(z) - \vartheta_2^2\vartheta_2^2(z) = \vartheta_4^2\vartheta_4^2(z),$$

which gives (13.5.16) when we set $z = 0$. There are three other identities of the same type as (13.5.17) joining the squares of three distinct theta functions.

In addition to the modulus k it is customary to consider the *complementary modulus* k' which satisfies the relation

$$(13.5.18) \qquad k^2 + k'^2 = 1.$$

If k is real and $0 < k < 1$, then k' is also real and, by definition, $0 < k' < 1$. By virtue of (13.5.15) and (13.5.16) we can define the positive square root of k' by

$$(13.5.19) \qquad (k')^{\frac{1}{2}} = \frac{\vartheta_4(0, q)}{\vartheta_3(0, q)}.$$

We have then

$$\frac{1 - (k')^{1/2}}{1 + (k')^{1/2}} = \frac{\vartheta_3(0, q) - \vartheta_4(0, q)}{\vartheta_3(0, q) + \vartheta_4(0, q)}.$$

But a simple calculation shows that

$$\vartheta_3(0, q) - \vartheta_4(0, q) = 2\vartheta_2(0, q^4),$$
$$\vartheta_3(0, q) + \vartheta_4(0, q) = 2\vartheta_3(0, q^4),$$

so that

$$\frac{1 - (k')^{1/2}}{1 + (k')^{1/2}} = \frac{\vartheta_2(0, q^4)}{\vartheta_3(0, q^4)}$$

and

(13.5.20) $$\varepsilon \equiv \frac{1}{2}\frac{1 - (k')^{1/2}}{1 + (k')^{1/2}} = \frac{q + q^9 + q^{25} + q^{49} + \cdots}{1 + 2q^4 + 2q^{16} + 2q^{36} + \cdots}.$$

The improvement in the rapidity of convergence of the series is striking: even if $|q| = 0.8$, five terms suffice to give numerator and denominator with errors $< 10^{-9}$. Once ε is known, we can of course easily obtain k' and k.

But the expansion may also be used for the opposite purpose. Normally k is known and q or τ is unknown. If k is known we may consider that ε is known. We have to assume that $|\varepsilon| < \frac{1}{2}$, which is the case as long as k' is not real negative. We have now to find q from the transcendental equation

$$q = \varepsilon + 2\varepsilon q^4 - q^9 + 2\varepsilon q^{16} - q^{25} + \cdots.$$

That this equation has a solution q which is a holomorphic function of ε for small values of ε, follows from the implicit function theorem of Section 9.4. We obtain

(13.5.21) $$q(\varepsilon) = \sum_{n=0}^{\infty} \delta_n \varepsilon^{4n+1},$$

where the δ_n's are positive integers and

$$\delta_0 = 1, \quad \delta_1 = 2, \quad \delta_2 = 15, \quad \delta_3 = 150, \quad \delta_4 = 1707.$$

This series is of interest from the point of view both of theory and of numerical computation. It converges for $|\varepsilon| \leq \frac{1}{2}$ and

$$\sum_{n=0}^{\infty} \delta_n 2^{-4n-1} = 1.$$

It should be noted that the function $\varepsilon = \varepsilon(q)$ defined by (13.5.20) has $|q| = 1$ as its natural boundary since the numerator and the denominator are both lacunary series satisfying Fabry's gap condition. We shall return to the inverse function later. For the present it is enough to call attention to the rapid convergence of (13.5.21). If k is real and $0 < k < 2^{-\frac{1}{2}}$ we have

$$0 < \varepsilon \leq \frac{1}{2}\frac{2^{1/4} - 1}{2^{1/4} + 1} = 0.043 \cdots.$$

For such values of ε the third term of (13.5.21) is already negligible and the first two terms give q with ten correct decimals.

There are interesting connections between the theta functions and the theory of heat conduction in mathematical physics. The *heat equation* is a partial differential equation of the form

$$(13.5.22) \qquad \frac{\partial T}{\partial t} = a \frac{\partial^2 T}{\partial x^2}.$$

Here x represents position on the real line, t is the time coordinate, a is a physical constant, and $T = T(x, t)$ is the temperature at the point x at the time t. Since an exponential function of the form

$$e^{iax - a\alpha^2 t}$$

is a solution of (13.5.22) for any choice of the number α, we see that each theta function is a solution of the heat equation provided that

$$z = x, \quad -i\tau = t, \quad a = \tfrac{1}{4}\pi.$$

In particular, if τ is purely imaginary we see that $t > 0$ and that we are dealing with a possible solution of a physical problem. Actually the function $\vartheta_3(x \mid it)$ occurs in some problems of heat conduction, for instance, in the theory of what is known as *Fourier's ring*. Here a thin insulated circular wire of length 2π is supposed to have an initial temperature $f(x)$, $-\pi \leq x \leq \pi$, where $f(x)$ is a continuous periodic function of period 2π. If $a = \pi$, which can be achieved by suitable choice of the physical units, then the temperature at the time t at the place x is given by the integral

$$(13.5.23) \qquad T(x, t; f) = \frac{1}{2\pi} \int_{-\pi}^{\pi} \vartheta_3[\tfrac{1}{2}(x - s) \mid it] f(s) \, ds.$$

Above we have considered the transformations which the theta functions undergo when z is replaced by $z + c$, where c is a period or half of a period. There is also a very rich theory concerned with transformations involving changes of τ. Here the important problem is the study of the transformations corresponding to replacing τ by an element of the modular group

$$\frac{a\tau + b}{c\tau + d},$$

where a, b, c, d are integers and

$$ad - bc = 1.$$

We take the particular case

$$a = d = 0, \quad b = -1, \quad c = 1$$

and ask for the relations holding between $\vartheta_j(z \mid -\tau^{-1})$ and $\vartheta_k(z \mid \tau)$. These results are of importance because they give insight into the behavior of the theta functions when the variable τ approaches zero and, thus, $q \to 1$. Our object is to prove the relations

$$(13.5.24) \qquad \vartheta_j(z \mid \tau) = \delta_j(-i\tau)^{-\frac{1}{2}} \exp\left(\frac{z^2}{\pi i \tau}\right) \vartheta_k\left(\frac{z}{\tau} \mid -\tau^{-1}\right).$$

Here $j = 1, 2, 3, 4$ corresponds to $k = 1, 4, 3, 2$ respectively, and $\delta_j = 1, j \neq 1$, while $\delta_1 = i$. Further, the square root is real positive when τ is purely imaginary. We start with $j = k = 3$. Set

$$\psi(z) = \exp\left(\frac{z^2}{\pi i \tau}\right) \vartheta_3\left(\frac{z}{\tau} \mid -\tau^{-1}\right) [\vartheta_3(z \mid \tau)]^{-1}.$$

It will be shown that this is a doubly-periodic function without poles and hence a constant whose value depends upon τ.

That $\psi(z)$ has the periods π and $\pi\tau$ follows from formulas (13.5.6) and (13.5.7); the verification is left to the reader. The possible poles of $\psi(z)$ are the zeros of $\vartheta_3(z \mid \tau)$, that is, the points

$$m\pi + n\pi\tau + \tfrac{1}{2}\pi + \tfrac{1}{2}\pi\tau.$$

But the second factor of $\psi(z)$ has zeros at the points

$$a\pi\tau - b\pi + \tfrac{1}{2}\pi\tau - \tfrac{1}{2}\pi,$$

where a and b are arbitrary integers just as m and n are. Thus, taking $a = n$, $b = -m - 1$, we see that poles and zeros cancel, so that $\psi(z)$ has no poles and is a constant, which we denote by $A(\tau)$. We have thus proved

$$A(\tau)\, \vartheta_j(z \mid \tau) = \delta_j \exp\left(\frac{z^2}{\pi i \tau}\right) \vartheta_k\left(\frac{z}{\tau} \mid -\tau^{-1}\right)$$

for the case $j = 3$. The other formulas are obtained by replacing z by $z + \tfrac{1}{2}\pi$, $z + \tfrac{1}{2}\pi\tau$, and $z + \tfrac{1}{2}\pi + \tfrac{1}{2}\pi\tau$. In the formulas with $j > 1$ we set $z = 0$. For $j = 1$ we differentiate with respect to z and set $z = 0$ in the result. This gives

$$A(\tau)\, \vartheta_1'(0 \mid \tau) = \frac{i}{\tau} \vartheta_1'(0 \mid -\tau^{-1}),$$

$$A(\tau)\, \vartheta_j(0 \mid \tau) = \vartheta_{6-j}(0 \mid -\tau^{-1}), \quad j > 1.$$

We now resort to formula (13.5.13), which will be proved at the end of this section. Then the first equation becomes

$$A(\tau)\prod_{j>1}\vartheta_j(0 \mid \tau) = \frac{i}{\tau}\prod_{j>1}\vartheta_j(0 \mid -\tau^{-1}).$$

In view of the other equations, this implies that

$$[A(\tau)]^2 = -i\tau,$$

so that $A(\tau)$ is a square root of $-i\tau$. Here we must choose that determination of the square root which is real positive when τ is purely imaginary since both $\vartheta_3(0\,|\,\tau)$ and $\vartheta_3(0\,|\,-\tau^{-1})$ are real positive in this case. This completes the proof.

The following consequence of (13.5.24) is of importance for the theory of Fourier's ring, mentioned above. Take $j = 3$, $z = x$ real, $\tau = it$, $t > 0$. Then

$$\vartheta_3(x\,|\,it) = t^{-\frac{1}{2}} \exp\left(-\frac{x^2}{\pi t}\right)\vartheta_3\left(\frac{x}{it}\,\Big|\,\frac{i}{t}\right)$$

(13.5.25)

$$= t^{-\frac{1}{2}} \exp\left(-\frac{x^2}{\pi t}\right)\sum_{n=-\infty}^{\infty} \exp\left[\frac{1}{t}(2nx - n^2\pi)\right].$$

Here the sum of the infinite series is bounded uniformly in x, $-\frac{1}{2}\pi \le x \le \frac{1}{2}\pi$, and in t, $0 < t \le 1$, and it tends to 1 as t decreases to 0. Thus we have

(13.5.26) $\quad 0 < \vartheta_3(x\,|\,it) < Mt^{-\frac{1}{2}} \exp\left(-\frac{x^2}{\pi t}\right), \quad |x| \le \frac{1}{2}\pi, 0 < t \le 1.$

This shows that $\vartheta_3(x\,|\,it) \to 0$ as $t \to 0$, uniformly in x, $0 < \delta \le |x| \le \frac{1}{2}\pi$. On the other hand, for $x = 0$ we have

(13.5.27) $\qquad\qquad 1 < t^{\frac{1}{2}}\vartheta_3(0\,|\,it) < M, \quad 0 < t \le 1.$

The last topic in the theory of theta functions which we shall discuss is that of the product representations. The desired formulas are the following:

(13.5.28)
$$\vartheta_1(z, q) = 2q_0q^{\frac{1}{4}} \sin z \prod_{n=1}^{\infty}(1 - 2q^{2n}\cos 2z + q^{4n}),$$

$$\vartheta_2(z, q) = 2q_0q^{\frac{1}{4}} \cos z \prod_{n=1}^{\infty}(1 + 2q^{2n}\cos 2z + q^{4n}),$$

$$\vartheta_3(z, q) = q_0\prod_{n=1}^{\infty}(1 + 2q^{2n-1}\cos 2z + q^{4n-2}),$$

$$\vartheta_4(z, q) = q_0\prod_{n=1}^{\infty}(1 - 2q^{2n-1}\cos 2z + q^{4n-2}),$$

where

(13.5.29) $\qquad\qquad\qquad q_0 = \prod_{n=1}^{\infty}(1 - q^{2n}).$

We shall also have occasion to consider the products

$$q_1 = \prod_{n=1}^{\infty}(1 + q^{2n}),$$

(13.5.30) $\qquad\qquad\qquad q_2 = \prod_{n=1}^{\infty}(1 + q^{2n-1}),$

$$q_3 = \prod_{n=1}^{\infty}(1 - q^{2n-1}).$$

It is clear that all these products are absolutely convergent since $|q| < 1$ and, in the case of the theta products, the convergence is uniform with respect to z in any bounded set.

We shall start with the function $\vartheta_4(z, q)$. We recall that its zeros are located at the points

$$\tfrac{1}{2}\pi\tau + m\pi + n\pi\tau,$$

where m and n run through all integers. For a fixed n, let us consider the functions

$$1 - q^{2n-1}e^{2iz} \quad \text{and} \quad 1 - q^{2n-1}e^{-2iz}.$$

Since $q = e^{\pi i\tau}$, we see that the zeros of the first function are the points

$$\tfrac{1}{2}\pi\tau + m\pi - n\pi\tau,$$

where m runs through all integers. Similarly the zeros of the second function are

$$-\tfrac{1}{2}\pi\tau + m\pi + n\pi\tau.$$

Let us now form the infinite product

$$f(z) \equiv \prod_{n=1}^{\infty} (1 - q^{2n-1}e^{2iz}) \prod_{n=1}^{\infty} (1 - q^{2n-1}e^{-2iz}).$$

This is clearly absolutely convergent, uniformly with respect to z on bounded sets. It is thus an entire function having the same zeros as $\vartheta_4(z, q)$. Further, $f(z)$ is periodic of period π since each factor has this property. But we have also

$$\begin{aligned}
f(z + \pi\tau) &= \prod_{n=1}^{\infty} (1 - q^{2n+1}e^{2iz}) \prod_{n=1}^{\infty} (1 - q^{2n-3}e^{-2iz}) \\
&= f(z)(1 - q^{-1}e^{-2iz})(1 - qe^{2iz})^{-1} \\
&= -q^{-1}e^{-2iz}f(z).
\end{aligned}$$

Thus, $f(z)$ undergoes the same transformations as $\vartheta_4(z, q)$ when z is replaced by $z + \pi$ or by $z + \pi\tau$. It follows that the ratio of the two functions is a doubly-periodic function without poles and hence a constant with respect to z. We denote this constant by q_0. This proves the fourth formula under (13.5.28). The other formulas are proved with the aid of (13.5.5).

It remains to find q_0. For this purpose we set

$$e^{2iz} = Z.$$

This function maps the strip $0 < \Re(z) < \pi$ conformally on the Z-plane slit along the positive real axis. We now have

$$\vartheta_3(z, q) = q_0 \prod_{n=1}^{\infty} (1 + q^{2n-1}Z) \prod_{n=1}^{\infty} (1 + q^{2n-1}Z^{-1}) \equiv \Theta(Z).$$

This is a single-valued function of Z holomorphic in $0 < |Z| < \infty$. Hence it can be expanded in a Laurent series in Z, and this series must coincide with the expansion

$$\sum_{n=-\infty}^{\infty} q^{n^2} Z^n$$

implied by (13.5.4). We set

$$\Theta_m(Z) = q_0 \prod_{n=1}^{m}(1 + q^{2n-1}Z) \prod_{n=1}^{m}(1 + q^{2n-1}Z^{-1})$$

$$= A_{m,0} + A_{m,1}(Z + Z^{-1}) + \cdots + A_{m,m}(Z^m + Z^{-m}).$$

Here, obviously,

$$A_{m,m} = q_0 q^{m^2}.$$

Our aim is to express $A_{m,0}$ in terms of $A_{m,m}$ and q. Since $A_{m,0} \to 1$ as $m \to \infty$, we can then expect to get q_0 expressed in terms of q. It is easy to verify that

$$(qZ + q^{2m})\Theta_m(q^2Z) = (1 + q^{2m+1}Z)\Theta_m(Z)$$

by straightforward computation. In other words, we have, with $A_{m,-j} = A_{m,j}$,

$$(qZ + q^{2m}) \sum_{j=-m}^{m} A_{m,j}(q^{2j}Z^j + q^{-2j}Z^{-j}) = (1 + q^{2m+1}Z) \sum_{j=-m}^{m} A_{m,j}(Z^j + Z^{-j}).$$

Collecting terms and equating the coefficient of Z^j to 0, we get the recurrence relation

$$A_{m,j-1}q^{2j-1}(1 - q^{2(m-j)+2}) = A_{m,j}(1 - q^{2(m+j)}), \quad j = 1, 2, \cdots, m.$$

We multiply these relations together, substitute the value of $A_{m,m}$, and note that $\sum_1^m (2j - 1) = m^2$. This gives

$$A_{m,0}\prod_1^m(1 - q^{2n}) = q_0\prod_1^m(1 - q^{2(m+j)}).$$

This holds for every m. We now let $m \to \infty$. Then $A_{m,0} \to 1$ and the product in the right member tends to 1. Thus q_0 has the value stated in formula (13.5.29), and this completes the proof of formulas (13.5.28).

We note that the theta null values can be expressed in terms of q_0, q_1, q_2, and q_3. We obviously get

(13.5.31)
$$\vartheta_2(0, q) = 2q^{\frac{1}{4}}q_0 q_1^2,$$
$$\vartheta_3(0, q) = q_0 q_2^2,$$
$$\vartheta_4(0, q) = q_0 q_3^2.$$

Further, if we divide both sides of the first formula under (13.5.28) by z and let $z \to 0$, we get

(13.5.32)
$$\vartheta_1'(0, q) = 2q^{\frac{1}{4}}q_0^3.$$

In all these formulas, by definition,

$$q^{\frac{1}{4}} = e^{\frac{1}{4}\pi i \tau}.$$

All the infinite products are absolutely convergent, so we can rearrange the factors in any way we please. This observation leads to the identity

$$(13.5.33) \qquad q_1 q_2 q_3 \equiv 1.$$

For

$$q_0 = \prod_1^\infty (1 - q^{2n}) = \prod_1^\infty (1 - q^{4m}) \prod_1^\infty (1 - q^{4m-2})$$

$$= \prod_1^\infty (1 - q^{2m}) \prod_1^\infty (1 + q^{2m}) \prod_1^\infty (1 + q^{2m-1}) \prod_1^\infty (1 - q^{2m-1})$$

$$= q_0 q_1 q_2 q_3,$$

and (13.5.33) follows.

We are finally prepared to give the long-postponed proof of the relation joining the theta null series, formula (13.5.13). We have now

$$\vartheta_2(0, q)\vartheta_3(0, q)\vartheta_4(0, q) = 2q^{\frac{1}{4}} q_0{}^3 (q_1 q_2 q_3)^2$$

$$= 2q^{\frac{1}{4}} q_0{}^3 = \vartheta_1{}'(0, q)$$

as asserted.

This completes our discussion of the theta functions, except for the modular functions, which are treated in the next section.

EXERCISE 13.5

1. Express k, k', and K as infinite products in terms of q.

2. Prove that for $0 < q < 1$

$$\frac{q_3}{q_2} = \frac{\prod(1 - q^{2n-1})}{\prod(1 + q^{2n-1})}$$

is a monotone decreasing function of q which goes from 1 to 0 as q goes from 0 to 1.

3. Use the preceding result to prove that the modulus k is a monotone increasing function of q which goes from 0 to 1 with q.

4. Express the function $\varepsilon(q)$ of (13.5.20) as an infinite product in terms of q.

5. Show that $\lim_{q \to 1} \varepsilon(q) = \frac{1}{2}$. (*Hint:* Use formulas (13.5.24) with τ replaced by 4τ.)

6. Let $a_0 = 1$, $a_n = 2(-1)^k$ if $n = k^2$ and $a_n = 0$ if n is not a square. Show that $\sum_0^\infty a_n$ is summable $(C, 1)$ to the sum zero. What is the relation to the theory of theta functions?

7. Verify that

$$\int_{-\pi/2}^{\pi/2} \vartheta_3(s \mid it) \, ds \equiv 2\pi.$$

What are the values of the corresponding integrals for the other theta functions?

8. Show that

$$\lim_{t \to 0} T(x, t; f) = f(x),$$

uniformly in x, if $T(x, t; f)$ is defined by (13.5.23) and $f(x)$ is continuous and periodic. Prove similar results under weaker assumptions on $f(x)$. (*Hint:* Use the preceding problem and (13.5.26).)

9. Prove the formula

$$\text{sn}\,(2Kx/\pi) = 2q^{\frac{1}{4}}k^{-\frac{1}{2}}\sin x \prod_{n=1}^{\infty}\left\{\frac{1 - 2q^{2n}\cos 2x + q^{4n}}{1 - 2q^{2n-1}\cos 2x + q^{4n-2}}\right\}$$

and obtain similar representations for cn u and dn u.

10. Since sn $(2Kx/\pi)$ is a continuous function of x of period 2π when x is real, this function can be expanded in a Fourier series. Show that

$$\text{sn}\,\frac{2Kx}{\pi} = \frac{2\pi}{kK}\sum_{n=0}^{\infty}\frac{q^{n+\frac{1}{2}}}{1 - q^{2n+1}}\sin\,(2n + 1)x.$$

This series converges in a strip of the complex plane. Determine the strip. Find similar formulas for cn u and dn u. (*Hint:* The Fourier coefficients may be evaluated by contour integration.)

11. Show that K and K' are $\neq 0$ for any choice of k.

13.6. Modular functions. There is an extensive theory of *elliptic modular functions* which was a forerunner to the theory of *automorphic functions*. We shall give just enough of the former to fill in the gaps in the preceding discussion of elliptic and theta functions and to lay the foundations for the proof of Picard's theorem.

DEFINITION 13.6.1.   *A function $f(\tau)$ is said to be a modular function if (1) $f(\tau)$ is holomorphic save for poles in the upper half-plane, and (2) there exists a subgroup $\mathfrak{G}$ of the modular group $\mathfrak{M}$ which leaves $f(\tau)$ invariant.*

We recall that $\mathfrak{M}$ is the set of all linear fractional transformations of the form

$$(13.6.1) \qquad\qquad T(\tau) = \frac{a\tau + b}{c\tau + d}$$

where a, b, c, d are integers and

$$(13.6.2) \qquad\qquad ad - bc = 1.$$

Condition (2) then asserts that

$$(13.6.3) \qquad\qquad f[T(\tau)] = f(\tau) \quad \text{if} \quad T \in \mathfrak{G}.$$

Our first concern will be with the modular group. Every $T \in \mathfrak{M}$ maps the real axis onto itself and leaves the upper and lower half-planes invariant. For if $T(\tau)$ is given by (13.6.1), then

$$(13.6.4) \qquad \mathfrak{I}[T(\tau)] = \frac{\mathfrak{I}(\tau)}{|\, c\tau + d\,|^2} \,,$$

so that $\mathfrak{I}[T(\tau)]$ and $\mathfrak{I}(\tau)$ have the same sign. Moreover,

$$|\, \mathfrak{I}[T(\tau)]\,| \lessgtr |\, \mathfrak{I}(\tau)\,|$$

according as

$$|\, c\tau + d\,| \gtrless 1.$$

If $c \neq 0$, the circle

$$|\, c\tau + d\,| = 1$$

is known as the *isometric circle* of T. Its center is on the real axis at $\tau = -d/c$, and its radius is $|\, c\,|^{-1} \leq 1$. Thus $|\, \mathfrak{I}[T(\tau)]\,| < |\, \mathfrak{I}(\tau)\,|$ as soon as τ lies outside of the isometric circle of T and, in particular, if $|\, \mathfrak{I}(\tau)\,| > |\, c\,|^{-1}$.

There are two elements S and U of $\mathfrak{M}$ which are of particular interest. Here

$$S(\tau) = \tau + 1, \quad U(\tau) = -\frac{1}{\tau}.$$

In fact, $\mathfrak{M}$ is generated by S and U, as we shall see.

S is a parabolic transformation (see Section 3.2 for the terminology) with $\tau = \infty$ as its fixed point, and its powers are the only transformations of $\mathfrak{M}$ with a fixed point at infinity. S generates a cyclic subgroup $\mathfrak{S}$ of $\mathfrak{M}$

$$\mathfrak{S}: \quad \{S^n \mid n = 0, \pm 1, \pm 2, \cdots\}.$$

$\mathfrak{S}$ divides the plane into period strips of width 1. We may take

$$-\tfrac{1}{2} \leq \mathfrak{R}(\tau) < \tfrac{1}{2}$$

as the basic strip. S^n maps this strip onto

$$-\tfrac{1}{2} + n \leq \mathfrak{R}(\tau) < \tfrac{1}{2} + n.$$

In particular, S takes the line $\mathfrak{R}(\tau) = -\tfrac{1}{2}$ into $\mathfrak{R}(\tau) = \tfrac{1}{2}$. We note that any rational function of $e^{2\pi i \tau}$ is a modular function according to our definition since such a function is meromorphic and also invariant under $\mathfrak{S}$.

U on the other hand is an elliptic transformation with fixed points $\pm i$. It is an involution

$$U^2 = I$$

so that U is its own inverse. U maps the unit circle onto itself and maps the interior of the unit circle onto the exterior and vice versa.

We should also note the transformation SU:

$$SU(\tau) = \frac{\tau - 1}{\tau}.$$

It is an elliptic transformation with fixed points at $\tau = \rho$, ρ^{-1} where ρ is $\exp(\tfrac{1}{3}\pi i)$. SU has period 3 in the sense that $(SU)^3 = I$. Thus it generates a subgroup of $\mathfrak{M}$ of order three. This is also a subgroup of the group of anharmonic ratios

$$(13.6.5) \qquad \mathfrak{A}: \quad \lambda, \; \frac{1}{1-\lambda}, \; \frac{\lambda-1}{\lambda}, \; \frac{1}{\lambda}, \; \frac{\lambda}{\lambda-1}, \; 1-\lambda.$$

We write λ instead of τ because later we want λ to be a function of τ. $\mathfrak{A}$ is not a subgroup of $\mathfrak{M}$ since it contains transformations of determinant -1. But any symmetric function of the six ratios is a modular function of λ. In particular, adding 1 to each ratio and taking the product of the results, we obtain

$$(13.6.6) \qquad F(\lambda) = -\frac{(\lambda+1)^2(\lambda-2)^2(2\lambda-1)^2}{\lambda^2(\lambda-1)^2}.$$

Another important case is

$$(13.6.7) \qquad J_0(\lambda) = 1 - \frac{1}{27} F(\lambda) = \frac{4}{27} \frac{(1-\lambda+\lambda^2)^3}{\lambda^2(\lambda-1)^2}.$$

This function will be encountered below. We note that it has triple zeros at $\lambda = \rho$ and ρ^{-1}. Further, $J_0(-1) = J_0(\tfrac{1}{2}) = J_0(2) = 1$.

If $T \in \mathfrak{M}$, $T \neq S^n$, the fixed points of T are either real or conjugate complex and all fixed points lie in the strip $-1 \leq \Im(\tau) \leq 1$. A simple calculation shows that the complex fixed points have ordinates of the form

$$\pm \frac{1}{2\,|c|} \quad \text{or} \quad \pm \frac{\sqrt{3}}{2\,|c|}.$$

All such transformations are elliptic and of period 2 in the first case and of period 3 in the second.

We now introduce the fundamental region R of $\mathfrak{M}$ in the upper half-plane. Its interior is given by

$$(13.6.8) \qquad \text{Int } R = [\tau \mid -\tfrac{1}{2} < \Re(\tau) < \tfrac{1}{2}, \; 1 < |\tau|, \; 0 < \Im(\tau)].$$

To get R itself, we add all limit points of R with $\Re(\tau) \leq 0$. Finally, let R^* be R less the points $\rho - 1$ and i (fixed points of US and U).

THEOREM 13.6.1. *If $T \in \mathfrak{M}$, $T \neq I$, then T maps R one-to-one and onto a region $T(R)$ and $T(R^*) \cap R^* = \emptyset$. If $T_1 \neq T_2$, then $T_1(R^*) \cap T_2(R^*) = \emptyset$. Finally, the union of all images $T(R)$, $\mathfrak{M}(R)$ say, covers the whole upper half-plane.*

Proof. That $T(R^*) \cap R^* = \emptyset$ is clearly true if $T = S^n$, $n \neq 0$, and it is also true for U, US, and S^2U since the points i and $\rho - 1$ are excluded from R^*. Suppose now that $T \neq S^n$, U, US, S^2U. Then T^{-1} is also distinct from these transformations. This implies that R^* is completely outside the isometric circles of T and T^{-1}. It follows from the discussion above that for τ, $\tau_1 \in R^*$ we have

$$0 < \Im[T(\tau)] < \Im(\tau), \quad 0 < \Im[T^{-1}(\tau_1)] < \Im(\tau_1).$$

Suppose that $T(\tau) = \tau_1 \in R^*$ for a $\tau \in R^*$. We would then have

$$0 < \Im(\tau_1) < \Im(\tau), \quad 0 < \Im(\tau) < \Im(\tau_1),$$

which is absurd. Hence $T(R^*) \cap R^* = \emptyset$.

Suppose now that $T_1 \neq T_2$. If $T_1(R^*) \cap T_2(R^*)$ is not void, then there would be points τ_1 and τ_2 in R^* such that $T_1(\tau_1) = T_2(\tau_2)$, thus $T_2^{-1}T_1(\tau_1) = \tau_2$. Since $T_2^{-1}T_1 \in \mathfrak{M}$, this is again a contradiction.

Finally it is required to show that

$$\mathfrak{M}(R) = [\tau \mid \Im(\tau) > 0].$$

We note first that $\mathfrak{M}(R)$ contains all the regions $S^n(R)$ and $S^n U(R)$ for all integers n. The union of $S^n(R)$ is the set of points τ for which $\mid \tau + n \mid > 1$ for all n plus the left half of each circular arc on the boundary. Each $S^n U(R)$ is a curvilinear triangle bounded by arcs of the circles $\mid \tau + k \mid = 1$ for $k = n - 1$, n, $n + 1$ plus part of the boundary. See Figure 12. We see that $\mathfrak{M}(R)$ at least covers the half-plane $\Im(\tau) \geq \frac{1}{2}\sqrt{3}$.

The transformation

$$S^n U S^{-n}(\tau) = n - \frac{1}{\tau - n}$$

maps the half-plane $\Im(\tau) \geq \delta > 0$ on a circular disk $D_n(\delta)$ in the upper half-plane. $D_n(\delta)$ is tangent to the real axis at $\tau = n$ and its diameter is δ^{-1}. It is clear that each such disk belongs to $\mathfrak{M}(R)$ provided it is known in advance that the half-plane $\Im(\tau) \geq \delta$ has this property. This is certainly the case for $\delta = \frac{1}{2}\sqrt{3}$. Each disk $D_n(\delta)$ overlaps with the half-plane $\Im(\tau) \geq \delta$ if $\delta \leq 1$, and for such values of δ adjacent disks also overlap. If $\delta \leq \frac{1}{2}\sqrt{3}$, then one verifies that $[\tau \mid \Im(\tau) \geq \delta] \cup \{\cup_n D_n(\delta)\}$ is simply-connected. Since the bounding circles of adjacent disks intersect at two points, the lower one of which has the ordinate

$$\delta_1 = \tfrac{1}{2}\delta[1 + (1 - \delta^2)^{\frac{1}{2}}]^{-1} < \tfrac{1}{2}\delta,$$

we conclude that $\mathfrak{M}(R)$ must contain the half-plane

$$\Im(\tau) \geq \delta_1.$$

We can then replace δ in the preceding argument by δ_1. Repeating this process we see that $\mathfrak{M}(R)$ cannot omit any point in the upper half-plane. This completes the proof.

COROLLARY.   *S and U generate $\mathfrak{M}$.*

This is implied by the preceding proof. For we know that $\cup[T(R) \mid T \in \mathfrak{M}]$ covers the upper half-plane without overlapping, since any two regions $T(R)$ have at most two boundary points in common. We started out with all transformations of the form S^k and $S^m U$. Their maps of R cover the half-plane $\Im(\tau) \geq \frac{1}{2}\sqrt{3}$. Next we applied the transformations $S^n U S^{-n}$ to what we already had. The union of

$$S^k(R), \quad S^m U(R), \quad S^n U S^{k-n}(R), \quad \text{and} \quad S^n U S^{k-m} U(R)$$

where k, m, n run through all integers independently of each other now covers the half-plane $\Im(\tau) \geq \frac{1}{6}\sqrt{3}$. Operating again and again with the transformations $S^n U S^{-n}$, we get the whole upper half-plane covered, and each transformation that appears in this process is of the form

(13.6.9) $$S^{j_1} U S^{j_2} U \cdots U S^{j_n}$$

where the j's are integers. Since we must get every transformation of $\mathfrak{M}$ in the exhaustion process, we see that all these transformations are of the form (13.6.9), that is, S and U generate $\mathfrak{M}$.

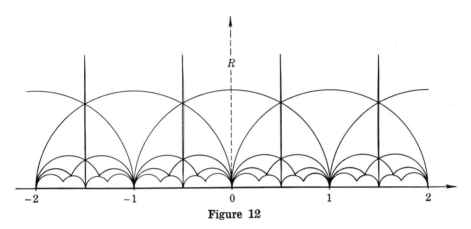

Figure 12

In Figure 12 we show the region R and part of the net obtained by mapping the boundary of R with the aid of the transformations of $\mathfrak{M}$. Each of the resulting regions can serve as fundamental region of $\mathfrak{M}$.

We come now to the construction of a modular function which is often referred to as *the* modular function. In view of the discussion in the preceding section, it is fairly obvious that such a function can be formed with the aid of the theta null values $\vartheta_2(0 \mid \tau)$, $\vartheta_3(0 \mid \tau)$, and $\vartheta_4(0 \mid \tau)$. If we apply the transformation S to these functions we obtain

$$e^{\frac{1}{4}\pi i}\vartheta_2(0 \mid \tau), \quad \vartheta_4(0 \mid \tau), \quad \vartheta_3(0 \mid \tau)$$

respectively. Applying U instead we get

$$(-i\tau)^{\frac{1}{2}}\vartheta_4(0 \mid \tau), \quad (-i\tau)^{\frac{1}{2}}\vartheta_3(0 \mid \tau), \quad (-i\tau)^{\frac{1}{2}}\vartheta_2(0 \mid \tau).$$

The least power of $\vartheta_2(0 \mid \tau)$ which is invariant under S is the eighth. It is clear that any rational symmetric function of ϑ_2^8, ϑ_3^8, ϑ_4^8 is invariant under S. The simplest functions of this kind are the sum

$$[\vartheta_2(0 \mid \tau)]^8 + [\vartheta_3(0 \mid \tau)]^8 + [\vartheta_4(0 \mid \tau)]^8$$

and the product

$$[\vartheta_2(0 \mid \tau)\,\vartheta_3(0 \mid \tau)\,\vartheta_4(0 \mid \tau)]^8.$$

If we now apply the transformation U, the sum is reproduced multiplied by τ^4, whereas the product is multiplied by τ^{12}. This suggests forming

$$(13.6.10) \qquad J(\tau) = \frac{1}{54} \frac{\{[\vartheta_2(0 \mid \tau)]^8 + [\vartheta_3(0 \mid \tau)]^8 + [\vartheta_4(0 \mid \tau)]^8\}^3}{[\vartheta_2(0 \mid \tau)\,\vartheta_3(0 \mid \tau)\,\vartheta_4(0 \mid \tau)]^8},$$

where the constant multiplier is a normalizing factor. Each of the functions $\vartheta_j(0 \mid \tau)$, $j = 2, 3, 4$, is holomorphic and different from 0 in $\Im(\tau) > 0$. Since S and U leave $J(\tau)$ invariant, every transformation of $\mathfrak{M}$ has this property, that is, $J(\tau)$ is a modular function.

$J(\tau)$ is closely related to the function $J_0(\lambda)$ of formula (13.6.7). In fact, we have

$$(13.6.11) \qquad\qquad J(\tau) = J_0(k^2(\tau))$$

where $k = k(\tau)$ is the corresponding modulus. This follows from (13.5.15) and (13.5.10). We have

$$\vartheta_2{}^8 + \vartheta_3{}^8 + \vartheta_4{}^8 = [k^4 + 1 + (1 - k^2)^2]\vartheta_3{}^8 = 2(1 - k^2 + k^4)\vartheta_3{}^8,$$
$$(\vartheta_2\vartheta_3\vartheta_4)^8 = k^4(1 - k^2)^2\vartheta_3{}^{24},$$

of which (13.6.11) is an immediate consequence.

We have now to determine the analytical properties of $J(\tau)$. This will also give information about $k^2(\tau)$, which is an algebraic function of $J(\tau)$. Some of the properties will follow from the invariance properties, others from explicit representations, and still others from (13.6.11) and special properties of $k^2(\tau)$.

We observe first that $J(\tau)$ takes on conjugate complex values at points which are symmetric with respect to the imaginary axis in the τ-plane. This follows from the fact that $q = e^{\pi i \tau}$ has this property, so that each $\vartheta_j(0 \mid \tau)$ has it, and hence also $J(\tau)$ and $k^2(\tau)$. This implies that $J(\tau)$ is real on the imaginary axis. But $J(\tau + 1) = J(\tau)$. This gives not merely that $J(\tau)$ is real on the lines $\Re(\tau) = n$ but also that it is real on the lines $\Re(\tau) = n + \frac{1}{2}$ where n is any integer. The same is obviously true for $k^2(\tau)$.

There is a difference in the behavior of $J(\tau)$ on these two sets of lines: on the lines $\Re(\tau) = n$ we find that $J(\tau)$ is positive, but on the other lines $J(\tau)$ changes its sign twice and is negative for $\Im(\tau) > \frac{1}{2}\sqrt{3}$ and for $\Im(\tau) < \frac{1}{6}\sqrt{3}$. That $J(\tau)$ is positive on the imaginary axis follows from the fact that $q = e^{\pi i \tau} > 0$, so that each $\vartheta_j(0 \mid \tau)$ is real.

To get further information we shall represent $J(\tau)$ as a function of q. With the aid of formulas (13.5.31) we can express $J(\tau)$ in terms of q, q_1, q_2, q_3. After some simplification we obtain

$$J(\tau) = 2^{-9}3^{-3}q^{-2}[q_2{}^{16} + q_3{}^{16} + 2^8 q^2 q_1{}^{16}]^3.$$

Here we substitute the infinite products for q_1, q_2, q_3 and obtain

$$(13.6.12)$$
$$J(\tau) = 2^{-9}3^{-3}q^{-2}\left\{\prod_1^\infty (1 - q^{2n-1})^{16} + \prod_1^\infty (1 + q^{2n-1})^{16} + 2^8 q^2 \prod_1^\infty (1 + q^{2n})^{16}\right\}^3.$$

Each of the infinite products is a holomorphic function of q for $|q| < 1$ and may be expanded in a convergent power series in q for such values. These power series are obtained by formal multiplication of the factors. When we collect terms inside the braces, we note that the third summand contributes only even powers of q with positive integral coefficients. The first summand will give rise to expressions of the form

$$C(-q^{j_1})(-q^{j_2}) \cdots (-q^{j_n}),$$

where C is a positive integer and the j's are positive *odd* integers. This is matched by an expression

$$Cq^{j_1}q^{j_2} \cdots q^{j_n}$$

coming from the second summand, so that the two expressions cancel if n is odd, and yield double the second expression if n is even. This implies that the power series for the function between the braces contains only even powers of q and that all coefficients are positive integers. The same then is true for the third power of this function. It follows that

$$(13.6.13) \qquad J(\tau) = 2^{-9}3^{-3}q^{-2}\left[2^3 + \sum_{n=1}^{\infty} a_n q^{2n}\right]$$

where every a_n is a non-negative integer.

Various conclusions may be drawn from this expansion. First we note that $J(\tau)$ becomes infinite with $\mathfrak{I}(\tau)$. Moreover, since $q^2 > 0$ when $\mathfrak{R}(\tau) = n$ and since $q^2 < 0$ when $\mathfrak{R}(\tau) = n + \frac{1}{2}$, we see that

$$J(n + i\beta) \to +\infty \quad \text{and} \quad J(n + \tfrac{1}{2} + i\beta) \to -\infty$$

when $\beta \to +\infty$. Since $J(-\tau^{-1}) = J(\tau)$, this implies that $J(n + i\beta) \to +\infty$ also when $\beta \to 0+$. Next we observe that the function

$$J(i\beta) = 2^{-9}3^{-3}\left[2^3 e^{2\pi\beta} + \sum_{n=1}^{\infty} a_n e^{-(n-1)2\pi\beta}\right]$$

has positive even-order derivatives. It follows that the graph of $\eta = J(i\beta)$ is concave upward and has a single positive minimum. This is located at $\beta = 1$ and $J(i) = 1$. For, differentiating the relation $J(\tau) = J(-\tau^{-1})$, we get

$$(13.6.14) \qquad J'(\tau) = \tau^{-2}J'(-\tau^{-1}),$$

so that $J'(i) = -J'(i)$ or $J'(i) = 0$. From (13.6.11) we get

$$J'(\tau) = 2k(\tau)k'(\tau)J_0'(k^2(\tau)).$$

But $k(\tau)$ is positive and monotone decreasing on the imaginary axis (see Problem 3 of Exercise 13.5). Thus $k^2(i)$ must be one of the zeros of $J_0'(\lambda)$, that is, $k^2(i)$ is restricted to one of the values ρ, ρ^2, -1, $\frac{1}{2}$, and 2. Since $k^2(i)$ is real positive, the choice is between $\frac{1}{2}$ and 2. Since $J_0(\frac{1}{2}) = J_0(2) = 1$, we see that $J(i) = 1$. Thus $J(\tau)$ assumes every real value ≥ 1 at least once in R.

We now go back to the relation $J(\tau) = J(-\tau^{-1})$. Let $\tau*$ be the point symmetric to τ with respect to the unit circle. Then $\tau*$ and $-\tau^{-1}$ are symmetric with respect to the imaginary axis and

$$J(\tau*) = \overline{J(\tau)}.$$

In particular, $J(\tau)$ is real on the unit circle and takes on the same value at the points $e^{i\theta}$ and $e^{i(\pi-\theta)}$. As θ goes from $\frac{1}{2}\pi$ to π, $J(\tau)$ goes from $+1$ to $-\infty$. Note that the transformation SU takes the arc from $\theta = \frac{2}{3}\pi$ to $\theta = \pi$ into the line segment $\Re(\tau) = -\frac{1}{2}$, $\Im(\tau) \geq \frac{1}{2}\sqrt{3}$, and that $J(\tau)$ goes to $-\infty$ on this segment and hence also on the arc. It follows that as τ describes the left half of the boundary of R, which belongs to R, then $J(\tau)$ takes on every real value ≤ 1 at least once. In particular, the value 0 is assumed. But from the definition of $J(\tau)$ every zero of $J(\tau)$ is a triple zero. Going back to (13.6.14), we see that $J'(\rho^2) = \rho^2 J'(\rho) = \rho^2 J'(\rho^2)$ since $J'(\tau + 1) = J'(\tau)$. Hence $J'(\rho^2) = 0$. Differentiating (13.6.14) we get similarly that $J''(\rho^2) = 0$. Thus, $J(\rho^2)$ is a triple value which, naturally, we suspect to be 0. This is indeed the case, but it is easier to prove it after we have studied the non-real values of $J(\tau)$. We shall prove

THEOREM 13.6.2. *The function $J(\tau)$ takes on every non-real value once and only once in* Int (R).

Proof. Let w be such a value and consider the integral

$$\frac{1}{2\pi i} \int \frac{J'(\tau)\, d\tau}{J(\tau) - w}$$

taken along the contour $ABCDEA$ where $A = \rho^2$, $B = i$, $C = \rho$, $D = \frac{1}{2} + ib$, $E = -\frac{1}{2} + ib$, following the boundary of R except for the crosspiece DE. If b is so large that $|J(\alpha + ib)| > |w|$, the integral will exist since $J(\tau)$ is real on the rest of the contour. We recall that the integral gives the number of times that $J(\tau)$ assumes the value w inside the contour; its value is consequently a positive integer or 0. Now the integrals along the vertical sides CD and EA cancel since $J(\tau) = J(\tau + 1)$. The integral along ABC, following the unit circle, turns out to be 0. For $J(\rho^2) = J(\rho)$, so the value of the integral equals $(2\pi)^{-1}$ times the change in the argument of $J(\tau) - w$ as we go from A to C. But $J(\tau)$ is real and $\Im(w) \neq 0$, so the argument must return to its original value. We have left the contribution from DE. Using (13.6.13), where we substitute $e^{\pi i \tau}$ for q, and differentiating termwise with respect to τ, we see that

$$\frac{J'(\alpha + ib)}{J(\alpha + ib) - w} = 2\pi i + O(e^{-2\pi b}).$$

Thus, this part contributes

$$1 + O(e^{-2\pi b})$$

to the integral. We conclude that the value of the integral is identically 1, and this completes the proof.

We can now settle the remaining questions concerning the distribution of the real values. In particular, we can prove that $J(\rho^2) = J(\rho) = 0$. We know that $J(\tau)$ must take on the value 0 somewhere in R and that each such point is a triple zero. Suppose that $\tau = \tau_0$ is such a point. Let w be a complex number of small absolute value. Then the equation $J(\tau) = w$ has three roots in a small neighborhood of $\tau = \tau_0$. If $\tau_0 \in \text{Int}\,(R)$, these three roots would also belong to Int (R). If $\tau \in \partial R$ but $\tau \neq \rho^2$, all three roots are not necessarily in Int (R). Counting homologous points under the transformations S or U, we see however that the equation $J(\tau) = w$ would have three roots in the interior of R. Since this is out of the question, we conclude that $J(\rho^2) = J(\rho) = 0$. In this case we get no contradiction, for, owing to the geometric nature of R at the vertices, only one root of the equation $J(\tau) = w$ belongs to R. We can use the same argument to conclude that for real values of w the equation $J(\tau) = w$ has only one root in R. From this we also see that $J(\tau)$ is monotone on the line segment $\Re(\tau) = -\frac{1}{2}, \Im(\tau) \geqq \frac{1}{2}\sqrt{3}$, where it goes steadily from 0 to $-\infty$. It is also monotone on the arc of the unit circle from $\tau = \rho^2$ to $\tau = i$, where it goes steadily from 0 to $+1$.

Thus we see that the function $w = J(\tau)$ maps the interior of R conformally on the w-plane cut along the real axis from $-\infty$ to $+1$. We can continue this function across the boundary of R by reflection in one of the straight lines or in the unit circle. This means continuation in the w-plane across one of the cuts by the adding of new copies of the w-plane cut along two of the three segments $(-\infty, 0), (0, 1), (1, +\infty)$. We note that τ always stays in the upper half-plane. Hence we have

THEOREM 13.6.3. *The inverse function $\tau = \mu(w)$ of the modular function is infinitely many-valued with branch points at $w = 0, 1, \infty$. The imaginary part of $\mu(w)$ is positive for every w and for every determination of $\mu(w)$.*

In the preceding discussion we have often referred to properties of $k^2(\tau)$. We shall now make a brief study of this function. It is a modular function with respect to a group $\Re$ which is a proper subgroup of $\mathfrak{M}$. We recall that

$$(13.6.15) \qquad k^2(\tau) = \left(\frac{\vartheta_2(0 \mid \tau)}{\vartheta_3(0 \mid \tau)}\right)^4.$$

Now the transformation S acting on τ replaces $[\vartheta_2(0 \mid \tau)]^4$ by its negative and interchanges $\vartheta_3(0 \mid \tau)$ and $\vartheta_4(0 \mid \tau)$. Since

$$[\vartheta_2(0 \mid \tau)]^4 + [\vartheta_4(0 \mid \tau)]^4 = [\vartheta_3(0 \mid \tau)]^4,$$

it follows that

$$S[k^2(\tau)] = S^{-1}[k^2(\tau)] = \frac{k^2(\tau)}{k^2(\tau) - 1}$$

and

$$S^2[k^2(\tau)] = k^2(\tau).$$

Further, U carries $[\vartheta_2(0 \mid \tau)]^4$ into $-\tau^2[\vartheta_4(0 \mid \tau)]^4$ and multiplies $[\vartheta_3(0 \mid \tau)]^4$ by $-\tau^2$ so that

$$U[k^2(\tau)] = 1 - k^2(\tau).$$

From these relations we get three more:

$$SU[k^2(\tau)] = \frac{k^2(\tau) - 1}{k^2(\tau)}, \quad US[k^2(\tau)] = \frac{1}{1 - k^2(\tau)}, \quad SUS[k^2(\tau)] = \frac{1}{k^2(\tau)}.$$

Thus, the effect of the modular group $\mathfrak{M}$ acting on $k^2(\tau)$ in the τ-plane is the same as that of the anharmonic group $\mathfrak{A}$ acting in the k^2-plane.

Further, we see that if a transformation $T \in \mathfrak{M}$ leaves $k^2(\tau)$ invariant, then T is of the form

$$S^{2j_1}US^{2j_2}U \cdots US^{2j_n},$$

where the j's are integers. It is clear that these transformations form a group $\mathfrak{R}$ which is generated by

$$(13.6.16) \qquad S^2(\tau) = \tau + 2, \quad US^2U(\tau) = \frac{\tau}{1 - 2\tau}, \quad (SU)^3(\tau) = \tau.$$

As fundamental region of $\mathfrak{R}$ we may take what is defined by

$$(13.6.17) \quad \text{Int } (B) = [\tau \mid \mid \tau + \tfrac{1}{2} \mid > \tfrac{1}{2}, \mid \tau - \tfrac{1}{2} \mid > \tfrac{1}{2}, -1 < \mathfrak{R}(\tau) < 1].$$

To get B we add that part of the boundary where $\mathfrak{R}(\tau) \leqq 0$. For a given value of w the equation $J(\tau) = w$ clearly has six roots in B.

THEOREM 13.6.4.   *The equation*

$$(13.6.18) \qquad\qquad\qquad k^2(\tau) = \lambda$$

is satisfied by one and only one value of τ in B for every choice of λ.

Proof.   If for a particular choice of λ the equation had two roots τ_1 and τ_2 in B, then by (13.6.11) we must have $J(\tau_1) = J(\tau_2)$. This requires that $\tau_2 = T(\tau_1)$ where T belongs to $\mathfrak{M}$ but not to $\mathfrak{R}$. We would then have $k^2(\tau_1)$ equal to one of the numbers

$$\frac{1}{1 - \lambda}, \quad 1 - \lambda, \quad \frac{\lambda}{\lambda - 1}, \quad \frac{\lambda - 1}{\lambda}, \quad \text{or} \quad \frac{1}{\lambda}$$

which arise when we apply the anharmonic group to λ. One of these numbers could coincide with λ if and only if λ had one of the values

$$0, -1, +1, \tfrac{1}{2}, 2, \rho, \rho^2.$$

It is a simple matter to handle these special cases, and we shall restrict ourselves to a few cases. For $\lambda = \tfrac{1}{2}$ we have $J_0(\lambda) = 1$, $J(\tau) = 1$. The equation $J(\tau) = 1$ has three double roots in B at the points i, $-1 + i$, and $-\tfrac{1}{2} + \tfrac{1}{2}i$. Here

$k^2(i) = \frac{1}{2}$, $k^2(-1 + i) = -1$, $k^2(-\frac{1}{2} + \frac{1}{2}i) = 2$. Thus the solution of (13.6.18) is unique for $\lambda = \frac{1}{2}$, 2, and -1. The other doubtful cases are handled in the same manner, and we conclude that the solution is unique if it exists.

To obtain the existence we have merely to solve the equation

$$J(\tau) = J_0(\lambda)$$

for τ. It has six roots in B, one and only one of which is also a root of (13.6.18). The actual computation of the root may be based on the rapidly converging series (13.5.21), which expresses $q = \exp(\pi i \tau)$ in terms of

$$\varepsilon = \frac{1}{2} \frac{1 - [1 - k^2(\tau)]^{1/4}}{1 + [1 - k^2(\tau)]^{1/4}}.$$

We note that $4\varepsilon^2 = k(4\tau)$.

For the mapping (13.6.18) compare Figure 13, where the symbols in parentheses are the corresponding values of λ.

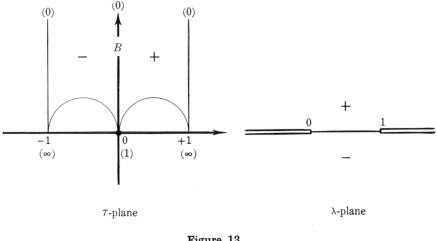

τ-plane   λ-plane

Figure 13

Once the right values of q and of τ have been found, corresponding to a given value of k^2, we can determine K and K', the former from (13.5.14), the latter from $K' = -i\tau K$. We can then pass over to the Jacobi functions themselves, for which we have representations as theta quotients or infinite products or Fourier series. See formulas (13.5.9), (13.5.11), and (13.5.12) and Problems 9 and 10 of Exercise 13.5.

We have referred to the corresponding problem for the Weierstrass $\wp$-function in Section 13.2. Here g_2 and g_3 are known, or equivalently e_1, e_2, e_3, and the problem is to determine ω_1 and ω_3.

THEOREM 13.6.5.   *Given two numbers a and b such that $a^3 - 27b^2 \neq 0$, then there exist numbers ω_1 and ω_3 such that $\Im(\omega_3/\omega_1) > 0$ and $g_2(\omega_1, \omega_3) = a$, $g_3(\omega_1, \omega_3) = b$.*

Here, as usual

$$g_2(\omega_1, \omega_3) = 60 \sum \sideset{}{'}\sum (2m\omega_1 + 2n\omega_3)^{-4},$$

$$g_3(\omega_1, \omega_3) = 140 \sum \sideset{}{'}\sum (2m\omega_1 + 2n\omega_3)^{-6}.$$

The condition $a^3 - 27b^2 \neq 0$ serves to ensure that the corresponding cubic has three distinct roots. For we have that the discriminant is

$$\Delta = (e_1 - e_2)^2 (e_2 - e_3)^2 (e_3 - e_1)^2 = \tfrac{1}{16}(g_2{}^3 - 27g_3{}^2).$$

A direct proof of the theorem would involve a repetition of the argument given above in discussing $J(\tau)$ and $k^2(\tau)$. We would then define $J(\tau)$ by

$$(13.6.19) \qquad\qquad J(\tau) = \frac{[g_2(1, \tau)]^3}{[g_2(1, \tau)]^3 - 27[g_3(1, \tau)]^2} .$$

A direct study of this function would show that it is identical with the function defined by (13.6.10). The first step in the solution of the problem posed in Theorem 13.6.5 would then be to determine the root of the equation

$$(13.6.20) \qquad\qquad J(\tau) = \frac{a^3}{a^3 - 27b^2}$$

which is located in R. This root is unique. In setting up the equation (13.6.20) we have used the homogeneity properties

$$g_2(\omega_1, \omega_3) = \omega_1{}^{-4} g_2(1, \tau), \quad g_3(\omega_1, \omega_3) = \omega_1{}^{-6} g_3(1, \tau), \quad \tau = \frac{\omega_3}{\omega_1} .$$

Using these again we see that

$$\omega_1{}^2 = \frac{a\, g_3(1, \tau)}{b\, g_2(1, \tau)} ,$$

whence

$$\omega_3 = \tau\omega_1.$$

Since τ is known from (13.6.20), ω_1 and ω_3 are uniquely determined.

The series defining $g_2(1, \tau)$ and $g_3(1, \tau)$ are slowly convergent and are not suitable for numerical computation. Actually we get a more satisfactory solution of the problem if we reduce it to the Jacobi case. Problem 10 of Exercise 13.4 shows how this is to be done. The relation between the Jacobi modulus k and the Weierstrass zeros e_1, e_2, e_3 with $e_1 + e_2 + e_3 = 0$ is simply

$$(13.6.21) \qquad\qquad k^2 = \frac{e_2 - e_3}{e_1 - e_3} .$$

If we form $J_0(k^2)$ with this choice of k^2, an elementary but lengthy calculation shows that $J(\tau)$ coincides with the right member of (13.6.19). We can then proceed as above and calculate τ from (13.6.20). The periods $2\omega_1$, $2\omega_3$ are now as easy to calculate as in the Jacobi case. Referring to Problem 10 once more we see that

$$(13.6.22) \qquad \omega_1 = \tfrac{1}{2}\pi(e_1 - e_3)^{-\frac{1}{2}}[\vartheta_3(0 \mid \tau)]^2,$$

where now e_1, e_2, e_3 are the roots of the cubic

$$4w^3 - aw - b = 0.$$

We have, of course, $\omega_3 = \tau\omega_1$. This shows that the problem admits of a neat and compact solution.

EXERCISE 13.6

1. Prove that the group $\Re$ contains all transformations (13.6.1) such that b and c are even integers.

2. Show that the transformations (13.6.1) such that b and c are divisible by a fixed integer n form a group.

3. The circles $\mid \tau \mid = 1$ and $\mid \tau + 1 \mid = 1$ divide the plane into three domains. Show that each of them with the addition of one of the bounding circular arcs can be taken as fundamental region of the group generated by US. If R_1 is the region, it is to be shown that no two points of R_1 can be congruent under the group and that the transforms of R_1 cover the plane exactly once.

4. Verify that the function $J(\tau)$ maps the left half of Int (R), that is, the domain $-\tfrac{1}{2} < \Re(\tau) < 0$, $\mid \tau \mid > 1$, conformally on the upper half-plane. Discuss the boundary and, in particular, the vertices.

5. Verify that $x = k^2(\tau)$ maps the domain $0 < \Re(\tau) < 1$, $\mid \tau - \tfrac{1}{2} \mid > \tfrac{1}{2}$, conformally on the upper half-plane. Discuss the boundary and, in particular, the vertices.

6. Given the function

$$w = \frac{1 - (1 - z)^{1/4}}{1 + (1 - z)^{1/4}},$$

where the fourth root equals $+1$ for $z = 0$. Show that this function maps the z-plane cut along the real axis from $z = +1$ to $z = +\infty$ on a lens-shaped region bounded by two arcs which intersect at right angles at $w = \pm 1$. The lens is symmetric with respect to the axes and its "thickness" is $2(\sqrt{2} - 1)$, so that it lies inside the unit circle except for the vertices. What is the bearing of this problem on (13.5.21)?

7. Prove that the hypergeometric differential equation

$$x(1 - x)y'' + (1 - 2x)y' - \tfrac{1}{4}y = 0$$

is unchanged if x is replaced by $1 - x$. It is also unchanged if x is replaced by $1/x$ or by $(x - 1)/x$ and y by $yx^{\frac{1}{2}}$ or if x is replaced by $1/(1 - x)$ or by $x/(x - 1)$ and y by $y(1 - x)^{\frac{1}{2}}$.

8. Prove by direct computation or otherwise that the periods of sn (z, k) as functions of $x = k^2$ satisfy the preceding differential equation and, more precisely, that

$$K(x) = \tfrac{1}{2}\pi F(\tfrac{1}{2}, \tfrac{1}{2}, 1; x), \quad K'(x) = \tfrac{1}{2}\pi F(\tfrac{1}{2}, \tfrac{1}{2}, 1; 1 - x),$$

where the prime does *not* denote differentiation with respect to x.

9. Prove that as $x \to 1-$

$$K(x) = \tfrac{1}{2} \log \frac{1}{1 - x} + O(1).$$

10. Prove that the inverse of the mapping function in Problem 5 is the quotient of two hypergeometric functions with parameters $\alpha = \beta = \tfrac{1}{2}, \gamma = 1$. A similar result holds for the inverse of the mapping function in Problem 4. Here $\alpha = \beta = \tfrac{1}{12}, \gamma = \tfrac{2}{3}$.

COLLATERAL READING

The standard text on elliptic functions is

TANNERY, J., and MOLK, J. *Éléments de la Théorie des Fonctions Elliptiques*, Vols. I–IV. Gauthier-Villars, Paris, 1893–1902. Second Edition. Chelsea Publishing Co., New York, 1972.

There are good accounts of the theory in a number of treatises as, for instance,

COPSON, E. T. *An Introduction to the Theory of Functions of a Complex Variable*, Chaps. XIII–XV. Oxford University Press, Inc., Fair Lawn, New Jersey, 1935.

HURWITZ, A., and COURANT, R. *Vorlesungen über allgemeine Funktionentheorie und elliptische Funktionen*, Second Edition, Part II. Springer-Verlag, Berlin, 1925.

SAKS, S., and ZYGMUND, A. *Analytic Functions*, Chap. 8. Monografie Matematyczne, Vol. 28, Warsaw and Wroclaw, 1952.

WHITTAKER, E. T., and WATSON, G. N. *A Course of Modern Analysis*, Fourth Edition, Chaps. 20–22. Cambridge University Press, New York, 1952.

In dealing with theta functions the reader is warned that different systems of notation are in use and that the same symbol may denote different functions. Our notation agrees with that of Whittaker and Watson.

The most complete monograph on modular functions is

KLEIN, F., and FRICKE, R. *Vorlesungen über die Theorie der elliptischen Modulfunktionen*, Vols. I, II. B. G. Teubner, Leipzig, 1890 and 1892.

A brief account is to be found in

FORD, L. R. *Automorphic Functions*, Second Edition, Chap VII. Chelsea Publishing Company, Inc., New York, 1951. (The reader is also referred to this work for an introduction to the theory of automorphic functions.)

14

ENTIRE AND MEROMORPHIC
FUNCTIONS

14.1. Order relations for entire functions. In Section 8.7 we laid the foundation of the theory of entire functions culminating in the factorization theorem of Weierstrass. In the present section we shall study some aspects of the theory of entire functions and, in particular, the relations which hold between the rate of growth of the maximum modulus, the rate of decrease of the coefficients, and the frequency of the zeros.

Let

$$(14.1.1) \qquad f(z) = \sum_{n=0}^{\infty} c_n z^n$$

be a transcendental entire function and let

$$(14.1.2) \qquad M(r;f) = \max_{0 \le \theta < 2\pi} |f(re^{i\theta})|$$

be its maximum modulus. Denote its zeros by $\{z_n\}$, where a k-fold zero is supposed to be repeated k times in the sequence. We suppose that the zeros are ordered according to increasing absolute values and write

$$(14.1.3) \qquad |z_n| = r_n,$$

so that

$$r_n \le r_{n+1}$$

for all n. It is, of course, possible for $f(z)$ to have only a finite number of zeros or none at all.

DEFINITION 14.1.1. *We call* $\rho = \rho(f)$ *the order of* $f(z)$ *if*

$$(14.1.4) \qquad \rho(f) = \limsup_{r \to \infty} \frac{\log \log M(r;f)}{\log r}.$$

We have $0 \le \rho(f) \le \infty$. *If* $0 < \rho(f) < \infty$, *we say that* $f(z)$ *is of type* $\tau = \tau(f)$ *of its order if*

$$(14.1.5) \qquad \tau(f) = \limsup_{r \to \infty} r^{-\rho(f)} \log M(r;f).$$

A function $f(z)$ *is said to be of minimal type of its order if* $\tau = 0$, *of normal type if* $0 < \tau < \infty$, *and of maximal type if* $\tau = \infty$.

182

The function $\exp (az^k)$, where $a > 0$ and k is a positive integer, is of order k and of type a of that order, and $\exp (e^z)$ is of order ∞. We shall see later that the theta-like functions

$$\prod_{n=1}^{\infty} (1 - q^n z) \quad \text{and} \quad \sum_{n=0}^{\infty} q^{n^2} z^n, \quad |q| < 1,$$

are entire functions of order zero. The following three functions:

(14.1.6) $$\prod_{n=2}^{\infty} \left\{ 1 + \frac{z}{n \log^2 n} \right\}, \quad e^{az}, \quad \frac{1}{\Gamma(z)},$$

which are of order one, are of type 0, a, ∞ respectively.

There are close relations between the order and the type of an entire function on the one hand and the infinitary behavior of the sequence $\{c_n\}$ formed by the coefficients of its Maclaurin series on the other hand. This is due to the fact that the maximum modulus $M(r; f)$ is essentially of the order of the maximal term in (14.1.1) for $|z| = r$. For each value of r, $r > 0$, there exists an index $\nu(r)$, known as the *central index*, such that

(14.1.7) $$|c_{\nu(r)}| r^{\nu(r)} \geq |c_k| r^k$$

for all k, and such that strict inequality holds for all $k > \nu(r)$. The term on the left is, by definition, the *maximal term* of the series for $|z| = r$. *The central index is a never decreasing function of r, and it becomes infinite with r.* These notions are used in the proof of the following lemma:

LEMMA 14.1.1.   *The entire function*

(14.1.8) $$F_\alpha(z) = \sum_{n=1}^{\infty} \left(\frac{n}{\alpha e} \right)^{-n/\alpha} z^n, \quad \alpha > 0,$$

is of order α and of type 1 of that order. More precisely, as $r \to \infty$,

(14.1.9) $$M(r; F_\alpha) = F_\alpha(r) = (2\pi)^{1/2} \alpha r^{\alpha/2} \exp (r^\alpha)\{1 + o(1)\}.$$

Proof.   The terms of $F_\alpha(r)$ are strictly increasing up to the maximal term and then strictly decreasing. Equality holds in (14.1.7) for at most one value of $k < \nu(r)$. To determine the maximal term, we use the methods of the calculus. We set

$$g(r, u) = u \log r - \frac{u}{\alpha} \log \frac{u}{\alpha e},$$

so that the nth term of the series is given by $\exp [g(r, n)]$. The maximum of $g(r, u)$ is r^α, and it is reached for

$$u = u_0 = \alpha r^\alpha.$$

Thus, for a sequence of values of r which tend to infinity, the maximal term equals exp (r^α). Now, in a power series with positive terms the maximum modulus clearly exceeds the maximal term. Hence

$$M(r; F_\alpha) > \exp(r^\alpha)$$

infinitely often, and this shows that $\rho(F_\alpha) \geq \alpha$. To establish equality and to prove the estimate (14.1.9), we cannot restrict ourselves to the maximal term. It turns out, however, that the terms in a certain vicinity of the maximal term determine the growth of $M(r; F_\alpha)$. This is due to the fact that for large values of r the graph of

$$y = g(r, u)$$

in the (u, y)-plane is comparatively flat over a considerable range containing the maximum, and, for u outside of this range, exp $[g(r, u)]$ is relatively insignificant.

To make this more precise we use Taylor's series with remainder and find that

$$g(r, u) = r^\alpha - (2\alpha u_1)^{-1}(u - u_0)^2,$$

where u_1 lies between u_0 and u. For

$$|u - u_0| < \omega(r) = o(u_0)$$

we can replace this by

(14.1.10) $g(r, u) = r^\alpha - (2\alpha u_0)^{-1}(u - u_0)^2 + R,$

where

$$|R| < \alpha^{-1}u_0^{-2}[\omega(r)]^3$$

for large values of r. A suitable choice of $\omega(r)$ is

$$\omega(r) = \left[\alpha u_0 \log \frac{u_0}{\alpha}\right]^{1/2} = [\alpha^3 r^\alpha \log r]^{1/2}.$$

With this choice of $\omega(r)$, we estimate the integral of exp $[g(r, u)]$ over the interval $I:\ [u_0 - \omega(r), u_0 + \omega(r)]$.

The integral equals

$$[1 + \delta(r)] \exp(r^\alpha) \int_I \exp[-(2\alpha u_0)^{-1}(u - u_0)^2]\, du,$$

where

$$|\delta(r)| < C(\alpha)u_0^{-1/2}\left(\log \frac{u_0}{\alpha}\right)^{3/2} = C_1(\alpha)r^{-\alpha/2}(\log r)^{3/2}.$$

Here the factor $1 + \delta(r)$ corresponds to the factor exp R of the integrand. Since $|R|$ is small, exp $R = 1 + O(R)$, and the remainder does not exceed $(\log 2)^{-1} \max |R|$ if max $|R| < \log 2$. This is certainly true for large values of

r since $\max |R| \to 0$ as $r \to \infty$. The numbers $C_j(\alpha)$ occurring here and in the following depend only upon α as soon as $r > 1$. Further,

$$\int_I \exp\left[-(2\alpha u_0)^{-1}(u - u_0)^2\right] du = (2\alpha u_0)^{1/2} \int_{-a}^{a} e^{-s^2} ds$$

where

$$a = (\tfrac{1}{2}\alpha \log r)^{1/2}.$$

But

$$\int_{-a}^{a} = \int_{-\infty}^{\infty} - \int_{-\infty}^{-a} - \int_{a}^{\infty}.$$

The first integral on the right equals $\pi^{1/2}$, and the second and third integrals together do not exceed

$$a^{-1}e^{-a^2} = C_2(\alpha)r^{-\alpha/2}(\log r)^{-1/2}.$$

Combining these computations, we see that

(14.1.11) $$\int_I \exp[g(r, u)] du = \alpha(2\pi)^{1/2}[1 + \delta_1(r)]r^{\alpha/2} \exp(r^\alpha),$$

where

(14.1.12) $$|\delta_1(r)| < C_3(\alpha)r^{-\alpha/2}(\log r)^{3/2}.$$

The same estimate holds for

$$\Sigma \exp[g(r, n)]$$

where the summation extends over those values of n which lie in the interval I. This is due to the fact that $\exp[g(r, u)]$ is increasing for $u < u_0$ and decreasing for $u_0 < u$. Expressing the sum as the area of rectangles and comparing this with the area under the curve $y = \exp[g(r, u)]$, we see that the difference between the integral and the corresponding sum cannot exceed four times the maximum of the integrand, that is, $4 \exp(r^\alpha)$. This means that

(14.1.13) $$\sum_{n \in I} \exp[g(r, n)] = \alpha(2\pi)^{1/2}[1 + \delta_2(r)]r^{\alpha/2} \exp(r^\alpha),$$

where $\delta_2(r)$ satisfies (14.1.12) for a suitable choice of $C_3(\alpha)$.

It remains to discuss the terms with index n outside of I. Again we can replace sums by integrals with an error which does not exceed twice the largest term involved. In the interval $(1, u_0 - \omega(r))$ the function $g(r, u)$ is increasing, and $g_u(r, u)$ is positive and assumes its least value at the upper end of the interval. We write c for $u_0 - \omega(r)$ and note that

$$\int_1^c \exp[g(r, u)] du < [g_u(r, c)]^{-1} \int_1^c d_u \exp[g(r, u)]$$

$$< [g_u(r, c)]^{-1} \exp[g(r, c)].$$

A simple calculation shows that

$$\exp\,[g(r,\,c)] < C_4(\alpha)r^{-\alpha/2}\exp\,(r^\alpha),$$
$$g_u(r,\,c) > C_5(\alpha)r^{-\alpha/2}(\log\,r)^{-1/2},$$

so that

$$\sum_{n<c}\exp\,[g(r,\,n)] < C_6(\alpha)(\log\,r)^{1/2}\exp\,(r^\alpha).$$

The same device shows that this estimate is also valid for the infinite sum in which $n > u_0 + \omega(r)$.

Combining these estimates we see that (14.1.9) holds. Moreover, we can replace the factor $1 + o(1)$ by $1 + \delta_3(r)$ where

(14.1.14) $$|\,\delta_3(r)\,| < C_8(\alpha)r^{-\alpha/2}\,(\log\,r)^{3/2}.$$

This completes the proof of the lemma.

We have used the proof of this lemma to familiarize the reader with important analytical techniques. For the application in the proof of the next theorem a less precise result, obtainable by cruder methods, would have sufficed.

THEOREM 14.1.1.   *The order of $f(z)$ is also given by*

(14.1.15) $$\rho(f) = \limsup_{n\to\infty}\frac{n\log n}{\log|\,c_n\,|^{-1}}\,.$$

Proof.   We denote the right member of (14.1.15) by σ, and we start by showing that $\rho(f) \leqq \sigma$. If σ is finite and $\varepsilon > 0$ is fixed, there exists an n_ε such that

$$\frac{n\log n}{\log|\,c_n\,|^{-1}} < \sigma + \varepsilon, \quad n_\varepsilon < n.$$

From this we get

$$|\,c_n\,| < n^{-n/(\sigma+\varepsilon)}, \quad n_\varepsilon < n.$$

Hence there exists a finite $M(\varepsilon)$ such that

$$|\,c_n\,| < M(\varepsilon)n^{-n/(\sigma+\varepsilon)}, \quad 1 \leqq n.$$

Thus, by the definition of $F_\alpha(z)$ we have

$$M(r;f) \leqq |\,c_0\,| + M(\varepsilon)F_\alpha[(\alpha e)^{1/\alpha}r], \quad \alpha = \sigma + \varepsilon.$$

Since the order of $F_\alpha(z)$ equals α, we see that $\rho(f) \leqq \sigma + \varepsilon$ for every $\varepsilon > 0$. Hence $\rho(f) \leqq \sigma$ provided σ is finite. But the inequality is trivially true if $\sigma = \infty$, so we have $\rho(f) \leqq \sigma$ regardless of the value of σ.

To prove the converse inequality we use Cauchy's estimates. For any $r > 0$ we have

$$|\,c_n\,| < r^{-n}M(r;f).$$

Suppose that $\rho(f)$ is finite. Then for any fixed $\varepsilon > 0$ there is a finite $M_1(\varepsilon)$ such that

$$M(r;f) \leq M_1(\varepsilon) \exp(r^{\rho+\varepsilon})$$

for all r. Now for a fixed n the minimum value of the expression

$$r^{-n} \exp(r^{\rho+\varepsilon})$$

is

$$\left\{\frac{n}{e(\rho+\varepsilon)}\right\}^{-n/(\rho+\varepsilon)},$$

which is reached for

$$r = \left\{\frac{n}{\rho+\varepsilon}\right\}^{1/\rho}.$$

It follows that

(14.1.16)
$$|c_n| \leq M_1(\varepsilon)\left\{\frac{n}{e(\rho+\varepsilon)}\right\}^{-n/(\rho+\varepsilon)},$$

whence

$$\sigma = \limsup_{n\to\infty} \frac{n \log n}{\log |c_n|^{-1}} \leq \rho + \varepsilon$$

for every $\varepsilon > 0$ or $\sigma \leq \rho$. This obviously is true also if $\rho = \infty$. Since the opposite inequality holds, we conclude that $\sigma = \rho$ as asserted.

In applying (14.1.15), we can clearly disregard the values of n for which $c_n = 0$ in forming the superior limit.

THEOREM 14.1.2.   *If $f(z)$ is of order ρ, $0 < \rho < \infty$, then the type of $f(z)$ is also given by*

(14.1.17)
$$\tau(f) = \frac{1}{e\rho} \limsup_{n\to\infty} n |c_n|^{\rho/n}.$$

Proof.   We use the same argument as above. We denote the right member of (14.1.17) by v and we see that if $v < \infty$ then

$$|c_n| \leq M(\varepsilon)\left\{\frac{n}{e\rho(v+\varepsilon)}\right\}^{-n/\rho}, \quad n \geq 1.$$

This gives

$$M(r;f) \leq |c_0| + M(\varepsilon)F_\rho[(v+\varepsilon)^{1/\rho}r].$$

Since $F_\rho(z)$ is of order ρ and type 1, we conclude that $\tau(f) \leq v$.

The converse inequality is obtained from Cauchy's estimates. We now have if $\tau < \infty$

$$|c_n| \leq M_1(\varepsilon)r^{-n} \exp[(\tau+\varepsilon)r^\rho]$$

for any choice of $r > 0$ and any fixed $\varepsilon > 0$. The minimum of the right-hand side is attained for

$$r = \left\{ \frac{n}{\rho(\tau + \varepsilon)} \right\}^{1/\rho}$$

and gives

(14.1.18) $$|c_n| \leqq M_1(\varepsilon) \left\{ \frac{n}{e\rho(\tau + \varepsilon)} \right\}^{-n/\rho},$$

whence

$$v = \frac{1}{\rho e} \limsup_{n \to \infty} n \, |c_n|^{\rho/n} \leqq \tau + \varepsilon$$

for every $\varepsilon > 0$ so that $v \leqq \tau$. Hence $v = \tau$ as asserted.

We shall now take up the relations between the frequency of the zeros and the maximum modulus. Since an entire function need not have any zeros, the implications go only one way: the order of $f(z)$ imposes an upper bound on the frequency of the zeros, and a given frequency of the zeros imposes a lower bound on the order.

We can measure the frequency in various ways, and several of these will be used in the following. The first is the *exponent of convergence* of the series

(14.1.19) $$\Sigma \, r_n^{-\alpha}, \quad r_n > 0,$$

that is, the lower bound σ of the values of α for which the series converges. We have $0 \leqq \sigma \leqq \infty$. The three cases

$$r_n = e^n, \quad n^{1/\gamma}, \quad \log n, \quad \text{with } n > 1,$$

give $\sigma = 0, \gamma, \infty$ respectively.

We can also study the frequency with the aid of an *enumerative function*. Let

(14.1.20) $$n(r) = n(r, 0; f)$$

denote the number of zeros z_n with $r_n \leqq r$. This is a never decreasing step function. For some purposes the function

(14.1.21) $$N(r) = N(r, 0; f) = \int_0^r [n(s) - n(0)] \frac{ds}{s}$$

is preferable.

THEOREM 14.1.3. *For any $\alpha > 0$ the following three integrals are equi-convergent:*

(14.1.22) $$\int_a^\infty r^{-1-\alpha} N(r) \, dr, \quad \int_a^\infty r^{-1-\alpha} n(r) \, dr, \quad \int_a^\infty r^{-\alpha} \, dn(r).$$

Here $a > 0$ is arbitrary. We can take $a = 0$ if $n(0) = 0$.

Proof.   We start by observing that

$$\int_a^\infty r^{-\alpha}\, dn(r) = \sum_{a \leq r_n} r_n^{-\alpha}$$

by the definition of the Stieltjes integral.

An integration by parts gives

$$\int_a^b r^{-\alpha}\, dn(r) = b^{-\alpha} n(b) - a^{-\alpha} n(a) + \alpha \int_a^b r^{-1-\alpha} n(r)\, dr.$$

It follows that if the left member tends to a finite limit as $b \to \infty$, so does the right; that is, the convergence of the third integral in (14.1.22) or, equivalently, of the series (14.1.19) implies the convergence of the second integral. Further,

(14.1.23) $$\lim_{r \to \infty} r^{-\alpha} n(r, 0; f) = 0,$$

for the limit must exist and be finite, and the integral on the right is convergent. Suppose next that the integral on the right tends to a finite limit as $b \to \infty$ while the integral on the left becomes infinite. But this leads to a contradiction, for it would imply that the limit in (14.1.23) is infinite, and this is not consistent with the convergence of the integral on the right. Thus, the second and third integrals are equiconvergent.

We can handle the first and second integrals in the same manner. If $n(0) > 0$, we replace $n(r)$ by $n(r) - n(0)$ in the second integral, and this does not affect convergence. Integrating by parts we get

$$\int_a^b r^{-1-\alpha}[n(r) - n(0)]\, dr = b^{-\alpha} N(b) - a^{-\alpha} N(a) + \alpha \int_a^b r^{-1-\alpha} N(r)\, dr$$

and argue as above. The details are left to the reader. We note that in the convergence case

(14.1.24) $$\lim_{r \to \infty} r^{-\alpha} N(r, 0; f) = 0.$$

This completes the proof.

The clue to the connection between the maximum modulus and the frequency of the zeros is furnished by Jensen's theorem (Theorem 9.2.5). Since this theorem is basic for the following, we shall restate it in terms of the enumerative functions.

THEOREM 14.1.4.   *Let $f(z)$ be meromorphic in the finite plane and let*

$$N(r, 0; f) \quad and \quad N(r, \infty; f)$$

be the enumerative functions for zeros and poles respectively. Let $f(0) \neq 0, \infty$. Then

(14.1.25) $$\frac{1}{2\pi} \int_0^{2\pi} \log |f(re^{i\theta})|\, d\theta = \log |f(0)| + N(r, 0; f) - N(r, \infty; f).$$

The function $N(r, 0; f)$ is defined by formula (14.1.21). If $n(r, \infty; f)$ denotes the number of poles p_n of $f(z)$ with $|p_n| \leq r$, then

$$(14.1.26) \qquad N(r, \infty; f) = \int_0^r [n(s, \infty; f) - n(0, \infty; f)] \frac{ds}{s} .$$

In order to verify (14.1.25) we have merely to show that

$$(14.1.27) \qquad N(r, 0; f) = \sum_{r_n \leq r} \log \frac{r}{r_n} , \quad N(r, \infty; f) = \sum_{|p_n| \leq r} \log \frac{r}{|p_n|} ,$$

for then the formula coincides with (9.2.9) where we replace R by r. But we now have $n(0, 0; f) = 0$ so that

$$\sum_{r_n \leq r} \log \frac{r}{r_n} = \int_0^r \log \frac{r}{s} \, dn(s, 0; f) = \int_0^r n(s, 0; f) \frac{ds}{s}$$
$$= N(r, 0; f),$$

and similarly for the second sum. Here we have expressed the finite sum as a Stieltjes integral, integrated by parts, using the fact that the integrated part vanishes at both ends of the interval $(0, r)$.

For an entire function, formula (14.1.25) reduces to

$$(14.1.28) \qquad \frac{1}{2\pi} \int_0^{2\pi} \log |f(re^{i\theta})| \, d\theta = \log |f(0)| + N(r, 0; f),$$

and this gives the basic inequality

$$(14.1.29) \qquad \log M(r; f) \geq \log |f(0)| + N(r, 0; f)$$

and the following theorem:

THEOREM 14.1.5.   *The three integrals in (14.1.22) converge for $\alpha > \rho(f)$ so that $\sigma \leq \rho(f)$.*

For

$$\log M(r; f) \leq r^{\rho + \varepsilon} + O(1),$$

and $N(r, 0; f)$ satisfies the same inequality by (14.1.29).

COROLLARY.   *Formulas (14.1.23) and (14.1.24) hold for $\alpha > \rho(f)$.*

By the definition of type, these relations hold also for $\alpha = \rho(f)$ if $\tau(f) = 0$. The integral

$$(14.1.30) \qquad \int_1^\infty r^{-1-\rho} \log M(r; f) \, dr$$

normally diverges even if $\tau(f) = 0$. If it converges, $f(z)$ is said to belong to the *convergence class* of its order. An example is given in the Exercise below.

EXERCISE 14.1

1. Determine the central index for the exponential series.

2. Prove that the central index $\nu(r)$ is a never decreasing step function which becomes infinite with r.

3. Determine order and type of $\sin z$.

4. Same question for the Bessel function

$$J_k(z) = \sum_{n=0}^{\infty} \frac{(-1)^n z^{k+2n}}{2^{k+2n} n! \, (n+k)!}.$$

5. If $\alpha > 0$, use Stirling's formula and Theorems 14.1.1 and 14.1.2 to determine order and type of the Mittag-Leffler function

$$E_\alpha(z) = \sum_{n=0}^{\infty} \frac{z^n}{\Gamma(1+n\alpha)}.$$

6. Prove that the Jacobi theta function $\vartheta_4(z, q)$ is an entire function of z of order two. If $q = e^{-\pi t}, t > 0$, show that the type is $(4\pi t)^{-1}$. Are the same results true for the other theta functions?

7. Use the technique of Lemma 14.1.1 to prove that for $a > 0, r > 0$,

$$\sum_{n=0}^{\infty} e^{-an^2} r^n = \left(\frac{\pi}{a}\right)^{\frac{1}{2}} [1 + o(1)] \exp\left\{\frac{\log^2 r}{4a}\right\}.$$

8. Determine the exponent of convergence of the zeros of the Weierstrass σ-function. What conclusion can be drawn concerning its order?

9. Same question for $1/\Gamma(z)$.

10. Knowing that for $a > 0$, $a \to 0+$ we have $\int_a^{\infty} r^{-2} \log (1 + r) \, dr = \log a^{-1} + O(1)$, prove that the entire function

$$\prod_{n=2}^{\infty} \left\{1 + \frac{z}{n(\log n)^{\alpha}}\right\}, \quad \alpha > 2,$$

is of order one and belongs to the convergence class of that order.

11. If $f_1(z)$ and $f_2(z)$ are entire functions of order ρ_1 and ρ_2 respectively, examine the orders of $f_1(z) + f_2(z)$ and $f_1(z)f_2(z)$. Pay special attention to the case $\rho_1 = \rho_2$.

12. If $f(z)$ is of order ρ, so is $f'(z)$. Prove! (*Hint:* An argument based on Cauchy's formula for $f'(z)$ shows that $\rho(f') \leq \rho(f)$ if $\rho(f)$ is finite, and this is trivially true if $\rho(f) = \infty$. Since $f(z) = f(0) + \int_0^z f'(t) \, dt$, the converse inequality follows.)

13. The differential equation $w'' - zw = 0$ has a solution $f(z)$ with $f(0) = 1$, $f'(0) = 0$. Show that $f(z)$ is an entire function of order $\frac{3}{2}$ and type $\frac{2}{3}$. The other solutions?

14.2. Entire functions of finite order. We plan to study the properties of entire functions of finite order in the light of the concepts developed in the preceding section. The main task is to prove Hadamard's factorization theorem, but many other problems will be broached.

We start with some extensions of Cauchy's integral where we integrate the real or the imaginary part of an analytic function instead of the function itself. Suppose that

$$(14.2.1) \qquad\qquad F(z) = U(z) + iV(z)$$

is holomorphic in $|z| < R$ and continuous in $|z| \leq R$ and, further, that

$$(14.2.2) \qquad\qquad F(z) = \sum_{n=0}^{\infty} c_n z^n.$$

By Cauchy's formulas we then have

$$c_n R^n = \frac{1}{2\pi} \int_0^{2\pi} F(Re^{i\varphi}) e^{-ni\varphi}\, d\varphi, \quad n = 0, 1, 2, \cdots.$$

We also have

$$0 = \frac{1}{2\pi} \int_0^{2\pi} F(Re^{i\varphi}) e^{ni\varphi}\, d\varphi, \quad n = 1, 2, 3, \cdots,$$

since $F(z)$ is holomorphic. In the second set of formulas we replace the integrands by their complex conjugates and add the result to the corresponding formula of the first set. Since $F + \bar{F} = 2U$, we get

$$(14.2.3) \qquad c_n R^n = \frac{1}{\pi} \int_0^{2\pi} U(Re^{i\varphi}) e^{-ni\varphi}\, d\varphi, \quad n = 1, 2, 3, \cdots.$$

For $n = 0$ the value of the integral is $c_0 + \bar{c_0}$. Similarly we get

$$(14.2.4) \qquad c_n R^n = \frac{i}{\pi} \int_0^{2\pi} V(Re^{i\varphi}) e^{-ni\varphi}\, d\varphi, \quad n = 1, 2, 3, \cdots,$$

and for $n = 0$ the integral equals $c_0 - \bar{c_0}$.

Let $z = re^{i\theta}$, $r < R$. We multiply by $(rR^{-1}e^{i\theta})^n$ and add the results for $n = 0, 1, 2, \cdots$. On the left we get $F(z) + \overline{F(0)}$ in the first case and similarly $F(z) - \overline{F(0)}$ in the second. On the right we have a geometric series which turns out to be the Cauchy kernel. Thus

$$(14.2.5) \qquad
\begin{aligned}
F(z) &= -\overline{F(0)} + \frac{1}{\pi i} \int_{|t|=R} \frac{U(t)\, dt}{t - z}, \\
F(z) &= \overline{F(0)} + \frac{1}{\pi} \int_{|t|=R} \frac{V(t)\, dt}{t - z}.
\end{aligned}$$

We can use these formulas just as we use the Cauchy formulas. Below we shall differentiate under the sign of integration to get formulas for the derivatives of $F(z)$. For the moment we are more interested in the maximum modulus. We define

$$(14.2.6) \quad \begin{aligned} A(r; F) &= \max_{0 \le \varphi < 2\pi} U(re^{i\varphi}), \quad a(r; F) = \min_{0 \le \varphi < 2\pi} U(re^{i\varphi}), \\ B(r; F) &= \max_{0 \le \varphi < 2\pi} V(re^{i\varphi}), \quad b(r; F) = \min_{0 \le \varphi < 2\pi} V(re^{i\varphi}). \end{aligned}$$

LEMMA 14.2.1.   *The functions $A(r; F)$, $-a(r; F)$, $B(r; F)$, and $-b(r; F)$ are increasing functions of r unless $F(z)$ is a constant.*

Proof.   The function $\exp[F(z)]$ has $\exp[A(r; F)]$ as its maximum modulus. The latter function is increasing unless $F(z)$ is a constant, so the same must hold for $A(r; F)$. The other functions are handled in the same way.

We can now derive analogues of the Cauchy estimates. We have

$$|c_n| r^n \le \frac{1}{\pi} \int_0^{2\pi} |U(re^{i\varphi})| \, d\varphi, \quad 2\Re(c_0) = \frac{1}{\pi} \int_0^{2\pi} U(re^{i\varphi}) \, d\varphi,$$

so that

$$|c_n| r^n + 2\Re(c_0) \le \frac{1}{\pi} \int_0^{2\pi} [|U(re^{i\varphi})| + U(re^{i\varphi})] \, d\varphi = \frac{2}{\pi} \int_0^{2\pi} U^+(re^{i\varphi}) \, d\varphi,$$

$$|c_n| r^n - 2\Re(c_0) \le \frac{1}{\pi} \int_0^{2\pi} [|U(re^{i\varphi})| - U(re^{i\varphi})] \, d\varphi = \frac{2}{\pi} \int_0^{2\pi} U^-(re^{i\varphi}) \, d\varphi.$$

Here and in the following we shall often write

$$a^+ = \max(0, a), \quad a^- = \max(0, -a).$$

The above inequalities then give

$$(14.2.7) \qquad |c_n| r^n + 2\Re(c_0) \le 4A^+(r; F),$$

$$(14.2.8) \qquad |c_n| r^n - 2\Re(c_0) \le 4a^-(r; F).$$

We obtain similar inequalities by using the imaginary parts instead.

From these inequalities we can get the analogue of the theorem of Liouville.

THEOREM 14.2.1.   *If $f(z)$ is an entire function and if one of the four functions $A(r; f)$, $-a(r; f)$, $B(r; f)$, and $-b(r; f)$ is dominated by an expression of the form $M_1 + M_2 r^\alpha$ for a sequence of values of r tending to infinity where M_1, M_2, α are nonnegative, then $f(z)$ is a polynomial of degree $\le \alpha$.*

Proof.   The assumptions imply that an inequality of the form

$$|c_n| r^n - 2|c_0| \le M_1 + M_2 r^\alpha$$

holds for every n and a sequence of values of r tending to infinity. Such an inequality cannot hold for $n > \alpha$ unless $c_n = 0$. Thus $f(z)$ is a polynomial of degree $\le \alpha$.

It is clear that the functions $M(r; f)$, $A(r; f)$, and so on are interrelated. The following theorem due to É. Borel throws some light on this question:

THEOREM 14.2.2.   *If $F(z)$ is holomorphic for $|z| \leq R$, then for $0 < r < R$ we have*

(14.2.9) $M(r; F) \leq |F(0)| + 2R(R - r)^{-1} \max [A(R; F), -a(R; F)].$

Proof.   We use the first half of formula (14.2.5), which gives

$$|F(z)| \leq |F(0)| + \frac{1}{\pi} \int_{|t|=R} \frac{|U(t)|}{|t-z|} |dt|$$
$$= |F(0)| + \frac{1}{\pi} \int_{|t|=R} [U^+(t) + U^-(t)] \frac{|dt|}{|t-z|}.$$

Here

$$U^+(t) + U^-(t) \leq \max [A(R; F), -a(R; F)].$$

The integral

$$\frac{1}{\pi} \int_{|t|=R} \frac{|dt|}{|t-z|}$$

can be worked out and is found to be

$$2F\left(\tfrac{1}{2}, \tfrac{1}{2}, 1; \left(\frac{r}{R}\right)^2\right),$$

which becomes logarithmically infinite as $r \to R-$. For our purposes it suffices to note that the denominator is $\geq R - r$, so that the integral does not exceed $2R/(R - r)$, whence the theorem follows. In (14.2.9), A and a may be replaced by B and b respectively.

In the opposite direction we observe that since

$$|F|^2 = U^2 + V^2,$$

we have

(14.2.10) $M(r; F) \geq \max [A(r; F), -a(r; F), B(r; F), -b(r; F)].$

We obtain formulas for the derivatives of $F(z)$ by formal differentiation of (14.2.5) under the sign of integration. If we apply the estimates given above to these integrals instead, we obtain

THEOREM 14.2.3.   *For any z with $|z| = r$, $0 < r < R$, and any k*

(14.2.11) $|F^{(k)}(z)| \leq 2k! R(R - r)^{-k-1} \text{Max} [A(R; F), -a(R; F)].$

We shall now turn to Hadamard's factorization theorem for entire functions. We recall that Weierstrass's factorization theorem states that every entire function can be factored:

$$f(z) = e^{g(z)} z^k \prod_{n=1}^{\infty} E\left(\frac{z}{z_n}, p_n\right).$$

Here $g(z)$ is an entire function, k a non-negative integer, and the Weierstrass prime factors $E(z, p)$ are defined by

$$E(z, 0) = 1 - z, \quad E(z, p) = (1 - z) \exp\left\{\frac{z}{1} + \frac{z^2}{2} + \cdots + \frac{z^p}{p}\right\}, \quad p > 0.$$

Finally, the p_n's are a sequence of non-negative integers. Hadamard's theorem states that if $f(z)$ is of finite order, then $g(z)$ is a polynomial and p_n can be chosen as a fixed integer p. To prove this theorem we shall first prove a couple of results concerning canonical products.

THEOREM 14.2.4. *Let $\{z_n\}$ be a sequence of complex numbers such that $z_n \neq 0$, $|z_n| = r_n$, $r_n \leq r_{n+1}$, and the series*

$$(14.2.12) \qquad \sum_{n=1}^{\infty} r_n^{-\alpha}$$

has a finite exponent of convergence σ. If σ is not an integer, let $p = [\sigma]$, the integral part of σ. If σ is an integer, we take $p = \sigma - 1$ or σ according as the series converges or does not converge for $\alpha = \sigma$. Let

$$(14.2.13) \qquad P(z) = \prod_{n=1}^{\infty} E\left(\frac{z}{z_n}, p\right).$$

Then $P(z)$ is an entire function of order σ.

Proof. Since the series converges for $\alpha = p + 1$ by our choice of p, we know that the infinite product converges and defines an entire function. By Theorem 14.1.5, $\rho(P) \geq \sigma$. It suffices then to prove the opposite inequality. We recall Theorem 8.7.2, according to which

$$E(z, p) = 1 - z^{p+1} \sum_{k=0}^{\infty} a_{k, p} z^k,$$

where $a_{k, p} \geq 0$ and $\sum_{k=0}^{\infty} a_{k, p} = 1$. This shows that

$$(14.2.14) \qquad |E(z, p) - 1| \leq |z|^{p+1}, \quad |z| \leq 1,$$

and

$$(14.2.15) \qquad \log |E(z, p)| < \log(1 + |z|^{p+1}) < |z|^{p+1}, \quad |z| \leq 1,$$

since $\log(1 + u) < u$ for $0 < u$. If $p = 0$, these estimates hold for all values of z. For $1 < |z|$, $0 < p$, we have

$$\log E(z, p) = \log(1 - z) + \Sigma_1^p \frac{z^k}{k}.$$

Hence

$$\log |E(z, p)| < \log(1 + |z|) + \sum_{k=1}^{p} \frac{|z|^k}{k}$$

$$< |z| + \left\{\Sigma_1^p \frac{1}{k}\right\} |z|^p < \left\{1 + \Sigma_1^p \frac{1}{k}\right\} |z|^p,$$

so that

$$(14.2.16) \qquad \log |E(z, p)| < (2 + \log p) |z|^p, \quad 1 < |z|, 0 < p.$$

If $p > 0$ we then have

$$\log |P(z)| = \{\Sigma' + \Sigma''\} \log \left| E\left(\frac{z}{z_n}, p\right) \right|,$$

where in the first sum $r_n < r = |z|$ and in the second $r \leq r_n$. We use (14.2.16) for terms in the first sum, (14.2.15) for terms in the second. Writing the resulting sums as Stieltjes integrals, we get

$$(14.2.17) \quad \log |P(z)| \leq (2 + \log p)\, r^p \int_0^r s^{-p}\, dn(s) + r^{p+1} \int_r^\infty s^{-p-1}\, dn(s).$$

Integration by parts gives

$$\log |P(z)| \leq (1 + \log p)\, n(r) + p(2 + \log p)\, r^p \int_0^r s^{-p-1} n(s)\, ds$$

$$(14.2.18)$$

$$+ (p+1) r^{p+1} \int_r^\infty s^{-p-2} n(s)\, ds.$$

The choice of p guarantees the convergence of the integrals over the infinite range. If the series (14.2.12) converges for $\alpha = \sigma$, we know that $n(s) = o(s^\sigma)$. It is then a simple matter to verify that each of the three terms on the right is $o(r^\sigma)$, so that

$$(14.2.19) \quad\quad\quad \log |P(z)| = o(r^\sigma), \quad |z| = r.$$

This estimate applies also if $\sigma = p + 1$, for then we have convergence of (14.2.12) for $\alpha = p + 1$, and the second integral in (14.2.18) tends to zero as $r \to \infty$ and the other terms are $o(r^{p+1})$.

In the remaining cases we have $p \leq \sigma < p + 1$ and

$$n(s) < M(\varepsilon)\, s^{\sigma + \varepsilon}$$

for every $\varepsilon > 0$. Substituting this estimate in (14.2.18), we get

$$(14.2.20) \quad\quad\quad \log |P(z)| < M(\varepsilon, p)\, r^{\sigma + \varepsilon}.$$

These estimates show that $\rho(P) \leq \sigma$. Since we already know that $\rho(P) \geq \sigma$, the theorem is proved for the case $p > 0$.

For $p = 0$, we use the first half of (14.2.15) for $1 < |z|$ and the second half for $|z| \leq 1$. This gives

$$(14.2.21) \quad\quad \log |P(z)| < \int_0^r \log\left(1 + \frac{r}{s}\right) dn(s) + r \int_r^\infty s^{-1}\, dn(s)$$

or, after integration by parts,

$$(14.2.22) \quad \log |P(z)| < (\log 2 - 1)\, n(r) + r \int_0^r \frac{n(s)\, ds}{s(r + s)} + r \int_r^\infty s^{-2} n(s)\, ds.$$

The second term on the right does not exceed $N(r)$ since $r < r + s$. The rest of the argument proceeds as in the case $p > 0$ and we are led to (14.2.19) or (14.2.20) according as the series (14.2.12) converges or diverges for $\alpha = \sigma$. This completes the proof.

CORROLLARY. $P(z)$ *is of minimal type of the order* σ *if the series* (14.2.12) *converges for* $\alpha = \sigma$.

Our next task is to obtain estimates for $\log | P(z) |$ from below. It is clear that we must keep away from the zeros, but it turns out that if we exempt certain small neighborhoods of the zeros, $\log | P(z) |^{-1}$ will also satisfy (14.2.20).

THEOREM 14.2.5.   *Let k be a fixed positive real number and omit from the complex plane all the' disks:*

$$(14.2.23) \qquad | z - z_n | < | z_n |^{-k}, \quad n = 1, 2, 3, \cdots.$$

In the remaining closed set we have for every $\varepsilon > 0$

$$(14.2.24) \qquad \log | P(z) | > -M(p, \varepsilon) | z |^{\sigma + \varepsilon}.$$

Proof.   We need some more inequalities for $\log | E(z, p) |$. From (14.2.14) we get

$$(14.2.25) \qquad \log | E(z, p) | > \log (1 - | z |^{p+1}), \quad | z | < 1.$$

Actually, in using this inequality we shall restrict $| z |$ still further. Let

$$(14.2.26) \qquad \beta = 2^{-1/(p+1)}.$$

We need also an inequality which is valid for $| z | > \beta$. We have obviously

$$\log | E(z, p) | > \log | 1 - z | - \sum_{k=1}^{p} \frac{| z |^k}{k}.$$

Now, for $| z | > \beta$ we have $| z |^k \leq \beta^{k-p} | z |^p$, so that

$$\sum_{k=1}^{p} \frac{| z |^k}{k} < \left(\sum_{k=1}^{p} \frac{1}{k} \beta^{k-p} \right) | z |^p$$

$$< \beta^{-p} \left(\sum_{k=1}^{p} \frac{1}{k} \right) | z |^p < 2(1 + \log p) | z |^p.$$

Hence for $p > 0$

$$(14.2.27) \quad \log | E(z, p) | > \log | 1 - z | - 2(1 + \log p) | z |^p, \quad | z | > \beta.$$

Let us take $p > 0$ and consider

$$\log | P(z) | = \{\Sigma' + \Sigma''\} \log \left| E\left(\frac{z}{z_n}, p\right) \right|,$$

where in the first sum n is such that $\beta r_n < r$ and the second sum takes in the rest. We now use (14.2.27) or (14.2.25) according as n belongs to the first or the second sum. We set

$$(14.2.28) \qquad R = \beta^{-1} r = 2^{1/(p+1)} r$$

and obtain

$$\log | P(z) | > \Sigma' \log \left| 1 - \frac{z}{z_n} \right| - 2(1 + \log p) r^p \int_0^R s^{-p} \, dn(s)$$

$$+ \int_R^\infty \log \left\{ 1 - \left(\frac{r}{s}\right)^{p+1} \right\} dn(s).$$

We integrate by parts and obtain

$$\log |P(z)| > \Sigma' \log \left| 1 - \frac{z}{z_n} \right| - (2 - \log 2 + 2 \log p) \, n(R)$$

$$- 2p(1 + \log p) \, r^p \int_0^R s^{-p-1} n(s) \, ds$$

$$- (p+1) r^{p+1} \int_R^\infty \frac{n(s) \, ds}{s(s^{p+1} - r^{p+1})}.$$

Our choice of β implies that $s^{p+1} - r^{p+1} > \frac{1}{2} s^{p+1}$ for $s > R$. We use this and replace the powers of r before the integrals by the corresponding powers of R. This gives the basic inequality

(14.2.29) $$B(r, p) > \log |P(z)| > L(z, R) - B(R, p)$$

where

(14.2.30)
$$B(r, p) = 2(1 + \log p) \, n(r) + 2p(1 + \log p) \, r^p \int_0^r s^{-p-1} n(s) \, ds$$
$$+ 2(p+1) \, r^{p+1} \int_r^\infty s^{-p-2} n(s) \, ds,$$

(14.2.31) $$L(z, R) = \Sigma' \log \left| 1 - \frac{z}{z_n} \right|,$$

and, we repeat, the summation extends over those values of n for which $r_n < R$. It is clear that the first half of the inequality (14.2.29) is implied by (14.2.18), which gives a slightly smaller upper bound.

The inequality (14.2.29) is also valid for $p = 0$. A suitable choice for $B(r, 0)$ is

(14.2.32) $$B(r, 0) = N(r) + 2r \int_r^\infty s^{-2} n(s) \, ds.$$

This will do as an upper bound by virtue of (14.2.22), even without the factor 2. For the lower bound, on the other hand, we can do without the term $N(R)$, but we need the factor 2 instead.

We know how the function $B(r, p)$, $p \geq 0$, behaves as $r \to \infty$. It satisfies an inequality of the form

(14.2.33) $$B(r, p) < M_0(\varepsilon, p) \, r^{\sigma + \varepsilon}$$

for every $\varepsilon > 0$. It remains to show that a similar estimate holds for $-L(z, R)$ outside of the disks (14.2.23). This is elementary. We have

$$-L(z, R) < (1 + k) \Sigma' \log r_n = (1 + k) \int_0^R \log s \, dn(s),$$

so that

(14.2.34) $$-L(z, R) < (1 + k)[n(R) \log R - N(R)].$$

If $k > -1$, the right-hand side is positive but is dominated by an expression of the form $M(\varepsilon)r^{\sigma+\varepsilon}$. Combining the last two estimates we see that (14.2.24) holds outside the disks (14.2.23).

On the face of it, the bound for $-L(z, R)$ looks very crude, and, in a way, so it is. However, if the zeros are distributed in clusters of ever increasing multiplicities, then it will be possible for "nearly all" the factors $|z - z_n|$ in $L(z, R)$ to be equal to r_n^{-k}, which is the underlying assumption for (14.2.34).

The most interesting case in connection with Theorem 14.2.5 is that in which $k > \sigma$, for then the sum of the diameters of the excluded circles forms a convergent series. Thus we can find a sequence $\{R_n\}$, $0 < R_n < R_{n+1} \to \infty$, such that (14.2.24) holds everywhere on the circle $|z| = R_n$.

We are now ready to prove the factorization theorem of Hadamard.

THEOREM 14.2.6.   *If $f(z)$ is an entire function of finite order ρ, then*

$$(14.2.35) \qquad f(z) = e^{g(z)}z^k P(z),$$

where $g(z)$ is a polynomial of degree $q \leq \rho$, k is a non-negative integer, and $P(z)$ is a canonical product of order $\sigma \leq \rho$. If ρ is not an integer, then $\sigma = \rho$ and $q \leq [\rho]$. If ρ is an integer, then at least one of the quantities q and σ equals ρ.

Proof. By Theorem 14.1.5 the exponent of convergence of the zeros of $f(z)$ satisfies $\sigma \leq \rho$. We can then form the corresponding canonical product

$$P(z) = \prod_{n=1}^{\infty} E\left(\frac{z}{z_n}, p\right),$$

where $p \leq \sigma \leq p + 1$. It is an entire function of order σ and it satisfies (14.2.24) outside of the circles (14.2.23). In particular, it satisfies such an inequality on a system of circles $|z| = R_n$ where $R_n \to \infty$. By the factorization theorem of Weierstrass we then have (14.2.35), but it remains to prove that $g(z)$ is actually a polynomial. We have

$$\Re[g(z)] = \log |f(z)| - \log |P(z)| - k \log |z|$$

$$< M_1(\varepsilon)r^{\rho+\varepsilon} + M_2(\varepsilon)r^{\sigma+\varepsilon}, \quad r = R_n.$$

By Theorem 14.2.1 such inequalities imply that $g(z)$ is a polynomial of degree $q \leq \rho$. The remaining assertions are fairly obvious. If ρ is not an integer, then $q < \rho$ and $\rho = \max(q, \sigma)$ implies that $\sigma = \rho$. On the other hand, if ρ is an integer, then we may have $q = \rho$ and $\sigma < \rho$, but if $q < \rho$ we have again $\sigma = \rho$. This completes the proof.

The larger of the two integers p and q is known as the *genus* of $f(z)$. This concept played an important role in the early literature on entire functions, but it is nowadays largely replaced by the more fundamental classification according to order and type. Finally we note the

COROLLARY.   *If $f(z)$ is of minimal type and of integral order, then $q \leq \rho - 1$.*

EXERCISE 14.2

1. Take q such that $0 < q < 1$ and let

$$f(z) = \prod_{n=0}^{\infty} (1 - q^n z).$$

Determine $n(r)$ and show that

$$\log M(r; f) < \frac{1}{2}\left(\log \frac{1}{q}\right)^{-1}[\log^2 (er) + 1].$$

2. Consider the canonical product with zeros at $z = n!$ $(n > 0)$. Show that $n(r) \sim \log r/(\log \log r)$ and estimate $\log M(r; f)$.

3. Determine $n(r)$ and $\log M(r; f)$ for the canonical product with zeros at $z = \exp n^{\alpha}$ $(0 \le n, \alpha > 1$ and fixed).

4. Construct a canonical product of order $\frac{1}{2}$ and minimal type.

5. Determine $n(r)$ for the function $1/\Gamma(z)$ and estimate $\log M(r; f)$. Improve the estimate with the aid of the expression for $\Gamma(z)\Gamma(1 - z)$ and Stirling's formula.

6. If $f(z)$ is a canonical product of genus p and if C_n: $|z| = R_n$ is a system of circles on which there are no zeros of $f(z)$ and such that $R_n \to \infty$, prove that

$$\lim_{n \to \infty} \frac{1}{2\pi i} \int_{C_n} z^{-p-1} \frac{f'(z)}{f(z)} \, dz$$

exists and find it.

7. If $f(z)$ is a canonical product of genus p, prove that

$$\frac{f'(z)}{f(z)} = \sum_{n=1}^{\infty} \frac{z^p}{z_n{}^p(z - z_n)}.$$

8. If $f(z) = i[C - \log (1 - z)]$, $C > 0$, then $|f(0)| = C$. Show that, for $r < 1$, $A(r; f)$ and $-a(r; f)$ are less than $\frac{1}{2}\pi$ while $M(r; f) \ge C - \log (1 - r)$. This example (due to E. C. Titchmarsh) shows that the estimate given in the proof of Theorem 14.2.2, namely,

$$M(r; f) \le |f(0)| + 2F\left(\frac{1}{2}, \frac{1}{2}, 1; \left(\frac{r}{R}\right)^2\right) \max [A(r; f), -a(r; f)],$$

is practically the best of its kind for

$$F(\tfrac{1}{2}, \tfrac{1}{2}, 1; x) = \frac{1}{\pi} \log \frac{1}{1 - x} + O(1) \quad \text{as} \quad x \to 1-.$$

9. If $|z| = r < R$, prove that

$$\frac{1}{\pi} \int_{|t|=R} |t - z|^{-k-1} |dt| = 2F\left[\tfrac{1}{2}(k + 1), \tfrac{1}{2}(k + 1), 1; \left(\frac{r}{R}\right)^2\right],$$

and reformulate Theorem 14.2.3 with the resulting sharper estimate.

14.3. Functions with real zeros.   For a real differentiable function of a real variable Rolle's theorem states that the derivative vanishes at least once between any two consecutive zeros of the function. This theorem of course applies to entire functions which are real on the real axis. In the case of polynomials Rolle's theorem implies that if all zeros of $f(z)$ are real, the same holds for $f'(z)$. As is shown by the example $f(z) = (z + 1)e^{z^2}$, such a conclusion does not hold for arbitrary entire functions. For polynomials with real zeros the roots of $f(z) = 0$ and of $f'(z) = 0$ separate each other. The example

$$f(z) = (1 - z^2)e^{5z^2 + z}$$

shows that this result may also break down. Here all three zeros of $f'(z)$ lie in $(-1, 1)$. There are, however, important classes of entire functions for which both corollaries of Rolle's theorem are valid.

THEOREM 14.3.1.   *Let $f(z)$ be a canonical product of genus p whose zeros are real. Then the zeros of $f'(z)$ are also real. The point $z = 0$ is a p-fold zero of $f'(z)$. All other zeros of $f'(z)$, multiple zeros of $f(z)$ excepted, are simple zeros, and there is one and only one zero of $f'(z)$ between any two consecutive zeros of $f(z)$ except for the interval containing $z = 0$.*

Proof.   Let the zeros of $f(z)$ be denoted by $\{x_n\}$ and consider the expansion given in Problem 7 of the preceding Exercise:

$$(14.3.1) \qquad F(z) \equiv z^{-p}\frac{f'(z)}{f(z)} = \sum_{n=1}^{\infty} x_n^{-p}(z - x_n)^{-1}.$$

Suppose that $f'(z) = 0$ for $z = c = a + ib$ where $b \neq 0$. Separating reals and imaginaries in $F(a + bi)$, we get

$$a \sum_{n=1}^{\infty} x_n^{-p} |c - x_n|^{-2} - \sum_{n=1}^{\infty} x_n^{-p+1} |c - x_n|^{-2} = 0,$$

$$b \sum_{n=1}^{\infty} x_n^{-p} |c - x_n|^{-2} = 0.$$

We see right away that the multiplier of b cannot be zero if either p is even or all the x_n's have the same sign. This would contradict the assumption that $b \neq 0$. But if p is odd and there are positive as well as negative zeros, then for $b \neq 0$ the equations imply that

$$\sum_{n=1}^{\infty} x_n^{-p+1} |c - x_n|^{-2} = 0,$$

and this is impossible since $p - 1$ is even. This proves that the zeros of $f'(z)$ are real. That $z = 0$ is a p-fold zero of $f'(z)$ follows from (14.3.1). Again, if p is even or if all the x_n's have the same sign, then $F'(x)$ keeps a constant sign when x is real, so that $F(x)$ is monotonic between consecutive zeros of $f(x)$. This implies that each such interval, provided it does not include the origin, contains one and only one zero of $f'(x)$. Therefore, if α and β were distinct zeros of $f'(z)$ between a pair of consecutive zeros of $f(z)$, p would have to be odd and there would have to be both positive and negative x_n's. But then we have

$$F(\alpha) + F(\beta) = (\alpha + \beta) \sum_{n=1}^{\infty} x_n^{-p}(\alpha - x_n)^{-1}(\beta - x_n)^{-1}$$

$$- 2 \sum_{n=1}^{\infty} x_n^{-p+1}(\alpha - x_n)^{-1}(\beta - x_n)^{-1} = 0,$$

$$F(\alpha) - F(\beta) = (\beta - \alpha) \sum_{n=1}^{\infty} x_n^{-p}(\alpha - x_n)^{-1}(\beta - x_n)^{-1} = 0,$$

so that

$$\sum_{n=1}^{\infty} x_n^{-p+1}(\alpha - x_n)^{-1}(\beta - x_n)^{-1} = 0.$$

This is impossible since $p - 1$ is even and $(\alpha - x_n)(\beta - x_n) > 0$ for all n. Hence, there can be at most one zero of $f'(z)$ in such an interval not containing the origin. The same type of argument excludes multiple zeros of $f'(z)$ except at $z = 0$ and at multiple zeros of $f(z)$.

The presence of an outside exponential factor usually complicates matters, but Theorem 14.3.1 holds, except for the reference to $z = 0$, for the following classes of entire functions:

(14.3.2) $$e^{bz}P(z),$$

(14.3.3) $$e^{-az^2 + bz}N(z).$$

Here b is real, $a > 0$, $P(z)$ and $N(z)$ are canonical products of genus 0 or 1, the zeros of $P(z)$ are real, and those of $N(z)$ are negative. It is a simple matter to verify that the imaginary part of the logarithmic derivative of such a function equals 0 if and only if z is real. Thus the zeros of $f'(z)$ are real. The discussion of the distribution of the zeros of $f'(z)$ on the various intervals requires only the usual methods of the calculus.

This study of the implications of Rolle's theorem for entire functions may be extended in several directions. We may enlarge the range of the zeros or change the operator. Thus we may ask: If the zeros of the entire function $f(z)$ are known to belong to some unbounded convex set, when do the zeros of $f'(z)$ lie in the same set? A couple of such extensions of the theorem of Gauss will be found in Exercise 14.3. In the other direction we may replace the operator d/dz by a more general differential operator and ask whether the operator takes entire functions with real zeros into functions of the same class. We shall devote some attention to the latter problem. The results given below are due to Edmond Laguerre (1834–1886).

In dealing with power series we have found the operator

(14.3.4) $$\vartheta = z \frac{d}{dz}$$

particularly appropriate since it simply multiplies z^n by n. In Section 11.2 we studied the operator $G(\vartheta)$, where $G(w)$ is an entire function which is at most of order one and minimal type. Such an operator is holomorphy preserving. In applying $G(\vartheta)$ to entire functions, we are not restricted to functions $G(w)$ of order one and minimal type. In general we cannot allow the order to exceed one, but $G(w)$ may very well be of normal type. Ordinarily, functions of maximal type do not give admissible operators. Thus

$$\{\Gamma(-\vartheta + \tfrac{1}{2})\}^{-1} e^z$$

is not an entire function. On the other hand, $\{\Gamma(\vartheta)\}^{-1}$ takes any Maclaurin series into an entire function; so, in particular, it applies to entire functions.

We recall that

$$(14.3.5) \qquad G(\vartheta)[f(z)] = \sum_{n=0}^{\infty} c_n G(n) z^n \quad \text{if} \quad f(z) = \sum_{n=0}^{\infty} c_n z^n.$$

We now ask: *If $f(z)$ is an entire function with real zeros only, when will the transform have the same property?* We shall give a partial answer to this question. As a preliminary step we observe that the operator $I - \alpha\vartheta$, $0 < \alpha < 1$, does not preserve the reality of zeros, even in the case of polynomials. For the polynomial $z^{k+2} - z^k$ has only real zeros, but the transform has two conjugate complex zeros if $\dfrac{1}{k+2} < \alpha < \dfrac{1}{k}$, as is easily seen by computation. This means that the entire function $G(w)$ cannot be allowed to have positive zeros. It will be convenient to restrict both $f(z)$ and $G(w)$, and this will be done in the following. The classes considered are not the largest to which the argument applies, but the choice made simplifies the exposition.

DEFINITION 14.3.1.   *An entire function is of class* **A** *if*

$$(14.3.6) \qquad f(z) = e^{bz+c} \prod_{n=1}^{\infty} \left(1 + \frac{z}{p_n}\right);$$

it is of class **B** *if*

$$(14.3.7) \qquad f(z) = e^{az+c} \prod_{n=1}^{\infty} [(1 + \delta_n z) e^{-\delta_n z}].$$

Here a and c are real, $b \geq 0$, $p_n > 0$, $\delta_n > 0$, $\Sigma\, p_n^{-1} < \infty$, and $\Sigma\, \delta_n^2 < \infty$.

THEOREM 14.3.2.   *If $\delta > 0$, if $f(z) \in$ **A**, and if $(I + \delta\vartheta)[f] \equiv f_1(z)$, then $f_1(z) \in$ **A**.*

Proof.   We have

$$Q(z) \equiv \frac{f_1(z)}{f(z)} = 1 + \delta b z + \delta z \sum_{n=1}^{\infty} \frac{1}{z + p_n}.$$

The imaginary part of $Q(z)$ equals

$$\delta y \left[b + \sum_{n=1}^{\infty} p_n \, |z + p_n|^{-2} \right].$$

This is 0 if and only if $y = 0$. Hence all zeros of $Q(z)$ and of $f_1(z)$ are real. Moreover, $Q(x) > 0$ if $x > 0$, so all zeros are negative. Now

$$Q'(x) = \delta b + \delta \sum_{n=1}^{\infty} p_n(x + p_n)^{-2} > 0,$$

so that $Q(x)$ is monotone in each of the intervals

$$\cdots, (-p_{n+1}, -p_n), \cdots, (-p_2, -p_1), (-p_1, 0).$$

Since $Q(x)$ goes from $+\infty$ to $-\infty$ as x passes through a point $-p_n$, we see that each of the intervals in question contains one and only one zero of $Q(x)$, that is, of $f_1(x)$. Denote the zero in $(-p_{n+1}, -p_n)$ by $-\xi_n$, that in $(-p_1, 0)$ by $-\xi_0$. Let σ be the exponent of convergence of the zeros of $f(z)$, so that $0 \leq \sigma \leq 1$. Then σ is also the exponent of convergence of the zeros of $f_1(z)$. Now the order of $f(z)$ equals 1 if $b > 0$ or $\sigma = 1$, otherwise it is $\sigma < 1$. Since the order of $f'(z)$ is the same as that of $f(z)$, we see that the order of $f_1(z)$ is at most equal to that of $f(z)$. But

$$M(r; f_1) = f_1(r) = f(r) + \delta r f'(r) > f(r),$$

whence it follows that the order of $f_1(z)$ is the same as that of $f(z)$. If $b = 0$, then $f(z)$ is at most of minimal type of order one since $\Sigma p_n^{-1} < \infty$. In this case $f_1(z)$ must have the same property. This means that $f_1(z)$ is a positive constant times the canonical product with the zeros $\{-\xi_n\}$. Suppose now that $b > 0$. Then $f(z)$ and $f_1(z)$ are both of order one and normal type. The type of $f(z)$ is b, and a simple calculation shows that that of $f_1(z)$ has the same value. Since $f_1(r) > f(r)$ and $f_1(0) = f(0)$, this gives

$$f_1(z) = e^{bz+c} \prod_{n=0}^{\infty} \left(1 + \frac{z}{\xi_n}\right),$$

that is, $f_1(z) \in \mathbf{A}$ as asserted.

This tells us that if $\delta_1, \delta_2, \cdots, \delta_n$ are positive numbers, then

$$\prod_{k=1}^{n}(1 + \delta_k \vartheta)[f] \in \mathbf{A}$$

when $f \in \mathbf{A}$. Here we can pass to the limit and obtain

THEOREM 14.3.3.   *If $G(w) \in \mathbf{B}$ and if $f(z) \in \mathbf{A}$, then*

(14.3.8) $G(\vartheta)[f] \in \mathbf{A}.$

Proof.   Without restricting the generality, we may suppose that $a = 0$ and $G(0) = 1$. For

$$\exp(a\vartheta)[f](z) = f(e^a z),$$

which belongs to $\mathbf{A}$ if $f(z)$ does, and changing the value of $G(0)$ merely multiplies the transform by a constant. Now set

$$G_m(w) = \prod_{k=1}^{m} \{(1 + \delta_k w)e^{-\delta_k w}\} = \prod_{k=1}^{m}(1 + \delta_k w) \cdot \exp\left[-\sum_{1}^{m}\delta_k w\right].$$

Then

$$F_m(z) \equiv G_m(\vartheta)[f](z) = \prod_{k=1}^{m}(1 + \delta_k\vartheta)f\left[z\,\exp\left(-\sum_1^m \delta_k\right)\right]$$

$$= \sum_{n=0}^{\infty} c_n G_m(n)z^n$$

where

$$f(z) = \sum_{n=0}^{\infty} c_n z^n.$$

If $f(z) \in \mathbf{A}$, then $F_m(z) \in \mathbf{A}$ by the preceding theorem. For $x > 0$

$$(1 + x)e^{-x}$$

is < 1 and monotone decreasing to 0 as $x \to \infty$. From this we conclude that

$$1 = G(0) > G(1) > \cdots > G(n) > \cdots,$$

and for each n

$$G_m(n) > G_{m+1}(n) > G(n).$$

Since $f(z) \in \mathbf{A}$, the coefficients c_n are non-negative. It follows that

$$F_m(r) > F_{m+1}(r) > F(r), \quad r > 0.$$

Thus the sequence $\{F_m(r)\}$ is monotone decreasing for each $r > 0$ and converges to a limit. The usual argument shows that this limit is $F(r)$ and that the convergence is uniform with respect to r for $0 \leq r \leq R < \infty$. Since

$$M(r;F_m) = F_m(r), \quad M(r;F) = F(r),$$

this implies that the sequence $\{F_m(z)\}$ converges to $F(z)$ uniformly in z for $|z| \leq R$.

We have to show that this limit function $F \in \mathbf{A}$. For this purpose we need a theorem due to Adolf Hurwitz.

THEOREM 14.3.4.   *Let $\{f_n(z)\}$ be a sequence of functions holomorphic in a domain D. Let $f_n(z)$ converge to $f(z)$ uniformly in D. Let $z = a \in D$ be a zero of $f(z)$. Then for each ε, $0 < \varepsilon < \varepsilon_0$, there exists an N_ε such that $f_n(z)$ for $n > N_\varepsilon$ has the same number of zeros in the disk $|z - a| < \varepsilon$ as $f(z)$ has.*

Proof.   We use the theorem of Rouché (Theorem 9.2.3). On the circle $|z - a| = \varepsilon$ the functions $f(z) - f_n(z)$ converge uniformly to 0. On the other hand, for sufficiently small values of ε, there exist a $\delta > 0$ such that $|f(z)| > \delta$ and an N_ε such that $|f(z) - f_n(z)| < \delta$ for $n > N_\varepsilon$ and $|z - a| = \varepsilon$. Then, since

$$f_n(z) = f(z) - [f(z) - f_n(z)],$$

it follows that the functions $f_n(z)$ and $f(z)$ have the same number of zeros in the disk $|z - a| < \varepsilon$ for $n > N_\varepsilon$.

Having this theorem at our disposal, we can easily complete the proof of Theorem 14.3.3. Since the sequence $\{F_m(z)\}$ converges uniformly to $F(z)$ in each disk $|z| < R$ and since each $F_m(z)$ has real negative zeros only, it follows that all zeros of $F(z)$ must be real and negative. Since the sequence $\{G(n)\}$ is bounded and $G(n) < 1$ for $n > 0$, we see that

$$M(r; F) = F(r) < f(r) = M(r; f).$$

This shows that the order of the entire function $F(z)$ is at most equal to that of $f(z)$, that is, at most 1. If $\rho(f) < 1$, then, a fortiori, $\rho(F) < 1$ and

$$F(z) = c_0 \prod_{n=1}^{\infty} \left(1 + \frac{z}{\xi_n}\right), \quad \Sigma \, \xi_n^{-1} < \infty.$$

If $\rho(f) = \sigma(f) = 1$, then $\rho(F) \leq 1$, $\sigma(F) \leq 1$. Suppose we have equality in both places. Then we must still have $\Sigma \, \xi_n^{-1} < \infty$, for if this series diverged, the canonical product formed with the zeros $\{-\xi_n\}$ would define an entire function of order one and maximal type. This contradicts $M(r; F) < M(r; f)$. Thus we have now

$$F(z) = c_0 e^{bz} \prod_{n=1}^{\infty} \left(1 + \frac{z}{\xi_n}\right), \quad \Sigma \, \xi_n^{-1} < \infty.$$

Here $b \geq 0$ since otherwise we would have $\lim_{r \to \infty} F(r) = 0$, and this is impossible since $F(z)$ is a power series with positive terms. If $\rho(F) = 1$, $\sigma(F) < 1$, there is nothing to change in the representation of $F(z)$. In this case $b > 0$. Thus $F(z) \in \mathbf{A}$ as asserted.

    COROLLARY.   *If $G(w) \in \mathbf{B}$, then*

(14.3.9) $$\sum_{n=0}^{\infty} \frac{G(n)}{n!} z^n \in \mathbf{A}.$$

    For this is the transform of e^z by $G(\vartheta)$ and $e^z \in \mathbf{A}$.

    If the sequence $\{p_n\}$ is sufficiently regular, the asymptotic behavior of the corresponding canonical product is a very simple one. This is shown by the following theorem due to G. Pólya and G. Szegö:

    THEOREM 14.3.5.   *If $f(z)$ is the canonical product with zeros $\{-p_n\}$ and if*

(14.3.10) $$n(r, 0; f) = Cr^\rho[1 + o(1)], \quad 0 < \rho < 1,$$

then

(14.3.11) $\log f(z) = C\pi \, \mathrm{cosec} \, \rho\pi \, z^\rho \, [1 + o(1)], \quad -\pi < \arg z < \pi.$

Here the logarithm and the power are real for z real positive, and the error term tends to zero as $z \to \infty$, uniformly in z in the sector $|\arg z| \leq \pi - \delta$.

Proof.   Suppose that z is not real negative. Giving the logarithm on the right its principal value, we have for $-\pi < \arg z < \pi$

$$\log f(z) = \int_0^\infty \log \left(1 + \frac{z}{s}\right) dn(s)$$

$$= z \int_0^\infty \frac{n(s)\, ds}{s(s + z)},$$

where the integration by parts is easily justified. Hence

$$\log f(z) = Cz \int_0^\infty \frac{s^{\rho-1}\, ds}{s + z} + z \int_0^\infty \frac{s^{\rho-1}\varepsilon(s)\, ds}{s + z}.$$

The first integral we can evaluate by contour integration. We turn the line of integration from the positive real axis so that it goes through the point z. Using a result on page 250 of Volume I, we get for the leading term

$$Cz^\rho \int_0^\infty \frac{t^{\rho-1}\, dt}{t + 1} = C\pi \csc \rho\pi \, z^\rho.$$

The absolute value of the error term does not exceed

$$r^\rho \int_0^\infty \frac{|\varepsilon(rt)|\, t^{\rho-1}}{|t + e^{i\theta}|} \, dt, \quad z = re^{i\theta}.$$

Here

$$|t + e^{i\theta}| > \cos \tfrac{1}{2}\theta \, (1 + t^2)^{\frac{1}{2}},$$

so that the error does not exceed

$$\sec \tfrac{1}{2}\theta \, r^\rho \int_0^\infty \frac{|\varepsilon(rt)|\, t^{\rho-1}}{(1 + t^2)^{1/2}} \, dt,$$

and this integral tends to zero as $r \to \infty$ since $|\varepsilon(rt)|$ is bounded and tends to zero uniformly on any fixed interval $0 < \varepsilon \leq t < \infty$. This completes the proof.

COROLLARY.   *If* $0 < \rho < \tfrac{1}{2}$, *then* $|f(z)| \to \infty$ *uniformly   in   the   sector* $|\arg z| \leq \pi - \delta, 0 < \delta.$

For functions of order $\rho < \tfrac{1}{2}$, it is possible to show that there is a sequence of values r tending to infinity through which the minimum modulus of $f(z)$ tends to infinity. We shall not elaborate on this point here.

In Chapter 11 use was made of entire functions of the form

$$(14.3.12) \qquad\qquad G(w) = \prod_{k=1}^\infty \left(1 - \frac{w^2}{a_k^2}\right),$$

where

$$(14.3.13) \qquad\qquad \frac{a_k}{k} \to \infty, \quad \min (a_{k+1} - a_k) \geq 1.$$

This class of functions was claimed to have properties such as (11.2.27) and (11.7.12). These are special cases of

THEOREM 14.3.6.   *Let $\{p_n\}$ be a sequence of positive numbers such that*

$$(14.3.14) \qquad \frac{p_n}{n^2} \to \infty, \quad p_{n+1} - p_n > c p_n^{\frac{1}{2}}.$$

Let $f(z)$ be the canonical product with the zeros $\{-p_n\}$. Then $f(z)$ is an entire function of order $\leq \frac{1}{2}$ and it is of minimal type if $\rho = \frac{1}{2}$. Let $h > 0$ be fixed and let I_h be the subset of the negative real axis outside the intervals $(-p_k - h, -p_k + h)$, $k = 1, 2, 3, \cdots$. Then

$$(14.3.15) \qquad \lim r^{-\frac{1}{2}} \log |f(-r)| = 0,$$

if $r \to \infty$ through a sequence of values such that $-r \in I_h$.

Proof. By the first part of (14.3.14), $n(r, 0; f) = n(r) = o(r^{\frac{1}{2}})$; the statements about order and type follow from this estimate. Theorem 14.2.5 gives the limit relation if $\rho(f) < \frac{1}{2}$ or if $n(r) = o[r^{\frac{1}{2}}(\log r)^{-1}]$. If these conditions do not hold, an elaborate argument is required. We observe first that for $r \neq p_k$

$$(14.3.16) \quad \log |f(-r)| = \int_0^\infty \log \left| 1 - \frac{r}{s} \right| dn(s) > \int_{r/2}^\infty \log \left| 1 - \frac{r}{s} \right| dn(s)$$

since the logarithm is positive for $0 < s < \frac{1}{2}r$. Here we can neglect the interval $(2r, \infty)$. True, it gives a negative contribution, but

$$-\int_{2r}^\infty \log \left| 1 - \frac{r}{s} \right| dn(s) < r \log 4 \int_{2r}^\infty \frac{dn(s)}{s} < r \log 4 \int_{2r}^\infty \frac{n(s)}{s^2} ds = o(r^{\frac{1}{2}}).$$

Thus we can restrict ourselves to the interval $(\frac{1}{2}r, 2r)$ and we need only a lower bound for $\log |f(-r)|$ since the upper bound is $o(r^{\frac{1}{2}})$.

We observe next that since $n(s) = o(s^{\frac{1}{2}})$ we have

$$n(2r) - n(\tfrac{1}{2}r) = o(r^{\frac{1}{2}}),$$

and this may be sharpened to

$$(14.3.17) \qquad \lim \frac{[n(v) - n(u)]^2}{v - u} = 0$$

for

$$(14.3.18) \qquad u \to \infty, \quad O(u^{\frac{1}{2}}) < v - u < 3u.$$

The second part of (14.3.14) shows that

$$p_{n+m} - p_n > Cm p_n^{\frac{1}{2}}, \quad m \geq 1.$$

We now choose m and n so that

$$p_{n-1} \leq u < p_n < p_{n+m} \leq v < p_{n+m+1}.$$

Hence

$$p_{n+m} - p_n < v - u, \quad n(u) = n - 1, \quad n(v) = n + m, \quad n(v) - n(u) = m + 1,$$

and

(14.3.19) $$\frac{v - u}{[n(v) - n(u)]^2} > \frac{1}{4} C \frac{p_n^{\frac{1}{2}}}{m} > \frac{1}{4} C \frac{u^{\frac{1}{2}}}{m}.$$

But we have seen that $m = n(v) - n(u) - 1 = o(u^{\frac{1}{2}})$ if $v \leq 4u$. It follows that the right member of (14.3.19) becomes infinite with u if (14.3.18) holds, and this implies (14.3.17).

Suppose now that $-r \in I_h$. Then the distance from r to the nearest p_n is at least h. Condition (14.3.14) shows that we can find an interval containing r and of length $\frac{1}{2}Cr^{\frac{1}{2}}$ which contains at most one point p_n. Let this interval be (a, b) where a and b are chosen so that

$$b - a = \tfrac{1}{2}Cr^{\frac{1}{2}}, \quad ab = r^2.$$

We can then write

$$\int_{r/2}^{2r} \log\left| 1 - \frac{r}{s} \right| dn(s) = \left\{ \int_{r/2}^{a} + \int_{b}^{2r} \right\} \log\left| 1 - \frac{r}{s} \right| dn(s) + O(\log r).$$

The last term is present if and only if there is a point p_n in (a, b). The contribution of this point to the value of the integral is

$$\log\frac{| p_n - r |}{p_n} > \log\frac{h}{p_n} > \log\frac{h}{a} = -O(\log r).$$

We can now integrate by parts and obtain five terms:

$$n(a) \log\left| 1 - \frac{r}{a} \right| - n(b) \log\left| 1 - \frac{r}{b} \right| - n(2r) \log 2$$

$$+ r \int_{r/2}^{a} \frac{n(s)\, ds}{s(r - s)} - r \int_{b}^{2r} \frac{n(s)\, ds}{s(s - r)}.$$

The first two terms are the biggest but they practically annihilate each other. If $n(b) = n(a)$, then their difference is

$$n(a) \log\frac{r}{a} = n(a)O(r^{-\frac{1}{2}}) = o(1).$$

If $n(b) = n(a) + 1$ instead, we have to add a term $O(\log r)$. The remaining integrated term $-n(2r) \log 2$ is $o(r^{\frac{1}{2}})$. The two integrals may be joined to a single one over the interval $\left(\frac{1}{2}, \frac{a}{r}\right)$ by setting $s = rt$ in the first integral and $s = r/t$ in the second. The result is

(14.3.20) $$\int_{1/2}^{a/r} \frac{n(rt) - tn(r/t)}{t(1 - t)}\, dt = \int_{1/2}^{a/r} \frac{n(rt) - n(r/t)}{t(1 - t)}\, dt + \int_{1/2}^{a/r} n\left(\frac{r}{t}\right)\frac{dt}{t}.$$

Here we can disregard the second integral in the right member since it is positive and $o(r^{\frac{1}{2}})$. The first integral may be written

$$-r^{\frac{1}{2}} \int_{1/2}^{a/r} \frac{n(r/t) - n(rt)}{(r/t - rt)^{1/2}} \left\{ \frac{1+t}{t^3(1-t)} \right\}^{\frac{1}{2}} dt.$$

Here we can use (14.3.17) with $u = rt$, $v = r/t$ for $r/t < 4rt$ in the interval $(\frac{1}{2}, 1)$, and $r/t - rt$ takes its minimum value $b - a = \frac{1}{2}Cr^{\frac{1}{2}}$ at $t = a/r$. Since the minimum value goes to infinity with r, we see that (14.3.17) applies. It follows that the integral in question is dominated by a constant times

$$r^{\frac{1}{2}} \max_t \frac{n(r/t) - n(rt)}{(r/t - rt)^{1/2}} = o(r^{\frac{1}{2}}).$$

This shows that

$$\log |f(-r)| > -\varepsilon(r)r^{\frac{1}{2}},$$

where $\varepsilon(r) \to 0$ as $r \to \infty$, $-r \in I_h$, and completes the proof of the theorem. Formulas (11.2.27) and (11.7.12) are obviously special instances of this theorem.

EXERCISE 14.3

1. If $f(z)$ is a canonical product of genus zero whose zeros are in the strip $a < y < b$, prove that the zeros of $f'(z)$ lie in the same strip.

2. Same question if the zeros of $f(z)$ lie in a sector of opening $\leq \pi$. (*Hint:* Prove first for a half-plane.)

3. Prove that the zeros of the Bessel function $J_\alpha(z)$, $\alpha > -1$, are real.

4. If $G(w) \in \mathbf{B}$, show that $G(\vartheta)e^{-z^2}$ has only real zeros.

5. If $G(w) \in \mathbf{B}$, show that $G(\vartheta)e^{z^2}$ has only purely imaginary zeros.

6. If $G(w) \in \mathbf{B}$ and k is an integer > 2, where are the zeros of $G(\vartheta)e^{z^k}$?

7. Find the distribution of the zeros of the solution of Problem 11 of Exercise 14.1.

8. The second part of (14.3.14) in the hypotheses for Theorem 14.3.6 may be replaced by the following condition: We have

$$n(\alpha r) \geq \alpha^k n(r), \quad 1 > \alpha \geq \tfrac{1}{2},$$

for some fixed positive integer k, all large values of r, and uniformly with respect to α. Verify! (*Hint:* The new hypothesis applies directly to the discussion of the left member of (14.3.20) with $\alpha = t^2$.)

9. If $f(z)$ satisfies the conditions of Theorem 14.3.6, prove that

$$\lim_{n \to \infty} p_n^{-\frac{1}{2}} \log |f'(-p_n)| = 0.$$

14.4. Characteristic functions. In the preceding sections we have made a study of entire functions based essentially on the two non-decreasing functions $\log M(r;f)$ and $n(r, 0;f)$. The resulting picture is somewhat incomplete inasmuch as we lack information about the distribution of other values than zeros. This gap in the theory could be filled by a study of the functions $n(r, a;f)$. A more serious defect would seem to be the fact that there is no obvious way of carrying over the theory to meromorphic functions. To be sure, the functions $n(r, a;f)$ still make sense, but $\log M(r;f)$ is no longer a suitable majorant for these enumerative functions. Here the decisive new ideas are due to Rolf Nevanlinna (1924), who placed the whole theory of entire and meromorphic functions on a new basis.

The point of departure is Theorem 14.1.4, Jensen's theorem. We define

$$(14.4.1) \qquad m(r, a;f) = \frac{1}{2\pi}\int_0^{2\pi} \log^+ |f(re^{i\theta}) - a|^{-1}\, d\theta, \quad a \neq 0,$$

$$(14.4.2) \qquad m(r, \infty;f) = \frac{1}{2\pi}\int_0^{2\pi} \log^+ |f(re^{i\theta})|\, d\theta,$$

$$(14.4.3) \qquad T(r;f) = m(r, \infty;f) + N(r, \infty;f) + n(0, \infty;f) \log r,$$

where the last term is zero unless $z = 0$ is a pole of $f(z)$. We can now rewrite formula (14.1.25) in several suggestive ways:

$$(14.4.4) \qquad m(r, 0; f) + N(r, 0; f) = T(r; f) - \log |f(0)|,$$

$$(14.4.5) \qquad T\left(r; \frac{1}{f}\right) = T(r;f) - \log |f(0)|.$$

If $f(0) = 0$ or ∞, the term $\log |f(0)|$ has to be replaced by $\log |c|$, where c is the coefficient of the lowest power in z actually present in the Laurent series of $f(z)$ at $z = 0$.

$T(r;f)$ is called the *characteristic function* of $f(z)$ or, more precisely, the *Nevanlinna characteristic* since another characteristic function will be considered later in this section. For $m(r, a;f)$ we shall use the term *proximity function*, which gives the sense of the usual German term *Schmiegungsfunktion*. $T(r;f)$ is the sum of two terms, the proximity function $m(r, \infty;f)$ which measures the proximity of $f(z)$ to ∞ on the circle $|z| = r$, and the enumerative function $N(r, \infty;f)$ which gives a weighted average of the number of infinitudes in the disk $|z| \leq r$. If $f(z)$ is an entire function, the second term is missing and we have

$$(14.4.6) \qquad T(r;f) = m(r, \infty;f) \leq \log M(r;f).$$

Similarly, $m(r, a;f)$ measures the proximity of $f(z)$ to the value $w = a$ on $|z| = r$, and $N(r, a;f)$ gives the weighted average of the number of a-values in the disk $|z| \leq r$.

With another vague but suggestive term we may say that the characteristic function $T(r;f)$ measures the *affinity* of $f(z)$ for the value $w = \infty$ in the disk $|z| \leq r$. Formula (14.4.4) now states that *the affinity of $f(z)$ to the value 0 differs from the affinity to the value ∞ by a fixed constant*. Formula (14.4.5) states that *the characteristic functions of $f(z)$ and $1/[f(z)]$ differ by a constant*. Both statements are capable of generalization, and this shows that the values 0 and ∞ do not play a special role.

For this purpose we replace $f(z)$ by $f(z) - a$, $a \neq 0, \infty$. We have obviously

$$m(r, 0; f - a) = m(r, a; f), \quad N(r, 0; f - a) = N(r, a; f),$$
$$N(r, \infty; f - a) = N(r, \infty; f).$$

In order to estimate the proximity function, we need two properties of the function

$$\log^+ p = \max (\log p, 0), \quad p > 0,$$

which we state without proof:

(14.4.7)
$$\log^+ (p_1 p_2 \cdots p_m) \leq \sum_{k=1}^{m} \log^+ p_k,$$
$$\log^+ \left(\sum_{k=1}^{m} p_k \right) \leq \sum_{k=1}^{m} \log^+ p_k + \log m.$$

The second of these relations gives:

$$m(r, \infty; f - a) \leq m(r, \infty; f) + \log^+ |a| + \log 2.$$

If, finally, we denote the coefficient of the lowest power of z in the Laurent expansion of $f(z) - a$ at the origin by $c(a)$, then the relation

$$m(r, 0; f - a) + N(r, 0; f - a)$$
$$= m(r, \infty; f - a) + N(r, \infty; f - a) - \log |c(a)|$$

implies that

(14.4.8) $\quad m(r, a; f) + N(r, a; f) \leq T(r;f) + \big| \log |c(a)| \big| + \log^+ |a| + \log 2.$

This is known as the *first fundamental theorem* of R. Nevanlinna. In view of its importance we restate it as follows:

THEOREM 14.4.1.   *Let $f(z)$ be meromorphic in $|z| < R \leq \infty$, and let*

$$f(z) - a = z^k \left[c(a) + \sum_{n=1}^{\infty} c_n z^n \right].$$

Then

(14.4.9) $\quad m(r, a; f) + N(r, a; f) = T(r;f) + S(r, a;f), \quad 0 \leq r < R,$

where

$$\big| S(r, a;f) \big| \leq \big| \log |c(a)| \big| + \log^+ |a| + \log 2.$$

Thus, if we measure the affinity of $f(z)$ for the value $w = a$ in $|z| \leq r$ by the left member of (14.4.9), then the theorem asserts that $f(z)$ has essentially the same affinity for all values, in the sense that the difference of any two affinities is a bounded function of r.

We shall see below that if $f(z)$ is meromorphic in the finite plane and if its characteristic function is bounded, then $f(z)$ is necessarily a constant. More generally, if $T(r;\dot f) = O[\log r]$, then $f(z)$ is a rational function.

We note the following inequalities which follow from the definitions and from (14.4.7):

$$
(14.4.10) \quad
\begin{cases}
m(r, \infty; f_1 f_2 \cdots f_m) \leq \sum_{k=1}^{m} m(r, \infty; f_k), \\[2mm]
N(r, \infty; f_1 f_2 \cdots f_m) \leq \sum_{k=1}^{m} N(r, \infty; f_k), \\[2mm]
T(r; f_1 f_2 \cdots f_m) \leq \sum_{k=1}^{m} T(r; f_k),
\end{cases}
$$

$$(14.4.11) \qquad T(r; f_1 + f_2 + \cdots + f_m) \leq \sum_{k=1}^{m} T(r; f_k) + \log m.$$

Further, if k is any constant, $k \neq 0$, then

$$(14.4.12) \qquad |\, T(r; kf) - T(r; f) \,| \leq |\, \log | k | \,|.$$

Similarly, as we saw above,

$$(14.4.13) \qquad |\, T(r; f - a) - T(r; f) \,| \leq \log^+ |\, a \,| + \log 2.$$

If we combine the last two relations with (14.4.5), we see that replacing $f(z)$ by a linear fractional transform

$$\frac{\alpha f(z) + \beta}{\gamma f(z) + \delta}, \quad \alpha\delta - \beta\gamma \neq 0,$$

changes the characteristic function by a bounded function of r.

$N(r, a; f)$ *is a monotone increasing function of* r *and a convex function of* $\log r$. The second property implies that for $0 < r < s$, $0 < \alpha < 1$, we have

$$(14.4.14) \qquad N(r^\alpha s^{1-\alpha}, a; f) \leq \alpha N(r, a; f) + (1 - \alpha) N(s, a; f).$$

This is proved as follows. To simplify the notation we omit a and f. We also write $r^\alpha s^{1-\alpha} = R$. Then

$$
N(R) = \int_0^R \log \frac{R}{t}\, dn(t) = \alpha \int_0^R \log \frac{r}{t}\, dn(t) + (1 - \alpha) \int_0^R \log \frac{s}{t}\, dn(t)
$$

$$
\leq \alpha \int_0^r \log \frac{r}{t}\, dn(t) + (1 - \alpha) \int_0^s \log \frac{s}{t}\, dn(t)
$$

$$
= \alpha N(r) + (1 - \alpha) N(s).
$$

The justification for replacing R by r in the upper limit of the first integral in the third member is that the integrand becomes negative for $t > r$. The rest is obvious. It should be noted that $N(r)$ is not a convex function of r, for in any interval which contains a discontinuity of $n(r, a; f)$ there are arcs of the curve

$$y = N(x, a; f)$$

which are above the chord as well as arcs which are below.

The function $m(r, a; f)$ is rather irregular, so it is somewhat surprising that $T(r; f)$ is an increasing convex function of $\log r$. The following proof of this fact is due to Henri Cartan (1929). It is based on

THEOREM 14.4.2.   *If $f(z)$ is meromorphic in $|z| < R \leq \infty$ and if $f(0) \neq \infty$,* *then*

(14.4.15) $$T(r; f) = \log^+ |f(0)| + \frac{1}{2\pi} \int_0^{2\pi} N(r, e^{i\theta}; f)\, d\theta.$$

This theorem in its turn rests on the following lemma:

LEMMA 14.4.1.   *We have*

(14.4.16) $$\frac{1}{2\pi} \int_{|t|=1} \log |c - t|\, |dt| = \log^+ |c|.$$

Proof of the Lemma. If $|c| > 1$, the function

$$\log |c - z|$$

is harmonic in the unit disk, and by the mean value theorem of Gauss [see formula (8.2.8)] the value of the integral equals the value of the function at the center, that is, $\log |c|$. If $|c| < 1$, the argument applies to the function

$$\log |1 - cz|$$

instead. This function has the mean value 0; on the other hand, the integral coincides with the left member of (14.4.16). This completes the proof of the lemma, since the case $c = 1$ follows by continuity.

Proof of Theorem 14.4.2. We apply Jensen's theorem to the function $f(z) - e^{i\theta}$, θ real, and obtain

$$\frac{1}{2\pi} \int_0^{2\pi} \log |f(re^{i\varphi}) - e^{i\theta}|\, d\varphi + N(r, \infty; f) = N(r, e^{i\theta}; f) + \log |f(0) - e^{i\theta}|.$$

Here we divide by 2π and integrate with respect to θ from 0 to 2π. The second term on the left remains unchanged. In the first double integral we can interchange the order of the two integrations and obtain

$$\frac{1}{2\pi} \int_0^{2\pi} d\varphi \, \frac{1}{2\pi} \int_0^{2\pi} \log |f(re^{i\varphi}) - e^{i\theta}|\, d\theta + N(r, \infty; f)$$

$$= \frac{1}{2\pi} \int_0^{2\pi} N(r, e^{i\theta}; f)\, d\theta + \frac{1}{2\pi} \int_0^{2\pi} \log |f(0) - e^{i\theta}|\, d\theta.$$

We now apply the lemma to the inner integral on the left and to the second integral on the right and obtain

$$\frac{1}{2\pi}\int_0^{2\pi}\log^+ |f(re^{i\varphi})|\,d\varphi + N(r,\infty;f) = \frac{1}{2\pi}\int_0^{2\pi} N(r,e^{i\theta};f)\,d\theta + \log^+ |f(0)|.$$

On the left we have $T(r;f)$; so the theorem is proved. It is possible to permit $f(0)$ to be infinite. We then have simply to replace $\log^+ |f(0)|$ by $\log^+ |c|$ where c is the coefficient of the algebraically lowest power in the Laurent expansion.

In (14.4.15) the function $N(r,e^{i\theta};f)$ is, for each fixed θ, a monotone increasing convex function of $\log r$. It follows that its mean value with respect to θ has the same property; that is, $T(r;f)$ is indeed a monotone increasing convex function of $\log r$ as asserted.

The formula permits an interesting geometric interpretation of the Nevanlinna characteristic. We have

$$\int_0^{2\pi} N(r,e^{i\theta};f)\,d\theta = \int_0^r \frac{dt}{t}\int_0^{2\pi} n(t,e^{i\theta};f)\,d\theta.$$

Now the function

$$w = f(z)$$

maps the disk $|z| \le t$ upon a Riemann surface $\mathfrak{R}_t$ over the w-plane. Here $\mathfrak{R}_t$ covers parts of the w-plane several times, in fact, $n(t,a;f)$ tells us how many times $\mathfrak{R}_t$ covers the point $w = a$. In particular, some arcs of the unit circle are covered by $\mathfrak{R}_t$, and

$$\int_0^{2\pi} n(t,e^{i\theta};f)\,d\theta \equiv L(t)$$

gives the total length of these arcs where each arc is measured as often as it is covered. Thus, we get

$$(14.4.17) \qquad T(r;f) = \log^+ |f(0)| + \frac{1}{2\pi}\int_0^r L(t)\frac{dt}{t},$$

which is the desired geometric interpretation. Since $L(t)$ is a never decreasing function of t, we see once more that $T(r;f)$ is a monotone increasing convex function of $\log r$.

The study of the properties of the surface $\mathfrak{R}_t$ adds to our insight into the properties of $f(z)$. For this purpose it is desirable to use stereographic projection, and we refer the reader to Section 2.5 for this notion. We recall that the chordal distance of two points in the w-plane is

$$\chi(w_1,w_2) = \frac{|w_1 - w_2|}{(1 + |w_1|^2)^{1/2}(1 + |w_2|^2)^{1/2}}.$$

This formula tells us that if a figure in the plane is projected on the sphere, then the linear magnification at the point w equals $(1 + |w|^2)^{-1}$ while the areal magnification is $(1 + |w|^2)^{-2}$.

We denote the stereographic projection of $\Re_t$ by $\mathfrak{S}_t$ and observe that its area is given by

$$(14.4.18) \qquad \pi A(t;f) = \int_{|z|<t} \frac{|f'(z)|^2 \, d\omega}{[1 + |f(z)|^2]^2},$$

where $d\omega$ is the surface element in the z-plane. We recall that the factor $|f'(z)|^2$ gives the areal magnification under the conformal mapping defined by $w = f(z)$.

For work on the sphere it is more natural to define the proximity function by

$$(14.4.19) \qquad m°(r, a; f) = \frac{1}{2\pi}\int_0^{2\pi} \log \frac{1}{\chi[f(re^{i\theta}), a]} \, d\theta - \log \frac{1}{\chi[f(0), a]},$$

where we assume that $f(0) \neq a$. A simple calculation shows that if we have also $f(0) \neq b$, then

$$m°(r, a; f) - m°(r, b; f) = \frac{1}{2\pi}\int_0^{2\pi} \log\left|\frac{f(re^{i\theta}) - b}{f(re^{i\theta}) - a}\right| \, d\theta - \log\left|\frac{f(0) - b}{f(0) - a}\right|.$$

This relation we differentiate with respect to r, under the sign of integration on the right. Now

$$\log \frac{w - b}{w - a} = \log\left|\frac{w - b}{w - a}\right| + i \arg \frac{w - b}{w - a} \equiv U + iV,$$

and

$$r\frac{\partial U}{\partial r} = \frac{\partial V}{\partial \theta}$$

by the Cauchy-Riemann equations. Hence

$$r\frac{d}{dr}[m°(r, a; f) - m°(r, b; f)] = \frac{1}{2\pi}\int_0^{2\pi} d_\theta \arg\frac{f(re^{i\theta}) - b}{f(re^{i\theta}) - a}$$

$$= n(r, b; f) - n(r, a; f)$$

by the principle of the argument. It follows that

$$\frac{d}{dr} m°(r, a; f) + \frac{1}{r} n(r, a; f) = \frac{d}{dr} m°(r, b; f) + \frac{1}{r} n(r, b; f),$$

whence

$$(14.4.20) \qquad m°(r, a; f) + N(r, a; f) = m°(r, b; f) + N(r, b; f)$$

since all four functions vanish for $r = 0$.

This is the "spherical form" of the first fundamental theorem. It suggests introducing the *spherical characteristic*

$$(14.4.21) \qquad T°(r; f) = m°(r, a; f) + N(r, a; f),$$

which is independent of a. In order that this shall be strictly true, it is necessary to define $m°(r, a; f)$ when $f(0) = a$ and (14.4.19) breaks down. We merely have to replace the constant term by

$$\log \frac{1 + |a|^2}{|c|} \quad (a \neq \infty), \qquad \log |c| \quad (a = \infty),$$

where in each case c is the coefficient of the leading term in the Laurent expansion at the origin.

$T°(r; f)$ has a very elegant geometrical meaning which we now determine. For this purpose we integrate both sides of (14.4.21) with respect to a over the whole sphere and divide by π, the area of the sphere. The left side is clearly invariant and we get

$$T°(r; f) = \frac{1}{\pi} \int \frac{m°(r, a; f)}{[1 + |a|^2]^2} \, d\omega + \frac{1}{\pi} \int \frac{N(r, a; f)}{[1 + |a|^2]^2} \, d\omega,$$

where we integrate over the plane using the spherical metric. Here the first integral does not depend upon r; in fact, if we substitute (14.4.19) and interchange the order of integration we see that the integral does not depend upon $f(z)$ at all and is an absolute constant (in fact, zero). We then differentiate with respect to r and obtain

$$r \frac{d}{dr} T°(r; f) = \frac{1}{\pi} \int \frac{n(r, a; f) \, d\omega}{[1 + |a|^2]^2}.$$

This is nothing but the area of $\mathfrak{S}_r$ divided by π, that is, $A(r; f)$ in the notation of (14.4.18). It follows that

(14.4.22) $$T°(r; f) = \int_0^r A(t; f) \frac{dt}{t}.$$

The discovery of the importance of the spherical characteristic was made in 1929 by Tatsujirô Shimizu and Lars Ahlfors, independently of each other. The elegant derivation given above is due to Ahlfors.

We have

$$m°(r, \infty; f) = \frac{1}{2\pi} \int_0^{2\pi} \log [1 + |f(re^{i\theta})|^2]^{\frac{1}{2}} \, d\theta - \log [1 + |f(0)|^2]^{\frac{1}{2}},$$

and here the integral lies between

$$m(r, \infty; f) \quad \text{and} \quad m(r, \infty; f) + \log 2.$$

From this it follows that the difference

$$T°(r; f) - T(r; f)$$

is a bounded function of r, so that the two characteristics are interchangeable.

From (14.4.22) we obtain

$$T^\circ(r;f) > A(1;f) \log r, \quad r > 1,$$

whence it follows that

(14.4.23) $$\liminf \frac{T^\circ(r;f)}{\log r} > 0$$

unless $f(z)$ is a constant.

Suppose that the left member of this inequality has a finite value c. Then $f(z)$ is a rational function and c is a positive integer. Since

$$N(r, a;f) > n(r^{1/k}, a;f) \int_{r^{1/k}}^{r} \frac{dt}{t} = \frac{k-1}{k} \, n(r^{1/k}, a;f) \log r$$

for each $k > 1$, the assumption implies that

$$n(r, a;f) \leq c$$

for every a and r. But such a property can hold only for a rational function, and in that case the limit in (14.4.23) is a positive integer.

Let us finally return to our starting point, the question of how to define order and type for meromorphic functions. This is now a simple matter. We define

(14.4.24) $$\rho(f) = \limsup_{r \to \infty} \frac{\log T^\circ(r;f)}{\log r},$$

and if $0 < \rho(f) < \infty$ we set

(14.4.25) $$\tau(f) = \limsup_{r \to \infty} r^{-\rho} T^\circ(r;f).$$

The type is normal if $0 < \tau(f) < \infty$, minimal if $\tau(f) = 0$, maximal if $\tau(f) = \infty$. In the case of an entire function this notion of order coincides with that given by Definition 14.1.1. Likewise the classification into minimal, normal, and maximal types is the same, but in the case of functions of normal type the numerical value of the type will depend upon which definition we use.

EXERCISE 14.4

1. Determine $m(r, a;f)$, $N(r, a;f)$, and $T(r;f)$ if $f(z) = e^z$. Note that $a = 0, \infty$ are exceptional values.

2. Same question for $f(z) = \int_0^z e^{-t^2}\, dt$ with exceptional values $\pm \tfrac{1}{2}\sqrt{\pi}, \infty$.

3. What values are exceptional if $f(z) = \int_0^z e^{-t^p}\, dt$, $p > 2$?

4. Give a similar discussion for $f(z) = \tan z$.

5. What is the nature of the exceptional values in the preceding problems? Note that they are also asymptotic values approached by the function in certain sectors. Determine these sectors.

6. Verify that e^z is of order one also according to (14.4.24) and compute the two types.

7. Determine order and type of the Gamma function.

8. Show that $\log (1 + |w|^2)^{\frac{1}{2}}$ satisfies the equation

$$\Delta F = \frac{\partial^2 F}{\partial u^2} + \frac{\partial^2 F}{\partial v^2} = 2e^{-4F}, \quad w = u + iv.$$

9. Use the preceding result to evaluate directly the integral (14.4.18). Take $g(z) = \log [1 + |f(z)|^2]^{\frac{1}{2}}$ and apply Green's theorem,

$$\iint_D \Delta G \, dS = \int_{\partial D} \frac{\partial G}{\partial n} \, ds,$$

where $D = D_\delta$ is the disk $|z| < t$ from which disks of radius δ about each of the poles of $f(z)$ have been deleted. Let $\delta \to 0$. Differentiation is taken in the direction of the outer normal in the right member.

10. Prove that

$$\iint \log [\chi(w, a)]^{-1}(1 + |a|^2)^{-2} \, d\omega,$$

extended over the plane, is independent of w and has the value $\frac{1}{2}\pi$. (*Hint:* Rotate the sphere so that w goes into ∞.)

14.5. Picard's and Landau's theorems. In 1879 Émile Picard proved that a meromorphic function which omits more than two values is a constant. This theorem was justly hailed as an outstanding achievement and for many years it stirred the imagination of many prominent mathematicians with results that were ultimately epoch making. Picard's proof was based on the elliptic modular function, a theory which was in the center of analysis at that time. Toward the end of the nineteenth century, mathematicians tried to replace Picard's method by more elementary considerations, and in 1896 Émile Borel succeeded in giving such a proof. In the able hands of Edmund Landau this proof gave in 1904 some astonishing results concerning the influence of the first two coefficients of a power series on the properties of the function defined by the series. These results were taken as a vindication of the elementary methods, but the triumph was brief, for in 1905 Constantin Carathéodory proved that the modular function played the role of extremal function in Landau's problem.

A fresh start was made in 1924 by André Bloch (1893–1948), who proved a remarkable theorem on conformal mapping: If $f(z)$ is holomorphic in the unit disk and if $|f'(0)| \geqq 1$, then the map of the disk covers a disk whose radius exceeds a positive absolute constant. The theorems of Picard and Landau follow from this result.

In this area, new prospects were opened by R. Nevanlinna's theory of the distribution of values of meromorphic functions, a theory whose basis is his second fundamental theorem. Picard's theorem is a simple and very special consequence of this theory.

In the present section we shall prove Picard's and Landau's theorems using the modular function, and the Nevanlinna theory will be discussed in the next two sections. Bloch's theorem and its implications will be proved in Section 17.7. We start with Picard's theorem.

THEOREM 14.5.1.   Let $f(z)$ be meromorphic in the finite plane and let $f(z) \neq a$, b, and c for all values of z. Then $f(z)$ is a constant.

Proof.   We may always assume that the three omitted values are $0, 1, \infty$, for if this is not the case at the outset, we replace $f(z)$ by

$$g(z) = \frac{c - b}{c - a} \frac{f(z) - a}{f(z) - b}.$$

This is an entire function without zeros and "ones."

We now go back to the modular function

(14.5.1) $w = k^2(\tau)$

of Section 13.6. We recall that this function maps the domain Int (B) defined by (13.6.17) conformally on the w-plane slit along the intervals $(-\infty, 0)$ and $(1, \infty)$ of the real axis. See Figure 13, page 177. It maps the upper half-plane on a Riemann surface $\mathfrak{R}$ having branch points over $w = 0, 1, \infty$, and the surface has no regular elements over these points since $w \neq 0, 1, \infty$ for $0 < \mathfrak{I}(\tau) < \infty$. The inverse function

$$\tau = \nu(w)$$

then is an infinitely many-valued function defined for every $w \neq 0, 1, \infty$, and it is locally holomorphic in the neighborhood of any such point. Further, for every determination of $\nu(w)$ we have $\mathfrak{I}[\nu(w)] > 0$.

We now form the composite function

(14.5.2) $G(z) = \nu[g(z)]$.

For this purpose, suppose that $g(0) = a_0$ and determine $\nu(a_0)$ in B. This can be done, and since $a_0 \neq 0, 1$, the point in question is not a vertex of B. Since $\nu(w)$ is holomorphic at a_0 and $g(z)$ is entire, it follows that $G(z)$ is definable as a holomorphic function of z in some disk $|z| < R$. But if Γ is any path in the z-plane starting at the origin, then $g(z)$ can be continued along this path, and since $g(z)$ omits the values $0, 1, \infty$, we can also continue $G(z)$ along the path without ever encountering a singularity of this function. Since the finite plane is simply-connected, the theorem of monodromy applies, and we see that $G(z)$ is holomorphic in the finite plane; that is, $G(z)$ is an entire function. Now $\mathfrak{I}[G(z)] > 0$, and an entire function cannot omit all values in a half-plane unless it reduces to a constant. Hence, $G(z)$ is a constant and, consequently, also $g(z)$, as asserted.

The same method gives the Landau-Carathéodory theorem.

THEOREM 14.5.2.   *If*

$$f(z) = a_0 + a_1 z + \cdots, \quad a_1 \neq 0,$$

is holomorphic and different from 0 *and* 1 *in the disk* $|z| < R$, *then*

(14.5.3) $$R \leq R(a_0, a_1) = \frac{2\,\Im[\nu(a_0)]}{|a_1|\,|\nu'(a_0)|}.$$

Proof.   Here we consider the auxiliary function

(14.5.4) $$\lambda(w) = \frac{\nu(w) - \nu(a_0)}{\nu(w) - \overline{\nu(a_0)}}.$$

It maps the Riemann surface $\Re$ on the disk $|\lambda| < 1$. For $\nu(w)$ and $\nu(a_0)$ are both points in the upper half-plane, while $\overline{\nu(a_0)}$ is in the lower half-plane. Hence $|\lambda(w)| < 1$, and it is easy to show that no point of this disk is omitted by $\lambda(w)$. Further, $\lambda(w)$ is a holomorphic function of w in a neighborhood of any point $w_0 \neq 0, 1$, and ∞. Thus the latter three points are the only singularities of $\lambda(w)$. Suppose now that

$$f(z) = a_0 + a_1 z + a_2 z^2 + \cdots + a_n z^n + \cdots$$

is holomorphic and different from 0 and 1 in $|z| < R$. Then the composite function $\lambda[f(z)]$ can also be defined in $|z| < R$, and it is seen to be holomorphic in this disk by the theorem of monodromy. We have

$$\lambda[f(z)] = A_1 z + A_2 z^2 + \cdots$$

where

(14.5.5) $$A_1 = \frac{a_1 \nu'(a_0)}{\nu(a_0) - \overline{\nu(a_0)}} = \frac{a_1 \nu'(a_0)}{2i\Im[\nu(a_0)]}.$$

Since $|\lambda[f(z)]| < 1$ for $|z| < R$, the estimates of Cauchy give

$$R\,|A_1| \leq 1,$$

and this is precisely (14.5.3). Here the inequality always holds unless every $A_n = 0$ for $n > 1$, that is,

$$\lambda[f(z)] = A_1 z.$$

This means that we have equality if and only if $f(z)$ is the function which maps the disk $|z| < R$ on $\Re$. This completes the proof.

This theorem can be generalized in a variety of ways. Thus, we can pre-assign other coefficients than the first two and we may permit more than three omitted values. The extremal functions are known for these problems. For the related theorem of Schottky, see Section 15.4.

EXERCISE 14.5

1. Prove that the expression for $R(a_0, a_1)$ is independent of the determination of $v(a_0)$.

2. If

$$a_0 = k^2 = \frac{e_2 - e_3}{e_1 - e_3}$$

and if $2\omega_1$, $2\omega_2$ is a pair of primitive periods of the corresponding $\wp$-function, prove that

$$R(a_0, a_1) = \frac{2A}{\pi} \mid e_1 - e_2 \mid \left| \frac{a_0}{a_1} \right|$$

where A is the area of the period parallelogram [Gaston Julia].

3. If $f(z)$ is prescribed to be of the form

$$f(z) = a_0 + a_n z^n + \cdots,$$

and if $f(z)$ is holomorphic and different from 0 and 1 in the disk $\mid z \mid < R$, prove that

$$R \leq \left\{ \frac{2\Im[v(a_0)]}{\mid a_n \mid \mid v'(a_0) \mid} \right\}^{\frac{1}{n}}.$$

When does equality hold?

4. In his original proof Picard formed the function $\exp[iG(z)]$. How is the proof concluded?

5. Picard also proved the following theorem: *If $f(z)$ is meromorphic in the punctured disk $0 < \mid z - z_0 \mid < R$ and omits three values a, b, c in this domain, then $z = z_0$ is a pole or a removable singularity.* Apply the argument used in proving Theorem 14.5.1 and see where it breaks down in the present case. Compare Theorem 15.4.2.

14.6. The second fundamental theorem. We shall now return to the study of value distribution in the case of meromorphic functions. The first fundamental theorem asserts that the sum of a proximity function and the corresponding enumerative function

$$m^\circ(r, a; f) + N(r, a; f)$$

is independent of a. The theorem fails to tell us which of the two addends is normally the more important one. It is fairly clear from the proof of formula (14.4.22) that $N(r, a; f)$ must dominate in general, for its average with respect to a over the sphere equals $T^\circ(r; f)$, whereas the average of $m^\circ(r, a; f)$ is 0. The second fundamental theorem gives the quantitative measure of this dominance, from which important conclusions may be drawn concerning the value distribution.

The proof of this theorem given by R. Nevanlinna in 1924 was based on an elaborate discussion of the proximity function $m(r, \infty; f'/f)$. It is fairly elementary but certainly not simple. A simple but far from elementary proof was given the following year by Nevanlinna's older brother Frithiof. This proof drew heavily on the theory of the differential equation

$$(14.6.1) \qquad \Delta F = e^{2F}$$

which plays an important role in the theory of automorphic functions and uniformization (compare Problems 8 and 9 of Exercise 14.4, where an equivalent equation is used in the study of the first fundamental theorem). A third proof is due to Ahlfors. It employs integration of the first fundamental theorem over the surface of the a-sphere with respect to a suitable measure. This particular measure is so chosen that its density has essentially the same properties as the particular solution of (14.6.1) used by F. Nevanlinna. This method is both simple and elementary. Further, it provides convenient means of studying other problems in value distribution theory. We shall follow this method below.

Suppose then that a mass distribution is given on the sphere such that the total mass is one and that a density exists which is positive and continuous almost everywhere. In other words, we assume the existence of a function $\rho(a)$ defined on the sphere such that

(1) $\rho(a)$ is positive and continuous almost everywhere on the sphere A;

$$(2) \qquad \int\!\!\int_A \rho(a)\, d\omega(a) = 1.$$

Here $d\omega(a)$ is the surface element on the sphere. We can integrate over the plane instead, noting that

$$(14.6.2) \qquad d\omega(a) = (1 + |a|^2)^{-2}\, d\omega.$$

With the aid of this density function $\rho(a)$ we construct what may be regarded as a potential function

$$(14.6.3) \qquad p(w) = \int\!\!\int_A \log \frac{1}{\chi(w, a)}\, \rho(a)\, d\omega(a).$$

Suppose now that $f(z)$ is a meromorphic function and let $N(r, a; f)$ have its usual meaning. With this function $f(z)$ and the density $\rho(a)$ we associate a proximity function

$$(14.6.4) \qquad m_\rho(r; f) = \frac{1}{2\pi} \int_0^{2\pi} p[f(re^{i\theta})]\, d\theta$$

and an enumerative function

$$(14.6.5) \qquad N_\rho(r; f) = \int\!\!\int_A N(r, a; f)\rho(a)\, d\omega(a).$$

We now go back to (14.4.21), which defines $T^\circ(r; f)$. We multiply both sides by $\rho(a)$ and integrate over the sphere. The result is the *generalized first fundamental theorem*:

(14.6.6) $$T^\circ(r;f) = m_\rho(r;f) + N_\rho(r;f).$$

We define

$$n_\rho(r;f) = rN_\rho'(r;f)$$

so that

$$n_\rho(r;f) = \int\int_A n(r,a;f)\rho(a)\,d\omega(a).$$

This is the integral of the density function over the Riemann surface $\mathfrak{S}_r$ which is the stereographic projection of the image of the disk $|z| < r$ under the mapping $w = f(z)$. From this interpretation it follows that

(14.6.7) $$n_\rho(r;f) = \int_0^r \int_0^{2\pi} \rho[f(te^{i\theta})] \frac{|f'(te^{i\theta})|^2 \, d\theta t\, dt}{[1 + |f(te^{i\theta})|^2]^2}.$$

Here we set

(14.6.8) $$\lambda(t;f) = \int_0^{2\pi} \rho[f(te^{i\theta})] \frac{|f'(te^{i\theta})|^2}{[1 + |f(te^{i\theta})|^2]^2}\, d\theta$$

and obtain

(14.6.9) $$N_\rho(r;f) - N_\rho(r_0;f) = \int_{r_0}^r \frac{ds}{s} \int_0^s \lambda(t;f)t\, dt,$$

where the lower bound $r_0 > 0$ is arbitrary.

In view of (14.6.6) we have

$$N_\rho(r;f) - N_\rho(r_0;f) = T^\circ(r;f) - T^\circ(r_0;f) - m_\rho(r;f) + m_\rho(r_0;f)$$
$$< T^\circ(r;f) + m_\rho(r_0;f).$$

This gives the important inequality

(14.6.10) $$T^\circ(r;f) > \int_{r_0}^r \frac{ds}{s} \int_0^s \lambda(t;f)t\, dt - m_\rho(r_0;f).$$

Our next task will be to find upper and lower bounds for $\lambda(r;f)$ in terms of $T^\circ(r;f)$. For the discussion of lower bounds we need the following lemma:

LEMMA 14.6.1.   *If $f(x)$ is a non-negative function which is integrable over the finite interval (α, β), then*

(14.6.11) $$\log\left\{\frac{1}{\beta - \alpha} \int_\alpha^\beta f(x)\, dx\right\} \geq \frac{1}{\beta - \alpha} \int_\alpha^\beta \log f(x)\, dx.$$

REMARK.   This is a special case of the inequality

(14.6.12) $$g\left\{\frac{1}{\beta - \alpha} \int_\alpha^\beta f(x)\, dx\right\} \geq \frac{1}{\beta - \alpha} \int_\alpha^\beta g[f(x)]\, dx,$$

valid if $g(u)$ is convex upward on the range of $f(x)$. The inequality is reversed if $g(u)$ is convex downward instead.

Proof.   The inequality states that

(14.6.13) $$\exp \left\{ \frac{1}{\beta - \alpha} \int_\alpha^\beta \log f(x) \, dx \right\} \leqq \frac{1}{\beta - \alpha} \int_\alpha^\beta f(x) \, dx.$$

The expression in the right member is the arithmetic mean value of $f(x)$ over the interval (α, β), and the expression on the left is the geometric mean. Thus the lemma is an extension to functions of the familiar inequality

$$(a_1 a_2 \cdots a_n)^{\frac{1}{n}} \leqq \frac{1}{n} (a_1 + a_2 + \cdots + a_n)$$

for positive numbers. Suppose for a moment that $f(x)$ is continuous in $[\alpha, \beta]$ and set

$$a_{j,n} = f\left(\alpha + j \frac{\beta - \alpha}{n} \right), \quad j = 1, 2, \cdots, n.$$

Then

$$\frac{1}{n} \sum_{j=1}^n a_{j,n} = \frac{1}{\beta - \alpha} \sum_{j=1}^n f\left(\alpha + j \frac{\beta - \alpha}{n} \right) \frac{\beta - \alpha}{n} \rightarrow \frac{1}{\beta - \alpha} \int_\alpha^\beta f(x) \, dx$$

as $n \rightarrow \infty$. On the other hand,

$$\log \left\{ \prod_{j=1}^n a_{j,n} \right\}^{\frac{1}{n}} = \frac{1}{n} \sum_{j=1}^n \log f\left(\alpha + j \frac{\beta - \alpha}{n} \right) \rightarrow \frac{1}{\beta - \alpha} \int_\alpha^\beta \log f(x) \, dx.$$

This shows that (14.6.13) holds for positive continuous functions. It then holds for arbitrary positive integrable functions, for the integral of such a function is the limit of integrals of continuous functions. This then completes the proof of the lemma.

Define

(14.6.14) $$\mu(r; f) = \frac{1}{2\pi} \int_0^{2\pi} \log \frac{|f'(re^{i\theta})|}{1 + |f(re^{i\theta})|^2} \, d\theta.$$

The lemma then gives

(14.6.15) $$2\mu(r; f) + \frac{1}{2\pi} \int_0^{2\pi} \log \rho[f(re^{i\theta})] \, d\theta \leqq \log \lambda(r; f) - \log 2\pi.$$

We also have

$$\mu(r; f) = \frac{1}{2\pi} \int_0^{2\pi} \log |f'(re^{i\theta})| \, d\theta - \frac{2}{2\pi} \int_0^{2\pi} \log [1 + |f(re^{i\theta})|^2]^{\frac{1}{2}} \, d\theta.$$

The second term on the right is

$$-2 \left\{ m^\circ(r, \infty; f) + \frac{1}{2\pi} \log [1 + |f(0)|^2]^{\frac{1}{2}} \right\}.$$

The first integral may be evaluated by differentiation with respect to r. By the

Cauchy-Riemann equations and the argument principle the derivative equals

$$\frac{1}{2\pi r} \int_0^{2\pi} d_\theta \arg f'(re^{i\theta}) = \frac{1}{r}[n(r, 0; f') - n(r, \infty; f')].$$

Thus

$$\mu'(r; f) = \frac{1}{r}[n(r, 0; f') - n(r, \infty; f')] - 2\frac{d}{dr} m^\circ(r, \infty; f).$$

Here the enumerative functions are connected with the multiple values of $f(z)$. If $z = z_0$ is a k-tuple a-value of $f(z)$, then $z = z_0$ is also a $(k-1)$-tuple zero of $f'(z)$. Multiple poles of $f(z)$ contribute to $n(r, \infty; f)$, but a k-tuple pole of $f(z)$ is a $(k+1)$-tuple pole of $f'(z)$.

Let us introduce an enumerative function $n_1(r; f)$ which registers the multiple values of $f(z)$ in the disk $|z| < r$, but in such a manner that a finite or infinite k-tuple value contributes $(k-1)$ to $n_1(r; f)$. We have then

$$(14.6.16) \qquad n_1(r; f) = n(r, 0; f') - n(r, \infty; f') + 2n(r, \infty; f).$$

Using this function and the first fundamental theorem we see that

$$\mu'(r; f) = \frac{1}{r} n_1(r; f) - 2T^{\circ\prime}(r; f),$$

and this gives upon integration

$$(14.6.17) \qquad \mu(r; f) = N_1(r; f) - 2T^\circ(r; f)$$

for the appropriate choice of the constant of integration.

Thus we have proved the inequality

$$(14.6.18) \qquad \frac{1}{2\pi} \int_0^{2\pi} \log \rho[f(re^{i\theta})] \, d\theta \leqq 4T^\circ(r; f) - 2N_1(r; f)$$
$$+ \log \lambda(r; f) - \log 2\pi.$$

This gives a lower bound for $\lambda(r; f)$ and may be used to extend the inequality (14.6.10).

Next we shall use (14.6.10) to get an upper bound for $\lambda(r; f)$. We consider only functions meromorphic in the finite plane, though corresponding estimates are also obtainable for functions meromorphic in a bounded disk. If $f(z)$ is of infinite order, the estimates hold outside of certain exceptional very sparse intervals, but if the order is finite there are no exceptions.

We set

$$(14.6.19) \qquad \int_0^r \lambda(t; f) t \, dt = L(r; f), \qquad \int_{r_0}^r L(s; f) \frac{ds}{s} = K(r; f).$$

Then (14.6.10) becomes

$$(14.6.20) \qquad T^\circ(r; f) \geqq K(r; f) - C,$$

where C is a constant.

We choose arbitrarily a number $k \geq 0$ and denote by Δ_{1k} the intervals $(\alpha_{1j}, \beta_{1j})$ where

$$\lambda(r;f) > r^{k-1}[L(r;f)]^2.$$

The measure of Δ_{1k} is finite; moreover the integral of r^k over Δ_{1k} is finite, for

$$\int_{\Delta_{1k}} r^k \, dr < \int_{\Delta_{1k}} \frac{r\lambda(r;f)}{[L(r;f)]^2} \, dr = \int_{\Delta_{1k}} \frac{dL(r;f)}{[L(r;f)]^2} < \frac{1}{L(\alpha_{11};f)} .$$

Similarly, we denote by Δ_{2k} the intervals $(\alpha_{2j}, \beta_{2j})$ where

$$L(r;f) > r^{k+1}[K(r;f)]^2.$$

Then

$$\int_{\Delta_{2k}} r^k \, dr = \int_{\Delta_{2k}} r^{k+1} \frac{dK(r;f)}{L(r;f)} < \int_{\Delta_{2k}} \frac{dK(r;f)}{[K(r;f)]^2} < \frac{1}{K(\alpha_{21};f)} .$$

Now set $\Delta_k = \Delta_{1k} \cup \Delta_{2k}$. Then for r outside of Δ_k, we have

$$\lambda(r;f) \leq r^{k-1}[L(r;f)]^2 < r^{3k+1}[K(r;f)]^4 < r^{3k+1}[T^\circ(r;f) + C]^4.$$

In particular we have

(14.6.21) $$\log \lambda(r;f) = O[\log r] + O[\log T^\circ(r;f)]$$

outside of Δ_k.

We are now ready to state and prove the *second fundamental theorem*.

THEOREM 14.6.1.   *Let $f(z)$ be meromorphic in the finite plane and let $a_1, a_2, \cdots, a_q$ be q given points, $q > 2$. Let k be a given real number, $k \geq 0$. Then*

(14.6.22) $$\sum_{j=1}^{q} m^\circ(r, a_j; f) < 2T^\circ(r;f) - N_1(r;f) + O[\log r]$$
$$+ O[\log T^\circ(r;f)]$$

for r outside an open set Δ_k such that $\displaystyle\int_{\Delta_k} r^k \, dr < \infty$. If $f(z)$ is of finite order, then (14.6.22) *holds for all large r without exception.*

Proof.   For the proof we have to choose a suitable density function $\rho(a)$. It is reasonable to choose a function which becomes infinite at the points a_j, but in such a manner that $\rho(a)$ remains integrable over the sphere A. A simple choice satisfying these requirements is

(14.6.23) $$\log \rho(a) = 2 \sum_{j=1}^{q} \log \frac{2}{\chi(a_j, a)} - \alpha \log \left\{ \sum_{j=1}^{q} \log \frac{2}{\chi(a_j, a)} \right\} - 2C.$$

Here α is fixed, $\alpha > 1$, and C is so chosen that $\displaystyle\int_A \rho(a) \, d\omega(a) = 1$. The condition on α suffices for integrability, for in the neighborhood of one of the points a_j the density becomes infinite as $t^{-2}\left(\log \dfrac{2}{t}\right)^{-\alpha}$. The surface element introduces a factor t, so $\rho(a)$ is integrable over a spherical cap with center at a_j if and only if $\int_0^1 t^{-1}\left(\log\dfrac{2}{t}\right)^{-\alpha} dt$ exists, that is, if and only if $\alpha > 1$. Here $t = \chi(a_j, a)$.

From (14.6.18) we now get

(14.6.24)
$$\sum_{j=1}^q m^\circ(r, a_j; f) - \frac{1}{2}\,\alpha\,\frac{1}{2\pi}\int_0^{2\pi} \log\left\{\sum_{j=1}^q \log\frac{1}{\chi[f(re^{i\theta}), a_j]}\right\} d\theta$$
$$\leq 2T^\circ(r;f) - N_1(r;f) + \tfrac{1}{2}\log\lambda(r;f) - \tfrac{1}{2}\log 2\pi + C.$$

The second term on the left may be simplified with the aid of Lemma 14.6.1. From this we get

$$\frac{1}{2\pi}\int_0^{2\pi}\log\left\{\sum_{j=1}^q\log\frac{1}{\chi[f(re^{i\theta}), a_j]}\right\} d\theta \leq \log\left(\sum_{j=1}^q m^\circ(r, a_j; f)\right);$$

for large values of r this does not exceed $\log T^\circ(r;f) + O(1)$. For $\log \lambda(r;f)$ we have the estimate (14.6.21) outside of Δ_k. Combining these inequalities, we see that (14.6.22) holds outside of Δ_k. If $f(z)$ is of infinite order, there is nothing to be added, but if the order is finite, then we have to show that (14.6.22) holds for all large values of r.

Since $\log T^\circ(r;f) = O(\log r)$ if $\rho[f] < \infty$, we can rewrite (14.6.22) as follows:

(14.6.25)　　　$(q - 2)T^\circ(r;f) + N_1(r;f) < \displaystyle\sum_{j=1}^q N(r, a_j; f) + C\log r.$

To start with, this holds outside of the set Δ_k. It should be noted that C depends upon k. We choose a fixed $k > \rho[f]$. Now the left member of (14.6.25) is an increasing function of r. Hence, if $r \in \Delta_k$ we can find an r_1, $r_1 > r$, such that $r_1 \notin \Delta_k$ and

$$(q - 2)T^\circ(r;f) + N_1(r;f) < (q - 2)T^\circ(r_1;f) + N_1(r_1;f)$$
$$< \sum_{j=1}^q N(r_1, a; f) + C\log r_1.$$

From the properties of Δ_k we conclude that it is possible to choose r_1 so that for large values of r

$$r_1 < r + r^{-k}.$$

This gives

$$\log r_1 < \log r + r^{-k-1}.$$

Further, for any a,

$$N(r_1, a; f) - N(r, a; f) = \int_r^{r_1} n(s, a; f)\,\frac{ds}{s}$$
$$< n(r_1, a; f)\log\frac{r_1}{r}$$
$$< r^{\rho + \varepsilon - k - 1} = o(1).$$

Here we have used the fact that

(14.6.26)

$$\limsup_{r \to \infty} \frac{\log T^\circ(r;f)}{\log r} \geq \limsup_{r \to \infty} \frac{\log N(r, a;f)}{\log r} = \limsup_{r \to \infty} \frac{\log n(r, a;f)}{\log r} .$$

Thus an inequality of type (14.6.25) holds also for $r \in \Delta_k$, the only difference being that we may have to replace C by a larger constant. Thus, if C is sufficiently large to start with, then (14.6.25) holds for all r. This completes the proof.

EXERCISE 14.6

1. Prove (14.6.26).

2. The second fundamental theorem holds also for functions meromorphic in a finite disk $|z| < R$. Prove that in (14.6.22) the first remainder term $O[\log r]$ is to be replaced by $(k + \varepsilon) \log \dfrac{1}{R - r}$ and the estimate is valid outside an open set Δ_k such that $\displaystyle\int_{\Delta_k} (R - r)^{-k} \, dr < \infty$. If $T^\circ(r;f) = O[(R - r)^{-\rho}]$, then there is no exceptional set.

14.7. Defect relations. We come now to the consequences of the second fundamental theorem. It is necessary to introduce a certain amount of new notation and terminology.

We recall the function $n_1(r;f)$ of (14.6.16). It keeps account of the multiple points of $f(z)$: for a k-tuple point it registers $k - 1$. Let the set of multiple values of $f(z)$ be denoted by B. It is a countable set, and its intersection with the disk $|z| < r$ is always finite. This is obvious, for $b \in B$ if and only if there exists a point z_0 such that $f(z_0) = b, f'(z_0) = 0$, and the set of zeros of $f'(z)$ in $|z| < r$ is finite. For each $b \in B$ we have also an enumerative function $n_1(r, b; f)$ giving $\Sigma (k - 1)$ summed over the multiple b-points in the disk $|z| < r$. Corresponding to this disk there is a Riemann surface $\mathfrak{S}_r$ over the sphere A. Now $n(r, b;f)$ gives the total number of sheets of this surface over the point $w = b$; on the other hand, $n_1(r, b;f)$ gives the sum of the orders of the branch points over b. For each function $n_1(r, b;f)$ there is a corresponding function

(14.7.1) $$N_1(r, b;f) = \int_0^r n_1(s, b;f) \frac{ds}{s} + n_1(0, b;f) \log r.$$

We can define $N_1(r, a;f)$ to be zero for any $a \notin B$. We then see that

(14.7.2) $$N_1(r;f) = \sum_a N_1(r, a;f)$$

where the sum on the right is actually finite.

DEFINITION 14.7.1.   *We set*

$$\delta(a;f) = \liminf_{r \to \infty} \frac{m^\circ(r,a;f)}{T^\circ(r;f)},$$

$$\Phi(f) = \liminf_{r \to \infty} \frac{N_1(r;f)}{T^\circ(r;f)},$$

(14.7.3)

$$\vartheta(a;f) = \liminf_{r \to \infty} \frac{N_1(r,a;f)}{T^\circ(r;f)},$$

$$\theta(a;f) = \liminf_{r \to \infty} \frac{m^\circ(r,a;f) + N_1(r,a;f)}{T^\circ(r;f)},$$

and refer to these numbers thus: $\delta(a;f)$ *is the defect of a with respect to* $f(z)$, $\Phi(f)$ *is the total ramification of* $f(z)$, $\vartheta(a;f)$ *is the ramification index of a, and* $\theta(a;f)$ *is the ramification of a with respect to* $f(z)$. *A value a is said to be completely ramified with respect to* $f(z)$ *if all the roots of the equation*

$$f(z) = a \quad \left(\frac{1}{f(z)} = 0 \text{ if } a = \infty\right)$$

are multiple roots.

We now have

THEOREM 14.7.1.   *Each of the numbers* $\delta(a;f)$, $\vartheta(a;f)$, *and* $\theta(a;f)$ *lies between 0 and 1 and is different from 0 for at most a countable set of values of a. Further,*

(14.7.4) $0 \leq \Phi(f) \leq 2, \quad \delta(a;f) + \vartheta(a;f) \leq \theta(a;f)$

and

(14.7.5) $\sum_a \delta(a;f) + \Phi(f) \leq 2, \quad \sum_a \delta(a;f) + \sum_a \vartheta(a;f) \leq 2, \quad \sum_a \theta(a;f) \leq 2.$

Proof.   The numbers in question are obviously non-negative. From the first fundamental theorem we get

$$m^\circ(r,a;f) < m^\circ(r,a;f) + N_1(r,a;f)$$
$$< m^\circ(r,a;f) + N(r,a;f) = T^\circ(r;f),$$

whence it follows that $\delta(a;f)$, $\vartheta(a;f)$, and $\theta(a;f)$ cannot exceed 1. This also gives $\delta(a;f) + \vartheta(a;f) \leq \theta(a;f)$. The remaining relations follow from the second fundamental theorem. In (14.6.22) we divide both sides by $T^\circ(r;f)$ and pass to the limit with r. Since we are concerned with inferior limits, the exceptional intervals may be neglected. This gives

$$\sum_{j=1}^{q} \delta(a_j;f) + \Phi(f) \leq 2$$

valid for any choice of the integer $q > 2$ and the points $a_1, a_2, \cdots, a_q$. This implies first that $0 \leq \Phi(f) \leq 2$ as asserted. Secondly we see that $\delta(a;f) \geq \dfrac{1}{m}$

can hold at most for $2m$ values of a, that is, the set of values a such that $\delta(a;f) > 0$ is countable. This implies the first relation under (14.7.5). For the other relations we use (14.7.2). This implies that

$$\sum_{j=1}^{q} m^{\circ}(r, a_j; f) + \sum_{j=1}^{q} N_1(r, a_j; f) < 2T^{\circ}(r;f) + O[\log r + \log T^{\circ}(r;f)].$$

Dividing through by $T^{\circ}(r;f)$ and passing to the limit with r we get the remaining relations.

There are a number of important consequences of this theorem.

CÔROLLARY 1.   *A meromorphic function which is not a constant can have at most two values of defect one.*

This is now obvious, but the statement implies Picard's theorem in a stronger form. It is clear that if $f(z)$ never takes on the value a, then $N(r, a; f) = 0$, $m^{\circ}(r, a; f) = T^{\circ}(r;f)$, and $\delta(a;f) = 1$. But the fact that $\delta(a;f) = 1$ does not imply that $f(z) \neq a$ for all z. It does imply, however, that the a-points are less frequent. This is worth stating as a separate corollary.

CÔROLLARY 2.   *Let $f(z)$ be a meromorphic function of finite order $\rho(f)$. Let $\{z_n(a)\}$ be the roots of the equation $f(z) = a$, where multiple roots are repeated according to their multiplicity. Let $\sigma(a)$ be the exponent of convergence of the series*

$$\Sigma \mid z_n(a) \mid^{-\alpha}.$$

Then $\sigma(a) \leq \rho(f)$ and $<$ holds for at most two values of a.

Suppose that $f(z)$ is an entire function. We have then $\delta(\infty;f) = 1$. If there exists a finite value a such that $\delta(a;f) = 1$, then

$$f(z) = a + P(z)e^{Q(z)}$$

where $P(z)$ is a canonical product of order σ and $Q(z)$ is a polynomial of degree q. The order of this function is $\rho = \max(\sigma, q)$. Since $\delta(a;f) = 1$ by assumption, we must have $\sigma \leq q$, and if $\sigma = q$, then $f(z)$ is of minimal type of its order. It follows that $\rho = q$; that is, if an entire function has a finite value a with $\delta(a;f) = 1$, then its order is necessarily an integer ≥ 1.

No function with a countable number of positive defects is known. Defects can have any value between 0 and 1 as is shown by examples given below.

CÔROLLARY 3.   *A meromorphic function has at most four completely ramified values.*

Proof.   If a is completely ramified with respect to $f(z)$, then the equation $f(z) = a$ either has no roots or all roots are multiple roots. In either case

$$m^{\circ}(r, a; f) + N_1(r, a; f) \geq \tfrac{1}{2}[m^{\circ}(r, a; f) + N(r, a; f)] = \tfrac{1}{2}T^{\circ}(r;f),$$

so that $\theta(a) \geq \tfrac{1}{2}$. Since $\Sigma \theta(a) \leq 2$, it follows that this situation can hold for at most four values of a. The limit four is sharp, for the $\wp$-function of Weierstrass admits e_1, e_2, e_3, and ∞ as completely ramified values.

This result also has a bearing on a theorem of Picard which was mentioned at the end of Section 12.4.

THEOREM 14.7.2.   *If* $F(z, w) = 0$ *is an algebraic curve of genus* $p > 1$, *then it is not possible to find two functions* $f_1(s)$ *and* $f_2(s)$ *which are meromorphic in the finite s-plane and are such that*

(14.7.6) $$F[f_1(s), f_2(s)] = 0.$$

Proof.   It is sufficient to deal with the hyperelliptic case in which the equation is reduced to the form

(14.7.7) $$w^2 = \prod_{n=1}^{2p+1} (z - a_n).$$

We recall that the general case may be reduced to this form by a birational transformation. Any parametric representation of the old coordinates by means of meromorphic functions would give rise to such a representation of the new coordinates and vice versa. But if

$$[f_2(s)]^2 = \prod_{n=1}^{2p+1} [f_1(s) - a_n],$$

where $f_1(s)$ and $f_2(s)$ are meromorphic in the finite s-plane, then the $2p + 2$ values $a_1, a_2, \cdots, a_{2p+1}$, and ∞ would have to be completely ramified with respect to $f_1(s)$ since otherwise $f_2(s)$ could not be single-valued. We have seen that this is impossible for $p > 1$.

If $w = a$ has a positive ramification index $\vartheta(a; f)$ with respect to $f(z)$, then this indicates that the Riemann surface of $f(z)$ contains infinitely many algebraic elements above $w = a$. On the other hand, the total ramification $\Phi[f]$ may be positive and still $\vartheta(a; f) = 0$ for each a. This simply indicates that there are infinitely many algebraic elements, but no point $w = a$ carries enough such elements to affect $\vartheta(a; f)$. Examples of this phenomenon are given below.

Most of the functions whose value distributions have been studied in detail have the property that an exceptional value a with $\delta(a; f) > 0$ is also an asymptotic value of $f(z)$, that is, there exists a path tending to infinity along which $f(z)$ approaches the value a. It has been shown, however, that this is not necessarily so; there exist functions having values of positive defect which are not asymptotic values of the function.

Finally we make a remark affecting the intrinsic character of concepts used in the theory of value distribution. Suppose that $w = a$ is an exceptional value of $f(z)$ and that the corresponding defect is $\delta > 0$. If c is a fixed complex number, is $w = a$ also an exceptional value of $f(z + c)$ and is the corresponding defect also δ? Unfortunately this is not necessarily the case for functions of infinite order or irregular growth. No discrepancy will arise as long as

(14.7.8) $$\lim_{r \to \infty} \frac{T^\circ(r + 1; f)}{T^\circ(r; f)} = 1.$$

EXERCISE 14.7

1. Determine the defects of the finite exceptional values of the function considered in Problem 2 of Exercise 14.4. Since $f(z)$ is odd, the two defects are equal.

The following four problems involve functions of the form $Ae^{az} + Be^{bz}$ and exhibit rather different features which illustrate remarks made in the text:

2. Consider $f(z) = \cos z$. Show that $w = \pm 1$ are completely ramified with $\theta(1) = \theta(-1) = \frac{1}{2}$. Show that $\delta(c) = 0$ for $c \neq \infty$.

3. Take $a = \alpha$, $b = 1$, $A = 1$, $B = -\alpha$ where α is irrational > 1. Show that $\delta(0; f) = 1/\alpha$, $\Phi(f) = (\alpha - 1)/\alpha$, $\vartheta(c; f) = 0$ for each c.

4. Same choice of a, b, A, B, but α is rational, $\alpha = p/q > 1$. Show that in this case there are $p - q$ points a_j, forming the vertices of a regular polygon, which carry the algebraic elements of the Riemann surface, and $\vartheta(a_j, f) = 1/p$,

$$\Phi(f) = \sum_{j=1}^{p-q} \vartheta(a_j; f).$$

5. Take $a = 1$, $b = i$, $A = 1$, $B = i$. Show that the double points carrying the algebraic elements converge to 0 and ∞, and that each has $\vartheta(c; f) = 0$ while $\Phi(f) = 2(\sqrt{2} - 1)$. Show also that $\delta(0; f) = 3 - 2\sqrt{2}$.

6. Let $0 < \rho < 1$ and define

$$f(z) = \prod_{n=1}^{\infty} \frac{n^{1/\rho} + z}{n^{1/\rho} - z}.$$

Show that $f(z)f(-z) \equiv 1$ and $|f(iy)| \equiv 1$. Show that

$$N(r, 0; f) = N(r, \infty; f) = \frac{1}{\rho} r^\rho [1 + o(1)].$$

7. Use Theorem 14.3.5 to show that for $0 < \theta < \frac{1}{2}\pi$

$$\log |f(re^{i\theta})| = \frac{\pi}{\gamma} \sin \rho(\tfrac{1}{2}\pi - \theta) r^\rho [1 + o(1)], \quad \gamma = \cos(\tfrac{1}{2}\pi\rho).$$

Use this to compute

$$m^\circ(r, \infty; f) = \frac{1 - \gamma}{\gamma} \frac{1}{\rho} r^\rho [1 + o(1)]$$

and find

$$\delta(0; f) = \delta(\infty; f) = 1 - \gamma = 1 - \cos(\tfrac{1}{2}\pi\rho).$$

8. Suppose that $f(z)$ is meromorphic in the unit circle and that

$$\liminf_{r \to 1} T^\circ(r; f) \left(\log \frac{1}{1-r} \right)^{-1} = \lambda \leqq \infty.$$

Define defects and ramification indices for such functions with the aid of (14.7.3) with $r \to \infty$ replaced by $r \to 1$. Prove that

$$\Sigma \, \delta(a; f) \leqq 2 + \frac{1}{\lambda}.$$

9. The modular function

$$f(z) = k^2\left(i\,\frac{1+z}{1-z}\right)$$

is meromorphic in the unit circle. Prove that $\lambda = 1$ and that $\delta(a;f) = 1$ for $a = 0, 1, \infty$.

COLLATERAL READING

For the first three sections consult

BOAS, R. P., JR. *Entire Functions*, Chapters 1–4. Academic Press, New York, 1954.

VALIRON, G. *Lectures on the General Theory of Integral Functions*. Édouard Privat, Toulouse, 1923. Chelsea Publishing Company, New York, 1949.

VIVANTI, G. and GUTZMER, A. *Theorie der eindeutigen analytischen Funktionen*, Sections 259–300. B. G. Teubner, Leipzig and Berlin, 1906.

For the rest of the chapter see

BLOCH, A. *Les Fonctions Holomorphes et Méromorphes dans le Cercle-unité*. Mémorial des Sciences Mathématiques, No. 20. Gauthier-Villars, Paris, 1926.

NEVANLINNA, R. *Le Théorème de Picard-Borel et la Théorie des Fonctions Méromorphes*. Gauthier-Villars, Paris, 1929. Chelsea Publishing Company, New York, 1973.

NEVANLINNA, R. *Eindeutige analytische Funktionen*, Second Edition, Chaps. 6, 8–10. Springer-Verlag, Berlin, 1953.

WITTICH, HANS. *Neuere Untersuchungen über eindeutige analytische Funktionen*. Ergebnisse der Mathematik, New Series, No. 8. Springer-Verlag, Berlin, 1955.

In connection with Section 14.5 the student should read the article

LINDELÖF,E. "Sur le théorème de M. Picard dans la théorie des fonctions monogènes," in *Compte Rendu du Congrès des Mathématiciens à Stockholm*. B. G. Teubner, Leipzig, 1910.

Section 14.6 is based on

AHLFORS, L. V. "Über eine Methode in der Theorie der meromorphen Funktionen," *Societatis Scientiarum Fennicæ, Commentationes Physico-Mathematicæ*, Vol. 8, No. 10 (1936), 14 pp.

15

NORMAL FAMILIES

15.1. Schwarz's lemma and hyperbolic measure. The main topic of this chapter is the theory of compactness of families of analytic functions and some of its applications. Further applications of this theory to conformal mapping will follow in Chapter 17. Questions concerning denseness of polynomials and of rational functions in certain function spaces will be considered later in Section 16.6.

The present section is devoted to a discussion of Schwarz's lemma and of hyperbolic measure, which are important tools in later considerations. The lemma in question appeared in 1869 in a paper by H. A. Schwarz dealing with conformal mapping.

THEOREM 15.1.1. *If $f(z)$ is holomorphic and less than 1 in absolute value in the unit disk and if $f(0) = 0$, then either*

$$(15.1.1) \qquad\qquad |f(z)| < |z|$$

for all z with $0 < |z| < 1$, or else

$$(15.1.2) \qquad\qquad f(z) = e^{i\alpha}z$$

where α is a real constant.

Proof. We consider the function

$$g(z) = \frac{f(z)}{z},$$

which is holomorphic in the unit disk if it is defined by its limit at $z = 0$. The maximum principle shows that for $|z| \leq r < 1$

$$|g(z)| \leq \frac{1}{r} \max_{\theta} |f(re^{i\theta})| \leq \frac{1}{r}.$$

Here r may be as close to 1 as we please. It follows that for any fixed z with $|z| < 1$ we must have

$$|g(z)| \leq 1.$$

Now the maximum principle offers two mutually exclusive alternatives: either $|g(z)| < 1$ for every z with $|z| < 1$ or $|g(z)| \equiv 1$ and $g(z) = e^{i\alpha}$, where α is a real constant. This proves the theorem.

COROLLARY. *Under the preceding assumptions either* $|f'(0)| < 1$ *or* $f(z) = e^{i\alpha}z$.

For $f'(0) = g(0)$.

The formulation of Schwarz's lemma seems to assign a special role to the origin of the two planes. Actually it is possible to give a more general version of the theorem which brings out both the geometric character of the inequalities and the nature of the extremal functions. For this purpose we need a result concerning the mapping of the unit disk onto itself which is easily proved with the aid of the special form of the theorem given above. We also need to introduce a hyperbolic non-Euclidean metric in the unit disk. We start with

THEOREM 15.1.2. *If $f(z)$ maps the unit disk in a one-to-one manner onto itself in such a way that $f(z_0) = 0$ and $\arg f'(z_0) = \beta$, then*

$$(15.1.3) \qquad\qquad f(z) = e^{i\beta} \frac{z - z_0}{1 - \bar{z}_0 z}.$$

Proof. A simple calculation shows that the function defined by this formula has the desired properties. We recognize that $w = f(z)$ is a Möbius transformation belonging to the group U of Section 3.1. It remains to show that the mapping function is unique. Suppose that $F(z)$ has the same properties. From the two equations

$$w = f(z), \quad \omega = F(z)$$

we can obtain w as a function of ω and vice versa:

$$w = f[\check{F}(\omega)] \equiv g(\omega), \quad \omega = F[\check{f}(w)] \equiv h(w)$$

with the notation for inverse functions introduced in Section 4.5. The functions $g(\omega)$ and $h(w)$ are holomorphic in the unit disks of their planes; further, note that $|g(\omega)| < 1$, $|h(w)| < 1$, and $g(0) = h(0) = 0$. Hence the conditions of Schwarz's lemma are satisfied by both these functions, so that

$$|g(\omega)| \le |\omega|, \quad |h(w)| \le |w|.$$

This says that $|w| \le |\omega|$ and $|\omega| \le |w|$, so that

$$|w| = |\omega| \quad \text{or} \quad |f(z)| = |F(z)|.$$

This requires

$$F(z) = e^{i\alpha}f(z)$$

where α is a real constant. But

$$F'(z_0) = e^{i\alpha}f'(z_0).$$

Hence $\alpha = 0$ and $F(z) \equiv f(z)$ so that the mapping function is unique.

Ways of satisfying the axioms of non-Euclidean geometry by means of suitably chosen configurations in Euclidean space have been known for a long time. In 1859 Arthur Cayley[1] conceived the idea of regarding metric Euclidean geometry as the invariant theory of the projective plane after adjunction of the *absolute*, the configuration

$$x_1{}^2 + x_2{}^2 + x_3{}^2 = 0$$

in homogeneous coordinates. He went farther and replaced the absolute by an arbitrary conic and the Euclidean group of motions by the collineation group which leaves the conic invariant. The systematic development of these ideas was given by Felix Klein, starting in 1871. Klein saw that, depending upon the nature of the conic, the resulting geometry coincided with the (hyperbolic) geometry of Lobachevski and Bolyai or the (parabolic) geometry of Euclid or the (elliptic) geometry of Riemann. A particularly elegant realization of the hyperbolic case in a half-plane was used by Henri Poincaré in his theory of automorphic functions from 1881 onward. It is the equivalent situation in a circular disk that we shall use in the following.

The "points" of this hyperbolic geometry are the points z with $|z| < 1$. The "straight lines" are the arcs of the circles orthogonal to the unit circle and interior to it. In hyperbolic geometry the axioms of Euclid are valid except for the parallel postulate: there are two parallels to a given line through a given point not on the line. Figure 14 shows that the lines through the point P fall into two classes according as they intersect the given line $Q_1Q_0Q_2$ or not. The lines separating the two classes are the two parallels of $Q_1Q_0Q_2$ through P. The unit disk and the "straight lines" are left invariant by the transformations (15.1.3) which form the group U.

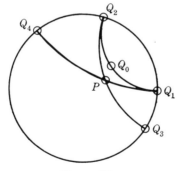

Figure 14

[1] Cayley (1821–1895) should probably be considered the most prominent British mathematician of the nineteenth century. He was certainly the most prolific. Starting in 1841, he produced a total of 887 papers. Some of his best work was done while he was reading for the bar and during the fourteen years that he practiced law as a conveyancer. From 1863 on he was Sadlerian professor of pure mathematics in the University of Cambridge. His nine memoirs on quantics (= forms) are famous; the ideas mentioned above occur in the sixth memoir. He made important contributions to group theory; he founded the theory of matrices and also, together with his good friend James Joseph Sylvester (1814–1897), the theory of invariants. There are Cayley numbers, a Cayley plane, and Cayley transforms.

We now introduce a metric on the disk, the "hyperbolic" metric, satisfying the following assumptions:

H_1. *For each pair of points z_1 and z_2 in the disk, there is a distance $\rho(z_1, z_2)$ which is a non-negative symmetric function of z_1 and z_2.*

H_2. *If $T \in U$, then*

$$(15.1.4) \qquad \rho(z_1, z_2) = \rho(Tz_1, Tz_2).$$

H_3. *If z_1, z_2, z_3 lie on the same hyperbolic straight line and if z_2 lies between z_1 and z_3, then*

$$(15.1.5) \qquad \rho(z_1, z_3) = \rho(z_1, z_2) + \rho(z_2, z_3).$$

H_4. *We have*

$$(15.1.6) \qquad \lim_{z \to 0} \frac{\rho(0, z)}{|z|} = 1.$$

These assumptions determine $\rho(z_1, z_2)$ uniquely. Since the transformation $z \to e^{i\beta}z$ belongs to U, we see that

$$(15.1.7) \qquad \rho(0, z) = \rho(0, |z|).$$

If $0 < r < r + \delta < 1$, we have by H_3

$$\rho(0, r + \delta) = \rho(0, r) + \rho(r, r + \delta).$$

Now the transformation

$$w = \frac{z - r}{1 - rz}$$

belongs to U. It takes $z = r$ into $w = 0$ and $z = r + \delta$ into

$$w = \frac{\delta}{1 - \delta r - r^2}.$$

It follows that

$$\rho(r, r + \delta) = \rho\left(0, \frac{\delta}{1 - \delta r - r^2}\right).$$

Let us define

$$(15.1.8) \qquad \rho(0, r) \equiv \rho(r), \quad 0 \leqq r < 1.$$

We have then

$$(15.1.9) \qquad \rho(r + \delta) - \rho(r) = \rho\left(\frac{\delta}{1 - \delta r - r^2}\right).$$

Here we divide by δ and use H_4, obtaining

$$\frac{1}{\delta}[\rho(r + \delta) - \rho(r)] = \frac{1}{1 - \delta r - r^2} \frac{1 - \delta r - r^2}{\delta} \rho\left(\frac{\delta}{1 - \delta r - r^2}\right) \to \frac{1}{1 - r^2}$$

as $\delta \to 0$. We have supposed $\delta > 0$, but the same result holds for $\delta < 0$. It follows that

$$(15.1.10) \qquad \rho(r) = \tfrac{1}{2} \log \frac{1+r}{1-r}.$$

It should be noted that $\rho(r)$ is a strictly increasing function of r. There exists an element of U which takes the two points z_1 and z_2, $z_1 \neq z_2$, into $w = 0$ and

$$w = \frac{z_2 - z_1}{1 - \bar{z}_1 z_2}$$

respectively. Hence

$$(15.1.11) \quad \rho(z_1, z_2) = \rho\left(\left|\frac{z_2 - z_1}{1 - \bar{z}_1 z_2}\right|\right) = \frac{1}{2} \log \frac{|1 - \bar{z}_1 z_2| + |z_1 - z_2|}{|1 - \bar{z}_1 z_2| - |z_1 - z_2|}.$$

We are now ready to give the invariantive formulation of Schwarz's lemma given by G. Pick in 1916.

THEOREM 15.1.3. *Suppose that $f(z)$ is holomorphic in $|z| < 1$ and that $|f(z)| < 1$. Then either*

$$(15.1.12) \qquad \rho[f(z_1), f(z_2)] < \rho(z_1, z_2)$$

for every z_1 and z_2, $z_1 \neq z_2$, or else

$$(15.1.13) \qquad \rho[f(z_1), f(z_2)] = \rho(z_1, z_2)$$

for all z_1, z_2, and $f(z)$ is one of the functions (15.1.3).

Proof. We take an arbitrary point z_1 with $0 < |z_1| < 1$ which we map into the origin by the transformation

$$t = \frac{z - z_1}{1 - \bar{z}_1 z}.$$

Also we consider

$$w = F(z) = \frac{f(z) - f(z_1)}{1 - \overline{f(z_1)} f(z)}.$$

Eliminating z between these two equations we get

$$(15.1.14) \qquad w = g(t),$$

where $g(t)$ is holomorphic in $|t| < 1$, $|g(t)| < 1$, and $g(0) = 0$. Hence by Schwarz's lemma

$$|g(t)| \leq |t|.$$

Here there are two alternatives. Either equality holds everywhere or else it holds nowhere. In the former case

$$(15.1.15) \qquad g(t) = e^{i\alpha} t,$$

where α is real. An elementary but laborious calculation shows that $f(z)$ is now of the form (15.1.3), where β and z_0 are expressible in terms of α, z_1, and $f(z_1)$. In this case (15.1.13) holds identically.

In the second case we have

$$| g(t) | < | t |$$

for every t, $| t | < 1$. Since $\rho(r)$ is strictly increasing,

$$\rho(0, g(t)) = \rho(| g(t)|) < \rho(| t |) = \rho(0, t)$$

for every t. We set $z = z_2$. Then by (15.1.11)

$$\rho(z_1, z_1) = \rho\left(\left| \frac{z_2 - z_1}{1 - \bar{z}_1 z_2} \right| \right)$$

and

$$\rho[f(z_1), f(z_2)] = \rho\left(\left| \frac{f(z_1) - f(z_2)}{1 - \overline{f(z_1)} f(z_2)} \right| \right) < \rho\left(\left| \frac{z_1 - z_2}{1 - \bar{z}_1 z_2} \right| \right),$$

as just proved. This implies the validity of (15.1.12) for all z_1, z_2.

EXERCISE 15.1

1. Suppose that $f(z)$ is holomorphic in $| z | < R$, $| f(z) | < M$, and $f(0) = 0$. Show that

$$| f(z) | \leqq \frac{M}{R} | z |$$

and discuss the case of equality.

2. Deduce Liouville's theorem (Theorem 8.2.2) from the preceding inequality.

3. If $f(z)$ is holomorphic in $| z | < 1$, $| f(z) | < 1$, and $f(z) = 0$ for $z = z_1$, $z_2, \cdots, z_n$, show that

$$| f(z) | \leqq \prod_{k=1}^{n} \left| \frac{z - z_k}{1 - \bar{z}_k z} \right|.$$

4. If the function $f(z)$ of the preceding problem has infinitely many zeros z_k in the unit disk, prove that either $\sum_1^\infty [1 - | z_k |] < \infty$ or $f(z) \equiv 0$. (This coincides with Problem 7 of Exercise 9.2. At the present juncture, use the preceding problem with z real positive.)

5. If $\sum [1 - | z_k |] < \infty$, prove that the infinite product

$$\prod_{k=1}^{\infty} \frac{\bar{z}_k}{| z_k |} \frac{z_k - z}{1 - \bar{z}_k z}$$

converges absolutely and defines a function $f(z)$ holomorphic in $| z | < 1$ such that $| f(z) | < 1$ and $f(z_k) = 0$ for all k (Wilhelm Blaschke, 1915).

6. In Poincaré's hyperbolic geometry the "points" are those of the upper half-plane, the "straight lines" are the (Euclidean) circular arcs orthogonal to the real axis which lie in the upper half-plane, and the transformation group is the set of Möbius transformations with real coefficients and positive determinant $ad - bc$. How should postulates H_1–H_4 be modified to introduce a metric? The distance function $D(z_1, z_2)$ should be invariant under real translations $z \to z + \alpha$, α real. This argument points to $D(i, iy)$ as a suitable replacement of $\rho(0, r)$ in the discussion and leads to the formula

$$D(z_1, z_2) = \frac{1}{2} \log \frac{|\overline{z_1} - z_2| + |z_1 - z_2|}{|\overline{z_1} - z_2| - |z_1 - z_2|}.$$

7. Verify that if (15.1.15) holds, then $f(z)$ is of the form (15.1.3).

8. The inequality $|g(t)| \leq |t|$ implies that $|g'(0)| \leq 1$. If $g(t)$ is defined by (15.1.14) and preceding formulas, compute $g'(0)$ and show that the inequality implies that

$$(1 - |z|^2)|f'(z)| \leq 1 - |f(z)|^2$$

for all z, and that equality holds if and only if $f(z)$ is given by (15.1.3).

9. Given a rectifiable arc γ in $|z| < 1$. Prove that γ also has a finite length in the hyperbolic metric and express the length by an integral. Let $w = f(z)$ be holomorphic and less than 1 in absolute value in the unit disk. Let Γ be the image of γ under this mapping. Prove that Γ also has finite hyperbolic length (abbreviated hl) and that $hl(\Gamma) \leq hl(\gamma)$, where equality holds if and only if $f(z)$ is of the form (15.1.3) (G. Pick, 1916).

15.2. Normal families.   We shall now consider the compactness properties of families of meromorphic functions. We start with a number of conventions, definitions, and lemmas.

The functions involved have definite, finite or infinite, values at all points of the domain of definition. We shall admit $f(z) \equiv \infty$, $z \in D$, as a function. It is convenient under these circumstances to use the chordal metric on the sphere in dealing with the range space. See Section 2.5 for the properties of $\chi(w_1, w_2)$. For the discussion of continuity, see also Section 4.1, where similar questions are treated, although there the spherical metric is used in the domain instead of in the range.

DEFINITION 15.2.1.   *A function $f(z)$ defined in a domain D is spherically continuous at $z = z_0$, $z_0 \in D$, provided that for any $\varepsilon > 0$ there exists a $\delta = \delta(\varepsilon, z_0)$ such that*

(15.2.1) $\chi[f(z), f(z_0)] < \varepsilon$ *if* $|z - z_0| < \delta$ *and* $z \in D$.

$f(z)$ is spherically continuous in D if it is spherically continuous at all points of D.

LEMMA 15.2.1.   *If $f(z)$ is spherically continuous in D and if S is a closed subset of D, then $f(z)$ is uniformly spherically continuous in S, that is, given any $\varepsilon > 0$ there exists a $\delta = \delta(\varepsilon)$ such that for all z, z_0 in S with $|z - z_0| < \delta$ we have*

$$(15.2.2) \qquad \chi[f(z), f(z_0)] < \varepsilon.$$

The proof is analogous to that of Theorem 4.1.1 and is left to the reader.

LEMMA 15.2.2.   *If $f(z)$ is spherically continuous in D, so is $1/f(z)$.*

This follows immediately from the relation

$$(15.2.3) \qquad \chi\left[\frac{1}{f(z_1)}, \frac{1}{f(z_2)}\right] = \chi[f(z_1), f(z_2)].$$

See Problem 9 on page 44 of Volume I.

LEMMA 15.2.3.   *A meromorphic function is spherically continuous in its domain of definition.*

Proof.   At a point z_0 where $f(z)$ is holomorphic, $f(z)$ is continuous in the ordinary sense, and from the inequality

$$(15.2.4) \qquad \chi[f(z), f(z_0)] < |f(z) - f(z_0)|$$

it follows that $f(z)$ is also spherically continuous. At a pole of $f(z)$ the same argument applies to $1/f(z)$ and we use Lemma 15.2.2 to pass back to $f(z)$.

DEFINITION 15.2.2.   *A family F of functions $f_\alpha(z)$ defined in a domain D is said to be spherically equicontinuous at a point z_0 of D if for each $\varepsilon > 0$ there exists a $\delta = \delta(\varepsilon, z_0)$ such that*

$$\chi[f_\alpha(z), f_\alpha(z_0)] < \varepsilon$$

for $z \in D$, $|z - z_0| < \delta$, and for every α. F is spherically equicontinuous in D if it has this property at every point of D.

LEMMA 15.2.4.   *A family of functions, equicontinuous in the ordinary sense on a domain D, is also spherically equicontinuous.*

Proof.   This follows from (15.2.4).

DEFINITION 15.2.3.   *A sequence of functions $f_n(z)$ defined on a set S converges spherically uniformly on S if, given any $\varepsilon > 0$, there exists an $N = N(\varepsilon)$ such that*

$$(15.2.5) \qquad \chi[f_m(z), f_n(z)] < \varepsilon, \quad m, n > N,$$

for every z in S.

LEMMA 15.2.5.   *If a sequence of spherically continuous functions $\{f_n(z)\}$ converges spherically uniformly on a closed set S, then it converges to a function $f(z)$ which is spherically continuous on S.*

Proof. It is clear that $f(z)$ is uniquely defined, but its value may very well be ∞ everywhere in S. That $f(z)$ is spherically continuous on S follows from the inequality

$$(15.2.6) \quad \chi[f(z), f(z_0)] \leq \chi[f(z), f_n(z)] + \chi[f_n(z), f_n(z_0)] + \chi[f(z_0), f_n(z_0)],$$

which holds for any n and for $z, z_0 \in S$. Given $\varepsilon > 0$, we can choose n so large that the first and third terms in the right member are each $< \varepsilon$. Now n is fixed and $f_n(z)$ is spherically continuous on S. Hence we can choose a $\delta = \delta(\varepsilon, z_0)$ such that

$$\chi[f_n(z), f_n(z_0)] < \varepsilon$$

for $|z - z_0| < \delta$. This shows that $f(z)$ is spherically continuous on S.

LEMMA 15.2.6.   *If a sequence $\{f_n(z)\}$ of spherically continuous functions converges spherically uniformly to a limit $f(z)$ on a set S, then the functions $f_n(z)$ are spherically equicontinuous in S.*

Proof. We have merely to interchange the left member and the second term on the right in (15.2.6) and apply the same argument as in the proof of the preceding lemma.

DEFINITION 15.2.4.   *A family F of functions $f_\alpha(z)$, meromorphic in a domain D, is said to be normal[1] in D if every sequence $\{f_n(z)\} \subset F$ contains a subsequence which converges spherically uniformly on every compact subset of D.*

THEOREM 15.2.1.   *A sequence of functions $\{f_n(z)\}$, meromorphic in a domain D, which converges spherically uniformly on each compact subset of D, converges to a meromorphic function $f(z)$. It is not excluded that $f(z)$ is actually holomorphic in D or reduces to a finite or infinite constant.*

Proof. By Lemma 15.2.5, $f(z)$ is spherically continuous in D, and by Lemma 15.2.6 the functions $f_n(z)$ are spherically equicontinuous on each fixed compact subset of D. In particular, they have this property in D_0 if D_0 is a bounded domain such that $\overline{D_0} \subset D$. Let $z_0 \in D_0$ and consider $f(z_0)$. There are three possibilities.

---

[1] This term is due to Paul Montel, whose work in this field started in 1907. It is now standard terminology in complex function theory though in other branches of analysis the term "compact" is preferred for similar concepts. The latter term was introduced by Maurice Fréchet in his Paris thesis of 1906. Montel's early work dealt with families of holomorphic functions. The extension to meromorphic functions was made by him in 1912. This development was strongly influenced by the new light shed on the theorem of Picard by the work of Landau and Carathéodory. The elegant formulation in terms of the spherical metric is due to Ostrowski in 1925. The notion of equicontinuity we owe to Giulio Ascoli (1843–1896) in 1884. Cesare Arzelà (1847–1912) proved in 1895 that equicontinuity and uniform boundedness of a family of continuous functions are necessary and sufficient for compactness. A corresponding theorem for holomorphic functions was proved in 1903 by Giuseppe Vitali (1875–1932). (See further below.)

First we may have $f(z_0)$ finite. By the equicontinuity there exists a neighborhood $N(z_0)$ of z_0 and a finite M such that $|f(z)| < M$ in $N(z_0)$. Further, the spherically uniform convergence implies the existence of an integer $m(z_0)$ such that $|f_n(z)| < 2M$ for $n > m(z_0)$ and $z \in N(z_0)$. But this implies that the sequence $\{f_n(z) \mid n > m(z_0)\}$ is uniformly bounded in $N(z_0)$ so that these functions are actually holomorphic in $N(z_0)$. Moreover, for $z \in N(z_0)$, $n > m(z_0)$ we have

$$|f(z) - f_n(z)| < (1 + 4M^2)\chi[f(z), f_n(z)]$$

(cf. page 68 of Volume I), whence it follows that the sequence converges uniformly in the Euclidean metric. Thus $f(z)$ is holomorphic in $N(z_0)$.

Secondly, if $f(z_0) = \infty$ there may be a neighborhood $N(z_0)$ of z_0 in which $f(z) \neq \infty$ except at $z = z_0$. We may assume this neighborhood to be small enough so that $f(z) \neq 0$ in $N(z_0)$. Then $1/f(z)$ is well defined in $N(z_0)$ and is 0 at $z = z_0$. Further, since

$$\chi\left[\frac{1}{f}, \frac{1}{f_n}\right] = \chi[f, f_n],$$

there exists an integer $m(z_0)$ such that the sequence $\{1/f_n \mid n > m(z_0)\}$ consists of functions holomorphic in $N(z_0)$ which converge uniformly to the limit $1/f$. Thus $1/f(z)$ is holomorphic in $N(z_0)$ and has a zero at $z = z_0$. It follows that $f(z)$ is meromorphic in $N(z_0)$ and has a pole at $z = z_0$.

The third possibility is that $f(z_0) = \infty$ and $f(z)$ takes on the value ∞ at points other than z_0 in every neighborhood of z_0. By the spherical continuity we must have $|f(z)| > M$ for $|z - z_0| < \delta(M)$, and the spherically uniform convergence shows that $|f_n(z)| > \frac{1}{2}M$ for $n > m(M)$. Thus, the functions $1/f_n(z)$, $n > m(M)$, are holomorphic in $|z - z_0| < \delta(M)$ and uniformly small. Again we have uniform convergence of $1/f_n(z)$ to $1/f(z)$ in this neighborhood. Hence $1/f(z)$ is holomorphic. But the point $z = z_0$ is a limit point of zeros of this function, and therefore $f(z)$ must be identically ∞ in the neighborhood under consideration.

Thus we see that for every point $z_0 \in D_0$ there is a neighborhood in which either $f(z)$ or $1/f(z)$ is holomorphic. We can clearly cover $\overline{D_0}$ by a finite number of these neighborhoods. If $1/f(z) \equiv 0$ in one of these neighborhoods, then it is also identically 0 in all overlapping neighborhoods. It follows that $1/f(z) \equiv 0$ in D_0 and, since D_0 is arbitrary, $\overline{D_0} \subset D$, the same conclusion holds in D. Thus, $f(z) \equiv \infty$ in D. In the other cases, the same type of reasoning shows that $f(z)$ is holomorphic except for poles in D. If there are no poles, then $f(z)$ is holomorphic in D and it may possibly reduce to a finite constant. This completes the proof.

THEOREM 15.2.2. *A necessary and sufficient condition for a family of meromorphic functions to be normal in a domain D is that it be spherically equicontinuous on compact subsets of D.*

Proof. I. *Necessity.* Suppose that the family $F \equiv \{f_\alpha(z)\}$ is normal. Lemma 15.2.6 tells us that any converging sequence belonging to F is spherically equicontinuous in D. Suppose that F as a whole is not spherically equicontinuous. This assumption implies that for a particular number $\varepsilon > 0$ we can find a point z_0 and a sequence of points $\{z_n\}$, all in D, together with a sequence $\{f_n(z)\}$ in F such that $z_n \to z_0$ but

(15.2.7) $\chi[f_n(z_0), f_n(z_n)] > \varepsilon$

for all n. On the other hand, since F is normal the sequence $\{f_n(z)\}$ contains a subsequence $\{f_{n_k}(z)\}$ which converges spherically uniformly on compact subsets of D. We may arrange that such a subset contains the points $z_n, n = 0, 1, 2, \cdots$. But by Lemma 15.2.6, $\{f_{n_k}(z)\}$ is spherically equicontinuous at $z = z_0$, and for $n = n_k$, k large, we have a contradiction of (15.2.7). Thus it is necessary that F be spherically equicontinuous in D.

II. *Sufficiency.* Suppose then that F is spherically equicontinuous in D. Let D_0 be a bounded domain such that $\overline{D_0} \subset D$ and consider all points in D_0 which are of the form

(15.2.8) $2^{-m}(n + ip)$,

where m runs through 0 and the positive integers, and where n and p are arbitrary integers. This is a countable set of points which we arrange into subsets according to increasing values of m. Within each subset we let $j_1 + ik_1$ precede $j_2 + ik_2$ if either $j_1 < j_2$ or $j_1 = j_2$ and $k_1 < k_2$. Finally we weed out all repetitions so that each point $2^{-m}(n + ip)$ occurs once and only once. This sequence we denote by $\{z_k\}$. It is dense in D_0.

We now consider the point set $\{f_\alpha(z_1)\}$ on the sphere. It has at least one limit point. Let w_1 be this limit point and let $\{f_{1,n}(z)\}$ be a sequence in F such that

$$\lim_{n \to \infty} f_{1,n}(z_1) = w_1.$$

Next we consider the point set $\{f_{1,n}(z_2)\}$ on the sphere. Let w_2 be a limit point of this sequence and let $\{f_{2,n}(z)\}$ be a subsequence of the sequence $\{f_{1,n}(z)\}$ which converges to w_2 at $z = z_2$. This subsequence also converges at $z = z_1$ to w_1. We then proceed to z_3, and so on. In this manner we obtain a sequence of sequences $\{f_{k,n}(z)\}$ such that

$$F \supset \{f_{1,n}(z)\} \supset \{f_{2,n}(z)\} \supset \cdots \supset \{f_{k,n}(z)\} \supset \{f_{k+1,n}(z)\} \supset \cdots$$

and

$$\lim_{m, n \to \infty} \chi[f_{k,m}(z_j), f_{k,n}(z_j)] = 0, \quad j = 1, 2, \cdots, k.$$

The diagonal sequence

$$\{f_{n,n}(z)\} \equiv \{f_n(z)\}$$

will then converge in the sense of the chordal metric at all points z_k. These points are everywhere dense in D_0, but at this stage of the process we have only a countable set of Cauchy sequences, and the structure of the limit function cannot be clearly discerned.

We shall prove, however, that the sequence $\{f_n(z)\}$ converges everywhere in D_0, spherically uniformly with respect to z. First we note that the sequence $\{f_n(z)\}$ is spherically equicontinuous in D_0 uniformly with respect to z. Hence if $\varepsilon > 0$ is given, we can find a $\delta = \delta(\varepsilon)$ such that

$$(15.2.9) \qquad \chi[f_n(Z_1), f_n(Z_2)] < \varepsilon$$

for any choice of Z_1 and Z_2 in D_0 provided $|\, Z_1 - Z_2 \,| < \varepsilon$. Choose an integer m_0 such that $2^{\frac{1}{2}-m_0} < \delta$ and let $z_1, z_2, \cdots, z_N$ be the finite subset of $\{z_k\}$ which results if we require that $m \leq m_0$ in (15.2.8). Then every point z of D_0 has a distance from this finite subset which is $< \delta$. Consider a particular point $z \in D_0$ and let z_k be a point in the finite subset such that $|\, z - z_k \,| < \delta$. Then

$$(15.2.10) \quad \chi[f_m(z), f_n(z)] \leq \chi[f_m(z), f_m(z_k)] + \chi[f_m(z_k), f_n(z_k)] + \chi[f_n(z_k), f_n(z)].$$

By (15.2.9) and the choice of z_k, the first and third terms on the right are each $< \varepsilon$. Since the sequence $\{f_n(z_k)\}$ converges for $k = 1, 2, \cdots, N$, there exists an integer $p = p(\varepsilon, N)$ such that

$$\chi[f_m(z_k), f_n(z_k)] < \varepsilon$$

for $m > p, n > p$ and any choice of $k \leq N$. Hence the right member of (15.2.10) does not exceed 3ε for such a choice of m and n. Since the estimate holds uniformly with respect to z in D_0, we conclude that the sequence $\{f_n(z)\}$ converges spherically uniformly in D_0. The limit function $f(z)$ by the preceding theorem is meromorphic in D_0.

We can proceed from D_0 to all of D by a second diagonal process. For this purpose we use a sequence of nested domains D_k, starting with D_0, such that $\overline{D_{k-1}} \subset D_k \subset D$ for each k and such that for each z in D there is an integer $m(z)$ such that $z \in D_k$ for $k > m(z)$. The first diagonal sequence $\{f_n(z)\}$ converges spherically uniformly in D_0. It contains a subsequence which converges spherically uniformly in D_1. In this manner we obtain a sequence of subsequences such that the kth one converges spherically uniformly in D_k. From these subsequences we select the diagonal sequence which, by virtue of its construction, converges spherically uniformly in each D_k. The limit function $f(z)$ consequently exists in D. It is a meromorphic function in the wide sense; that is, it may possibly be the constant function ∞, or it may be holomorphic or a finite constant. This completes the proof.

We shall now give some examples of normal families.

THEOREM 15.2.3. *A family F of functions $f_\alpha(z)$ holomorphic in a domain D is normal in D if for each compact subset S of D there is a finite number $M(S)$ such that*

$$|\, f_\alpha(z) \,| \leq M(S), \quad z \in S,$$

for each α. The limit functions are then holomorphic in D.

Proof. Let $D_0 \subset D_1 \subset \overline{D_1} \subset D$. We suppose that the boundary of D_1 is a rectifiable curve C_1 and that the distance from D_0 to C_1 is $\delta > 0$. Then for z_1, z_2 in D_0 we have

$$f_\alpha(z_1) - f_\alpha(z_2) = (z_1 - z_2) \frac{1}{2\pi i} \int_{C_1} \frac{f_\alpha(t)\, dt}{(t - z_1)(t - z_2)},$$

so that

$$(15.2.11) \qquad | f_\alpha(z_1) - f_\alpha(z_2) | \leqq | z_1 - z_2 | \frac{M(\overline{D_1})l(C_1)}{2\pi\delta^2}$$

and the functions $f_\alpha(z)$ are uniformly equicontinuous in D_0. By Lemma 15.2.4 we have then also spherical equicontinuity, and F is normal.

In this case spherically uniform convergence on compact subsets of D implies ordinary uniform convergence on such sets, and thus each limit function must be holomorphic in D. This completes the proof. We note the following

COROLLARY. *If F is a family of functions holomorphic and uniformly bounded in a domain D, then F is equicontinuous in D.*

Another consequence of (15.2.11) is

THEOREM 15.2.4. *If the functions $f_\alpha(z)$ are holomorphic in D and the family is normal in D, then the family $\{f_\alpha'(z)\}$ is also normal in D.*

For we have

$$(15.2.12) \qquad | f_\alpha'(z) | \leqq \frac{M(\overline{D_1})l(C_1)}{2\pi\delta^2}, \qquad z \in D_0.$$

The boundedness condition used above may be introduced in a less obvious fashion. We shall give a couple of examples in which for the sake of simplicity we choose D to be the unit disk.

THEOREM 15.2.5. *The family of power series*

$$(15.2.13) \qquad f_\alpha(z) = \sum_{n=0}^{\infty} a_{n\alpha} z^n$$

such that $| a_{n\alpha} | \leqq M$ for all n and α, is normal in the unit disk.

Proof. We have

$$| f_\alpha(z) | \leqq \frac{M}{1 - r}, \qquad | z | \leqq r < 1,$$

so we are back to the case considered above.

THEOREM 15.2.6. *The family of functions $f_\alpha(z)$ holomorphic in the unit disk and such that*

$$(15.2.14) \qquad \int_0^{2\pi} \int_0^1 | f_\alpha(re^{i\theta}) |^2 r\, dr\, d\theta \leqq M$$

is normal in the unit disk.

Proof. The value of the integral is

$$\pi \sum_{n=0}^{\infty} \frac{|a_{n\alpha}|^2}{n+1},$$

if $f_\alpha(z)$ is given by (15.2.13). By Cauchy's inequality

$$|f_\alpha(z)| \le \sum_{n=0}^{\infty} |a_{n\alpha}| r^n \le \left\{\frac{1}{r}\sum_{n=0}^{\infty} |a_{n\alpha}|^2 \frac{r^{n+1}}{n+1}\right\}^{\frac12}\left\{\sum_{n=0}^{\infty} (n+1)r^n\right\}^{\frac12}$$

$$\le \left(\frac{M}{\pi}\right)^{\frac12} \frac{1}{1-r},$$

so once more we are back to boundedness on compact subsets.

In these cases all functions $f_\alpha(z)$ omit a fixed spherical cap when z is restricted to a compact subset of D and all these caps are neighborhoods of the north pole, $w = \infty$. Since a rotation of the sphere does not change chordal distances, we conclude that Theorem 15.2.3 may be generalized as follows:

THEOREM 15.2.7. *F is normal in D if there exists a fixed point a and for every compact subset S of D a neighborhood $N(a, S)$ of a on the sphere which is omitted by every $f_\alpha(z)$ in F when z is restricted to S.*

The following theorem, due to Montel, lies much deeper.

THEOREM 15.2.8. *A family F of functions $f_\alpha(z)$ meromorphic in D is normal in D if there are three fixed numbers a, b, c such that none of the equations*

$$(15.2.15) \qquad f_\alpha(z) = a, \quad f_\alpha(z) = b, \quad f_\alpha(z) = c$$

has a solution in D.

Proof. This result is closely related to Picard's theorem, and we proceed as in the proof of the latter up to a certain point. First we normalize by setting

$$g_\alpha(z) = \frac{c-b}{c-a}\frac{f_\alpha(z)-a}{f_\alpha(z)-b}.$$

This gives a family G which omits the three values $0, 1, \infty$. If G is normal, so is F, and vice versa. We then take the inverse modular function $\nu(w)$. It is infinitely many-valued but each branch satisfies $\Im[\nu(w)] > 0$. Each branch is holomorphic in the neighborhood of an arbitrary point $w_0 \neq 0, 1$. We choose a point z_0 in D and take for $\nu[g_\alpha(z_0)]$ that uniquely determined value which belongs to the fundamental region B defined by (13.6.17) and the statement that follows it. With this initial value we can then define $\nu[g_\alpha(z)]$ uniquely in some neighborhood $N(z_0)$ of z_0. Let D_0 be a simply-connected domain such that $N(z_0) \subset D_0 \subset \overline{D_0} \subset D$. Then $\nu[g_\alpha(z)]$ can be continued indefinitely in D_0 without any singularity being encountered, for $g_\alpha(z)$ is holomorphic in D_0 and never assumes the values $0, 1,$ and ∞, which are singularities of $\nu(w)$. Since D_0

is simply-connected, the theorem of monodromy applies so that each of the functions $v[g_\alpha(z)]$ is holomorphic in D_0. Thus the family $N \equiv \{v[g_\alpha(z)]\}$ consists of holomorphic functions in D_0 and these functions have positive imaginary parts so they omit a fixed half of the sphere. By the preceding theorem N is a normal family in D_0. But D_0 was an arbitrary simply-connected subdomain of D. It follows that N is normal in D.

From this fact we have to conclude that G is also normal. Suppose that $\{h_n(z)\} = \{v[g_n(z)]\}$ is a sequence which converges spherically uniformly on compact sets to a limit function $h(z)$. Here $h(z)$ is meromorphic in the wide sense and its imaginary part is ≥ 0. The latter property prevents the presence of any poles, for in a small neighborhood of a pole a function is free to take on all large values, without limitation on the imaginary part. Thus $h(z)$ is either holomorphic in D or identically ∞. Further, we note that $h(z)$ cannot take on a real value in D unless it is a real constant. Thus either $\Im[h(z)] > 0$ in D or $\Im[h(z)] \equiv 0$. Actually the only real values which can occur are 0 and ± 1. For $h_n(z_0)$ is chosen in B so that $h(z_0)$ must be in $\bar{B}$ and the only real points in $\bar{B}$ are 0 and ± 1. See Figure 13, page 177.

Now if $h(z) \equiv 0$ and D_0 is chosen as above, then there exists an $m(\varepsilon)$ such that

$$| h_n(z) | < \varepsilon \quad \text{for} \quad z \in D_0, \ m(\varepsilon) < n.$$

We have

$$g_n(z) = k^2[h_n(z)]$$

in the notation of Section 13.6 since $k^2[v(w)] = w$. When $h_n(z) \to 0$, $g_n(z) \to 1$. The first limit holds uniformly with respect to z in D_0 and, thus, also the second. Similarly, if $h_n(z) \to 1$ or -1

$$g_n(z) = k^2[h_n(z)] \to \infty,$$

and this holds uniformly in the chordal metric. Further, if $h_n(z) \to \infty$ then $k^2[h_n(z)] \to 0$ uniformly in D_0.

Suppose now that $h(z)$ does not have one of these three exceptional constant values. Then $h(z)$ is holomorphic in D_0 and in this domain it is bounded away from infinity and from the real axis. We have now ordinary uniform convergence of $h_n(z)$ to its limit $h(z)$ and from this it follows that

$$(15.2.16) \qquad g_n(z) = k^2[h_n(z)] \to k^2[h(z)] \equiv g(z)$$

uniformly in D_0. Here D_0 is arbitrary, so the conclusion extends to any compact subset of D. This shows that any spherically uniformly convergent sequence in N gives rise to a spherically uniformly convergent sequence in G. On the other hand, if we have any sequence in G, then we can map it on a sequence in N by the transformation $w \to v(w)$. From the latter we select a spherically uniformly convergent subsequence which we then map back into G by the transformation $v \to k^2(v)$. The final sequence obtained in this manner is a subsequence of the original one and it converges spherically uniformly. Thus G is also normal. Since this implies that F is normal, the theorem is proved.

We give one more theorem, but without proof.

THEOREM 15.2.9.   *A family of functions $f_\nu(z)$ meromorphic in D is normal in D if there are three fixed numbers a, b, c such that the roots of the equations (15.2.15) have multiplicities α, β, γ respectively where*

(15.2.17)
$$\frac{1}{\alpha} + \frac{1}{\beta} + \frac{1}{\gamma} < 1.$$

The search for normal families has been much aided by the following heuristic principle:

A family of holomorphic (meromorphic) functions which have a property P in common in a domain D is (apt to be) a normal family in D if P cannot be possessed by non-constant entire (meromorphic) functions in the finite plane.

In Theorem 15.2.3 the property P is boundedness, and the theorem of Liouville asserts that a non-constant entire function cannot be bounded. In Theorem 15.2.8, P is the property of omitting three fixed values, and Picard's theorem shows that no non-constant meromorphic function can have this property. In Theorem 15.2.9 we are dealing with three fixed values which are completely ramified and whose ramifications would add up to at least

$$\left(1 - \frac{1}{\alpha}\right) + \left(1 - \frac{1}{\beta}\right) + \left(1 - \frac{1}{\gamma}\right) > 2.$$

This would violate Theorem 14.7.1 if the functions were meromorphic in the finite plane.

EXERCISE 15.2

1. If $m(r, \infty; f)$ is defined by (14.4.2), use the inequality

$$\log |f(re^{i\theta})| \leq \frac{R + r}{R - r} m(R, \infty; f), \quad r < R,$$

to prove that a family of functions which are holomorphic in the unit circle and which satisfy $m(r, \infty; f_\alpha) < M$ for $0 < r < 1$ is normal.

2. Consider the following special case of Theorem 15.2.9:

$$a = 0, \alpha = 3; \quad b = 1, \beta = 2; \quad c = \infty, \gamma = \infty,$$

that is, holomorphic functions with triple zeros and double "ones." Give a proof along the lines of Theorem 15.2.8, replacing $\nu(w)$ by the inverse of the function $J(\tau)$ of Section 13.6.

3. A sequence of spherically continuous functions is supposed to converge spherically to $f(z)$ in a domain D. The convergence is said to be *spherically continuous* at $z = z_0$ if $z_n \to z_0$ implies

$$\lim \chi[f(z_0), f_n(z_n)] = 0.$$

Show that the sequence converges spherically uniformly to $f(z)$ if the stated condition holds for each z_0 in D.

15.3. Induced convergence. A striking example of induced convergence
is given by Theorem 7.10.1. That result shows that uniform convergence in the
interior of a "scroc" C of a sequence of functions holomorphic inside and on C
follows from the uniform convergence of the sequence on C itself. The theorem,
a case of what may be called "inward propagation" of convergence, was discovered by Carl Runge (1856–1927) in 1884. The first example of outward
propagation of convergence occurs in Stieltjes' posthumous memoir on continued
fractions (1894–95). There it is shown that a uniformly bounded sequence of
functions holomorphic in a domain D converges uniformly in D provided the
sequence converges uniformly in some subdomain of D. Keeping the boundedness condition, Osgood was able in 1901 to replace uniform convergence in a
subdomain by simple convergence in a set dense in D. In 1904 Porter reduced
the condition still further, first to a set dense on a "scroc" in D and later to a set
having a limit point in D. The result had already been proved by Vitali, however, and is commonly known as Vitali's theorem.[1]

Vitali formulated his theorem for holomorphic functions. The extension to
meromorphic functions goes back to Montel.

THEOREM 15.3.1.   *Let the functions $f_n(z)$ be meromorphic and spherically
equicontinuous in a domain D, and let the sequence $\{f_n(z)\}$ converge spherically in a
point set $\{z_k\}$ which has a limit point z_0 in D. Then spherical convergence takes
place everywhere on D and is spherically uniform on compact sets contained in D.*

Proof.   This can be based on Theorem 15.2.2. The family $\{f_n(z)\}$ is normal
and contains at least one subsequence which converges spherically uniformly on
compact subsets to a function $f(z)$ which is meromorphic in the wide sense in D.
If the whole sequence does not converge spherically uniformly to $f(z)$, then we
can find a point a in D where the sequence $\{f_n(a)\}$ has two distinct limit points,
w_1 and w_2. We may suppose that $f(a) = w_1$. We can also find a subsequence
which converges to a function $g(z)$ such that $g(a) = w_2$. Both functions $f(z)$ and
$g(z)$ are meromorphic in the wide sense. Since

$$\lim_{n \to \infty} f_n(z_k)$$

exists for $k = 1, 2, 3, \cdots$, we must have

(15.3.1) $f(z_k) = g(z_k), \quad k = 1, 2, 3, \cdots.$

If $f(z)$ is holomorphic in some domain D_0 which contains $z = z_0$, then we see

---

[1] William Fogg Osgood (1864–1943) was, together with Maxime Bôcher (1867–1918),
the founder of the Harvard school of analysis. He made important contributions to the
theory of functions of one and of several complex variables, including early results on the
behavior of the conformal mapping function on the boundary. Milton Brockett Porter
(1869–1960) was one of Bôcher's brightest pupils; his early work on Bessel functions was
highly praised. Moreover, he also discovered "overconvergence." That phenomenon was
rediscovered twice before Porter was given credit for the original observation.

that $g(z)$ must also be holomorphic in D_0, and the identity theorem (the Corollary of Theorem 8.1.3) shows that $f(z)$ and $g(z)$ coincide in D_0 and, hence, by analytic continuation, everywhere in D. In particular, we then have also $f(a) = g(a)$, in conflict with our assumption that these numbers are distinct. The same conclusion holds if $f(z) \equiv \infty$. We have then also $g(z) \equiv \infty$, and $f(a) = g(a)$. Finally, suppose that $z = z_0$ is a pole of order m of $f(z)$. We can then find a domain D_0 containing $z = z_0$ in which $(z - z_0)^m f(z)$ is holomorphic. It is clear that

$$(z_k - z_0)^m f(z_k) = (z_k - z_0)^m g(z_k)$$

for all k. Since the left member tends to a limit $\neq 0$ when $k \to \infty$, the right member must have the same limit. From this we conclude that $(z - z_0)^m g(z)$ is also holomorphic in D_0 and that $g(z)$ has a pole of order m at $z = z_0$. Again the identity theorem may be applied, this time to the functions $(z - z_0)^m f(z)$ and $(z - z_0)^m g(z)$. From the fact that these two functions are holomorphic in D_0 and identical there, it follows that they are identical in all of D. Hence $f(z)$ and $g(z)$ are again identical, and $f(a) = g(a)$. This contradiction shows that the sequence $\{f_n(z)\}$ converges everywhere in D and spherically uniformly on compact subsets.

In view of the great importance of this theorem it is desirable to have a more direct and constructive proof. Such a proof was given by E. Lindelöf in 1913 for the case of holomorphic functions. It does not use the diagonal process or the compactness of the set $\{f_\alpha(z)\}$ for fixed z. The limit function is constructed in a neighborhood of $z = z_0$, and the extension to the rest of D is by analytic continuation. If we place the limit point at the origin the basic theorem takes the following form:

THEOREM 15.3.2.   *Given a sequence of power series*

(15.3.2)
$$f_n(z) = \sum_{p=0}^{\infty} a_{np} z^p, \quad n = 1, 2, 3, \cdots,$$

which converge for $|z| < R$ and which satisfy $|f_n(z)| < M$ for such values of z. Suppose there is a sequence $\{z_k\}$ with $|z_k| < R$ and $\lim z_k = 0$ such that

$$\lim_{n \to \infty} f_n(z_k)$$

exists for each k. Then

(15.3.3)
$$\lim_{n \to \infty} a_{np} \equiv a_p$$

exists for each p, the series

(15.3.4)
$$\sum_{p=0}^{\infty} a_p z^p \equiv f(z)$$

converges for $|z| < R$, and for each positive $R_1 < R$ we have

(15.3.5)
$$\lim_{n \to \infty} f_n(z) = f(z)$$

uniformly with respect to z in $|z| \leq R_1$.

Proof. Since $|f_n(z)| < M$ for $|z| < R$ and all n, the estimates of Cauchy show that

(15.3.6) $$|a_{np}| < MR^{-p}$$

for all n and p. Next we note that

$$|f_n(z) - f_n(0)| < 2M,$$

and with the aid of Schwarz's lemma this may be strengthened to

(15.3.7) $$|f_n(z) - f_n(0)| < 2\,\frac{M}{R}\,|z|.$$

It follows that

$$|f_n(z_k) - a_{n0}| < 2\,\frac{M}{R}\,|z_k|.$$

Now, for any choice of k, we have

$$|a_{m0} - a_{n0}| \leq |f_m(z_k) - a_{m0}| + |f_m(z_k) - f_n(z_k)| + |f_n(z_k) - a_{n0}|,$$

so that

$$\limsup_{m,\,n \to \infty} |a_{m0} - a_{n0}| \leq 4\,\frac{M}{R}\,|z_k|.$$

Since k is arbitrary and $\lim z_k = 0$, we see that the right member can be made as small as we please. It follows that $\{a_{n0}\}$ is a Cauchy sequence; we denote its limit by a_0. We note that $|a_0| \leq M$.

We now introduce new sequences

(15.3.8) $$f_{np}(z) \equiv \sum_{\mu=p}^{\infty} a_{n\mu} z^{\mu-p}, \quad n, p = 0, 1, 2, \cdots,$$

where $f_{n0}(z) = f_n(z)$. We note that

$$f_{np}(z) = a_{np} + z f_{n,\,p+1}(z),$$

and, in particular,

$$f_{np}(z_k) = a_{np} + z_k f_{n,\,p+1}(z_k)$$

for all k. We may suppose that $z_k \neq 0$ for all k. We then see that if we know the existence of

$$\lim_{n \to \infty} f_{np}(z_k) \quad \text{and} \quad \lim_{n \to \infty} a_{np} \equiv a_p,$$

then

$$\lim_{n \to \infty} f_{n,\,p+1}(z_k)$$

will also exist. Suppose that we have verified the existence of these limits for $p \leq q$. By (15.3.6)

(15.3.9) $$|a_p| \leq MR^{-p}$$

will hold for $p \leq q$. Suppose further that we have shown that for $|z| < R, p \leq q$,

(15.3.10) $$|f_{np}(z)| \leq (p+1)MR^{-p}.$$

Then by Schwarz's lemma

$$| z | | f_{n,\,p+1}(z) | = | f_{np}(z) - a_{np} | \leq (p + 2)MR^{-p-1} | z |,$$

so that (15.3.10) holds also for $p = q + 1$. One more use of the lemma gives

$$| f_{n,\,q+1}(z) - a_{n,\,q+1} | \leq (q + 3)MR^{-q-1} | z |.$$

We can then complete the argument as above. For any k

$$| a_{m,\,q+1} - a_{n,\,q+1} | \leq | f_{m,\,q+1}(z_k) - a_{m,\,q+1} | + | f_{m,\,q+1}(z_k) - f_{n,\,q+1}(z_k) |$$
$$+ | f_{n,\,q+1}(z_k) - a_{n,\,q+1} |$$
$$\leq 2(q + 3)MR^{-q-1} | z_k | + | f_{m,\,q+1}(z_k) - f_{n,\,q+1}(z_k) |,$$

whence it follows that $\lim_{n \to \infty} a_{np} \equiv a_p$ exists also for $p = q + 1$. Since the induction hypothesis holds for $q = 0$ we see that (15.3.3) and (15.3.9) hold for all p.

Thus $f(z)$ is well defined, and the series converges for $| z | < R$. Suppose that $R_1 < R$, and let δ be a positive number. We can then find an integer m such that

$$\frac{2MR_1{}^m}{R^{m-1}(R - R_1)} \leq \delta.$$

With this m and for $| z | \leq R_1$ we have

$$| f_n(z) - f(z) | \leq \sum_{p=0}^{m-1} | a_{np} - a_p | | z |^p + \sum_{p=m}^{\infty} (| a_{np} | + | a_p |) | z |^p$$
$$\leq \sum_{p=0}^{m-1} | a_{np} - a_p | R_1{}^p + 2M \sum_{p=m}^{\infty} (R_1/R)^p.$$

Hence

$$\limsup_{n \to \infty} | f_n(z) - f(z) | \leq \delta$$

by the choice of m, and this shows that the sequence $\{f_n(z)\}$ converges uniformly to $f(z)$ for $| z | \leq R_1 < R$.

This completes the proof of Theorem 15.3.2, but we shall briefly indicate how the argument extends to an arbitrary domain D in which the sequence $\{f_n(z)\}$ is holomorphic and uniformly bounded. It is no restriction to suppose that the sequence $\{z_k\}$ converges to $z = 0$. Suppose now that D_0 is a subdomain of D such that $\{0\} \subset D_0 \subset \overline{D_0} \subset D$. We can cover $\overline{D_0}$ by a finite number of circular disks $C_0, C_1, C_2, \cdots, C_m$ such that: (i) $f_n(z)$ is holomorphic in $\cup C_k$, (ii) the center of C_0 is $z = 0$, and (iii) the center ζ_k of C_k also belongs to a disk C_j with $j < k$ where $k = 1, 2, \cdots, m$. Theorem 15.3.2 now shows that the sequence converges in C_0 and uniformly on any concentric disk interior to C_0. It follows that the sequence converges in some small neighborhood of $z = \zeta_1$ which lies in C_0. Using the theorem once more, we see that this fact induces convergence in all of C_1. The center of C_2, $z = \zeta_2$, lies either in C_0 or in C_1. In either case we now have convergence in some small neighborhood of ζ_2, which then induces convergence in all of C_2. In a finite number of steps every point $z \in \overline{D_0}$ can be covered, and thus $\overline{D_0}$ belongs to a region of uniform convergence of $\{f_n(z)\}$. This proves Vitali's theorem for holomorphic functions.

Vitali's theorem is based on the identity theorem of Weierstrass, that is, the fact that a function $g(z)$ which is holomorphic in a domain D and which vanishes on an infinite point set $\{z_k\}$ in D is identically 0 if this set has a limit point in D. Any other condition on the set which leads to the same conclusion could be used for an extension of Vitali's theorem. At the time when Vitali wrote, no other condition was known, but new conditions became available about ten years later. The best known of these is the theorem of Blaschke (1915). This theorem has figured in several of the Exercises; we state it here for future reference. The proof is left to the reader if he has not already worked it out.

THEOREM 15.3.3.   Let $g(z)$ be holomorphic in the unit disk where it has zeros at the points $\{z_k\}$ and satisfies

$$(15.3.11) \qquad m(r, \infty; g) \equiv \frac{1}{2\pi} \int_0^{2\pi} \log^+ |g(re^{i\varphi})| \, d\varphi \le M$$

for $0 < r < 1$. Then $g(z) \equiv 0$ if and only if

$$(15.3.12) \qquad \sum_{k=1}^{\infty} [1 - |z_k|] = \infty.$$

From this theorem we get immediately the following result:

THEOREM 15.3.4.   Let the sequence $\{f_n(z)\}$ consist of functions holomorphic in the unit disk and suppose they satisfy

$$(15.3.13) \qquad m(r, \infty; f_n) \le M$$

for all n and $0 < r < 1$. If the sequence converges on a set of points $\{z_k\}$ satisfying condition (15.3.12), then it converges uniformly on each disk $|z| \le R < 1$.

Proof.   The sequence is normal by Problem 1 of Exercise 15.2. Thus there exists at least one limit function $f(z)$ which also satisfies (15.3.13). If the whole sequence should not converge to $f(z)$, uniformly on $|z| < R$, there would be a subsequence converging to a holomorphic limit $g(z)$ which satisfies (15.3.13), and there would be a point a where $g(a) \ne f(a)$. On the other hand, we have $f(z_k) - g(z_k) = 0$, whence it follows that $g(z) \equiv f(z)$, and the theorem is proved.

In his original paper Blaschke assumed $g(z)$ to be bounded. Condition (15.3.11) is due to Ostrowski.

Another condition can be read off from Theorem 11.3.3 (Carlson). Here some difficulty is encountered in formulating growth conditions which are preserved by the limit process. The following theorem is a sample of what can be done along such lines:

THEOREM 15.3.5. *Let the sequence $\{f_n(z)\}$ consist of functions holomorphic in the half-plane $\Re(z) > -\frac{1}{2}$ and such that*

$$(15.3.14) \quad |f_n(-\tfrac{1}{2} + re^{i\theta})| \leq (1 + r)^{-1} \exp[h(\theta)r], \quad -\tfrac{1}{2}\pi \leq \theta \leq \tfrac{1}{2}\pi.$$

Here $h(-\tfrac{1}{2}\pi) \leq \pi, h(\tfrac{1}{2}\pi) \leq \pi$, and $h(\theta)$ is the function of support of a convex region in the open strip $-\pi < v < \pi, w = u + iv$. If the sequence $\{f_n(k)\}$ converges for each positive integer k, then $\{f_n(z)\}$ converges everywhere in $\Re(z) > -\frac{1}{2}$, uniformly on compact sets.

Proof. The sequence $\{f_n(z)\}$ is obviously bounded on compact sets and hence normal in the half-plane $\Re(z) > -\frac{1}{2}$. Any limit function $f(z)$ must also satisfy (15.3.14). If there were two limit functions, their difference would satisfy the conditions of Theorem 11.3.3 and hence be identically zero. This completes the proof.

We shall encounter other uniqueness theorems for functions holomorphic in a half-plane in Section 19.2. They give rise to similar convergence theorems.

EXERCISE 15.3

1. Given the factorial series

$$\sum_{n=0}^{\infty} \frac{a_n n!}{z(z+1)\cdots(z+n)},$$

where $a_n \geq 0$. Prove that if the series converges for $z = 1$, then it converges uniformly for $\Re(z) \geq 1$. Would your method work if 1 were replaced by some other real number a?

2. A sequence of functions $\{f_n(z)\}$ is defined in the upper half-plane by

$$f_n(z) = z - \frac{1}{f_{n-1}(z)}, \quad f_0(z) = z.$$

Show that the values of these functions lie in the upper half-plane. Verify that the sequence converges for $z = iy, y > 1$, and use this to prove convergence in the upper half-plane. What is the limit function? (The functions $f_n(z)$ are the *convergents of a continued fraction*.)

3. Verify that the functions $\tan nz$ form a normal family in the upper half-plane. Find the unique limit function.

4. Given a continuous function $f(x)$ defined on $[0, \infty)$ and a sequence $\{f_n(z)\}$ of functions holomorphic in $\Re(z) > -\frac{1}{2}$ and such that

$$f_n(k2^{-n}) = f(k2^{-n}), \quad k, n = 0, 1, 2, 3, \cdots.$$

Under what conditions on the rate of growth of $f_n(z)$ can you conclude that these functions are identical and give the analytic continuation of $f(x)$ to a half-plane?

15.4. Applications. Normal families and Vitali's theorem are extremely useful tools in analytic function theory. We shall give some examples here. Other applications will occur later.

We start with an elementary question. It is often laborious to prove that an analytic expression defines an analytic function in a certain domain. In a situation of this kind the theorem of Vitali is apt to be helpful. The following theorem deals with a frequently occurring situation:

THEOREM 15.4.1.   *Given a function $F(t, z)$ defined for t on a curve C and z in a domain D such that* (i) $| F(t, z) | \leq M$ *for all t and z,* (ii) $F(t, z)$ *is a continuous function of t on C for each fixed z in D, and* (iii) $F(t, z)$ *is a holomorphic function of z in D for each fixed t on C. Given a function $g(t)$ defined and of bounded variation on C. Then*

$$(15.4.1) \qquad\qquad f(z) \equiv \int_C F(t, z) \, dg(t)$$

defines a holomorphic function of z in D.

Proof.   This follows almost directly from the definition of the integral as a limit of Riemann-Stieltjes sums. That definition gives a sequence of functions

$$f_n(z) = \sum_{k=1}^{n} F(t_{k, n}, z)[g(t_{k, n}) - g(t_{k-1, n})]$$

which converges to $f(z)$ for each z in D. Here $\{t_{k, n}\}$ is a set of partitions of C, independent of z. Each $f_n(z)$ is clearly holomorphic in D by condition (iii), and $| f_n(z) | \leq M V[g]$, where $V[g]$ is the total variation of $g(t)$ on C. Thus Vitali's theorem applies and the desired conclusion follows.

Instead of continuity we may assume that $F(t, z)$ is integrable with respect to $g(t)$.

Even if the theorem does not apply directly to a given situation, it may still be useful as a means of constructing an approximating sequence of functions to which Vitali's theorem does apply. Thus an integral of the form

$$(15.4.2) \qquad\qquad \int_0^{\infty} F(t, z) \, dt$$

may not satisfy the conditions of the theorem, but it may happen that the theorem applies to each of the integrals

$$f_n(z) = \int_0^{n} F(t, z) \, dt.$$

If now the sequence $\{f_n(z)\}$ is made up of functions holomorphic and uniformly bounded in a domain D and if the sequence converges in a point set with a limit point in D, then Vitali's theorem does apply, and we can conclude that (15.4.2) defines a function holomorphic in D.

We now turn to more sophisticated applications of the theory of normal families. Montel and Julia have shown that this theory may be used for a study of the behavior of an analytic function in a partial or full neighborhood of a singular point. The first theorem which we shall prove is the so-called *great theorem* of Picard.

THEOREM 15.4.2.   *If $f(z)$ is holomorphic in the punctured circular disk $0 < |z| < R$ and has an essential singular point at $z = 0$, then there is at most one value a ($\neq \infty$) such that the equation*

$$f(z) = a$$

has only a finite number of roots in the disk.

Proof.   Suppose that there were two such values, a and b. Without restricting the generality, we may assume that the equations

$$f(z) = a, \quad f(z) = b$$

have no roots in the punctured disk. Let us now consider the sequence of functions

(15.4.3) $$f_n(z) = f(2^{-n}z), \quad n = 0, 1, 2, \cdots,$$

in the annulus

$$A_0: \quad 2^{-2}R < |z| < 3 \cdot 2^{-2}R.$$

The values that $f_n(z)$ assumes in A_0 coincide with the values of $f(z)$ in the annulus

$$A_n: \quad 2^{-2-n}R < |z| < 3 \cdot 2^{-2-n}R.$$

Since A_n and A_{n+1} overlap, each value assumed by $f(z)$ in the disk $0 < |z| < 3 \cdot 2^{-2}R$ is assumed by at least one of the functions $f_n(z)$ in A_0. By Theorem 15.2.8 the sequence $\{f_n(z)\}$ is normal in A_0. There is consequently a subsequence $\{f_{n_k}(z)\}$ which converges spherically uniformly on compact subsets of A_0 to a limit function. In particular we have convergence on the circle $|z| = \frac{1}{2}R$. There are two possibilities.

(1) The limit function is holomorphic in A_0. It is then bounded on the circle $|z| = \frac{1}{2}R$, and this implies that the functions $f_{n_k}(z)$ are uniformly bounded on the circle. Thus, we can find a finite M such that

$$|f_{n_k}(\tfrac{1}{2}Re^{i\theta})| \leqq M.$$

Hence

$$|f(2^{-1-n_k}Re^{i\theta})| \leqq M, \quad k = 1, 2, 3, \cdots.$$

This asserts the existence of a sequence of concentric circles converging to the origin on which $f(z)$ is bounded, $|f(z)| \leqq M$. Since $f(z)$ is holomorphic in each annulus bounded by two consecutive circles in this family, the maximum principle shows that $|f(z)| \leqq M$ also in the annulus. This implies

(15.4.4) $$|f(z)| \leqq M, \quad 0 < |z| < 2^{-1-n_1}R.$$

We have now a contradiction, for in any neighborhood of an essential singular point the function is necessarily unbounded.

(2) The limit function is identically ∞. In this case there exists a subsequence $\{f_{n_k}(z)\}$ such that $1/f_{n_k}(z)$ converges uniformly to 0 on the circle $|z| = \frac{1}{2}R$. Using the same argument as above, we see that (15.4.4) holds with $f(z)$ replaced by $1/f(z)$. This also contradicts the Casorati-Weierstrass theorem: if $z = 0$ is an essential singularity, $1/f(z)$ cannot be bounded in a neighborhood of $z = 0$.

It follows that we cannot have two omitted values a and b. This is the best possible result since the function $e^{1/z}$ does omit one value, 0, in every neighborhood of the essential singularity at $z = 0$.

If $f(z)$ is supposed to be meromorphic instead of holomorphic in the disk, we can use the same argument to show that there are at most two omitted values, one of which may be infinity. As a matter of fact, we can always reduce the general case in which a, b, c are omitted to the special case in which 0, 1, ∞ are omitted and then we are back to Theorem 15.4.2.

It was first observed by Gaston Julia that the proof of Picard's great theorem given above permits a stronger conclusion, which we shall now bring out. The proof shows that if $z = 0$ is an essential singular point of the function $f(z)$ which is holomorphic in the disk $0 < |z| < R$, then the family $\{f_n(z)\}$ defined by (15.4.3) cannot be normal in the annulus A_0. This implies that there is at least one point $z = z_0$ of A_0 and a small disk D_0: $|z - z_0| < \varepsilon$, in A_0 such that the sequence is not normal in D_0. With this disk we also consider the homothetic disks

$$(15.4.5) \qquad D_n\colon \quad |z - 2^{-n}z_0| < 2^{-n}\varepsilon, \quad n = 1, 2, 3, \cdots.$$

The values taken on by $f_n(z)$ in D_0 are simply the values taken on by $f(z)$ in D_n. Let a and b be any two numbers. We assert that at least one of the equations

$$f(z) = a, \quad f(z) = b$$

has a root in infinitely many of the disks D_n.

Suppose this were not so. Then there would exist an integer k such that

$$f(z) \neq a, \quad f(z) \neq b, \quad z \in D_n, \quad k \leq n.$$

Theorem 15.2.8 now applies to the sequence $\{f_n(z) \mid k \leq n\}$ and shows that this sequence is normal in D_0. Since this implies that the original sequence $\{f_n(z)\}$ is normal in D_0, we have a contradiction. Thus we have proved Julia's theorem:

THEOREM 15.4.3.   *If $z = 0$ is an essential singular point of $f(z)$, then there exists a sequence of homothetic disks (15.4.5) in which $f(z)$ assumes every value infinitely often with at most one exception.*

COROLLARY.   *There exists at least one ray $\arg z = \theta$ such that $f(z)$ assumes every value with at most one exception in each sector $\theta - \delta < \arg z < \theta + \delta$ no matter how small the positive number δ is.*

Such a ray is known as a *direction of Julia* or *direction J*. This concept is particularly important in the theory of entire functions where the essential singularity is at $z = \infty$. We can also introduce this notion in the theory of meromorphic functions. It is then necessary to admit two exceptional values.

The technique used in proving Theorem 15.4.2 may also be used to prove other results. The following is a theorem discovered by E. Lindelöf. The proof is due to Montel.

THEOREM 15.4.4. *Let $f(z)$ be holomorphic in the sector S: $\alpha < \arg z < \beta$, $0 < |z| < R$, and suppose that $f(z)$ omits two values when z is restricted to S. Suppose further that there is a γ such that $\alpha < \gamma < \beta$ and*

$$(15.4.6) \qquad \lim_{r \to 0} f(re^{i\gamma}) = c$$

exists where c may be finite or infinite. Then

$$(15.4.7) \qquad \lim_{r \to 0} f(re^{i\theta}) = c$$

for $\alpha < \theta < \beta$, uniformly in any fixed interior sector.

Proof. We introduce the domains

$$(15.4.8) \qquad D_n: \quad \alpha < \arg z < \beta, \quad 2^{-n-2}R < |z| < 3 \cdot 2^{-n-2}R,$$

and consider the sequence of functions

$$(15.4.9) \qquad f_n(z) = f(2^{-n}z), \quad n = 0, 1, 2, \cdots,$$

in D_0. We observe that the values of $f_n(z)$ in D_0 are those of $f(z)$ in D_n. Since $f(z)$ omits two fixed values, the sequence $\{f_n(z)\}$ is normal in D_0. By (15.4.6) the sequence converges to c on the line segment

$$\arg z = \gamma, \quad \tfrac{1}{4}R < |z| < \tfrac{3}{4}R.$$

Vitali's theorem applies and shows that the sequence converges in D_0 and, in fact, uniformly on compact sets. If $c = \infty$, convergence and uniform convergence are understood in the spherical sense, otherwise in the usual sense. This fact implies the validity of (15.4.7).

COROLLARY. *If $f(z)$ is holomorphic in the sector S and if there are two rays $\arg z = \gamma_1$ and $\arg z = \gamma_2$ with $\alpha < \gamma_1 < \gamma_2 < \beta$ such that*

$$\lim_{r \to 0} f(re^{i\gamma_1}) = c_1, \quad \lim_{r \to 0} f(re^{i\gamma_2}) = c_2,$$

then either $c_1 = c_2$ or $f(z)$ omits at most one value in the sector $\gamma_1 < \arg z < \gamma_2$.

The following theorem is a variation on the same theme:

THEOREM 15.4.5. *Let $f(z)$ satisfy the assumptions of the preceding theorem, except for (15.4.6), which is to be replaced by*

$$(15.4.10) \qquad |f(re^{i\gamma})| \leq M, \quad 0 < r \leq \tfrac{5}{8}R.$$

If S_0 is any fixed interior sector $\alpha < \alpha_1 \leq \arg z \leq \beta_1 < \beta$, then there exists a finite $M(S_0)$ such that

$$(15.4.11) \qquad |f(z)| \leq M(S_0), \quad z \in S_0.$$

Proof. We define the sequence $\{f_n(z)\}$ as above and introduce the regions Q_n defined by

$$(15.4.12) \qquad \alpha_1 \leq \arg z \leq \beta_1, \quad 2^{-n-2}R \leq |z| \leq 5 \cdot 2^{-n-3}R.$$

The region Q_0 is shown in Figure 15. These regions cover S_0 since any two consecutive regions overlap. The sequence $\{f_n(z)\}$ is normal in D_0 and consequently spherically equicontinuous in Q_0. We want to prove ·that condition

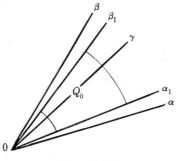

Figure 15

(15.4.10) implies that the sequence is actually uniformly bounded in Q_0. That this is indeed the case is shown by the following argument:

Suppose that $\max |f_n(z)| = M_n$ if z is restricted to Q_0. If this sequence is unbounded, we can find a subsequence $\{M_{n_k}\}$ which becomes infinite. The corresponding sequence $\{f_{n_k}(z)\}$ is normal in D_0, and we may suppose that it converges to a limit function $g(z)$ in D_0. Since the original sequence is made up of holomorphic functions, $g(z)$ is either holomorphic in D_0 or identically infinity. The second alternative is excluded by (15.4.10). It follows that $g(z)$ is bounded in Q_0. By assumption we can find a sequence of points $\{z_k\}$ in Q_0, having a limit point z_0 in Q_0, such that

$$\lim |f_{n_k}(z_k)| = \infty.$$

This implies that

$$\lim \chi[g(z_0), f_{n_k}(z_k)] = [1 + |g(z_0)|^2]^{-\frac{1}{2}} \neq 0.$$

This contradicts the spherically uniform convergence, so we conclude that the sequence $\{M_n\}$ must be bounded, and this in turn implies (15.4.11).

This type of argument can also be used to prove Schottky's theorem of 1904 (Friedrich Schottky, 1851–1935).

THEOREM 15.4.6. *Let $a_0 \neq 0, 1, \infty$ and let $R, 0 < R < \infty$, be given. Let F be a family of functions $f(z)$ holomorphic in the disk $|z| < R$ which do not assume the value 0 or 1, and let $f(0) = a_0$ for each $f(z)$ in F. Let θ be a fixed number, $0 < \theta < 1$. Then there exists a finite positive number $M(a_0, \theta)$ such that*

$$(15.4.13) \qquad |f(z)| \leq M(a_0, \theta) \quad for \quad |z| \leq \theta R$$

and every choice of $f(z)$ in F.

Proof. We set $g(z) \equiv f(zR)$ for $|z| < 1$. Then the functions $g(z)$ so obtained form a subset of the family G of all functions $g(z)$ holomorphic in the unit disk such that $g(0) = a_0$ and $g(z) \neq 0, 1$. This family G is normal in the unit disk and hence spherically equicontinuous in the disk $|z| \leq \theta$. Since $g(0) = a_0$ for every $g(z)$ in G, the argument used above shows that G is actually uniformly bounded in the disk $|z| \leq \theta$. The bound is clearly a function of a_0 and θ; that is, (15.4.13) holds.

Schottky's theorem was inspired by that of Landau (Theorem 14.5.2), and each theorem is obtainable from the other. The best value of $M(a_0, \theta)$ may be found by considering the properties of the mapping function (14.5.4).

EXERCISE 15.4

1. A function $f(z)$ holomorphic in $|z| < 1$ and continuous in $|z| \leq 1$ which is zero on an arc of the unit circle is identically zero. (*Hint:* See pages 186 and 187 of Volume I, the reflection principle of Schwarz. Actually the result is true if $f(z) = 0$ on a set of positive measure and the unit circle may be replaced by an arbitrary "scroc.")

2. Let $\{f_n(z)\}$ be a sequence of functions holomorphic in $|z| < 1$ and continuous in $|z| \leq 1$. If the sequence converges on an arc of the unit circle, show that it converges uniformly on every disk $|z| < R < 1$.

3. If $\Re(\alpha) > -\frac{1}{2}$, prove that the integral

$$\int_{-1}^{1} (1 - t^2)^{\alpha - \frac{1}{2}} \cos zt \, dt$$

is an entire function of z.

4. Show how Vitali's theorem can be used to show that Euler's integral for the Gamma function defines a holomorphic function of z in the right half-plane.

5. Prove that the Julia directions of e^z and of $\tan z$ are given by the values $\arg z = \pm \frac{1}{2}\pi$ and $\arg z = 0, \pi$ respectively.

6. Find the Julia directions of $f(z) = \int_0^z e^{-t^2} \, dt$.

7. Let $f(z)$ be holomorphic in a half-strip $a < x < b, \, 0 < y, \, z = x + iy$. Prove analogues of Theorems 15.4.4 and 15.4.5 assuming that $f(z)$ tends to a limit or is bounded on a line $x = c, \, a < c < b$.

8. Under the assumptions of Theorem 15.4.6 prove that for $|z| < \theta R$ we also have

$$|f(z)| \geq \left\{ M\left(\frac{1}{a_0}, \theta\right) \right\}^{-1} \quad \text{and} \quad |f(z) - 1| \geq \left\{ M\left(\frac{1}{1 - a_0}, \theta\right) \right\}^{-1}.$$

COLLATERAL READING

As a general reference see

MONTEL, PAUL. *Leçons sur les Familles Normales de Fonctions Analytiques et leurs Applications.* Gauthier-Villars, Paris, 1927. Chelsea Publishing Co., New York, 1973.

See also

LEAU, L. *Les Suites de Fonctions en Général (Domaine Complexe).* Mémorial des Sciences Mathématiques, No. 59. Gauthier-Villars, Paris, 1932.

VALIRON, G. *Familles Normales et Quasi-Normales de Fonctions Méromorphes.* Mémorial des Sciences Mathématiques, No. 38. Gauthier-Villars, Paris, 1932.

In connection with Section 15.1 see

GOLUSIN, G. M. *Geometrische Funktionentheorie,* Chap. 8. Deutscher Verlag der Wissenschaften, Berlin, 1957.

JULIA, GASTON. *Principes Géométriques d'Analyse.* Gauthier-Villars, Paris, 1930.

SCHWARZ, H. A. "Zur Theorie der Abbildung," in *Gesammelte Mathematische Abhandlungen,* Vol. II, pp. 108–132 (especially p. 110). Verlag von Julius Springer, Berlin, 1890. Chelsea Publishing Co., New York, 1972.

The use of the chordal metric in Section 15.2 is from

OSTROWSKI, ALEXANDER. "Über Folgen analytischer Funktionen und einige Verschärfungen des Picardschen Satzes," *Mathematische Zeitschrift,* Vol. 24 (1925), pp. 215–258.

Ostrowski's paper also deals with Julia directions and non-normal sequences, for which consult

JULIA, GASTON. *Leçons sur les Fonctions Uniformes à Point Singulier Essentiel Isolé.* Gauthier-Villars, Paris, 1924.

16

LEMNISCATES

16.1. Chebichev polynomials. The present chapter is not a discourse on the theory of lemniscates in all its ramifications, but lemniscates form a recurrent theme. We use the term "lemniscate" to denote a curve of the form

(16.1.1)
$$| P(z) | = C,$$

where $P(z)$ is a polynomial, and we call

(16.1.2)
$$| P(z) | \leq C$$

a lemniscatic region. We are interested in the remarkable properties of approximation of such curves and regions and of the corresponding polynomials for suitable choices of $P(z)$.

We start with the properties of a class of extremal polynomials known as Chebichev polynomials.[1] In 1859 Chebichev posed and solved the following problem:

Let $\mathbf{P}_n$ *be the class of polynomials*

(16.1.3)
$$P(z) = z^n + a_1 z^{n-1} + \cdots + a_n.$$

Find the element of $\mathbf{P}_n$ *that differs the least from 0 in a given interval* $[a, b]$.

In modern terminology, the problem is to determine the polynomial $P(z)$ of least norm as an element of the space $C[a, b]$. For the interval $[-1, 1]$ the unique solution is given by

(16.1.4)
$$T_n(z) = 2^{-n}\{[z + (z^2 - 1)^{\frac{1}{2}}]^n + [z - (z^2 - 1)^{\frac{1}{2}}]^n\}$$
$$= 2^{1-n} \cos (n \text{ arc cos } z), \quad \| T_n \| = 2^{1-n}.$$

The general case $[a, b]$ is easily reduced to this particular case. We note that for the interval $[a, b]$ the norm of the nth extremal polynomial equals

(16.1.5)
$$[\tfrac{1}{4}(b - a)]^{n-1}.$$

The polynomials (16.1.4) are known as Chebichev polynomials, but the same name is also given to the general class of polynomials which give the solutions of similar extremal problems.

---

[1] Pafnutiy Livovich Chebichev (1821–1894), professor at the University of St. Petersburg, was the founder of the Russian school of analysts. He contributed to prime-number theory, continued fractions, theory of approximations, probability, mechanical quadratures, and mechanics. He had a number of pupils who became famous in their own right. The name is frequently transliterated "Tchebycheff."

THEOREM 16.1.1.   *Let E be a closed bounded set in the complex plane. Then there exists an element $T_n(z)$ of $\mathbf{P}_n$ such that*

(16.1.6) $$\max_{z \in E} |T_n(z)| = \min_{P \in \mathbf{P}_n} \max_{z \in E} |P(z)|.$$

The solution is unique, for a particular n, if the set contains at least n points. If E is infinite, the solution is unique for all n.

Proof.   We shall prove the existence of an extremal polynomial but we do not give a complete uniqueness proof. If $P(z) \in \mathbf{P}_n$, then $|P(z)|$ reaches its maximum value $M(P)$ on E since E is closed. The set $\{M(P) \mid P \in \mathbf{P}_n\}$ has a non-negative infimum which we denote by M. We can find a sequence $\{P(z, k)\} \subset \mathbf{P}_n$ such that

$$\lim_{k \to \infty} M[P(\cdot, k)] = M.$$

We can arrange the n zeros of $P(z, k)$ according to increasing real and imaginary parts so that it makes sense to speak of the jth zero of $P(z, k)$ where $j = 1, 2, \cdots, n$. We may suppose that all the zeros of $P(z, k)$ are located in $H(E)$, the convex hull of E, for a simple argument shows that if $Q(z) \in \mathbf{P}_n$ and has a zero outside $H(E)$, then we can find a $P(z) \in \mathbf{P}_n$ which has all its zeros in $H(E)$ and is such that $\max\limits_{z \in E} |P(z)| < \max\limits_{z \in E} |Q(z)|$. Since the zeros of $P(z, k)$ form a bounded point set, we can find a subsequence $\{P(z, k_m)\}$ such that the first zero of $P(z, k_m)$ tends to a limit z_1, then a subsequence for which the second zeros converge also, etc., and finally a diagonal sequence such that all zeros converge to limits $z_1, z_2, \cdots, z_n$. The resulting polynomial

$$T_n(z) = (z - z_1)(z - z_2) \cdots (z - z_n)$$

evidently belongs to $\mathbf{P}_n$ and

$$\max_{z \in E} |T_n(z)| = M.$$

The zeros of $T_n(z)$ are located in $H(E)$.

It is clear that if E consists of only k points and if $n > k$, then $M = 0$. Any polynomial of degree n vanishing on the k points of E is an extremal polynomial; thus there is no uniqueness. On the other hand, if $k \geq n$ and, in particular, if E is infinite, then $T_n(z)$ is uniquely determined. We do not need this fact below, so the proof is omitted except for the following remark:

If $T_n(z)$ is an extremal polynomial with respect to E in $\mathbf{P}_n$ and if $|T_n(z)|$ reaches its maximum value M at n distinct points of E, then $T_n(z)$ is the unique solution. In fact, suppose that $|T_n(z)| = M$ for $z = \zeta_{n,1}, \zeta_{n,2}, \cdots, \zeta_{n,n}$ and that $U_n(z)$ is another element of $\mathbf{P}_n$ with $\max\limits_{z \in E} |U_n(z)| = M$. Then for any α with $0 < \alpha < 1$

$$\alpha T_n(z) + (1 - \alpha) U_n(z) \in \mathbf{P}_n,$$

and the maximum of its absolute value in E cannot exceed M. It would actually be less than M unless

$$T_n(\zeta_{n,j}) = U_n(\zeta_{n,j}), \quad j = 1, 2, \cdots, n.$$

But if these relations hold, then $T_n(z) - U_n(z)$ is a polynomial of degree $\leq n - 1$ which has n distinct zeros and which, hence, is identically zero. This shows that such an extremal polynomial is indeed unique. To complete the uniqueness proof it suffices to prove that an extremal polynomial of degree n assumes the maximum of its absolute value at least n times in E. This is indeed the case, but we omit the proof.

This theorem goes back to Leonida Tonelli (1885–1946), who proved a special case in 1908. Most of the results given below and in the next section are due to Mihály Fekete (1886–1957). These results center around the notion of transfinite diameter introduced by Fekete in 1923. Later work in this direction is due to G. Pólya, G. Szegö, and F. Leja.

THEOREM 16.1.2. *Let E be an infinite closed bounded point set in the complex plane. Let $\{T_n(z; E)\}$ be the corresponding sequence of Chebichev polynomials, and set* max $| T_n(z; E) | = M_n(E)$. *Then*

$$(16.1.7) \qquad \lim_{n \to \infty} [M_n(E)]^{1/n} \equiv \rho(E) < \infty.$$

Proof. If the diameter of E equals δ, then δ is also the diameter of $H(E)$. Since the zeros of $T_n(z; E)$ lie in $H(E)$, it follows that

$$M_n(E) = M_n = \max_{z \in E} \prod_{j=1}^{n} | z - z_{n,j} | \leq \delta^n,$$

so that the sequence $\{(M_n)^{1/n}\}$ is bounded. Denote its lim inf and lim sup by α and β respectively. Then $0 \leq \alpha \leq \beta \leq \delta$, so it suffices to prove that $\beta \leq \alpha$. Given an $\varepsilon > 0$, we can find an $n = n(\varepsilon)$ such that $(M_n)^{1/n} < \alpha + \varepsilon$, and, hence,

$$| T_n(z) | < (\alpha + \varepsilon)^n, \quad z \in E.$$

If k and m are positive integers and $1 \leq m \leq n - 1$, then there exists a $K = K(\varepsilon, E)$ such that

$$| z^m [T_n(z)]^k | \leq K(\alpha + \varepsilon)^{nk+m}, \quad z \in E,$$

for all k and all admissible m. But the polynomial in the left member belongs to $\mathbf{P}_{nk+m}$, whence

$$M_{nk+m} \leq K(\alpha + \varepsilon)^{nk+m}$$

for all k and m. Extracting the root of order $nk + m$ and passing to the limit with k, we see that $\beta \leq \alpha + \varepsilon$. It follows that $\beta \leq \alpha$ and, hence, that $\alpha = \beta$. The theorem is proved.

The non-negative set function $\rho(E)$ has many names. In the present context it is often referred to as the *Chebichev constant* of E. In the next section it will be called the *transfinite diameter* of E, and in the potential theoretic considerations of Section 16.4 it becomes the *logarithmic capacity* of E or its *Robin constant*. Finally, in Section 17.3 the same quantity will figure as the *exterior mapping radius* of E when E is a continuum.

As a by-product of the foregoing proof we see that the Chebichev constant of a set does not exceed its diameter in the point-set sense. Formula (16.1.5) shows that the Chebichev constant of an interval is $\frac{1}{4}$ of its length. It is a simple matter to show that for a circle or a circular disk the constant in question equals the radius. The following theorem is a useful reformulation of already familiar results:

THEOREM 16.1.3.　*The infinite bounded closed set E is contained in the lemniscatic region*

(16.1.8)　　　　　　　　L_n:　$|\, T_n(z; E)\,| \leqq M_n$,

and the boundary of L_n has at least n points in common with E.

The last assertion involves the uniqueness property of $T_n(z; E)$. In (16.1.8) it is only the extremal property that matters.

A lemniscatic region L like L_n above has at most n separate components, each bounded by an oval of ∂L. Each oval contains in its interior at least one zero of the corresponding polynomial $P(z)$ by the principle of the maximum, and for the same reason the ovals are exterior to each other. If an oval does not have multiple points, that is, if $P'(z) \neq 0$ on the oval, and if the oval contains k zeros of $P(z)$ in its interior, then it contains $k - 1$ zeros of $P'(z)$. This is a special case of a theorem which in principle goes back to Riemann (1857) but which is often named after H. M. Macdonald, who rediscovered it in 1898.

EXERCISE 16.1

1. Prove that for the circle $|\, z\,| = R$ as well as for the disk $|\, z\,| \leqq R$ we have $T_n(z; E) = z^n$ and $\rho(E) = R$.

2. If E is a closed convex region, prove that for all n

$$T_n(z; E) = T_n(z; \partial E).$$

3. Verify that (16.1.4) actually is a polynomial of degree n and that the two expressions for $T_n(z)$ agree. Show also that $T_n(z)$ reaches its maximum at $n + 1$ points in $[-1, 1]$.

4. If E is an ellipse with foci at $z = \pm 1$ and major axis $2a > 2$, then it may be shown that $T_n(z; E)$ is given by (16.1.4). Assuming this, determine M_n, find how often $|\, T_n(z; E)\,| = M_n$ on E, and find $\rho(E)$.

5. If $P(z) \in \mathbf{P}_k$ and E is the lemniscatic region $|\, P(z)\,| \leqq R^k$, prove that $\rho(E) = R$.

6. Set $\alpha_n = \dfrac{1}{n} \log M_n(E)$ and prove that

$$(m + n)\alpha_{m+n} \leqq m\alpha_m + n\alpha_n.$$

Use this subadditive inequality to prove that $\lim \alpha_n$ exists.

7. Prove the Riemann-Macdonald theorem: *If $f(z)$ is holomorphic inside and on a simple curve C on which $|f(z)| = constant$ and $f'(z) \neq 0$, and if $f(z)$ has k zeros inside C, then $f'(z)$ has $k - 1$ zeros there.* (*Hint:* Express the numbers in question as contour integrals along C using Theorem 9.2.1. Note that C is analytic and hence rectifiable and has a continuously turning tangent. The difference of the integrals involved is expressible in terms of the change of direction of the tangent when z describes C.)

8. Use the preceding theorem to show that the lemniscate $|P(z)| = C$ consists of a single oval if and only if $|P(z)| < C$ at the zeros of $P'(z)$.

16.2. The transfinite diameter. Fekete's original problem was an algebraic one involving discriminants of algebraic equations whose roots satisfy certain conditions. In trying to maximize such discriminants, he was led to a set function which can be identified with the Chebichev constant.

Let E be a bounded closed infinite point set in the complex plane, and let $z_1, z_2, \cdots, z_n$ be any n, $n > 1$, distinct points of E. Define a number $\delta_n = \delta_n(E)$ by

$$(16.2.1) \qquad [\delta_n(E)]^{\frac{1}{2} n(n-1)} = \max_{z_j \in E} \prod_{1 \leqq j < k \leqq n} |z_j - z_k|.$$

The product

$$(16.2.2) \qquad \prod_{1 \leqq j < k \leqq n} (z_j - z_k) = V_n(z_j) = \det | \quad 1 \quad z_j \quad z_j^{\,2} \quad \cdots \quad z_j^{\,n-1} \, |$$

is the Vandermonde determinant formed with $z_1, z_2, \cdots, z_n$. Its square is the discriminant of the algebraic equation whose roots are $z_1, z_2, \cdots, z_n$. It is clear that $\delta_2(E) = \delta(E)$, the diameter of E, and that for all n

$$(16.2.3) \qquad \delta_n(E) \leqq \delta(E).$$

THEOREM 16.2.1.   *The sequence $\{\delta_n(E)\}$ is monotone decreasing and*

$$(16.2.4) \qquad \lim_{n \to \infty} \delta_n(E) = \rho(E).$$

Proof. Since E is a closed bounded set, it is clear that the maximum in (16.2.1) is assumed. Suppose, in particular, that $n + 1$ numbers $z_1, z_2, \cdots, z_n, z_{n+1}$ have been found such that

$$(16.2.5) \qquad [\delta_{n+1}(E)]^{\frac{1}{2} n(n+1)} = \prod_{1 \leqq j < k \leqq n+1} |z_j - z_k|.$$

The right member equals

$$| z_1 - z_{n+1} | \, | z_2 - z_{n+1} | \cdots | z_n - z_{n+1} | \prod_{1 \leq j < k \leq n} | z_j - z_k |,$$

which does not exceed

$$| z_1 - z_{n+1} | \, | z_2 - z_{n+1} | \cdots | z_n - z_{n+1} | \, [\delta_n(E)]^{\frac{1}{2} n(n-1)}.$$

In the same manner one shows that

$$(16.2.6) \qquad [\delta_{n+1}(E)]^{\frac{1}{2} n(n+1)} \leq \prod_{j \neq k} | z_j - z_k | \, [\delta_n(E)]^{\frac{1}{2} n(n-1)}$$

holds for each fixed $k, k = 1, 2, \cdots, n + 1$. We multiply these inequalities together and note that the product of the factors involving $| z_j - z_k |$ is the square of the right-hand side of (16.2.5). Hence

$$[\delta_{n+1}(E)]^{\frac{1}{2} n(n+1)^2} \leq [\delta_{n+1}(E)]^{n(n+1)} [\delta_n(E)]^{\frac{1}{2} n(n^2 - 1)},$$

which reduces to

$$\delta_{n+1}(E) \leq \delta_n(E).$$

This is the first assertion and it shows that $\lim \delta_n(E) \equiv \delta_0(E)$ exists.

To prove that $\delta_0(E) = \rho(E)$ is harder. Let us first prove that $\rho(E) \leq \delta_0(E)$. To this end we choose numbers $z_1, z_2, \cdots, z_n, z_{n+1}$ in E such that

$$[\delta_n(E)]^{\frac{1}{2} n(n-1)} = \prod_{1 \leq j < k \leq n} | z_j - z_k |.$$

Then

$$[\delta_{n+1}(E)]^{\frac{1}{2} n(n+1)} \geq \prod_{j=1}^{n} | z_{n+1} - z_j | \prod_{1 \leq j < k \leq n} | z_j - z_k |$$
$$\geq M_n(E) [\delta_n(E)]^{\frac{1}{2} n(n-1)}.$$

Extracting the nth root we obtain

$$[\delta_{n+1}(E)]^{\frac{1}{2}(n+1)} \geq [M_n(E)]^{\frac{1}{n}} [\delta_n(E)]^{\frac{1}{2}(n-1)}$$
$$\geq [M_n(E)]^{\frac{1}{n}} [\delta_{n+1}(E)]^{\frac{1}{2}(n-1)},$$

whence

$$(16.2.7) \qquad \delta_{n+1}(E) \geq [M_n(E)]^{\frac{1}{n}}.$$

This inequality gives $\delta_0(E) \geq \rho(E)$.

In proving the reversed inequality, we can use the properties of Vandermonde determinants. We go back to formula (16.2.2) and replace n by $n + 1$. The value of the determinant is not changed if we add to the last column any linear combination of the preceding columns. In particular, we can choose multipliers such that the last column becomes

$$T_n(z_1), T_n(z_2), \cdots, T_n(z_{n+1}),$$

where $T_n(z) = T_n(z; E)$ is the nth Chebichev polynomial of E. Hence

$$\prod(z_j - z_k) = \det | 1 \quad z_j \quad z_j^2 \quad \cdots \quad z_j^{n-1} \quad T_n(z_j) |.$$

If we expand this determinant according to the elements of the last column, we obtain an expression of the form

$$\sum_{j=1}^{n+1} (-1)^{n+1+j} V_{n,\,j} T_n(z_j),$$

where $V_{n,\,j}$ is a Vandermonde determinant in n of the z_k's omitting z_j. Here we substitute the upper bound for such determinants and the upper bound for $|T_n(z)|$ on E and obtain the inequality

$$[\delta_{n+1}(E)]^{\frac{1}{2}n(n+1)} \leq (n+1) M_n(E)[\delta_n(E)]^{\frac{1}{2}n(n-1)},$$

or, extracting the nth root,

$$(16.2.8) \qquad [\delta_{n+1}(E)]^{\frac{1}{2}(n+1)} \leq (n+1)^{\frac{1}{n}} [M_n(E)]^{\frac{1}{n}} [\delta_n(E)]^{\frac{1}{2}(n-1)}.$$

Here we multiply together the first m inequalities, simplify, and extract the mth root. Omitting the E's we get

$$(\delta_2 \cdots \delta_m)^{\frac{1}{2m}}(\delta_{m+1})^{\frac{m+1}{2m}} \leq [2 \cdot 3^{\frac{1}{2}} \cdots (m+1)^{\frac{1}{m}}]^{\frac{1}{m}} [M_1 M_2^{\frac{1}{2}} \cdots M_m^{\frac{1}{m}}]^{\frac{1}{m}}.$$

At this stage we need

LEMMA 16.2.1.   *If $\varepsilon_n > 0$ and $\lim \varepsilon_n = \varepsilon \geq 0$, then*

$$(16.2.9) \qquad \lim_{n \to \infty} (\varepsilon_1 \cdot \varepsilon_2 \cdots \varepsilon_n)^{\frac{1}{n}} = \varepsilon.$$

If $\varepsilon > 0$, this follows from Theorem 5.1.10 upon setting $w_n = \log \varepsilon_n$. The case $\varepsilon = 0$ is easily handled directly.

The lemma applies to each of the three products above whose limits are $[\delta_0(E)]^{\frac{1}{2}}$, 1, and $\rho(E)$ respectively. Since the second factor on the left also tends to $[\delta_0(E)]^{\frac{1}{2}}$, we see that $\delta_0(E) \leq \rho(E)$. The opposite inequality has already been proved. Therefore, we have $\delta_0(E) = \rho(E)$, and the theorem is proved.

Using the terminology of Fekete we shall call $\delta_0(E) = \rho(E)$ the *transfinite diameter* of E. This is a non-negative set function which has a number of interesting properties. Some of the basic ones are listed in

THEOREM 16.2.2.   (i) *$E_1 \subset E_2$ implies that $\rho(E_1) \leq \rho(E_2)$.* (ii) *If a is a fixed complex number and if $E_1 = (w \mid w = az, z \in E)$, then $\rho(E_1) = |a| \rho(E)$.* (iii) *If E_ε is the set of points within distance $\leq \varepsilon$ from E, then $\lim_{\varepsilon \to 0} \rho(E_\varepsilon) = \rho(E)$.* (iv) *If E^* is the set of roots of the equations*

$$Q(z) \equiv z^k + a_1 z^{k-1} + \cdots + a_k = w$$

when w ranges over E, then $\rho(E^*) = [\rho(E)]^{1/k}$.

REMARKS.   Fekete proved conversely that a non-negative set function which does not vanish identically, which is defined for all bounded closed sets E, and which has properties (i)–(iv), must coincide with the transfinite diameter of

E. We shall not prove this, and we also omit the proof of (iv), which will not be used in the following.

Proof. (i) If $E_1 \subset E_2$, then

$$M_n(E_2) = \max_{z \in E_2} |T_n(z; E_2)|$$

is not less than the maximum of the same polynomial on the subset E_1, and the latter is $\geq M_n(E_1)$. It follows that $M_n(E_1) \leq M_n(E_2)$, and this implies that $\rho(E_1) \leq \rho(E_2)$.

(ii) It is obvious that $\delta_n(E_1) = |a| \delta_n(E)$ for each n, and from this the desired equality follows directly.

(iii) From (i) we get $\rho(E) \leq \rho(E_\varepsilon)$. Suppose that $\zeta_1, \zeta_2, \cdots, \zeta_n$ are chosen in E_ε in such a manner that

$$[\delta_n(E_\varepsilon)]^N = \Pi |\zeta_j - \zeta_k|, \quad N = \tfrac{1}{2}n(n-1).$$

We can then find n numbers $z_1, z_2, \cdots, z_n$ in E, not necessarily distinct, such that

$$\zeta_k = z_k + \omega_k \varepsilon, \quad |\omega_k| \leq 1.$$

This gives

$$[\delta_n(E_\varepsilon)]^N \leq \Pi[|z_j - z_k| + 2\varepsilon]$$
$$= \Pi |z_j - z_k| + \{\Pi[|z_j - z_k| + 2\varepsilon] - \Pi |z_j - z_k|\}.$$

For any choice of j and k we have $|z_j - z_k| \leq \delta(E)$, the diameter of E. Hence the expression between the braces does not exceed

$$[\delta(E) + 2\varepsilon]^N - [\delta(E)]^N,$$

as is seen by expanding the two expressions and comparing terms, so that

$$[\delta_n(E_\varepsilon)]^N \leq [\delta_n(E)]^N + \{[\delta(E) + 2\varepsilon]^N - [\delta(E)]^N\}.$$

We extract the Nth root on both sides and use the inequality

$$(a + b)^{1/N} \leq a^{1/N} + b^{1/N}, \quad 0 < a, 0 < b,$$

to obtain

$$(16.2.10) \qquad \delta_n(E_\varepsilon) \leq \delta_n(E) + \delta(E)\left\{\left[1 + \frac{2\varepsilon}{\delta(E)}\right]^N - 1\right\}^{1/N}.$$

Here we have $\rho(E_\varepsilon) \leq \delta_n(E_\varepsilon)$. Further, given any $\eta > 0$, we can choose n so large that

$$\delta_n(E) < \rho(E) + \eta.$$

Now n is fixed, and hence also N, and ε is at our disposal. It is clear that we can choose an $\varepsilon(N, \eta)$ so small that for $\varepsilon < \varepsilon(N, \eta)$ the second term on the right in (16.2.10) is $< \eta$. For such ε

$$\rho(E) < \rho(E_\varepsilon) < \rho(E) + 2\eta,$$

and this proves the continuity property (iii).

We return to formula (16.2.1). Suppose that the n points $z_{n,1}, z_{n,2}, \cdots,$ $z_{n,n}$ are so chosen that

(16.2.11) $$[\delta_n(E)]^N = \prod_{1 \leq j < k \leq n} |z_{n,j} - z_{n,k}|, \quad z_{n,j} \in E,$$

holds. We form

(16.2.12) $$F_n(z; E) = \prod_{j=1}^{n} (z - z_{n,j}).$$

This is a sequence of polynomials associated with the set E that is, normally, distinct from the sequence $\{T_n(z; E)\}$ introduced before. These polynomials are uniquely determined if and only if the choice of $\{z_{n,j}\}$ is unique for each n. We disregard questions of uniqueness and list merely properties of $F_n(z; E)$ which are of importance for the following.

We set

(16.2.13) $$K_n = K_n(E) = \max_{z \in E} |F_n(z; E)|,$$

(16.2.14) $$L_n: \quad |F_n(z; E)| \leq K_n.$$

THEOREM 16.2.3.   *We have $E \subset L_n$ and $\partial L_n \cap E \neq \emptyset$. Further,*

(16.2.15) $$\lim_{n \to \infty} [K_n(E)]^{1/n} = \rho(E).$$

Proof. The first assertions follow from the definitions. To prove (16.2.15) we note first that

$$K_n(E) \geq M_n(E)$$

by the extremal property of $T_n(z; E)$, whence

$$\liminf_{n \to \infty} [K_n(E)]^{1/n} \geq \lim_{n \to \infty} [M_n(E)]^{1/n} = \rho(E).$$

To prove the reverse inequality, we note that in the proof of (16.2.7) we may replace $M_n(E)$ by $K_n(E)$ throughout. This gives

$$\limsup_{n \to \infty} [K_n(E)]^{1/n} \leq \lim_{n \to \infty} \delta_{n+1}(E) = \delta_0(E) = \rho(E),$$

and this completes the proof.

We shall use the term "Fekete polynomial" in referring to $F_n(z; E)$. Actually, these Fekete polynomials are more useful to us than the polynomials $T_n(z; E)$. The polynomials

(16.2.16) $$F_{n,k}(z) = \frac{F_n(z)}{(z - z_{n,k})F_n'(z_{n,k})}$$

are known as the *basic polynomials* for the nth Fekete abscissas and expand to

$$F_{n,k}(z) = \frac{(z - z_{n,1}) \cdots (z - z_{n,k-1})(z - z_{n,k+1}) \cdots (z - z_{n,n})}{(z_{n,k} - z_{n,1}) \cdots (z_{n,k} - z_{n,k-1})(z_{n,k} - z_{n,k+1}) \cdots (z_{n,k} - z_{n,n})}.$$

THEOREM 16.2.4.   *For each k and n,*

(16.2.17) $$\underset{z \in E}{\text{Max}} \mid F_{n,k}(z) \mid = F_{n,k}(z_{n,k}) = 1.$$

Proof. This follows immediately from the extremal properties of the points $z_{n,k}$. For if there were a point $\zeta \in E$ for which the product of the distances from the $n-1$ points $z_{n,2}, z_{n,3}, \cdots, z_{n,n}$ exceeded the corresponding product for $z_{n,1}$, then we could replace $z_{n,1}$ by ζ in (16.2.11) and this would increase the right member. This is absurd in view of the extremal property of the set $\{z_{n,k}\}$.

We note that $F_{n,k}(z_{n,k}) = 1$, and it is clear that there are points in any neighborhood of $z = z_{n,k}$ where $\mid F_{n,k}(z) \mid > 1$. A point with this property cannot be in E, whence it follows that all points $z_{n,k}$ lie on the boundary of E. We can make this more precise.

The complement of E, $\mathbf{C}(E) = G$, is an open set which is the union of a finite or countably infinite number of disjoint domains. One of these, D_∞, has the point at infinity as a limit point. The common boundary of D_∞ and of E, that is,

$$\overline{D_\infty} \cap E,$$

we call the *outer boundary* of E. If G has any other component, D say, then the boundary of D belongs entirely to E, and D is surrounded by E. We have now

THEOREM 16.2.5.   *The points $\{z_{n,k}\}$ which maximize the right member of* (16.2.11) *all lie on the outer boundary of E.*

Proof. If such a point $z_{n,k}$ belonged to the boundary of D, $D \neq D_\infty$, then the corresponding basic polynomial $F_{n,k}(z)$ would take on values in D which exceed 1 in absolute value. By the maximum principle, $\mid F_{n,k}(z) \mid > 1$ for some $z \in \bar{D} \cap E$; but this contradicts the preceding theorem. It follows that every point $z_{n,k}$ must belong to $\overline{D_\infty}$, that is, to the outer boundary of E.

Our final observation in this section concerns interpolation at the Fekete abscissas.

THEOREM 16.2.6.   *If $f(z)$ is holomorphic in some domain D containing the closed bounded set E and if $\mid f(z) \mid \leq M$ for $z \in E$, then the polynomial*

(16.2.18) $$L_{n-1}(z;f) \equiv \sum_{k=1}^{n} f(z_{n,k}) F_{n,k}(z;E)$$

has the same values as $f(z)$ at the n zeros $z_{n,k}$ of the nth Fekete polynomial $F_n(z;E)$ and

(16.2.19) $$\mid L_{n-1}(z;f) \mid \leq nM \quad for \quad z \in E.$$

This follows from (16.2.16) and (16.2.17).

We shall resume the study of the transfinite diameter, now in the role of the logarithmic capacity of a set, in Section 16.4, and we shall discover it in another disguise in Section 17.3, where the polynomials $F_n(z;E)$ will also play a fundamental role.

EXERCISE   16.2

1. For the segment $[-1, 1]$

$$F_n(z; E) = \frac{n!}{(2n-2)!} \frac{d^{n-2}}{dz^{n-2}} (z^2 - 1)^{n-1}.$$

Verify this for $n = 4$ and 5, using the fact that the zeros of $F_n(z; E)$ must be symmetric with respect to the origin and should include $z = \pm 1$.

2. For the unit disk we may take $F_n(z; E) = z^n - 1$. Verify, determine the value of K_n, and find how often $|F_n(z; E)|$ reaches this value. Is $F_n(z; E)$ uniquely defined in this case?

3. Show that if a non-negative set function is not identically zero and satisfies the four conditions of Theorem 16.2.2, then its value for the closed unit disk is 1.

4. In the notation of Theorem 16.2.2, prove that $\rho(E^*) \leq [\rho(E)]^{1/k}$ by a discussion of the polynomial $T_n[Q(z); E]$ which is in $\mathbf{P}_{nk}$.

5. To prove the reversed inequality, let $z_1(w), \cdots, z_k(w)$ be the roots of the equation $Q(z) = w$ and form

$$\prod_{j=1}^{k} T_n[z_j(w); E^*] \equiv Q_n(w).$$

By a consideration of symmetric functions, prove that $Q_n(w)$ is a polynomial in w of degree n and leading coefficient $+1$ or -1. Use this to prove that

$$M_n(E) \leq \max_{w \in E} |Q_n(w)| \leq \prod_{j=1}^{k} [\max |T_n[z_j(w); E^*]|] \leq [M_n(E^*)]^k,$$

from which the desired inequality follows.

6. For a closed infinite point set E on the sphere we may define an "elliptic" transfinite diameter $\rho_e(E)$ by replacing Euclidean distance $|z_j - z_k|$ in (16.2.1) by chordal distance $\chi(z_j, z_k)$ and passing to the limit. Work out the details.

7. Same question for the "hyperbolic" transfinite diameter $\rho_h(E)$. Here E is confined to the open disk and Euclidean distance is replaced by hyperbolic. For problems 6 and 7 see pp. 89–96 of Tsuji's treatise cited on p. 318.

8. Suppose that in Theorem 16.2.6 the function $f(z)$ is a polynomial in z of degree k. Show that $L_{n-1}(z; f) \equiv f(z)$ for $n > k$.

16.3. Additive set functions; Radon-Stieltjes integrals. As a preparation for the discussion in Section 16.4 we shall give a brief summary of properties of completely additive set functions and the related Radon-Stieltjes integrals.[1] The presentation is geared to present needs and no attempt is made to reach great generality.

Let E be a bounded closed set in the complex plane. Let $\mathscr{B}$ be the ring of

---

[1] Named after the Austrian mathematician Johann Radon (1887–1956).

Borel sets in E. Thus, $\mathscr{B}$ contains the closed subsets of E and is closed under the formation of countable unions and differences. We recall that the difference $S_1 \ominus S_2$ is the subset of S_1 whose elements do not belong to S_2.

Let $\sigma(S)$ be a function of sets having the following properties:

(i) $\sigma(S)$ *is defined for* $S \in \mathscr{B}$;

(ii) $\sigma(S)$ *is a bounded real-valued function; and*

(iii) *if* $\{S_n\}$ *is a sequence of disjoint sets in* $\mathscr{B}$, *then*

$$(16.3.1) \qquad \sigma(\textstyle\bigcup_1^\infty S_n) = \sum_{n=1}^\infty \sigma(S_n),$$

and the series is absolutely convergent.

We say that $\sigma(S)$ is a "completely additive set function," and we denote the class of all such functions by $\mathscr{S}$. A subset of $\mathscr{S}$ is the class $\mathscr{M}$ of all *normalized measures* $\mu(S)$ on E. Here (ii) is replaced by

(ii′) $\qquad\qquad\qquad 0 \leq \mu(S) \leq 1, \quad \mu(E) = 1.$

A non-negative set function $\sigma(S)$ with $\alpha = \sigma(E)$ is a constant multiple of a measure: $\sigma(S) = \alpha\mu(S)$. A set function which takes on negative as well as positive values can be written as the difference of two non-negative set functions. This can be done in infinitely many ways, but there is a unique minimal decomposition

$$(16.3.2) \qquad \sigma(S) = \sigma^+(S) - \sigma^-(S), \quad |\sigma|(S) \equiv \sigma^+(S) + \sigma^-(S),$$

such that

$$(16.3.3) \qquad\qquad |\sigma|(E) = \sigma^+(E) + \sigma^-(E) = \min.$$

The non-negative set functions $\sigma^+(S)$, $\sigma^-(S)$, and $|\sigma|(S)$ are known as the *upper*, *lower*, and *total variations* of $\sigma(S)$ respectively.

A positive set function $\sigma(S)$ has a *point mass* at $z = z_0$ if $\sigma(\{z_0\}) = \mu_0 > 0$. In this case

$$(16.3.4) \qquad \lim_{\rho \to 0} \sigma(N_\rho) = \mu_0, \quad N_\rho\colon \ |z - z_0| \leq \rho.$$

Let E and F be two closed bounded sets in the complex plane, not necessarily distinct, let $\mathscr{B}_1 = \mathscr{B}(E)$ and $\mathscr{B}_2 = \mathscr{B}(F)$ be the corresponding Borel sets, and let $\mu(S_1) \in \mathscr{M}(E)$, $\nu(S_2) \in \mathscr{M}(F)$ be two normalized measures defined on $\mathscr{B}_1$ and $\mathscr{B}_2$ respectively. The "product space" $E \times F$ is defined as

$$(16.3.5) \qquad E \times F = [(z_1, z_2) \,|\, z_1 \in E, z_2 \in F].$$

The family $\mathscr{B}_3 = \mathscr{B}(E \times F)$ of Borel sets of the product space is defined as the family of sets generated by sets of the form

$$S_1 \times S_2 = [(z_1, z_2) \,|\, z_1 \in S_1, z_2 \in S_2]$$

through the operations of forming countable unions and differences. We define a "product measure" $\lambda = \mu \times \nu$ on $E \times F$ by setting

$$(16.3.6) \qquad\qquad \lambda(S_1 \times S_2) = \mu(S_1)\nu(S_2),$$

and we extend this to $\mathscr{B}_3$ with the aid of property (iii).

We come next to integration with respect to a set function. It is enough to define integrals with respect to measures $\mu(S)$. Let $\mu(S) \in \mathcal{M}$ and let $f(z)$ be a bounded real-valued Borel measurable function defined on E. In other words, for every choice of α and β, $\alpha < \beta$, the set

$$S(\alpha, \beta) = [z \mid \alpha \leq f(z) < \beta] \in \mathcal{B}.$$

Let the range of $f(z)$ be confined to the interval $[A, B]$ and let π be a partition of this interval,

$$\pi: \quad A = \alpha_{n, 0} < \alpha_{n, 1} < \cdots < \alpha_{n, n} = B.$$

We form the corresponding sum

(16.3.7)
$$\Sigma_n = \sum_{k=0}^{n-1} \alpha_{n, k} \mu[S(\alpha_{n, k}, \alpha_{n, k+1})],$$

which lies between A and B since $\mu(E) = 1$. Using the technique developed in Section C.3 of Volume I, we can show that the sums converge to a unique limit J when the norm of π tends to zero. This limit is known as the Radon-Stieltjes integral of $f(z)$ with respect to $\mu(S)$, and J is denoted indiscriminately by one of the symbols

(16.3.8)
$$\int_E f(z) \, d\mu, \quad \int_E f(z) \, d\mu(z), \quad \int_E f(z) \, \mu(dz).$$

The extension to positive set functions is immediate. Thus, if $\sigma(S) = \alpha\mu(S)$ we set

$$\int_E f(z) \, d\sigma(z) = \alpha \int_E f(z) \, d\mu(z).$$

If $\sigma(S)$ takes on both negative and positive values, we use the decomposition (16.3.2) and set

(16.3.9)
$$\int_E f(z) \, d\sigma(z) = \int_E f(z) \, d\sigma^+(z) - \int_E f(z) \, d\sigma^-(z).$$

Extension to complex-valued functions is also immediate. If $f(z)$ is real, bounded below but not above, we set

(16.3.10)
$$[f(z)]_N = \begin{cases} f(z), & f(z) \leq N, \\ N, & f(z) > N. \end{cases}$$

If $\mu(S) \in \mathcal{M}$, then

$$\lim_{N \to \infty} \int_E [f(z)]_N \, d\mu$$

always exists but may be $+\infty$. If the limit is finite, it is, by definition, the value of the integral of $f(z)$ with respect to $\mu(S)$. If $f(z)$ is unbounded below as well as above, we consider

$$f^+(z) = \max [f(z), 0], \quad f^-(z) = \max [-f(z), 0]$$

separately, and $f(z)$ is integrable if and only if each of these functions is integrable, that is, if and only if $|f(z)|$ is integrable. The integral then equals

$$(16.3.11) \qquad \int_E f(z)\, d\mu(z) = \int_E f^+(z)\, d\mu(z) - \int_E f^-(z)\, d\mu(z).$$

The extension to product spaces proceeds as in the classical Lebesgue theory. In particular, if $f(z_1, z_2)$ is a bounded Borel measurable function on $E \times F$ and if $\mu(S_1) \in \mathscr{M}(E)$, $\nu(S_2) \in \mathscr{M}(F)$ generate the product measure $\lambda = \mu \times \nu$, then the double integral

$$\int_E \int_F f(z_1, z_2)\, d\lambda(z_1, z_2)$$

exists and the Fubini theorem holds in the sharp form. That is, each of the integrals

$$\int_F f(z_1, z_2)\, d\nu(z_2) \quad \text{and} \quad \int_E f(z_1, z_2)\, d\mu(z_1)$$

exists for all values of the free variable and

$$\int_E \left\{ \int_F f(z_1, z_2)\, d\nu(z_2) \right\} d\mu(z_1) = \int_F \left\{ \int_E f(z_1, z_2)\, d\mu(z_1) \right\} d\nu(z_2),$$

the common value being that of the double integral.

The set $\mathscr{S}$ can be made into a normed space by taking either $\sup |\sigma(S)|$ or $|\sigma|(E)$ as the norm of σ. The following theorem states an important property of this space ("compactness in the weak star topology" in modern parlance). We state the result for $\mathscr{M}$ rather than for $\mathscr{S}$ and omit the proof.

THEOREM 16.3.1.   *Every infinite subfamily of $\mathscr{M}$ contains a convergent sequence $\{\mu_n(S)\}$ such that there exists an element $\mu(S)$ of $\mathscr{M}$ with*

$$(16.3.12) \qquad \lim_{n \to \infty} \int_E f(z)\, d\mu_n(z) = \int_E f(z)\, d\mu(z)$$

for every real $f(z)$ continuous on E.

We define the *carrier* or *support* of a normalized measure $\mu(S) \in \mathscr{M}(E)$ as the subset of E each point of which has a neighborhood whose μ-measure is positive. It is a closed point set. We shall need the following complement to the preceding theorem:

THEOREM 16.3.2.   *Let $\{\mu_n(S)\} \subset \mathscr{M}(E)$, where for each n the support of $\mu_n(S)$ is confined to a subset F of E, and let (16.3.12) hold; then the support of $\mu(S)$ is also a subset of F.*

Proof.   We may suppose that F is closed and is a proper subset of E. Then $U = E \ominus F$ is not void and is open with respect to E. There exists then a sequence of closed sets $\{K_m\}$ in E such that

$$K_1 \subset K_2 \subset \cdots \subset K_m \subset \cdots \subset U, \quad \mathsf{U}_1^\infty K_m = U.$$

Hence $\{\mu(K_m)\}$ is a never decreasing sequence whose limit is $\mu(U)$. For a given integer m we can find a continuous function $f_m(z)$ which equals 1 on K_m, 0 on F, while $0 < f_m(z) < 1$ elsewhere (see Problem 5 on page 279 of Volume I for the construction of such a function). We have then for every n

$$\int_E f_m(z)\, d\mu_n(z) = \int_F f_m(z)\, d\mu_n(z) = 0.$$

On the other hand, by (16.3.12)

$$\lim_{n \to \infty} \int_E f_m(z)\, d\mu_n(z) = \int_E f_m(z)\, d\mu(z) \geq \mu(K_m),$$

so that $\mu(K_m) = 0$ for each m. Hence $\mu(U) = 0$ and thus the support of $\mu(S)$ is a subset of F as asserted.

The following necessary and sufficient condition plays an important role in uniqueness questions:

THEOREM 16.3.3.   *If $\sigma(S) \in \mathscr{S}$ and if*

$$(16.3.13) \qquad \int_E z^m \bar{z}^n\, d\sigma(z) = 0, \quad m, n = 0, 1, 2, \cdots,$$

then $\sigma(S) = 0$ for all $S \in \mathscr{B}$.

Proof.   If $z = x + iy$, conditions (16.3.13) imply that

$$\int_E x^j y^k\, d\sigma(z) = 0, \quad j, k = 0, 1, 2, \cdots,$$

so that $d\sigma(z)$ is orthogonal to all polynomials in x and y. By the theorem of Weierstrass, polynomials are dense in the space of continuous functions. To fix the ideas, suppose that E is located in the rectangle

$$R: \quad a \leq x \leq b, \quad c \leq y \leq d,$$

and let $f(x, y)$ be a function of (x, y) continuous in this rectangle. Then the theorem asserts that, given any $\varepsilon > 0$, there exists a polynomial $P_\varepsilon(x, y)$ such that

$$\sup_{(x, y) \in R} |f(x, y) - P_\varepsilon(x, y)| \leq \varepsilon.$$

Since $P_\varepsilon(x, y)$ is orthogonal to $d\sigma(z)$ over E, this means that

$$\left| \int_E f(x, y)\, d\sigma(z) \right| \leq \varepsilon |\sigma|(E),$$

that is, arbitrarily small. It follows that $d\sigma(z)$ is orthogonal to all continuous functions defined on E.

This fact we can utilize as follows: If $\sigma(S)$ is not identically 0, it has a decomposition (16.3.2): $\sigma(S) = \sigma^+(S) - \sigma^-(S)$. Then E breaks up into three disjoint subsets E_+, E_0, and E_-. On the subsets of E_+ we have $\sigma^+(S) > 0$, $\sigma^-(S) = 0$; on those of E_0 we have $\sigma^+(S) = \sigma^-(S) = 0$; and on those of E_- we have $\sigma^+(S) = 0$, $\sigma^-(S) > 0$. Further, $\sigma^+(E_+) = \sigma^-(E_-) \equiv a > 0$. Given an $\varepsilon > 0$, we can then find two closed sets F_1 and F_2 such that

$$F_1 \subset \dot{E}_+, \sigma^+(F_1) > a - \varepsilon, \quad F_2 \subset E_-, \sigma^-(F_2) > a - \varepsilon,$$

and the distance from F_1 to F_2 is positive.

Using the same construction as in the preceding theorem we can find a continuous function, defined in the whole plane, which is 1 on F_1 and 0 on F_2 and which satisfies $0 < F(z) < 1$ everywhere else. Then, on the one hand,

$$\int_E F(z)\, d\sigma(z) = 0.$$

On the other hand, this integral is the sum of five integrals taken over the sets

$$F_1, F_2, E_+ \ominus F_1, E_0, \text{ and } E_- \ominus F_2$$

respectively. The integral over F_1 equals

$$\sigma^+(F_1) > a - \varepsilon;$$

the integrals over F_2 and E_0 are 0. The integral over $E_+ \ominus F_1$ is at most equal to $\sigma^+(E_+ \ominus F_1) < \varepsilon$, and the integral over $E_- \ominus F_2$ is negative but exceeds $-\sigma^-(E_- \ominus F_2) > -\varepsilon$. Thus we find that

$$\int_E F(z)\, d\sigma(z) > a - 2\varepsilon.$$

This contradicts our previous result, so we conclude that $\sigma(S)$ must be identically 0, and the theorem is proved.

EXERCISE 16.3

1. State and prove the analogues of the properties (C.3.6)–(C.3.11) for the Radon-Stieltjes integral. See pages 294 and 295 of Volume I.

2. Prove that if $\sigma(S) \in \mathscr{S}(E)$, then

$$\sup |\sigma(S)| \leq |\sigma|(E) \leq 2 \sup |\sigma(S)|.$$

3. Let $\sigma(S) \in \mathscr{S}(E)$ and set

$$f(z) = \frac{1}{2\pi i} \int_E \frac{d\sigma(t)}{t - z}.$$

Show that $f(z)$ is holomorphic in each of the domains which are components of the complement of E. Find $\lim z f(z)$ if $z \to \infty$.

4. If $f(z) \equiv 0$ for large values of $|z|$, show that $\sigma(S) \equiv 0$. (*Hint:* Use $\overline{f(\bar{z})}$ as well as $f(z)$ and reduce to Theorem 16.3.3.)

16.4. Logarithmic capacity. With our previous notation, let $\mu(S)$ be any normalized measure in $\mathscr{M} = \mathscr{M}(E)$ and form the so-called *energy integral*

$$(16.4.1) \qquad I[\mu] = \int_E \int_E \log | z_1 - z_2 |^{-1} d\mu(z_1) \, d\mu(z_2).$$

This integral is defined as the limit of

$$I_N[\mu] = \int_E \int_E [\log | z_1 - z_2 |^{-1}]_N \, d\mu(z_1) \, d\mu(z_2)$$

as $N \to \infty$, where the bracket notation is that of formula (16.3.10). The limit always exists but may be $+\infty$.

There are two cases: either $I[\mu] = +\infty$ for every $\mu(S) \in \mathscr{M}(E)$ or

$$(16.4.2) \qquad V = V(E) \equiv \inf I[\mu] < \infty.$$

We shall see later that $V > -\infty$.

DEFINITION 16.4.1.   *E is said to be of logarithmic capacity 0 if $I[\mu] = +\infty$ for all $\mu(S)$, and the logarithmic capacity equals*

$$(16.4.3) \qquad C(E) = e^{-V}$$

if $V < \infty$. $V(E)$ is known as Robin's constant.[1]

The function

$$(16.4.4) \qquad U(z; \mu) = \int_E \log | z - t |^{-1} d\mu(t)$$

is the harmonic function associated with the mass distribution $\mu(S)$ on E. It certainly exists for z in the complement G of E. $U(z; \mu)$ is harmonic in each of the domains which are the components of G. The only exception is the point at infinity, where

$$(16.4.5) \qquad U(z; \mu) = \log | z |^{-1} + U^*(z; \mu).$$

Here $U^*(z; \mu)$ is harmonic and tends to a finite limit as $z \to \infty$. If $I[\mu] < \infty$, the Fubini theorem applied to (16.4.1) shows that $U(z; \mu)$ exists almost everywhere on E and

$$(16.4.6) \qquad I[\mu] = \int_E U(z; \mu) \, d\mu(z).$$

$U(z; \mu)$ is a logarithmic potential function. So are the functions $U(z; \sigma)$ obtained by replacing $\mu(S)$ by an element $\sigma(S)$ of $\mathscr{S}$. We shall also have occasion to consider the corresponding energy integrals $I[\sigma]$. In addition it is desirable to study certain classes of *generalized potentials* introduced by Marcel Riesz and studied by him and by his pupil Otto Frostman (1935). We restrict ourselves to the case

$$(16.4.7) \qquad U_\alpha(z; \sigma) \equiv \int_E | z - t |^{-\alpha} d\sigma(t).$$

---

[1] (Victor) Gustave Robin (1855–1897), French applied mathematician.

For z fixed in G, the integral exists and defines an entire function of the parameter α. For α real positive, this is a subharmonic function of z since

(16.4.8) $$\Delta r^{-\alpha} = \alpha^2 r^{-\alpha-2} > 0, \quad r^2 = x^2 + y^2.$$

The integral may still exist on E. We shall see below that if $\sigma(E) = 0, I[\sigma] < \infty$,
and if $0 < \alpha \leq 1$, then $U_\alpha(z; \sigma)$ exists for almost all $z \in E$. The same holds for
$\Re(\alpha) < 0$ without restrictions on $\sigma(S)$, since in this case $| z - t |^{-\alpha}$ is continuous.

THEOREM 16.4.1. *If there exists an α, not 0 or a negative integer, such that
$U_\alpha(z; \sigma) = 0$ for all large $| z |$, then*

$$\sigma(S) \equiv 0.$$

Proof. We shall assume α real. Actually, $\alpha = 1$ is the only case used in
the following. Let R be so large that E is confined to the disk $| t | < R$. For
such values of t

$$| Re^{i\theta} - t |^{-\alpha} = R^{-\alpha}\left(1 - \frac{t}{R}e^{-i\theta}\right)^{-\alpha/2}\left(1 - \frac{\bar{t}}{R}e^{i\theta}\right)^{-\alpha/2}$$

$$= R^{-\alpha}\sum_{m=0}^{\infty}\sum_{n=0}^{\infty}(-1)^{m+n}\binom{-\alpha/2}{m}\binom{-\alpha/2}{n}t^m \bar{t}^n e^{-i(m-n)\theta}R^{-m-n}.$$

It follows that for any integer k

$$\int_0^{2\pi} e^{-ik\theta}U_\alpha(Re^{i\theta}; \sigma)\, d\theta$$

$$= (-1)^k 2\pi R^{-\alpha}\sum_{m=0}^{\infty}\binom{-\alpha/2}{m}\binom{-\alpha/2}{m+k}R^{-2m-k}\int_E t^m \bar{t}^{m+k}\, d\sigma(t).$$

Since the left member is identically zero by assumption and since R is arbitrary,
we conclude that each integral in the right member is zero for all k and m, and
this implies the desired result.

THEOREM 16.4.2. *Suppose that* (i) $\sigma(S) \in \mathscr{S}$, (ii) $\sigma(E) = 0$, *and* (iii) *the
energy integral $I[\sigma]$ is absolutely convergent. Then $U_1(z; \sigma)$ exists for almost all z
in E and*

(16.4.9) $$I[\sigma] = \frac{1}{2\pi}\int [U_1(z; \sigma)]^2\, d\omega$$

*where the integral is extended over the whole plane. Thus, $I[\sigma] \geq 0$ and equals 0 if
and only if $\sigma(S) \equiv 0$.*

Proof. Let R be a large positive number and let z_1 and z_2 be fixed, $z_1 \neq z_2$.
They will ultimately be confined to E. Form

(16.4.10) $$I(z_1, z_2; R) \equiv \frac{1}{2\pi}\int_{|t| \leq R} | t - z_1 |^{-1} | t - z_2 |^{-1}\, d\omega.$$

In this integral we make the substitution

$$t = z_1 + u(z_2 - z_1).$$

This leads to an integral of the same nature where, however, the integration is extended over a circular disk D_1 whose center is not at the origin. We can replace D_1 by a disk D_2 with center at the origin and radius $R \mid z_1 - z_2 \mid^{-1}$. Since the integrand is $O(\mid u \mid^{-2})$ as $u \to \infty$, and since $(D_1 \cup D_2) \ominus (D_1 \cap D_2)$ has an area which is $O(R)$, we conclude that the resulting error is at most $O(R^{-1})$, that is,

$$I(z_1, z_2; R) = I(0, 1; R \mid z_1 - z_2 \mid^{-1}) + O(R^{-1}).$$

But for $b > 1$

$$(16.4.11) \quad I(0, 1; b) = \frac{1}{2\pi} \int_0^b \int_0^{2\pi} \mid 1 - re^{i\theta} \mid^{-1} d\theta \, dr = C_0 + \log b + O(b^{-2}),$$

where C_0 is a constant. Hence

$$\log \mid z_1 - z_2 \mid^{-1} = -\log R - C_0 + \frac{1}{2\pi} \int_{|t| \le R} \mid t - z_1 \mid^{-1} \mid t - z_2 \mid^{-1} d\omega + O\left(\frac{1}{R}\right).$$

Here we multiply both sides by $d\sigma(z_1) \, d\sigma(z_2)$ and integrate over $E \times E$, using the fact that $\sigma(E) = 0$ so that the first two terms on the right give no contribution. Hence

$$I[\sigma] = \frac{1}{2\pi} \int_E \int_E \int_{|t| \le R} \mid t - z_1 \mid^{-1} \mid t - z_2 \mid^{-1} d\omega \, d\sigma(z_1) \, d\sigma(z_2) + O\left(\frac{1}{R}\right).$$

The assumption that $I[\sigma]$ is absolutely convergent is equivalent to $I[\mid \sigma \mid] < \infty$. This means that the triple integral is absolutely convergent and we can apply the Fubini theorem, obtaining

$$I[\sigma] = \frac{1}{2\pi} \int_{|t| \le R} [U_1(t; \sigma)]^2 \, d\omega + O(R^{-1}).$$

From this we conclude first that $U_1(t; \sigma)$ exists for almost all t in E and, secondly, that the integral tends to a finite limit when $R \to \infty$. This proves formula (16.4.9), and the rest follows from the preceding theorem.

We now return to the quantity $V(E)$ of (16.4.2).

THEOREM 16.4.3.   If $V(E) < +\infty$, then $V(E) > -\infty$, and there exists a unique element $v(S)$ of $\mathcal{M}$ such that $I[v] = V(E)$.

Proof.   Let δ be the diameter of the set E, which is bounded by assumption. Then

$$I[\mu] = \int_E \int_E \log \frac{\delta}{\mid z_1 - z_2 \mid} \, d\mu(z_1) \, d\mu(z_2) - \log \delta \ge -\log \delta,$$

so that $V(E) \ge -\log \delta$. This says that $C(E) \le \delta(E)$.

The existence of an extremal function $v(S) \in \mathcal{M}$ with $I[v] = V(E)$ follows from Theorem 16.3.1. First, by the definition of the infimum we can find a sequence $\{\mu_n(S)\} \subset \mathcal{M}$ such that

$$\lim_{n \to \infty} I[\mu_n] = V(E).$$

From this sequence we choose a (weakly) converging subsequence which we relabel $\{\mu_n(S)\}$. We can do this so that simultaneously

$$\lim_{n \to \infty} \int_E f(z) \, d\mu_n(z) \equiv \int_E f(z) \, d\nu(z)$$

exists for every $f(z)$ continuous on E, and

$$\lim_{n \to \infty} \int_E \int_E f(z_1, z_2) \, d(\mu_n \times \mu_n)(z_1, z_2) \equiv \int_E \int_E f(z_1, z_2) \, d\lambda(z_1, z_2)$$

exists for every $f(z_1, z_2)$ continuous on $E \times E$. We have to show that $\lambda = \nu \times \nu$. To this end we take an $f(z_1, z_2)$ of the form $f_1(z_1)f_2(z_2)$, where $f_1(z)$ and $f_2(z)$ are continuous on E. Then

$$\int_E \int_E f_1(z_1)f_2(z_2) \, d\lambda(z_1, z_2) = \lim_{n \to \infty} \int_E f_1(z_1) \, d\mu_n(z_1) \cdot \lim_{n \to \infty} \int_E f_2(z_2) \, d\mu_n(z_2)$$

$$= \int_E f_1(z_1) \, d\nu(z_1) \cdot \int_E f_2(z_2) \, d\nu(z_2)$$

$$= \int_E \int_E f_1(z_1)f_2(z_2) \, d(\nu \times \nu)(z_1, z_2).$$

Hence $d[\lambda - (\nu \times \nu)]$ is orthogonal to every continuous function of the form $f_1(z_1)f_2(z_2)$. Since linear combinations of such functions are dense in the space of continuous functions of two variables, we conclude that $[\lambda - (\nu \times \nu)](S) = 0$ for every set $S \in \mathscr{B}_3$. Hence $\lambda = \nu \times \nu$ as asserted.

We have now

$$I[\nu] = \lim_{N \to \infty} \int_E \int_E [\log |z_1 - z_2|^{-1}]_N \, d\nu(z_1) \, d\nu(z_2)$$

$$= \lim_{N \to \infty} \lim_{n \to \infty} \int_E \int_E [\log |z_1 - z_2|^{-1}]_N \, d\mu_n(z_1) \, d\mu_n(z_2).$$

But for each n and N

$$\int_E \int_E [\log |z_1 - z_2|^{-1}]_N \, d\mu_n(z_1) \, d\mu_n(z_2)$$

$$\leq \int_E \int_E \log |z_1 - z_2|^{-1} \, d\mu_n(z_1) \, d\mu_n(z_2) = I[\mu_n],$$

whence it follows that

$$I[\nu] \leq \liminf_{n \to \infty} I[\mu_n] = V(E).$$

Since $V(E)$ is the infimum of $I[\mu]$, we must have $I[\nu] = V(E)$.

To prove uniqueness, we use the identity

$$(16.4.12) \qquad I[\tfrac{1}{2}(\mu_1 + \mu_2)] + I[\tfrac{1}{2}(\mu_1 - \mu_2)] = \tfrac{1}{2}\{I[\mu_1] + I[\mu_2]\}.$$

Suppose there were two measures μ_1 and μ_2 such that $I[\mu_1] = I[\mu_2] = V$. Then $\tfrac{1}{2}(\mu_1 + \mu_2) \in \mathscr{M}$ and $\tfrac{1}{2}(\mu_1 - \mu_2) \in \mathscr{S}$. Further, $[\tfrac{1}{2}(\mu_1 - \mu_2)](E) = 0$, and, by

Theorem 16.4.2, $I[\frac{1}{2}(\mu_1 - \mu_2)] \geq 0$. Substituting these data in (16.4.12), we obtain

$$I[\tfrac{1}{2}(\mu_1 + \mu_2)] = V, \quad I[\tfrac{1}{2}(\mu_1 - \mu_2)] = 0;$$

that is, $\mu_1(S) = \mu_2(S)$ and uniqueness follows.

The function $\nu(S)$ is known as the *equilibrium, or natural, distribution*, and $U(z; \nu)$ is the *equilibrium potential*. The reason for the names is the analogy with the three-dimensional Newtonian case typified by the distribution of electricity on a conductor. Later theorems will bear out this analogy. Our next step is to identify the logarithmic capacity with the transfinite diameter. The following argument is due to G. Pólya and G. Szegö (for the case of absolutely continuous measures).

THEOREM 16.4.4. $C(E) = \rho(E) = \delta_0(E)$.

Proof.   We shall work with $C(E)$ and $\delta_0(E)$. Suppose that n points $z_{n,1}, z_{n,2}, \cdots, z_{n,n}$ are chosen in E so as to maximize the discriminant. Then

$$n(n - 1) \log \frac{1}{\delta_n} = \sum_{j \neq k} \log \mid z_{n,j} - z_{n,k} \mid^{-1}.$$

We recall that these points are located on the outer boundary of E. Let us place a mass $1/n$ at each of these n points. This defines a measure $\mu_n(S) \in \mathcal{M}$. We have $\mu_n(S) = k/n$ if S contains k of the points $z_{n,j}$ where $k = 0, 1, 2, \cdots, n$. We have

$$(16.4.13) \qquad\qquad I[\mu_n] = \frac{n - 1}{n} \log \frac{1}{\delta_n} > V(E).$$

Letting $n \to \infty$, we get

$$(16.4.14) \qquad\qquad \log [\delta_0(E)]^{-1} \geq V(E) \quad \text{or} \quad \delta_0(E) \leq C(E).$$

By Theorem 16.3.1 there exists a subsequence $\{\mu_{n_k}(S)\}$ which converges weakly to a limit $\mu_0(S)$, and

$$I[\mu_0] = \log [\delta_0(E)]^{-1}.$$

On the other hand, for any choice of n distinct points $z_1, z_2, \cdots, z_n$ in E we have

$$\sum_{j \neq k} \log \mid z_j - z_k \mid^{-1} \geq n(n - 1) \log \delta_n^{-1}.$$

We multiply both sides of this inequality by

$$d\mu(z_1) \, d\mu(z_2) \cdots d\mu(z_n)$$

and integrate over $E \times E \times \cdots \times E$. Here $\mu(S)$ is an arbitrary element of $\mathcal{M}$. Thus

$$\int_E \int_E \cdots \int_E \left\{ \sum_{j \neq k} \log \mid z_j - z_k \mid^{-1} \right\} d\mu(z_1) \, d\mu(z_2) \cdots d\mu(z_n) \geq n(n - 1) \log \delta_n^{-1}.$$

In the left member we have $n(n - 1)$ integrals all of the same type. Each integral has an integrand which depends upon only two variables, so we can carry out the integration with respect to the remaining $n - 2$ variables. This leaves us with $n(n - 1)$ double integrals which are identical except for the lettering of the variables of integration. Hence

$$\int_E \int_E \log | z_1 - z_2 |^{-1} d\mu(z_1) \, d\mu(z_2) \geq \log \delta_n^{-1}.$$

This holds also for the equilibrium distribution, so that

$$V(E) = I[\nu] \geq \log \delta_n^{-1}$$

for all n. It follows that

(16.4.15) $$V(E) = I[\nu] \geq \log [\delta_0(E)]^{-1} = I[\mu_0].$$

But $I[\nu]$ is the minimum value of the energy integral in $\mathcal{M}$, whence it follows that $I[\nu] = I[\mu_0]$, and the theorem is proved.

There are some important consequences of this theorem.

COROLLARY 1.   *The equilibrium distribution $\nu(S)$ of E is the weak limit of the sequence of point distributions $\mu_n(S)$ associated with the zeros of the Fekete polynomials $F_n(z; E)$ of (16.2.12).*

Above it was shown only that $\nu(S)$ is the limit of a subsequence extracted from $\{\mu_n(S)\}$. Since every such limit must satisfy (16.4.15) and the solution of $I[\mu] = V(E)$ is unique, we have

(16.4.16) $$\lim \int_E f(z) \, d\mu_n(z) = \int_E f(z) \, d\nu(z)$$

for every continuous $f(z)$ as asserted.

COROLLARY 2.   *The support of $\nu(S)$ is restricted to the outer boundary of E.*

This follows from (16.4.16) together with Theorem 16.3.2.

THEOREM 16.4.5.   *If E_1 and E_2 are two closed bounded sets having the same outer boundary Γ, then*

(16.4.17) $$C(E_1) = C(E_2) = C(\Gamma),$$

and the three sets have the same equilibrium distribution and the same equilibrium potential.

Proof.   Since $E_1 \supset \Gamma$, $E_2 \supset \Gamma$, we have $C(E_j) \geq C(\Gamma)$, $j = 1, 2$, by Theorem 16.2.2 (i). On the other hand, with obvious notation,

$$V(E_j) = \int_\Gamma \int_\Gamma \log | z_1 - z_2 |^{-1} d(\nu_j \times \nu_j) \geq V(\Gamma)$$

by the definition of $V(\Gamma)$ as the infimum of energy integrals over $\Gamma \times \Gamma$. Hence $C(E_j) \leq C(\Gamma)$, and this proves (16.4.17). The uniqueness of the minimizing measure shows that $\nu_1(S) \equiv \nu_2(S)$ and hence also that the corresponding potentials coincide.

This theorem shows that adjunction of a set to E or suppression of a subset of E does not change capacity and equilibrium potential as long as the outer boundary Γ is unchanged.

THEOREM 16.4.6. *Let $e \subset E$ where $C(e) = 0$, $C(E) > 0$. If $\mu(S) \in \mathcal{M}(E)$ and $I[\mu] < \infty$, then $\mu(e) = 0$.*

Proof. Without restricting the generality we may suppose that the diameter of E does not exceed 1 and that e is closed. We have then obviously $\log |z_1 - z_2|^{-1} \geq 0$ in E, so that

$$\int_e \int_e \log |z_1 - z_2|^{-1} d(\mu \times \mu) \leq \int_E \int_E \log |z_1 - z_2|^{-1} d(\mu \times \mu) < \infty.$$

If $\mu(e) > 0$, then $\mu(S)/\mu(e)$ would be a measure on e with finite energy integral; but this contradicts $C(e) = 0$. Thus $\mu(e) = 0$. In particular, this applies to Lebesgue measure: *any set in the plane whose logarithmic capacity is 0 has Lebesgue measure 0.* The converse is not true. Thus, a line segment has planar measure 0, a Cantor set on the segment has linear measure 0, but both have positive capacity.

COROLLARY 1. *If $C(E_1) = C(E_2) = 0$, then $C(E_1 \cup E_2) = 0$.*

For if $\mu(S)$ is any measure on $E \supset E_1 \cup E_2$ and if $I[\mu] < \infty$, then

$$\mu(E_1 \cup E_2) \leq \mu(E_1) + \mu(E_2) = 0.$$

This requires that $C(E_1 \cup E_2) = 0$. The same argument gives

COROLLARY 2. *If $E = \bigcup_1^\infty E_n$ and if $C(E_n) = 0$ for all n, then $C(E) = 0$.*

We shall also need a *maximum principle* for potential functions $U(z; \mu)$ where $\mu(S) \in \mathcal{M}(E)$. The following theorem is due to A. J. Maria (1934).

THEOREM 16.4.7. *If $U(z; \mu) \leq C$ on E, then the inequality holds in the whole plane.*

Proof. Let $G = \mathbf{C}(E)$ and let D be one of the components of G. Since $U(z; \mu)$ is harmonic in D, it suffices to prove that

(16.4.18) $$\limsup_{z \to a} U(z; \mu) \leq C, \quad z \in D, a \in \overline{D} \cap E.$$

Let N_ρ: $|z - a| < \rho$ and let $D_\rho = D \cap N_\rho$, $E_\rho = E \cap N_\rho$. If any point of E carried a positive mass, then $U(z; \mu)$ would necessarily become infinite at such a point. This is excluded since $U(z; \mu) \leq C$ on E. This implies that, given an $\varepsilon > 0$, we can find a ρ so small that $\mu(E_\rho) < \varepsilon$. We now choose points z and z_1 in such a manner that $z \in D_\rho$, $z_1 \in E_\rho$, and $|z_1 - z| \leq |z - t|$ when $t \in E_\rho$. Then

$$|z_1 - t| \leq |z_1 - z| + |z - t| \leq 2|z - t|,$$

and

$$\log |z - t|^{-1} \leq \log 2 + \log |z_1 - t|^{-1}.$$

This gives

$$\int_{E_\rho} \log |z - t|^{-1} d\mu(t) \leq \mu(E_\rho) \log 2 + \int_{E_\rho} \log |z_1 - t|^{-1} d\mu(t)$$

$$\leq \varepsilon \log 2 + \int_E \log |z_1 - t|^{-1} d\mu(t) - \int_{E \ominus E_\rho} \log |z_1 - t|^{-1} d\mu(t),$$

so that

$$\int_{E_\rho} \log |z - t|^{-1} d\mu(t) + \int_{E \ominus E_\rho} \log |z_1 - t|^{-1} d\mu(t) \leq C + \varepsilon \log 2.$$

Since $z_1 - z \to 0$ when $z \to a$ and the second integral is a continuous function of z_1 for $z_1 \in E_\rho$, we can replace z_1 by z, committing an error which does not exceed ε for $|z - a|$ sufficiently small. Hence

$$U(z; \mu) \leq C + (1 + \log 2)\varepsilon$$

for $|z - a|$ small. This implies (16.4.18) and proves the theorem.

We come now to the fundamental theorem of Frostman.

THEOREM 16.4.8.　*Let E be a closed bounded set of positive capacity whose complement is connected. Then $U(z; v) \leq V(E)$ in the whole plane and $U(z; v) = V(E)$ for z in E except possibly in a set e of capacity zero.*

Proof.　The support F of $v(S)$ is confined to Γ, the outer boundary of E. The complement of F is made up of domains in each of which $U(z; v)$ is harmonic, a fortiori continuous. We do not know that $U(z; v)$ is continuous on F, but at any rate it is lower semicontinuous since $U(z; v)$ is the limit of a never decreasing sequence of continuous functions

$$U_N(z; v) = \int_F [\log |z - t|^{-1}]_N \, dv(t).$$

We recall that a lower semicontinuous function assumes its infimum on a closed set, that the sets $[z \mid U \leq \alpha]$ are closed, and that, if $z_n \to z_0$, then

$$\liminf_{n \to \infty} U(z_n; v) \geq U(z_0; v).$$

We prove first that $U(z; v) \geq V$ on E, neglecting at most a subset of capacity 0. We set

$$A = [z \mid z \in E, U(z; v) < V]$$

and assume that $C(A) > 0$. Let A_n be the subset of A where

$$U(z; v) \leq V - \frac{1}{n}.$$

Then $A = \bigcup_1^\infty A_n$ and $C(A_n) > 0$ for all large n since otherwise $C(A) = 0$ by Corollary 2 of Theorem 16.4.6. Since A_n is closed, we conclude that a positive ε exists such that the set

$$E_1 = [z \mid z \in E, U(z; v) \leq V - 2\varepsilon]$$

is closed and $C(E_1) > 0$.

On the other hand, the relation

$$\int_E U(z; \nu)\, d\nu(z) = V, \quad \nu(E) = 1,$$

shows that the set

$$E_2 = [z \mid z \in E,\ U(z; \nu) > V - \varepsilon]$$

cannot be void, and, in fact, must have positive ν-measure. Let

$$\nu(E_2) = m.$$

Since $C(E_1) > 0$, we can find a positive measure $\mu(S)$ on E_1 with finite energy integral $I[\mu; E_1]$. We then define a set function $\sigma(S) \in \mathscr{S}(E)$ such that

$$\sigma(S) = m\mu(S), \quad S \subset E_1,$$

$$\sigma(S) = -\nu(S), \quad S \subset E_2, \quad \text{and}$$

$$\sigma(S) = 0 \quad \text{if} \quad S \cap [E_1 \cup E_2] = \emptyset.$$

Finally, we set $\rho(S) = \nu(S) + h\sigma(S)$ where $0 < h < 1$. Then $\rho(S) \in \mathscr{M}(E)$, so that

$$I[\rho] - I[\nu] \geq 0.$$

Actually the difference of the energy integrals equals

$$2h \int_E U(z; \nu)\, d\sigma(z) + h^2 I[\sigma] \leq 2h[-(V - \varepsilon)m + (V - 2\varepsilon)m] + h^2 I[\sigma]$$

$$= -2h\varepsilon m + h^2 I[\sigma].$$

Here $I[|\,\sigma\,|] < \infty$ since $I[\mu; E_1] < \infty$ and $I[\nu] < \infty$. It follows that for small values of h the difference $I[\rho] - I[\nu]$ is really negative. This contradiction shows that $C(E_1) = 0$ and, since ε is arbitrary, also that $C(A) = 0$. In other words, $U(z; \nu) \geq V$ for $z \in E$ outside of a set of capacity 0.

Next we prove that $U(z; \nu) \leq V$ for $z \in F$. If there were a point $a \in F$ such that $U(a; \nu) > V$, then the lower semicontinuity of $U(z; \nu)$ would imply that there is a neighborhood $N(a)$ of $z = a$ where $U(z; \nu) > V + \varepsilon$. Since $a \in F$ we have $\nu[N(a)] > 0$. Set

$$F_1 = F \cap N(a), \quad F_2 = F \ominus F_1.$$

Then

$$V = \int_{F_1} U(z; \nu)\, d\nu(z) + \int_{F_2} U(z; \nu)\, d\nu(z)$$

$$\geq (V + \varepsilon)\nu(F_1) + V[1 - \nu(F_1)] = V + \varepsilon\nu(F_1) > V,$$

which is absurd. It follows that $U(z; \nu) \leq V$ for $z \in F$.

Combining this with the previous inequality $U(z; \nu) \geq V$, valid for $z \in E$, outside of a set of capacity 0, we see that $U(z; \nu) = V$ for $z \in F \ominus e$ where e is void or of capacity 0.

We can now apply the maximum principle of Theorem 16.4.7, according to which

$$U(z; v) \leq V$$

in the whole plane. This implies that $U(z; v) = V$ for all $z \in E$ except possibly for a set $e \in F$ with $C(e) = 0$. The exceptional set is not necessarily void. Thus, every isolated point of E belongs to e. This completes the proof.

EXERCISE 16.4

1. Verify (16.4.11). (*Hint:* Split the interval of integration $(0, b)$ into $(0, 1)$ and $(1, b)$. The first gives a constant value which may be written as a convergent series. For the second use the technique employed in the proof of Theorem 16.4.1.)

2. In the proof of Theorem 16.4.3 it is asserted that linear combinations of functions of the form $f_1(z_1)f_2(z_2)$ are dense in the space of continuous functions of two complex variables. Assume the validity of Weierstrass's approximation theorem for functions of four real variables and verify the assertion.

3. Let $\mu_n(S)$ be the point distribution defined in the proof of Theorem 16.4.4 and let $U(z; \mu_n)$ be the corresponding potential function. Modify the technique used in the proof to show that

$$U(z; \mu_n) > \log \frac{1}{\delta_n} \quad \text{for} \quad z \in E.$$

(*Hint:* Use $n + 1$ points, namely, z and the roots of the nth Fekete polynomial. Note that $\delta_n > \delta_{n+1}$.)

4. If $\mu(S) \in \mathcal{M}(E)$ and $U(z; \mu)$ is the corresponding potential function, show that

$$V(E) \geq \min U(z; \mu), \quad z \in E.$$

(*Hint:* Evaluate $\int_E \int_E \log |z_1 - z_2|^{-1} d\mu(z_1) \, dv(z_2)$ in two different ways and discuss the resulting identity.)

5. If $E_1, E_2, \cdots, E_n$ are closed sets in a fixed circle of radius $\leq \frac{1}{2}$ and if E is their union, prove that

$$\left\{ \log \frac{1}{C(E)} \right\}^{-1} \leq \sum_{j=1}^{n} \left\{ \log \frac{1}{C(E_j)} \right\}^{-1}.$$

(*Hint:* If $v_j(S)$ is the equilibrium distribution for E_j and if $\delta_j \geq 0$, $\Sigma \, \delta_j = 1$, then $\mu(S) \equiv \Sigma \, \delta_j v_j(S) \in \mathcal{M}(E)$. If $z \in E_k$, then

$$U(z; \mu) = \Sigma \, \delta_j U(z; v_j) \geq \delta_k V(E_k).$$

Thus, $\min U(z; \mu) \geq \min_k [\delta_k V(E_k)]$, so that by the result of the preceding problem $V(E) \geq \min_k [\delta_k V(E_k)]$. Choose δ_k so that all products $\delta_j V(E_j)$ are equal. This gives the desired inequality in view of (16.4.3).)

16.5. Green's function; Hilbert's theorem.   For the applications which we have in mind it is desirable to remove the last ambiguity in Theorem 16.4.8, namely, the possible existence of an exceptional set of capacity 0 on which $U(z; \nu) < V(E), z \in E$. This can be done at the cost of restricting the admissible sets E. As a first step we prove a mean value theorem which holds for all finite values of z. It should be compared with formula (8.2.8), which holds where $U(z)$ is harmonic.

THEOREM 16.5.1.   *Let* $\sigma(S) \in \mathscr{S}(E)$, $\sigma(S) \geqq 0$, *and let* $U(z; \sigma)$ *be the corresponding potential function. Let*

$$(16.5.1) \qquad \mathfrak{M}(z_0; U, r) \equiv (\pi r^2)^{-1} \int_{D_r} U(z; \sigma)\, d\omega$$

be the mean value of $U(z; \sigma)$ *over the disk* D_r: $|z - z_0| \leqq r$. *Then*

$$
\begin{aligned}
\mathfrak{M}(z_0; U, r) = & \int_{E \ominus D_r} \log \frac{1}{|z_0 - t|}\, d\sigma(t) + \sigma(D_r) \log \frac{1}{r} \\
& + \frac{1}{2} \int_{D_r} \left[1 - \frac{|z_0 - t|^2}{r^2} \right] d\sigma(t),
\end{aligned}
$$
$(16.5.2)$

which does not exceed

$$U(z_0; \sigma) + \tfrac{1}{2}\sigma(D_r).$$

Proof.   Our point of departure is Lemma 14.4.1, according to which

$$\frac{1}{2\pi} \int_{|t|=1} \log |c - t|\, |dt| = \log^+ |c|.$$

This gives

$$\frac{1}{2\pi} \int_0^{2\pi} \log |\rho e^{i\theta} - \zeta|^{-1}\, d\theta = \begin{cases} \log \dfrac{1}{|\zeta|}, & |\zeta| > \rho, \\[2mm] \log \dfrac{1}{\rho}, & |\zeta| < \rho, \end{cases}$$

and shows that

$$(\pi r^2)^{-1} \int_{D_r} \log \frac{1}{|z - t|}\, d\omega = (\pi r^2)^{-1} \int_0^r \int_0^{2\pi} \log |\rho e^{i\theta} + z_0 - t|^{-1}\, d\theta\, \rho\, d\rho$$

has the values

$$\log \frac{1}{|z_0 - t|} \quad \text{if} \quad |z_0 - t| > r$$

and

$$\log \frac{1}{r} + \frac{1}{2}\left\{ 1 - \frac{|z_0 - t|^2}{r^2} \right\} \quad \text{if} \quad |z_0 - t| \leqq r.$$

We write $\mathfrak{M}(z_0; U, r)$ as a double integral and interchange the order of integration to obtain

$$\mathfrak{M}(z_0; U, r) = \int_E \left\{ \frac{1}{\pi r^2} \int_{D_r} \log \frac{1}{|z - t|}\, d\omega \right\} d\sigma(t).$$

Here we write the integral over E as the sum of two integrals, one over $E \ominus D_r$ and the other over D_r, and substitute the appropriate expression for the inner integrals. The result is (16.5.2). If E and D_r do not intersect, the second and third expressions in the right member of (16.5.2) are zero and the first becomes $U(z_0; \sigma)$. This result follows directly from (8.2.8). If $E \cap D_r \neq \emptyset$, the first two expressions have a sum $< U(z_0; \sigma)$ and the third is $< \frac{1}{2}\sigma(D_r)$. This completes the proof.

We shall now restrict the set E to satisfy the following conditions:

C_1. The complement of E is connected, and E is bounded by a finite number of simple closed curves Γ_j.

C_2. At each point z_0 on the boundary, there exists a sector

$$S_r: \quad 0 \leq |z - z_0| \leq r, \quad \alpha \leq \arg(z - z_0) \leq \beta,$$

where $2\pi > \beta - \alpha \geqq 2\theta\pi$, θ is independent of z_0, $0 < \theta < 1$, such that S_r lies completely in E if $r < r(z_0)$.

In analogy with the three-dimensional case we refer to C_1 and C_2 as the conditions of Poincaré. The following result is due to Frostman (for generalized potentials, the prototype of the theorem for Newtonian potentials is due to C. F. Gauss [1840]).

THEOREM 16.5.2.   *If E satisfies the conditions of Poincaré, then the function $U(z; v) = V(E)$ everywhere on E, and $U(z; v)$ is continuous in the finite plane.*

Proof.   Let z_0 be an arbitrary point on one of the curves Γ_j and consider two sectors S_r and S_R where $0 < r < R < r(z_0)$. We also note the two disks D_r and D_R with center at z_0 and radii r and R respectively. If the sectors have been chosen in conformity with condition C_2, the area of S_r is at least θ times the area of D_r, and the same relation holds for S_R and D_R. We now take an $\varepsilon > 0$ and choose r and R subject to the following conditions:

(i) $R < \frac{1}{2}$ and $v(D_r) < \theta\varepsilon$.

(ii) If $\sigma_1(S)$ is the restriction of $v(S)$ to D_R and $U_1(z)$ is the corresponding potential function, then $U_1(z_0) < \theta\varepsilon$.

(iii) If $\sigma_2(S)$ is the restriction of $v(S)$ to $E \ominus D_R$ and $U_2(z)$ is the corresponding potential function, then for all z in D_r we shall have $U_2(z) < U_2(z_0) + \varepsilon$.

The assumption $R < \frac{1}{2}$ serves to ensure that $U_1(z) \geqq 0$. Since

$$U_1(z_0) = \int_{D_R} \log \frac{1}{|z_0 - t|} \, dv(t)$$

tends to zero with R, we can clearly satisfy (ii). Here we have tacitly used the fact that $v(S)$ does not have any point masses, and this also makes it possible to satisfy (i). Finally, $U_2(z)$ is continuous in D_r, so it is possible to satisfy (iii). We note that

$$U(z; v) = U_1(z) + U_2(z).$$

We shall now consider various mean values. We denote by $\mathfrak{M}(U, S)$ the mean value of the function $U(z)$ in the Borel set S, that is, the integral of $U(z)$ over S divided by the Lebesgue measure of S. Since $S_r \subset E$ and $U(z; \nu) = V$ almost everywhere in E, we have

$$V = \mathfrak{M}(U, S_r) = \mathfrak{M}(U_1, S_r) + \mathfrak{M}(U_2, S_r)$$
$$< \mathfrak{M}(U_1, S_r) + U_2(z_0) + \varepsilon$$

by (iii). Now $U_1(z) \geq 0$, so the integral of $U_1(z)$ over D_r at least equals the integral over S_r. This shows that

$$\mathfrak{M}(U_1, S_r) \leq \frac{1}{\theta}\, \mathfrak{M}(U_1, D_r),$$

and by Theorem 16.5.1 together with (i) and (ii)

$$\mathfrak{M}(U_1, D_r) \leq U_1(z_0) + \tfrac{1}{2}\nu(D_r) < 2\theta\varepsilon.$$

Combining we get

$$V < U_2(z_0) + 3\varepsilon,$$

whence

$$U(z_0; \nu) \geq U_2(z_0) > V - 3\varepsilon.$$

It follows that $U(z_0; \nu) \geq V$. Since the converse inequality holds, we conclude that $U(z_0; \nu) = V$ everywhere in E without exception.

$U(z; \nu)$ is evidently continuous in $\mathbf{C}(E)$ and in Int (E). It remains to prove continuity for z on ∂E. Here the lower semicontinuity gives

$$\liminf U(z_n; \nu) \geq U(z_0; \nu) = V, \quad z_n \to z_0.$$

Since $U(z_n; \nu) \leq V$ we have

$$\limsup U(z_n; \nu) \leq V$$

and finally

$$\lim U(z_n; \nu) = U(z_0; \nu) = V.$$

This completes the proof.

THEOREM 16.5.3.　　*Under the assumptions* C_1 *and* C_2 *we have* $U(z; \nu) < V$ *everywhere in the complement of* E. *If* E *is connected, each level line*

$$(16.5.3) \qquad\qquad \Gamma_a: \quad U(z; \nu) = a < V$$

is a simple closed curve containing E *in its interior. If* $\varepsilon > 0$ *is given, there exists a* $\delta = \delta(\varepsilon)$ *such that every point of* Γ_a *is at a distance* $< \varepsilon$ *from* E *provided we have* $V - \delta < a < V$.

Proof.　　The first assertion follows from the maximum principle for harmonic functions. Since $U(z; \nu) = V$ on E and tends to $-\infty$ as $z \to \infty$, it follows that the continuous function $U(z; \nu)$ assumes every value between $-\infty$, and V. Now $U(z; \nu)$ is harmonic in $G = \mathbf{C}(E)$ and hence analytic in the real variables x

and y. This implies, among other things, that the curve Γ_a consists of a finite number of closed separate branches, that each branch can intersect itself only a finite number of times, and that each arc joining multiple points is analytic. In particular, no value a can be assumed at an isolated point $z = z_0$. The implicit function theorem shows that there are always one or more *real* branches going through the point. If a is large negative, then Γ_a consists of a single simple closed curve; if a is close to V and the number of components of E is finite, then there is one simple closed curve surrounding each component of E.

If E is connected, every Γ_a is a simple closed analytic curve surrounding E. Any other possibility leads to contradictions. Thus if Γ_a is not connected, then Γ_a separates the plane into three or more regions. One of these must be unbounded, one contains E in its interior, but in any other region G_0 obtained in this manner, $U(z; \nu)$ is harmonic. Then if $U(z; \nu) = c$ on the boundary of G_0 we must have $U(z; \nu) \equiv a$ in G_0 and hence also in the whole plane, which is absurd. It follows that Γ_a is connected, and the same argument shows that Γ_a is simple and surrounds E.

Finally, since $U(z; \nu)$ is uniformly continuous on compact sets and equals V on $\Gamma = \partial E$, it follows that Γ_a is close to Γ if a is close to V. This can be made more precise by introducing the Fréchet distance (French: *écart*) between two curves C_1 and C_2. This distance $\eta(C_1, C_2)$ is defined as the supremum of the distance from z to C_2 when z runs through C_1. It is a distance inasmuch as $\eta(C_1, C_2) \geqq 0$ and is equal to 0 if and only if $C_1 = C_2$. Further it is symmetric in C_1 and C_2 and satisfies the triangle axiom. We note that $\eta(C_1, C_2)$ is greater than or equal to the ordinary distance from C_1 to C_2.

Thus $\eta(a) \equiv \eta(\Gamma, \Gamma_a)$ is a well-defined function of a for $-\infty < a \leqq V$. Owing to the uniform continuity of $U(z; \nu)$, the function $\eta(a)$ is continuous and tends to zero when a tends to V. This completes the proof.

DEFINITION 16.5.1.   *If E is a bounded closed set of positive capacity, if $U(z; \nu)$ is the corresponding equilibrium potential, and if $G = \mathbf{C}(E)$, then*

(16.5.4) $$g(z; \infty) \equiv V(E) - U(z; \nu)$$

is called Green's function for G with respect to $z = \infty$.

If E satisfies the conditions of Poincaré, then $g(z; \infty)$ takes the value 0 on $\Gamma = \partial E$ and is harmonic in G except at infinity, where

(16.5.5) $$g(z; \infty) = \log |z| + V + O\left(\frac{1}{z}\right).$$

If D is any domain and $z_0 \in D$, then we can define a Green's function of D with respect to z_0 as a function $g(z; z_0, D)$ which is harmonic in D except at $z = z_0$, where it becomes positively infinite but in such a way that

$$g(z; z_0, D) + \log |z - z_0|$$

is still harmonic, and which approaches the value 0 as z approaches any point of the boundary of D with the possible exception of a subset of capacity 0.

We can prove the existence of such a function. In fact the transformation

$$\zeta = \frac{1}{z - z_0}$$

takes $z = {}^0z$into $\zeta = \infty$, maps D onto a domain G whose complement E is a bounded closed set, takes functions harmonic in D into functions harmonic in G, and sends sets of capacity 0 into sets of capacity 0. For G we know the existence of a Green's function with respect to infinity, namely $g(\zeta; \infty, G)$. Hence

$$g\left(\frac{1}{z - z_0} ; \infty, G\right)$$

is harmonic in D, becomes logarithmically infinite at $z = z_0$, and tends to zero as z approaches any point of ∂D with the possible exception of a point set of capacity 0. These are the desired properties of Green's function for D with respect to $z = z_0$, so we define

(16.5.6) $$g(z; z_0, D) = g\left(\frac{1}{z - z_0} ; \infty, G\right).$$

We shall see later that $g(z; z_0, D)$ is uniquely defined by its properties, but for the time being formulas (16.5.5) and (16.5.6) serve as the definition.

We turn now to David Hilbert's lemniscate theorem of 1897, which we formulate as follows:

THEOREM 16.5.4.   *If C_1 and C_2 are two simple closed curves, C_1 interior to C_2, then there exists a lemniscate*

$$L: \quad |P(z)| = R$$

which is interior to C_2 and contains C_1 in its interior.

Proof.   Suppose that the distance between C_1 and C_2 is d (> 0). Let us consider a covering of C_1 by circular disks of radius δ, $\delta < d$. There exists a finite subcovering, and with the aid of the circular arcs which bound the remaining disks we can construct a simple closed curve Γ with the following properties: (i) Γ lies in the annulus between C_1 and C_2; (ii) Γ contains C_1 in its interior and has a distance $\geq d - \delta$ from C_2; and (iii) Γ is the union of a finite number of circular arcs which turn their concave sides toward C_1.

We consider the closed connected region E bounded by Γ. It is clear from (iii) that E satisfies the conditions of Poincaré. Let $U(z; \nu)$ be the equilibrium potential and $V(E)$ the Robin constant of E. By the preceding theorems $U(z; \nu) = V(E)$ on E and the equipotential lines Γ_a are simple closed curves surrounding Γ and having the property

$$\eta(\Gamma, \Gamma_a) \to 0 \quad \text{as} \quad a \to V.$$

In the present case the boundary of E is a rectifiable curve whose length we denote by l. We choose arbitrarily a point t_0 on Γ and count the length of arc in the positive sense from t_0. We set

$$\Gamma: \quad t = t(s), \quad 0 \leq s \leq l.$$

The arc of Γ between $t(0)$ and $t(s)$ is a member of the set $\mathscr{B}(E)$. Let its ν-measure be denoted by $\nu(s)$. Then $\nu(s)$ is a monotone increasing continuous function of s such that

$$\nu(0) = 0, \quad \nu(l) = 1.$$

We now have

$$(16.5.7) \qquad U(z; \nu) = \int_0^l \log | z - t(s) |^{-1} \, d\nu(s).$$

We can approximate this Riemann-Stieltjes integral arbitrarily closely by Riemann sums. Let n be an integer to be determined later and divide Γ into n arcs, each having the same ν-measure $1/n$. Let the division points be

$$\zeta_0 = t_0, \zeta_1, \cdots, \zeta_n = t_0.$$

Then

$$(16.5.8) \qquad U_n(z) \equiv \frac{1}{n} \sum_{j=1}^{n} \log | z - \zeta_j |^{-1}$$

is an approximation to $U(z; \nu)$.

To make this precise, let M be the maximum of $U(z; \nu)$ on C_2 and choose **two** numbers V_1 and V_2 such that

$$M < V_1 < V_2 < V.$$

The two equipotential lines

$$\Gamma': \quad U(z; \nu) = V_1, \qquad \Gamma'': \quad U(z; \nu) = V_2$$

bound an annulus interior to the annulus bounded by C_1 and C_2. Let ε be given, $0 < 2\varepsilon < V_2 - V_1$, and choose n so large that

$$| U(z; \nu) - U_n(z) | < \varepsilon$$

for all z between Γ' and Γ''. Choose λ so that

$$V_1 + \varepsilon < \lambda < V_2 - \varepsilon$$

and define

$$(16.5.9) \qquad\qquad L: \quad U_n(z) = \lambda.$$

On L we have

$$V_1 < \lambda - \varepsilon < U(z; \nu) < \lambda + \varepsilon < V_2,$$

that is, L is located in the annulus bounded by Γ' and Γ''. But the equation of L can be written

$$\prod_{j=1}^{n} | z - \zeta_j | = e^{-n\lambda}.$$

This shows that L is a lemniscate with the desired properties.

COROLLARY. *If C is any simple closed curve and $\varepsilon > 0$ is given, then there exists a lemniscate L_2 surrounding C and a lemniscate L_1 surrounded by C, such that*

$$\eta(C, L_1) < \varepsilon, \quad \eta(C, L_2) < \varepsilon.$$

The existence of L_2 is immediate. In the proof of the theorem we take $C = C_1$, we choose δ^* small, and we construct C_2 in the same way as Γ was constructed, using $\delta = 2\delta^*$. This forces $\eta(C, L)$ to be small and we take $L_2 = L$. For the case of L_1, we take $C = C_2$ and construct a curve C_1 interior to C_2 and close to it. This we can do, for instance, by covering C_2 by small circles and using arcs of these circles interior to C_2 to form C_1. We can obviously do this in such a manner that the distance between C_1 and C_2 has a preassigned small value ρ and so that $\eta(C_1, C_2) \to 0$ with ρ. In this case C_1 is sufficiently regular so that it can be used directly as the curve Γ in constructing $U(z; \nu)$.

Another corollary which is sufficiently important to deserve statement as a separate theorem is the following:

THEOREM 16.5.5. *A simply-connected domain can be exhausted by a nested sequence of lemniscatic domains.*

Proof. Let D be a simply-connected domain and let $\{C_k\}$ be a sequence of simple closed curves in D such that (i) C_k lies in the interior of C_{k+1} for each k, and (ii) for each point z of D there is an integer $n(z)$ such that z is interior to C_k for $k > n(z)$. Such curves can always be found, even if D is unbounded. We can then use Theorem 16.5.3 to intercalate a lemniscate L_k in the annulus between C_k and C_{k+1} for each k. These lemniscates bound lemniscatic regions which have the desired property.

The two preceding theorems can be generalized to the multiply-connected case. Thus if we have n pairs of curves

$$C_{1,1}, C_{1,2}; \quad C_{2,1}, C_{2,2}; \quad \cdots; \quad C_{n,1}, C_{n,2}$$

such that $C_{j,1}$ is interior to $C_{j,2}$ for each j while $C_{j,2}$ and $C_{k,2}$ are mutually exterior to each other, then we can find a polynomial $P(z)$ and an R such that the lemniscate

$$|P(z)| = R$$

consists of n ovals, one for each of the annuli determined by the n pairs $C_{j,1}$, $C_{j,2}$. The proof of Theorem 16.5.3 carries over to this case without essential modification. This means that we can approximate simultaneously n simple closed curves from the outside by the ovals of the same fixed lemniscate, provided the curves are completely outside each other. The method breaks down if some of the given curves are interior to each other, and the principle of the maximum shows that no single lemniscate can give a solution in such a case.

These remarks show that if E is any bounded closed set, then there exists a lemniscate L such that (i) E is interior to L, and (ii) $\eta(L, \partial E)$ is as small as we please. Earlier in this chapter we have considered lemniscatic regions such as

$$| T_n(z; E) | \leq M_n(E), \quad | F_n(z; E) | \leq K_n(E)$$

which contain E as a subset. It is intuitively plausible if not obvious that these regions contract to E when $n \to \infty$. We shall give a formal proof for the Fekete polynomials since this result will be basic in Section 17.3. The argument is due to Fekete (1933). It is based on Theorem 16.6.5 of the next section.

THEOREM 16.5.6.   *Let E be a bounded closed set whose complement is connected. Let $F_n(z; E)$ be the corresponding Fekete polynomial and let $K_n(E)$ be the maximum of its absolute value in E. For a given $\varepsilon > 0$, let E_ε be the set of points within a distance ε from E. Then there is an integer $N = N_\varepsilon$ such that*

$$(16.5.10) \qquad \Lambda_n \equiv [z \,|\, | F_n(z; E) | \leq K_n(E)] \subset E_\varepsilon, \quad N_\varepsilon < n.$$

Proof.   Let $0 < \varepsilon < 1$ and let F be the closure of the set $E_1 \ominus E_\varepsilon$. It will be sufficient to show that $F \cap \Lambda_n = \emptyset$ for $n > N_\varepsilon$. We can cover F by a finite number of circular disks, all exterior to $E_{\varepsilon/2}$. Let $D: |z - z_0| \leq \rho$ be one of these and let C be a simple closed curve which separates D from E so that E is interior to C and D is exterior. Let a be any point of D. We now apply Theorem 16.6.5 to the function

$$\frac{1}{z - a}, \quad a \in D, z \in E.$$

According to the theorem there exist numbers $A(a)$ and $\vartheta(a)$ and polynomials $P_n(z, a)$ of degree $\leq n$ in z such that

$$\left| \frac{1}{z - a} - P_n(z, a) \right| \leq A(a)[\vartheta(a)]^n, \quad z \in E.$$

The proof shows that $A(a)$ and $\vartheta(a)$ are uniformly bounded for $a \in D$. More precisely, we can find numbers A and ϑ such that

$$A(a) \leq A < \infty, \quad \vartheta(a) \leq \vartheta, \quad 0 < \vartheta < 1,$$

so that

$$(16.5.11) \qquad \left| \frac{1}{z - a} - P_n(z, a) \right| \leq A\vartheta^n, \quad a \in D, z \in E.$$

Consider now $F_{n+1}(z; E)$ and construct the Lagrange interpolation polynomial of degree n which interpolates

$$\frac{1}{z - a} - P_n(z, a)$$

at the zeros of $F_{n+1}(z)$. This polynomial is evidently linear in the two functions involved. This observation leads to the following form for the polynomial:

$$(16.5.12) \qquad L_n(z) \equiv \frac{F_{n+1}(a) - F_{n+1}(z)}{F_{n+1}(a)(z - a)} - P_n(z, a).$$

For the fraction is a polynomial in z of degree n which coincides with $(z - a)^{-1}$

at the zeros of $F_{n+1}(z)$, and an interpolation polynomial of degree n which agrees with $P_n(z, a)$ at $n + 1$ points must coincide with $P_n(z, a)$ (see Problem 8 of Exercise 16.2). Hence

$$(16.5.13) \qquad \frac{F_{n+1}(z)}{F_{n+1}(a)} = (z - a)\left\{\left[\frac{1}{z - a} - P_n(z, a)\right] - L_n(z)\right\}.$$

By (16.5.11) combined with Theorem 16.2.6 the right member does not exceed

$$A \mid z - a \mid (n + 2)\vartheta^n.$$

It follows that

$$(16.5.14) \qquad \left|\frac{F_{n+1}(z)}{F_{n+1}(a)}\right| \leq AM(n + 2)\vartheta^n, \quad z \in E, a \in D,$$

where M is an upper bound for $\mid z - a \mid$ for $z \in E$, $a \in F$.

It is only ϑ that depends upon D in (16.5.14). We now recall that F is covered by a finite number of disks D. For each of them (16.5.14) holds, provided we take the appropriate value of ϑ for the disk under consideration. But the maximum value of ϑ is still < 1, and with this choice of ϑ the estimate holds uniformly with respect to a in F and z in E. There exists an $N = N_{\varepsilon, \theta}$ such that the right member of (16.5.14) is $< \theta^{n+1}$ for $n \geq N_{\varepsilon, \theta}$ where θ is a fixed number $\vartheta < \theta < 1$. Hence

$$(16.5.15) \qquad \mid F_n(z; E) \mid < \theta^n \mid F_n(a; E) \mid, \quad a \in F, z \in E, N_{\varepsilon, \theta} < n.$$

There exist values of z in E such that the left member reaches its maximum value $K_n(E)$. This gives the final inequality

$$(16.5.16) \qquad K_n(E) < \theta^n \mid F_n(a; E) \mid, \quad a \in F, N_{\varepsilon, \theta} < n.$$

This asserts, in particular, that for all large n the points of F cannot belong to Λ_n, and this proves (16.5.10) and the theorem. If $\rho(E)$ is the transfinite diameter of E, then

$$K_n(E) > [\rho(E) - \varepsilon]^n$$

for all large values of n. Hence there exists a $\delta > 0$ such that

$$(16.5.17) \qquad [\rho(E) + \delta]^n < \mid F_n(a; E) \mid, \quad a \in F, N < n.$$

This says that for all large n the set F is exterior to the lemniscates

$$\mid F_n(z; E) \mid = [\rho(E) + \delta]^n.$$

EXERCISE 16.5

1. Find $\mathfrak{M}(1; \nu, r)$ for the equilibrium potential on $[z \mid \mid z \mid \leq 1]$.

2. Verify that Green's function for (**a**) the unit disk, (**b**) the right half-plane is given by

$$\mathbf{a.}\ \log \frac{\mid 1 - \bar{z}_0 z \mid}{\mid z - z_0 \mid}\ ; \quad \mathbf{b.}\ \log \frac{\mid z + \bar{z}_0 \mid}{\mid z - z_0 \mid}.$$

3. If in 2(**b**) we replace z by z^2 and z_0 by z_0^2, we obtain Green's function for the sector $| \arg z | \leq \frac{1}{4}\pi$. Verify and generalize!

4. Prove Theorem 7.4.3 (the extended form of Cauchy's integral with several boundaries), assuming merely continuity of $f(z)$ on the boundary.

5. Show that an inversion takes sets of capacity 0 into sets of capacity 0.

16.6. Runge's theorem. In Section 10.1 we raised the question of the nature of the functions that can appear as limits of a sequence of polynomials or of rational functions. This question will now be answered. The basic results are due to Carl Runge (1885). The following basic proposition is suggested by Cauchy's formula and the definition of the Riemann-Stieltjes integral as the limit of a sequence of sums.

THEOREM 16.6.1. *Let D be a bounded domain whose boundary $\partial D = C$ is made up of a finite number of simple closed rectifiable curves $C_0, C_1, \cdots, C_m$. Suppose that $f(z)$ is holomorphic in D and continuous in D. Then there exists a series*

$$(16.6.1) \qquad \sum_{n=1}^{\infty} R_n(z),$$

the terms of which are rational functions whose poles lie on C such that the series converges to $f(z)$ in D, uniformly with respect to z on compact sets.

Proof. Our point of departure is Cauchy's formula in the generalized form of Theorem 7.4.3:

$$f(z) = \frac{1}{2\pi i} \int_C \frac{f(t)\,dt}{t-z}, \quad C = C_0 - C_1 - \cdots - C_m.$$

Here C_0 is the outer boundary of D, and $C_1, C_2, \cdots, C_m$ are interior to C_0 and exterior to each other. By Problem 4 of Exercise 16.5 this formula is valid when $f(z)$ is merely continuous on the boundary. We restrict z to a closed set B whose distance from ∂D equals $\delta > 0$, and we divide C into subarcs, $\Gamma_1, \cdots, \Gamma_p$, of length not exceeding a preassigned positive number η, and we choose on Γ_k a point t_k. Let

$$\frac{1}{2\pi i} \int_{\Gamma_k} f(t)\,dt = A_k$$

and consider the rational function

$$\sum_{k=1}^{p} \frac{A_k}{t_k - z} \equiv R(z; \eta).$$

For $z \in B$ we have

$$| f(z) - R(z; \eta) | = \left| \sum_{k=1}^{p} \frac{1}{2\pi i} \int_{\Gamma_k} \left\{ \frac{1}{t-z} - \frac{1}{t_k - z} \right\} f(t)\,dt \right|$$

$$\leq \sum_{k=1}^{p} \frac{1}{2\pi} \int_{\Gamma_k} \frac{| t - t_k | \, | f(t) | \, | dt |}{| t - z | \, | t_k - z |} \leq \frac{Ml(C)}{2\pi\delta^2} \eta,$$

where $M = \max |f(t)|$ on C and $l(C)$ is the length of C. Here we choose $\eta = 2^{-n}$, $n = 1, 2, 3, \cdots$, and set

$$R_1(z) = R(z; 2^{-1}), \quad R_n(z) = R(z; 2^{-n}) - R(z; 2^{-n+1}), \quad n > 1.$$

These functions are clearly rational and have their poles on C. We use them to form the series (16.6.1). Since

$$(16.6.2) \qquad |f(z) - R(z; 2^{-n})| \leq \frac{Ml(C)}{2\pi\delta^2} 2^{-n},$$

we see that the series converges everywhere in D and uniformly in B. This completes the proof.

The theorem may be extended in various directions. Thus, it is desirable to have a corresponding approximation theorem for more general domains, and, even for the restricted domains considered, it may be worth while to be able to move the poles away from the boundary. Using the method of the preceding theorem we shall next prove

THEOREM 16.6.2.   *If D is an arbitrary bounded domain and $f(z)$ is holo-morphic in D, then there exists a sequence of rational functions $\{S_n(z)\}$ converging to $f(z)$ in D, uniformly on compact sets.*

Proof.   We can exhaust D by a sequence of domains $\{D_n\}$ such that

$$\overline{D_n} \subset D_{n+1} \subset D$$

and that each point $z \in D$ lies in D_n for $n > n_z$. We may assume that D_n is bounded by a finite number of simple closed rectifiable curves. We write $\partial D_n = C_n$. Let δ_n be the distance of D_{n-1} from C_n, let λ_n be the length of C_n, and let $M_n = \max |f(z)|$ on C_n. We remind the reader that such a sequence of domains D_n may be obtained by introducing a Cartesian grid of mesh 2^{-m} in the plane and allotting to D_m a set of squares of the grid which are all completely in D and have sides in common so that the resulting (open) set is connected. Repeated bisection gives grids of finer mesh and allows the adjunction of contiguous smaller squares to those of D_m. It is clear that this process leads to a sequence of domains D_n of the desired nature. In this case C_n is made up of simple closed polygons. We suppose that D_n is constructed in this manner.

Using the preceding theorem we can now find rational functions $S_n(z)$ with poles on C_n such that for prescribed $\eta_n > 0$

$$(16.6.3) \qquad |f(z) - S_n(z)| \leq \frac{M_n\lambda_n}{2\pi\delta_n^2} \eta_n, \quad z \in D_{n-1}.$$

We can choose η_n so that the right member tends to zero with any desired rapidity. This sequence of rational functions evidently converges to $f(z)$ in D and uniformly in every fixed domain D_n.

It is clear that the poles of $S_n(z)$ are only apparent singularities of $f(z)$. It is possible, however, to avoid even apparent singularities and to approximate $f(z)$ by rational functions whose poles are outside of D. The artifice used to this end is also due to Runge. For the sake of simplicity we restrict ourselves to domains D of finite connectivity.

THEOREM 16.6.3.   *Suppose that D is bounded and $\mathbf{C}(D)$ has $m + 1$ components, $m > 0$, where F_0 is the component which contains the point at infinity and the other components $F_1, F_2, \cdots, F_m$ are bounded. Let $z_0, z_1, \cdots, z_m$ be given points, $z_k \in F_k$, z_0 finite. Let $f(z)$ be holomorphic in D. Then there exists a sequence of rational functions $\{T_n(z)\}$ such that $T_n(z)$ converges to $f(z)$ in D, uniformly on compact subsets, and the poles of $T_n(z)$ are located at the given points $\{z_k\}$.*

Proof.   To start with, we proceed as in the proof of the preceding theorem. We obtain rational functions $S_n(z)$ with poles on C_n such that (16.6.3) holds where the right-hand side equals 2^{-n}, for instance. We may suppose that n is so large that C_n already consists of $m + 1$ simple closed curves $C_{0,n}, C_{1,n}, \cdots, C_{m,n}$ where $C_{0,n}$ contains F_0 in its exterior and $C_{k,n}$ contains F_k in its interior for $k \geq 1$. There will now be poles of $S_n(z)$ on each of these $m + 1$ contours, and we shall try to replace the poles on $C_{k,n}$ by a single pole located at $z = z_k$ in such a manner that the degree of approximation in D_{n-1} is not materially affected.

To fix the ideas we suppose $m \geq 1$ and take $k = 1$. A typical term in $S_n(z)$ is $A(z - a)^{-1}$ where $a \in C_{1,n}$ in this case. Suppose that $b \neq a$, ∞ but otherwise is arbitrary. Then for any choice of p

$$\frac{1}{z - a} = \frac{1}{z - b} + \frac{a - b}{(z - b)^2} + \cdots + \frac{(a - b)^{p-1}}{(z - b)^p} + \frac{(a - b)^p}{(z - a)(z - b)^p},$$

so that

$$(16.6.4) \qquad \left| \frac{1}{z - a} - \sum_{j=1}^{p} \frac{(a - b)^{j-1}}{(z - b)^j} \right| = \frac{|a - b|^p}{|z - a||z - b|^p} < \delta_n^{-1} 2^{-p}$$

if $z \in D_{n-1}$ and if we choose b such that

$$(16.6.5) \qquad\qquad 2|b - a| < |z - b|.$$

This is always possible and shows that we can make the left member of (16.6.4) as small as we please for $z \in D_{n-1}$ provided p is sufficiently large. In other words, a simple pole at $z = a$ on C_n may be replaced by a pole at some other point $z = b$ provided the order of this pole is sufficiently high. If we can satisfy (16.6.5) with $b = z_1$ for all poles $z = a$ of $S_n(z)$ on $C_{1,n}$, then we are, in principle, out of the difficulties. Normally, we have to be prepared for a less favorable result, and we have to proceed step by step.

We note first that similar approximation formulas, which may be obtained by differentiation, hold for higher powers. As a result any polynomial in $(z - a)^{-1}$

may be replaced by a polynomial in $(z - b)^{-1}$ of sufficiently high degree and such that in any bounded domain which omits $z = a$ and $z = b$ the difference is arbitrarily small. This is the basis for the procedure.

We now return to the point $z = a$ on $C_{1,n}$ and join it to $z = z_1$ by a polygonal line L which does not intersect D_{n-1}. On L we choose points

$$b_0 = a, b_1, b_2, \cdots, b_N = z_1$$

such that

$$|b_j - b_{j-1}| < \tfrac{1}{2}\delta,$$

where δ is the distance from L to D_{n-1}. For $z \in D_{n-1}$ and $j = 1, 2, \cdots, N$ we have then

$$|b_j - b_{j-1}| < \tfrac{1}{2}|z - b_j|.$$

We have seen that the term $A(z - a)^{-1}$ of $S_n(z)$ may be replaced by a polynomial in $(z - b_1)^{-1}$, say $Q_1(z)$, such that

$$\left| \frac{A}{z - a} - Q_1(z) \right|$$

is uniformly small in D_{n-1}. Next we can replace each term $B(z - b_1)^{-\beta}$ of $Q_1(z)$ by a polynomial in $(z - b_2)^{-1}$, and so on. After N steps we have replaced $A(z - a)^{-1}$ by a polynomial in $(z - z_1)^{-1}$. This process we repeat for each term of $S_n(z)$ with poles on $C_{1,n}$. It should be noted that the number of steps required changes from one pole to the next, but the number of steps has a finite least upper bound.

We repeat the process for the poles on the other contours $C_{j,n}$, replacing z_1 by z_j. As a final result we obtain a rational function $T_n(z)$ whose poles are located at the points $z_0, \cdots, z_m$ such that

$$|S_n(z) - T_n(z)| < \rho_n, \quad z \in D_{n-1},$$

where $\{\rho_n\}$ is a given sequence of positive numbers converging to 0. In particular, we can take $\rho_n = 2^{-n}$ and obtain thus

$$|f(z) - T_n(z)| < 2^{1-n}, \quad z \in D_{n-1}$$

for each n. This completes the proof.

We note that the functions $S_n(z)$ have only simple poles, whereas the $T_n(z)$ have poles of higher order and the order presumably becomes infinite with n. We recall that in the case of a circular annulus, $0 < R_1 < |z - a| < R_2 < \infty$, the representation furnished by the Laurent expansion usually involves powers of $z - a$ of all orders, positive and negative. In this case there are two poles, one at $z = a$, the other at $z = \infty$. Here $m = 1$, $z_1 = a$, $z_0 = \infty$.

It is of course possible to modify the construction used above so as to allow more poles than the minimum number $m + 1$. It should be emphasized that it is necessary to place poles in each of the $m + 1$ components of $\mathbf{C}(D)$. The case $m = 0$ is of particular interest.

THEOREM 16.6.4. *If $f(z)$ is holomorphic in a bounded simply-connected domain D, then there exists a sequence of polynomials $\{P_n(z)\}$ which converges to $f(z)$ in D, uniformly on compact sets.*

Proof. The preceding theorem shows merely that it is possible to approximate $f(z)$ by means of polynomials in a variable $(z - b)^{-1}$ where b is exterior to D. We can get the desired result, however, with the aid of a suitable fractional linear transformation. We choose

$$ w = \frac{1}{z - a}, \quad a \in \mathbf{C}(D), $$

and set

$$ f(z) = F(w). $$

The transformation maps D conformally in a one-to-one manner on a simply-connected domain Δ, and $F(w)$ is holomorphic in Δ. Here Δ is bounded and $w = 0$ is exterior to Δ. Using the preceding theorem we can find a polynomial in $1/w$ such that

$$ \left| F(w) - P_n\left(\frac{1}{w}\right) \right| < \eta_n, \quad w \in \Delta_{n-1}, $$

where Δ_n is the image of the domain D_n used in the proof of Theorem 16.6.3. Hence

$$ |f(z) - P_n(z - a)| < \eta_n, \quad z \in D_{n-1}, $$

and the theorem is proved.

In all approximation problems for analytic functions in the complex plane, the central problem is that of approximating

$$ \frac{1}{z - b} $$

in a preassigned domain D. If the domain is simply-connected and we want to approximate by polynomials, we face a question which is closely related to the discussion of Sections 11.4 and 11.5. There we encountered a number of methods of summing the geometric series in a star-shaped domain, usually the Mittag-Leffler principal star of $(1 - z)^{-1}$. Any such method can be used to find a sequence of polynomials converging to $(1 - z)^{-1}$ in the principal star. Either the method gives such polynomials directly or they are obtained by taking partial sums of sufficiently high order of a sequence of power series. By a simple transformation of coordinates this leads to a sequence of polynomials in a variable $z - a$, $a \neq b$, which converges to $(z - b)^{-1}$ in any domain located in the plane cut from b to ∞ along the continuation of the line segment (a, b). If D is a convex domain and $f(z)$ is holomorphic in D and continuous in $\bar{D}$, then we can use this observation to give an alternate proof of Theorem 16.6.4.

We attach the discussion to formula (16.6.2). Each term of the rational function $R(z; 2^{-n})$ is of the form

$$\frac{A_k}{t_k - z}, \quad t_k \in C, z \in B.$$

We may suppose that B is also convex and take a fixed point a in B. The continuations of the line segments $(a, t_k]$ lie entirely outside of C, so that B is in the "right" position with respect to each of the points t_k. Hence there is a polynomial in $z - a$, given by the method of summability in question, which approximates $A_k(t_k - z)^{-1}$ in B with any desired degree of accuracy. Adding these polynomials for $k = 1, 2, \cdots, p$, we get a polynomial $P_n(z)$ such that

$$| R(z; 2^{-n}) - P_n(z) | < \eta_n, \quad z \in B,$$

and hence also

$$|f(z) - P_n(z) | < \eta_n^*, \quad z \in B$$

where η_n and η_n^* are at our disposal.

The methods used above give us no idea of the degree of approximation in terms of the degree of the approximating polynomial. This question is normally attacked with the aid of conformal mapping, but in 1927 G. Szegö showed that elementary methods suffice. We reproduce his proof.

THEOREM 16.6.5. *Let C be a simple closed rectifiable curve and let $f(z)$ be holomorphic inside C and continuous on C. Let B be a closed set interior to C and at a distance ρ from C. Then there exists an $A, 0 < A, a\vartheta, 0 < \vartheta < 1$, and a sequence of polynomials $\{P_n(z)\}$ where $P_n(z)$ is of degree n or less, such that for all $z \in B$*

$$(16.6.6) \qquad |f(z) - P_n(z) | \leq A\vartheta^n.$$

Here ϑ depends only upon the geometric configuration, that is, C and ρ, but A also depends upon $f(z)$.

Proof. For $z \in B$ we have

$$f(z) = \frac{1}{2\pi i} \int_C \frac{f(t)\, dt}{t - z}.$$

We partition C into a number of subarcs $C_1, \cdots, C_k$ such that on each arc

$$(16.6.7) \qquad \left| \frac{1}{t' - z} - \frac{1}{t'' - z} \right| \leq \frac{1}{4\delta}, \quad t', t'' \in C_\nu, z \in B,$$

where δ is the diameter of C. On C_ν we choose a point t_ν and use Theorem 16.6.4 to determine a polynomial $p_\nu(z)$ such that

$$\left| \frac{1}{t_\nu - z} - p_\nu(z) \right| \leq \frac{1}{4\delta}, \quad z \in B, \nu = 1, 2, \cdots, k.$$

Hence, by (16.6.7), for any $t \in C_\nu$ we have

$$\left| \frac{1}{t-z} - p_\nu(z) \right| \leq \frac{1}{2\delta} .$$

Since $|t - z| < \delta$ for $t \in C$, $z \in B$, we obtain

$$| 1 - (t - z)p_\nu(z) | \leq \tfrac{1}{2}.$$

Next we define

$$(16.6.8) \qquad Q_m(z) = \sum_{\nu=1}^{k} \frac{1}{2\pi i} \int_{C_\nu} \frac{1 - [1 - (t - z)p_\nu(z)]^m}{t - z} f(t)\, dt.$$

This is actually a polynomial; if the degree of $p_\nu(z)$ is d_ν, and if max $d_\nu = d$, then the degree of $Q_m(z)$ is $\leq (d + 1)m - 1$.

It follows that for $z \in B$

$$f(z) - Q_m(z) = \sum_{\nu=1}^{k} \frac{1}{2\pi i} \int_{C_\nu} \frac{[1 - (t - z)p_\nu(z)]^m}{t - z} f(t)\, dt,$$

so that

$$| f(z) - Q_m(z) | \leq \frac{1}{2^m} \frac{1}{2\pi} \int_C \frac{|f(t)|}{|t - z|} | dt | \leq \frac{Ml(C)}{2\pi\rho} 2^{-m},$$

where $M = \max | f(t) |$ on C, ρ has already been defined as the distance from B to C, and $l(C)$ is the length of C.

If we replace m in the definition of $Q_m(z)$ by the largest integer $\leq n/(d + 1)$, we obtain a polynomial of degree $\leq n$, which we define as $P_n(z), n = 1, 2, 3, \cdots$. This polynomial satisfies (16.6.6) with

$$(16.6.9) \qquad A = \frac{Ml(C)}{\pi\rho}, \quad \vartheta = 2^{-1/(d+1)}.$$

It is clear that these quantities have the desired properties. When ρ decreases, d will normally increase, so that $A \to \infty$, $\vartheta \to 1$ when $\rho \to 0$.

If D is the disk $| z | < R$, then the partial sums of the Maclaurin series of $f(z)$ furnish the desired approximation of $f(z)$ by polynomials. If $f(z)$ is continuous or at least bounded, $| f(t) | \leq M$, on $| t | = R$, then

$$\frac{| f^{(k)}(0) |}{k!} \leq MR^{-k}$$

by Cauchy's estimates. Hence, on $| z | = r < R$ we have

$$\left| f(z) - \sum_{k=0}^{n} \frac{f^{(k)}(0)}{k!} z^k \right| \leq \frac{M}{R - r} \left(\frac{r}{R} \right)^n ,$$

an estimate of the same character as (16.6.6).

This is the case of a circular domain. Similar results hold for lemniscatic domains, as is shown by the following theorem, due to C. G. J. Jacobi and published, after his death, in 1857.

THEOREM 16.6.6. *Let $P(z)$ be a polynomial of exact degree m and let L be the open set*

$$| P(z) | < R,$$

where R is so large that L is actually a domain. Let $f(z)$ be holomorphic in L and continuous in $\overline{L}$. Then $f(z)$ can be represented by a series of the form

(16.6.10) $$f(z) = \sum_{n=0}^{\infty} \left(\sum_{j=0}^{m-1} a_{j,\,n} z^j \right) [P(z)]^n,$$

convergent in L and uniformly convergent in any region

$$L_1: \quad | P(z) | \leqq R_1 < R.$$

The representation is unique.

Proof.   As is usual in such cases, the problem reduces to representing the Cauchy kernel $(t - z)^{-1}$ by such a series. Suppose that t lies on $C \equiv \partial L$ and restrict z to L_1. We note that

$$\frac{P(t) - P(z)}{t - z} = \sum_{k=1}^{m-1} a_k \sum_{j=0}^{k} z^j t^{k-j}$$

is a polynomial in z and t of degree $m - 1$, symmetric in the two variables. It follows that

(16.6.11)
$$\begin{aligned}
\frac{1}{t - z} &= \frac{1}{P(t) - P(z)} \frac{P(t) - P(z)}{t - z} \\
&= \sum_{n=0}^{\infty} \frac{[P(z)]^n}{[P(t)]^{n+1}} \left\{ \sum_{k=1}^{m-1} a_k \sum_{j=0}^{k} t^j z^{k-j} \right\}.
\end{aligned}$$

This series converges absolutely and uniformly with respect to t on C and z in L_1 since

$$\frac{| P(z) |}{| P(t) |} \leqq \frac{R_1}{R} < 1.$$

It follows that we can multiply by $f(t)$ and integrate termwise along C. The result is (16.6.10), and we find that

(16.6.12) $$a_{j,n} = \sum_{k=j}^{m-1} a_k \frac{1}{2\pi i} \int_C \frac{t^{k-j} f(t)}{[P(t)]^{n+1}} dt.$$

It is a simple matter to estimate these coefficients. Consider the Nth partial sum of the series (16.6.10), that is, the sum of the terms with $n \leqq N$. This is a polynomial in z of degree not exceeding $(m + 1)N - 1$. In L_1 the difference of $f(z)$ and this polynomial satisfies an inequality of type (16.6.6) with

(16.6.13) $$\vartheta = \left(\frac{R_1}{R} \right)^{1/(m+1)}$$

The remaining assertions of the theorem follow from this inequality.

Actually we can use Theorem 16.6.6 to prove Theorem 16.6.4 if we invoke

Hilbert's theorem. By Theorem 16.5.5 we can exhaust a bounded simply-connected domain by lemniscatic domains:

$$L_1 \subset L_2 \subset \cdots \subset L_n \subset L_{n+1} \subset \cdots \subset D,$$

such that each point $z \in D$ lies in L_n for $n > n_z$. We may suppose that L_n is given by

$$L_n: \quad |Q_n(z)| \leqq R_n$$

and that

$$L_{n-1} \subset [z \mid |Q_n(z)| \leqq R_n{}^*], \quad R_n{}^* < R_n.$$

By Theorem 16.6.6 we can represent $f(z)$ in L_n by a series of the form

$$\sum_{m=0}^{\infty} U_m(z)[Q_n(z)]^m,$$

where the degree of $U_m(z)$ is less than that of $Q_n(z)$. A suitably chosen partial sum of this series gives a good approximation of $f(z)$ in L_{n-1}, that is, we can determine a sequence of polynomials which converges to $f(z)$ in D, uniformly on each fixed L_{n-1}.

There is still another mode of approach to Theorem 16.6.4 which is of interest from the point of view of principle. First we shall give a slightly more general formulation to the topic of interpolation polynomials, mentioned briefly in Section 16.2. Suppose there is given a triangular array of points in the plane

$$z_{1,n}, z_{2,n}, \cdots, z_{n,n}, \quad n = 1, 2, 3, \cdots.$$

If all these points are located in a simply-connected domain D and if $f(z)$ is holomorphic in D, then we can form the interpolation polynomials of Lagrange

(16.6.14) $$L_{n-1}(z;f) \equiv P_n(z) \sum_{k=1}^{n} \frac{f(z_{k,n})}{P'(z_{k,n})(z - z_{k,n})},$$

where

$$P_n(z) = \prod_{k=1}^{n}(z - z_{k,n}).$$

These polynomials agree with $f(z)$ at the points $z_{k,n}$ and, naturally, the question arises whether the sequence $\{L_{n-1}(z;f)\}$ converges to $f(z)$ when $n \to \infty$. It is known that this is normally not the case unless the point set $\{z_{k,n}\}$ satisfies certain limitations. Several convergence theorems are known for the case in which the points $\{z_{k,n}\}$ are distributed on a simple closed curve C. For the case of the unit circle it suffices that the points $z_{k,n}$ are the roots of unity. See Problem 5 of Exercise 16.6. If the points $z_{k,n}$ are located on a simple closed curve C and if the function which maps the exterior of C conformally on the exterior of the unit circle, maps the points $z_{k,n}$ into the roots of unity, then, as proved by L. Fejér (1918), the corresponding interpolation polynomials for functions $f(z)$ holomorphic inside and on C converge to $f(z)$ inside and on C. We shall not prove this, but shall give another result due to M. Fekete (1926) where the polynomials $F_n(z; E)$ of (16.2.12) and the corresponding interpolation polynomials (16.2.18) are used.

THEOREM 16.6.7.   *Let E be a bounded closed set whose boundary C is a simple closed curve. Let $f(z)$ be holomorphic in E. Let $L_{n-1}(z;f)$ be defined by (16.2.18). Then, uniformly with respect to z in E,*

$$(16.6.15) \qquad\qquad \lim_{n\to\infty} L_{n-1}(z;f) = f(z).$$

Proof.   Let D be a simply-connected domain containing E such that $f(z)$ is holomorphic in D. Let $\rho(E)$ be the transfinite diameter of E. Theorem 16.5.6 shows that, if ε is sufficiently small, then for all large values of n

$$E \subset L_n(\varepsilon) \subset L_n(2\varepsilon) \subset D$$

where

$$L_n(\delta): \quad |F_n(z;E)| \leq [\rho(E) + \delta]^n.$$

Furthermore, for such n we can find a fixed simple closed rectifiable curve Γ which is contained in D and contains $L_n(2\varepsilon)$ in its interior. Let $l(\Gamma)$ be the length of Γ, η its distance from E, and let M be the maximum of $|f(t)|$ on Γ. A simple application of the residue theorem shows that

$$(16.6.16) \qquad f(z) - L_{n-1}(z;f) = F_n(z;E) \frac{1}{2\pi i} \int_\Gamma \frac{f(t)\, dt}{F_n(t;E)(t-z)}.$$

It follows that

$$|f(z) - L_{n-1}(z;f)| \leq \left(\frac{\rho(E)+\varepsilon}{\rho(E)+2\varepsilon}\right)^n \frac{Ml(\Gamma)}{2\pi\eta}, \quad z \in E,$$

which tends to zero as $n \to \infty$, uniformly with respect to z in E. This completes the proof.

EXERCISE 16.6

1. Let the simply-connected domain D be bounded by arcs of p circles C_k: $|z - z_k| = R_k$, $k = 1, 2, \cdots, p$, D being exterior to all the circles. If $f(z)$ is holomorphic in D, show that for $z \in D$

$$f(z) = \sum_{n=0}^{\infty} \sum_{k=1}^{p} A_{k,n}(z - z_k)^{-n}.$$

Determine the coefficients and verify the convergence. Show that the series also converges if z is in the infinite domain exterior to all circles, its sum being zero (P. Appell, 1882).

2. Find an interpolation polynomial which osculates $f(z)$. More precisely formulated, given m points $z_1, z_2, \cdots, z_m$ and m non-negative integers $\alpha_1, \alpha_2, \cdots, \alpha_m$, find the polynomial of lowest order which coincides with $f(z)$ and its first α_k derivatives at the m points. Suppose that $f(z)$ is holomorphic in a simply-connected domain containing the points in question and use a contour integral to represent the polynomial. What is the degree of the latter? This is a generalization of Lagrange's formula due to Charles Hermite (1878).

3. Show that Theorem 16.6.4 holds also for unbounded domains.

4. In (16.6.10) suppose that the zeros of $P(z)$ are at the m distinct points z_1,

$z_2, \cdots, z_m$. Take the difference between $f(z)$ and the Nth partial sum of the series and show that this function vanishes together with its first N derivatives at each of the points z_k.

5. Suppose that $f(z)$ is holomorphic in the disk $|z| < R$. Show that the sequence of interpolation polynomials $L_{n-1}(z; f)$ based on the unit roots converges to $f(z)$ in any disk $|z| < R_1 < R$.

6. Let $P_n(z)$ be a polynomial of degree n whose n zeros divide the line segment $[-1, +1]$ into $n - 1$ equal parts. If the leading coefficient is 1, show that

$$\lim_{n \to \infty} \log |P_n(z)|^{\frac{1}{n}} = \frac{1}{2} \int_{-1}^{1} \log |z - t| \, dt \equiv U(z),$$

and evaluate this integral explicitly. (*Hint:* It is easier to work with $\log(z - t)$ and to take the real part of the result.)

7. Let $f(z)$ be holomorphic in the domain

$$U(z) < U_1 \quad \text{where} \quad U_1 > \log 2 - 1.$$

Prove that the corresponding sequence of interpolation polynomials converges to $f(z)$ in this domain. (*Hint:* Use the analogue of formula (16.6.16) and the result of the preceding problem.)

8. Base a proof of Theorem 16.6.7 on the following propositions: (i) $f(z)$ can be approximated by a polynomial $P_n(z)$ satisfying (16.6.6). (ii) $L_{p-1}(z; g)$ is linear in g. (iii) Problem 8 of Exercise 16.2. (iv) Theorem 16.2.6 (Fekete, 1926).

16.7. Overconvergence. The phenomenon of overconvergence was discovered by M. B. Porter in 1906. It was rediscovered by Robert Jentzsch in 1914 and again by Alexander Ostrowski, who made a thorough investigation of the problem, starting in 1921.

Porter's example is a special case of the following one. Let $P(z)$ be a polynomial of degree $k + 1$, $k > 1$,

$$P(z) = z(z - z_1)(z - z_2) \cdots (z - z_k),$$

where at least one $z_j \neq 0$. Let $\{m_n\}$ be a sequence of positive integers such that

(16.7.1) $$m_{n+1} > (k + 1)m_n, \quad m_1 = 1,$$

and form the polynomial series (special Jacobi series)

(16.7.2) $$\sum_{n=1}^{\infty} a_n [P(z)]^{m_n}.$$

Here the coefficients a_n are so chosen that the corresponding Hadamard gap series

$$\sum_{n=1}^{\infty} a_n w^{m_n}$$

has a positive finite radius of convergence R. The series (16.7.2) then represents a holomorphic function of z in each of the components of the lemniscatic open set

(16.7.3) $| P(z) | < R,$

and the boundaries of these components form the natural boundaries of the corresponding analytic functions.

   If there are several components, let D be the one which contains the origin. Then there exists an analytic function $f(z)$ holomorphic in D where the function in question is defined by (16.7.2). Let C be the boundary of D and let C have the distance R_0 from the origin. It is fairly obvious that C cannot be a circle. Thus the disk $| z | < R_0$ is a proper subset of D and in this disk $f(z)$ is represented by a power series in z convergent for $| z | < R_0$. On the circle $| z | = R_0$ there is at least one point which is on C and which, hence, is a singular point of $f(z)$. Thus R_0 is the exact radius of convergence of the power series. This series is obtained by expanding the terms of (16.7.2)

$$a_n[P(z)]^{m_n} \equiv \sum_{p=m_n}^{(k+1)m_n} c_p z^p,$$

summing for n, and collecting terms. But condition (16.7.1) shows that there is no overlapping: a given power of z occurs in at most one of the terms of (16.7.2). It follows that

(16.7.4) $f(z) = \sum_{p=1}^{\infty} c_p z^p.$

   As already observed, the radius of convergence of this series is R_0, that is, the sequence of partial sums converges in $| z | < R_0$ and in no larger disk. On the other hand, if $N = (k+1)m_n$, then the Nth partial sum of (16.7.4) equals the nth partial sum of (16.7.2). Thus there exists a subsequence of partial sums of (16.7.4) which converges in the domain D of which the disk $| z | < R_0$ is a proper subset. This is the phenomenon of overconvergence. Porter took

$$P(z) = z(z+1).$$

   Overconvergence is not restricted to power series. A simple example of an overconvergent Dirichlet series is given by

(16.7.5)
$$\sum_{n=2}^{\infty} a_n e^{-\lambda_n z}.$$
$$a_{2k} = 1, \quad \lambda_{2k} = k, \quad a_{2k+1} = -1, \quad \lambda_{2k+1} = k + e^{-k^2}.$$

Here the series converges for $\Re(z) > 0$ but not for any z with $\Re(z) \leq 0$ since the terms do not tend to zero. On the other hand, the partial sums of odd order converge for all values of z, so the series represents an entire function.

   We return to power series. There is a peculiar relationship between over-convergence and gap series. An overconvergent series is not necessarily lacunary; for example, (16.7.4) need not be a gap series. Also, we can always add a power series of a larger radius of convergence to an overconvergent lacunary series so that the resulting series still exhibits the phenomenon of overconvergence but is

no longer lacunary. Actually every overconvergent power series is of this
nature: it is the sum of a lacunary series and a power series whose radius of
convergence exceeds that of the given one. The gaps have to satisfy certain
conditions about which the following theorem due to Ostrowski gives some
information.

THEOREM 16.7.1.   *Given a power series*

$$\sum_{n=0}^{\infty} c_n z^n \equiv f(z)$$

whose radius of convergence equals 1. *Given also two sequences of positive integers*
$\{m_k\}$ *and* $\{p_k\}$ *such that*

(16.7.6)
$$m_k < p_k \leqq m_{k+1}, \quad (1+\eta)m_k < p_k,$$
$$c_n = 0, \quad m_k < n < p_k, \quad k = 1, 2, 3, \cdots,$$

where η *is a fixed positive number. Then the partial sums*

$$S_{m_k}(z) = \sum_{n=0}^{m_k} c_n z^n$$

converge in a full neighborhood of each point of the unit circle at which $f(z)$ *is*
holomorphic.

Proof.   The following argument is due to G. Szegö (1922). It is based on a
device which we have already encountered in the proof of Theorem 16.6.5.

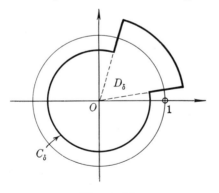

Figure 16

Suppose that $f(z)$ is holomorphic in the closure of the domain D_δ shown in
Figure 16. D_δ is the union of two circular sectors

$$|z| < 1 - \delta, \qquad\qquad |z| < R,$$
$$\theta_2 < \arg z < \theta_1 + 2\pi, \qquad \theta_1 < \arg z < \theta_2.$$

Here $\delta > 0$ is arbitrarily small and $R > 1$. We denote the boundary of D_δ
by C_δ.

For $z \in D_\delta$ we consider the difference

$$f(z) - S_{m_k}(z) = \frac{1}{2\pi i} \int_{C_\delta} \frac{f(t)}{t - z} \left(\frac{z}{t}\right)^{m_k + 1} dt.$$

Since at least $[\eta m_k] \equiv p$ terms are missing after the m_kth one, we see that

$$\frac{f(z) - S_{m_k}(z)}{z^{m_k + 1}} = \frac{1}{2\pi i} \int_{C_\delta} \frac{f(t)}{t^{m_k + 1}} \left[\frac{1}{t - z} - \sum_{j=1}^{p} \frac{\alpha_j}{t^j}\right] dt,$$

where the coefficients α_j are completely arbitrary. This gives

$$(16.7.7) \qquad \frac{f(z) - S_{m_k}(z)}{z^{m_k + 1}} = \frac{1}{2\pi i} \int_{C_\delta} \frac{f(t)}{t^{m_k + 2}} \left[\frac{1}{1 - \dfrac{z}{t}} - P\left(\frac{1}{t}\right)\right] dt$$

where $P(u)$ is any polynomial of degree p or less. We shall choose $P(u)$ to best suit our purposes.

Let B be a domain interior to C_δ in which

$$a < \left|1 - \frac{z}{t}\right| < b, \quad z \in B, t \in C_\delta,$$

where a and b are fixed positive numbers. Note that B is chosen first, a and b afterward. By Theorem 16.6.4 we can find a polynomial $R(1/t)$ of $1/t$ such that

$$(16.7.8) \qquad \left|\frac{1}{1 - \dfrac{z}{t}} - R\left(\frac{1}{t}\right)\right| < \frac{1}{2b}, \quad z \in B, t \in C_\delta.$$

We use the transformation $u = 1/t$ and consider the function $(1 - zu)^{-1}$ for fixed z in B as a function of u in the interior and on the boundary of the image of C_δ under the transformation. The theorem applies to the function and can be made to give the desired result. Let v be the degree of R and let

$$(16.7.9) \qquad P\left(\frac{1}{t}\right) = \frac{1 - \left[1 - \left(1 - \dfrac{z}{t}\right) R\left(\dfrac{1}{t}\right)\right]^m}{1 - \dfrac{z}{t}},$$

where m is the largest integer such that $m(v + 1) \leq p$. This is seen to be a polynomial in $1/t$ of degree $\leq p$ and is the polynomial to be used in (16.7.7). We have, in fact,

$$\left|\frac{1}{1 - \dfrac{z}{t}} - P\left(\frac{1}{t}\right)\right| \leq a^{-1} 2^{-m}, \quad z \in B, t \in C_\delta.$$

This gives

$$|f(z) - S_{m_k}(z)| \leq \frac{MR}{a} 2^{-m} \left(\frac{|z|}{1 - \delta}\right)^{m_k + 1}.$$

Here we extract the m_kth root and note that

$$\frac{m}{m_k} \to \frac{\eta}{\nu+1} \equiv \mu.$$

Hence

$$\limsup_{k\to\infty} |f(z) - S_{m_k}(z)|^{\frac{1}{m_k}} \leq (1-\delta)^{-1} |z| 2^{-\mu}.$$

This holds for every $\delta > 0$, so we must actually have

(16.7.10) $$\limsup_{k\to\infty} |f(z) - S_{m_k}(z)|^{\frac{1}{m_k}} \leq |z| 2^{-\mu}.$$

This holds for z in B. It is now desirable to be more specific in the choice of B. Suppose that $1 < R_1 < R$, $0 < \varepsilon < \frac{1}{2}(\beta - \alpha)$, and let B be the set

$$|z| \leq R_1, \quad \alpha + \varepsilon \leq \arg z \leq \beta - \varepsilon.$$

Then choose a and b in the manner indicated. This determines ν and hence also μ. Now $2^\mu > 1$ and we can find a θ, $0 < \theta < 1$, so that also $\theta 2^\mu > 1$. If we restrict z still further so that

(16.7.11) $$|z| \leq \min(\theta 2^\mu, R_1), \quad z \in B,$$

then the right member of (16.7.10) does not exceed θ, and for such values of z we have

$$\lim_{k\to\infty} |f(z) - S_{m_k}(z)| = 0$$

uniformly in z. This completes the proof.

COROLLARY [HADAMARD'S GAP THEOREM]. *If $p_k = m_{k+1}$, then the unit circle is the natural boundary of $f(z)$.*

For in this case the set of partial sums $\{S_{m_k}(z)\}$, which are potentially overconvergent, is identical with the set of all partial sums which converge only in the circle of convergence. This shows that there cannot be any points on the circle of convergence where $f(z)$ is holomorphic, that is, every point on the unit circle is singular and the circle is the natural boundary of $f(z)$.

For the class of functions defined by Theorem 16.7.1 the partial sums $S_{m_k}(z)$ provide the analytic continuation across the unit circle wherever such continuation is possible. Ordinarily we obtain no information concerning how far this method of continuation will be effective. There is one important case, however, in which it gives the complete domain of existence of the function. For the proof we have to anticipate the so-called *two constants theorem*, which will be proved later (the case $n = 2$ of Theorem 18.3.2). In a less precise form, sufficient for present needs, it may be formulated as follows:

Let $f(z)$ be holomorphic and bounded in a domain D bounded by a finite number of simple closed curves C and let $|f(z)| \leq M$ in D. Suppose that there exists a subarc Γ of C (or subarcs) such that when z approaches any point of Γ we have

$$(16.7.12) \qquad \qquad \limsup |f(z)| \leq m < M.$$

Then there exists a function $\lambda(z)$ with $0 < \lambda(z) < 1$ such that, for each $z \in D$,

$$(16.7.13) \qquad \qquad \log |f(z)| \leq \lambda(z) \log m + [1 - \lambda(z)] \log M.$$

Moreover, on compact subsets of D the function $\lambda(z)$ is bounded away from 0 and 1.

We can now prove

THEOREM 16.7.2.   If the assumptions on the gaps made in the preceding theorem are replaced by

$$(16.7.14) \qquad \qquad \frac{p_k}{m_k} \to \infty,$$

then the partial sums $S_{m_k}(z)$ converge uniformly to $f(z)$ in the neighborhood of any regular point of $f(z)$. The partial sums converge in the complete domain of existence $D[f]$ of $f(z)$, and $D[f]$ is simply-connected. Further, $f(z)$ is single-valued in $D[f]$.

Proof.   Suppose that D_0 is a bounded subdomain of $D[f]$ with $\overline{D_0} \subset D[f]$. See Figure 17. We also suppose that D_0 is contained in the disk $|z| < R$,

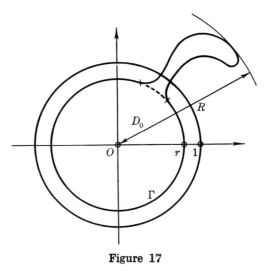

Figure 17

that it contains the disk $|z| < r < 1$, and that an arc Γ of the circle $|z| = r$ forms part of the boundary of D_0. We now consider

$$F_k(z) \equiv f(z) - S_{m_k}(z), \quad z \in D_0.$$

Since $f(z)$ is bounded in D_0 and $S_{m_k}(z)$ is a polynomial of degree m_k whose coefficients c_n satisfy $|c_n| = \exp[o(n)]$, we conclude that for any $R_1 > R$ and any large value of k

$$|F_k(z)| < R_1^{m_k}.$$

On the other hand, for $|z| = r$ and C_1: $|t| = r_1$, $r < r_1 < 1$,

$$
\begin{aligned}
F_k(z) &= \frac{1}{2\pi i} \int_{C_1} \left(\frac{z}{t}\right)^{m_k+1} \frac{f(t)}{t-z}\, dt \\
&= \frac{1}{2\pi i} \int_{C_1} \left(\frac{z}{t}\right)^{m_k+1} f(t) \left[\frac{1}{t-z} - \sum_0^{p_k-m_k-2} \frac{z^j}{t^{j+1}}\right] dt \\
&= \frac{1}{2\pi i} \int_{C_1} \left(\frac{z}{t}\right)^{p_k} \frac{f(t)}{t-z}\, dt,
\end{aligned}
$$

since the terms inserted in the third member give no contribution to the value of the integral. It follows that for $z \in \Gamma$

$$|F_k(z)| \leq \frac{r_1 M(r_1)}{r_1 - r}\left(\frac{r}{r_1}\right)^{p_k} \equiv A(1-\delta)^{p_k}.$$

Hence we have

$$M \leq R_1^{m_k}, \quad m \leq A(1-\delta)^{p_k},$$

and the two constants theorem gives

$$\log|F_k(z)| \leq \lambda(z)[\log A + p_k \log(1-\delta)] + [1 - \lambda(z)]m_k \log R_1, \quad z \in D_0.$$

By condition (16.7.14) the term involving p_k dominates for large values of k. Since $\log(1-\delta) < 0$, we conclude that

$$\lim F_k(z) = 0$$

uniformly on compact subsets of D_0.

Suppose now that $f(z)$ can be continued analytically along a path Π leading from $z = 0$ to $z = a$. We can then embed Π in a domain D_0 of the type considered above and find that $S_{m_k}(z)$ converges to $f(z)$ uniformly along Π. This means that $S_{m_k}(z)$ converges to $f(z)$ everywhere in the domain of existence $D[f]$ of $f(z)$. Since the domain of convergence of a sequence of polynomials is necessarily simply-connected and the corresponding limit function is single-valued, it follows that $D[f]$ is simply-connected and $f(z)$ is single-valued. In particular, $f(z)$ can have no isolated singularities and the boundary of $D[f]$ is connected and forms a singular line of $f(z)$.

The following theorem may be considered as a converse of the preceding one.

THEOREM 16.7.3. *Let D be a simply-connected domain containing the origin but not the point at infinity and having at least two boundary points. Then there exists a function $f(z)$ such that* (i) *D is the domain of existence of $f(z)$, and* (ii) *the Maclaurin series of $f(z)$ has a sequence of partial sums which converges to $f(z)$ in D, uniformly on compact sets.*

Proof. We use Hilbert's theorem. By Theorem 16.5.5 we can exhaust D by means of a sequence of lemniscatic domains:

$$L_1 \subset L_2 \subset \cdots \subset L_n \subset L_{n+1} \subset \cdots \subset D$$

such that each point z of D belongs to L_n for $n > n_z$. We suppose that $z = 0$ belongs to every L_n. We denote the boundary of L_n by Λ_n and suppose that

$$\Lambda_n: \quad |P_n(z)| = 1, \qquad L_n: \quad |P_n(z)| < 1.$$

It is understood that $\Lambda_n \subset L_{n+1}$. We denote the boundary of D by Γ, its distance from the origin by ρ. Let the largest distance of a point on Λ_n from $z = 0$ be ρ_n so that $\rho_n < \rho_{n+1}$. Let m_n be the degree of $P_n(z)$ and set

$$\min_{z \in \Gamma} |P_n(z)| = A_n, \qquad \max_{z \in \Lambda_{n-1}} |P_n(z)| = M_n$$

so that $M_n < 1 < A_n$. Let $\Sigma\, a_k$ be a convergent series with positive terms.

We determine a sequence of positive integers $\{i_k\}$ in the following manner. Choose i_1 so that

$$(M_1)^{i_1} < a_1.$$

If $i_1, i_2, \cdots, i_k$ have been chosen, take i_{k+1} so that

$$i_{k+1} > 2ki_k m_k, \quad \rho_k^{2ki_k m_k}(M_{k+1})^{i_{k+1}} < a_{k+1},$$

$$(16.7.15) \qquad 1 < \rho_k^{2ki_k m_k}(A_{k+1})^{i_{k+1}}.$$

Since $M_n < 1 < A_n$, this is always possible to attain. With this choice of $\{i_k\}$, we form the series

$$(16.7.16) \qquad f(z) \equiv \sum_{k=1}^{\infty} z^{2ki_k m_k}[P_{k+1}(z)]^{i_{k+1}}.$$

Suppose that z lies on Λ_n. Then $|z| \leq \rho_n$ and

$$|P_{k+1}(z)| \leq M_k \quad \text{for} \quad n \leq k.$$

Hence

$$|z^{2ki_k m_k}[P_{k+1}(z)]^{i_{k+1}}| \leq \rho_k^{2ki_k m_k}(M_k)^{i_{k+1}} < a_{k+1}$$

for $k \geq n$. It follows that the series (16.7.16) converges uniformly on Λ_n and hence also inside Λ_n. Since n is arbitrary, we see that the series converges everywhere in D, uniformly on compact sets. Further, on Γ and outside (if this makes sense), the third condition under (16.7.15) implies that the terms of the series are ≥ 1 in absolute value. It follows that D is the exact domain of convergence of the series.

In the kth term of the series the highest power of z has the degree

$$2ki_k m_k + i_{k+1}m_{k+1} < 2i_{k+1}m_{k+1},$$

while the least power of z in the $(k+1)$st term is of degree

$$2(k+1)i_{k+1}m_{k+1}.$$

It follows that there is no overlapping: a given power of z occurs in at most one term of the series. Hence the Maclaurin series of $f(z)$ is obtained by writing consecutively the expansions of the terms of the series (16.7.16). The resulting power series is a lacunary one and, as we have just seen, the length of the kth gap exceeds $k + 1$ times the order of the preceding partial sum. It follows that Theorem 16.7.2 applies so that the partial sums of (16.7.16) converge everywhere in $D[f]$, that is, $D[f] = D$. This completes the proof as well as our discussion of overconvergence.

EXERCISE 16.7

1. Prove the assertions concerning (16.7.5).

2. Prove that if $P(z) = 0$ has at least two distinct roots, then no component of the lemniscate $|P(z)| = R$ can be a circle.

3. Verify the assertions concerning (16.7.9).

4. Let a sequence of positive integers $\{k_p\}$ be given satisfying the conditions: $k_1 = 1$, $k_{p+1} \geq 2(p + 1)k_p$ for $p \geq 1$, and define

$$f(z) \equiv \sum_{p=1}^{\infty} \left\{ z \left(1 + \frac{z}{p} \right)^p \right\}^{k_p}.$$

Show that the series converges in the domain D: $|z| < e^{-x}$, $z = x + iy$, and that the boundary of D is the natural boundary of $f(z)$.

COLLATERAL READING

The subject matter of Chapter 16 comes from widely scattered sources, and no single reference gives adequate coverage of the whole field.

For Chebichev polynomials see

WALSH, J. L. *Interpolation and Approximation by Rational Functions in the Complex Domain*, Revised Edition. Colloquium Publications, Vol. 20. American Mathematical Society, Providence, Rhode Island, 1956.

The basic facts concerning transfinite diameter and capacity, together with a bibliography up to 1930 and extensions to three dimensions, are to be found in

PÓLYA, G., and SZEGÖ, G. "Über den transfiniten Durchmesser (Kapazitätskonstante) von ebenen und räumlichen Punktmengen," *Journal für die reine und angewandte Mathematik*, Vol. 165 (1931), pp. 4–49.

Later work in this field is largely dominated by F. Leja, who has also considered extensions to metric spaces. We note

LEJA, F. "Sur les suites de polynômes, les ensembles fermés et la fonction de Green," *Annales de la Société Polonaise de Mathématique*, Vol. 12 (1933), pp. 57–71.

An abstract notion of capacity, recently still further formalized by N. Bourbaki, was studied in great detail by

CHOQUET, GUSTAVE. "Theory of Capacities," *Annales de l'Institut Fourier*, Vol. 5 (1953–54), pp. 131–295.

For the theory of additive set functions see

HALMOS, P. R. *Measure Theory.* D. Van Nostrand Company, Inc., Princeton, New Jersey, 1950.

RADON, J. "Theorie und Anwendungen der absolut additiven Mengenfunktionen," *Sitzungsberichte der Akademie der Wissenschaften zu Wien*, Vol. 122 (1913), pp. 1295–1438.

SAKS, S. *Theory of the Integral*, trans. by L. C. YOUNG. Monografie Matematyczne, Vol. 7, Warsaw and Wrocław, 1937.

For potential theory consult

FROSTMAN, O. *Potentiel d'Équilibre et Capacité des Ensembles avec quelques Applications à la Théorie des Fonctions.* Thèse, Lund, 1935.

NEVANLINNA, R. *Eindeutige analytische Funktionen*, Second Edition. Springer-Verlag, Berlin, 1953.

ROBIN, G. "Sur la Distribution de l'Électricité à la Surface des Conducteurs Fermés et des Conducteurs Ouverts," *Annales de l'École Normale Supérieure*, Series 3, Vol. 3 (1886), Supplement, pp. S.1–S.58. (Paris thesis, 1886.)

TSUJI, M. *Potential Theory in Modern Function Theory.* Maruzen, Tokyo, 1959.

Robin's paper is mainly of historical interest. Tsuji's book starts *in medias res* and is harder to read than Nevanlinna's treatise, but it supplements the latter and gives excellent coverage of a wide range of ideas.

For Runge's theorem and related questions see Walsh's treatise cited above. The closely related question of expansion in polynomial series is treated from two different points of view in

BOAS, R. P., JR., and BUCK, R. C. *Polynomial Expansions of Analytic Functions.* Ergebnisse der Mathematik, New Series, No. 19. Springer-Verlag, Berlin, 1958.

MONTEL, P. *Leçons sur les Séries de Polynômes d'une Variable Complexe.* Gauthier-Villars, Paris, 1910.

Hilbert's lemniscate theorem was published in

HILBERT, D. "Über die Entwickelung einer beliebigen analytischen Funktion einer Variablen in eine unendliche nach ganzen rationalen Funktionen fortschreitende Reihe," in *Göttinger Nachrichten*, 1897. Reprinted in *Gesammelte Abhandlungen*, Vol. 3, pp. 3–9. Berlin, 1935. Chelsea Publishing Company, New York, 1966.

A simpler proof is given in Montel's treatise, cited above, pp. 45–49.

For overconvergence, see the following articles, the first of which is expository:

OSTROWSKI, A. "On Representation of Analytic Functions by Power Series," *Journal of the London Mathematical Society*, Vol. 1 (1926), pp. 251–263.

———. "Zur Theorie der Überkonvergenz," *Mathematische Annalen*, Vol. 103 (1930), pp. 15–27.

See also the excellent treatise

GOLUSIN, G. M. *Geometrische Funktionentheorie*, pp. 311–315. Deutscher Verlag der Wissenschaften, Berlin, 1957.

Golusin speaks of *Hyperkonvergenz*. The reader may also consult Chapter 7 of Golusin's book for Sections 16.1–16.4.

17

CONFORMAL MAPPING

17.1. Riemann's mapping theorem. The theory of conformal representation goes back to Riemann's inaugural dissertation (Göttingen, 1851) entitled *Grundlagen für eine allgemeine Theorie der Functionen einer veränderlichen complexen Grösse*, in which he considered, among other things, the correspondence between two Riemann surfaces. The special mapping theorem which carries his name, he formulated as follows:

> Zwei gegebene einfach zusammenhängende ebene Flächen können stets so auf einander bezogen werden, dass jedem Punkte der einen Ein mit ihm stetig fortrückender Punkt der andern entspricht und ihre entsprechenden kleinsten Theile ähnlich sind; und zwar kann zu Einem innern Punkte und zu Einem Begrenzungspunkte der entsprechende beliebig gegeben werden; dadurch aber ist für alle Punkte die Beziehung bestimmt.[1]

Riemann gave a proof of this theorem based upon *Dirichlet's principle* according to which the problem of minimizing the integral

$$(17.1.1) \qquad \int\int_D \left\{ \left(\frac{\partial u}{\partial x}\right)^2 + \left(\frac{\partial u}{\partial y}\right)^2 \right\} dx\, dy,$$

under suitable side conditions, has a solution. This principle was generally accepted in the days of Riemann until Weierstrass pointed out that similar problems in the calculus of variations need not have solutions. The fact that the integral has an infimum does not imply that there is a function in the class under consideration whose Dirichlet integral equals the infimum.

In view of this situation it became necessary to give new proofs for the mapping theorem. To start with, the efforts were concentrated on Dirichlet's problem, that is, the question of the existence of a function $U(x, y)$, harmonic in a given domain and taking on preassigned values on the boundary. Riemann had taken this for granted and had used the Green's function (see Section 16.5)

---

[1] See *Mathematische Werke* (B. G. Teubner, Leipzig, 1876), pp. 3–47 (quotation from p. 40). It is asserted that any two given simply-connected plane domains D and R can be brought into one-to-one correspondence which is continuous and locally a similitude, that is, R is the conformal map of D in the sense of Definition 4.6.2. Here one point of D and one point of the boundary of D may be paired off with one point of R and one point of the boundary of R respectively. These four points may be chosen arbitrarily and they determine the correspondence uniquely.

320

of the domain to construct its mapping function. Through the efforts of a number of mathematicians, including Carl Neumann (1832–1925), H. A. Schwarz, and H. Poincaré, the Dirichlet problem was solved for fairly general domains and boundary values by the end of the nineteenth century. As a consequence, the mapping problem was solved for domains bounded by piecewise analytic curves.

It should be noted that the discredited principle of Dirichlet was resurrected and justified under mild restrictions by D. Hilbert (1904), who in 1909 used it in an outline of a solution of the general problem of mapping a Riemann surface on a slit plane. This was carried out in more detail in the dissertation of Richard Courant (1910).

In 1913 W. F. Osgood and E. H. Taylor also used potential theoretic methods to verify earlier conjectures by Osgood concerning the mapping on the boundary.

By this time the mapping problem had changed character, and function theoretic methods had been successfully introduced. The boundary of a domain can be a very complicated point set, and the normalization of the mapping function, chosen by Riemann, involving boundary points, was found to lead to unnecessary difficulties. The mapping is conformal in the interior and usually not on the boundary. It is then better to fix the mapping function by conditions in the interior alone. One can prescribe corresponding points and directions in the two domains; that is, at a point $z_0 \in D$ the values of $f(z_0)$ and of $\arg f'(z_0)$ are given, and they are found to determine the mapping function uniquely. Further, it is enough to solve the problem for the case in which R is the unit disk in the w-plane.

Function theoretic methods for the study of the mapping problem were introduced by Paul Koebe (1882–1945) and by Constantin Carathéodory (1873–1950) and perfected by a number of writers, among whom should be mentioned Ludwig Bieberbach, T. H. Gronwall (1877–1932), E. Lindelöf, and P. Montel. In Koebe's method of 1908, the domain D is supposed to be interior to the unit circle, and the boundary of D is pushed out to the circle with the aid of a sequence of elementary transformations involving the extraction of square roots only. This simple idea reappears in later proofs and a trace of it will also be found in the proof below.

Modern proofs of the Riemann mapping theorem usually involve *a minimax principle* combined with the theory of normal families. The mapping function is characterized as that element of a family of functions holomorphic and univalent in D, satisfying the initial condition at $z = z_0$, for which a certain functional takes on an extreme value. The existence of an infimum or a supremum of the functional, as the case may be, is guaranteed by its definition, and the convergence of a minimizing or maximizing sequence of elements of the family is concluded from Montel's theorem.

After these preliminaries we proceed to a formulation and proof of the mapping theorem.

THEOREM 17.1.1.   *Given a simply-connected domain D whose boundary contains at least two distinct points and given a point $z_0 \in D$, then there exists a unique function $f(z)$ which maps D conformally on the disk $|w| < 1$ in such a manner that*

$$(17.1.2) \qquad\qquad f(z_0) = 0, \quad \arg f'(z_0) > 0.$$

Proof.   It is clear that neither the extended plane nor the punctured plane can be mapped conformally on a finite disk. Thus the assumption that D has at least two boundary points is necessary. If there are two boundary points, then there are automatically infinitely many since the complement of D is a continuum. We can use this fact for a preliminary simplification of the problem. We may suppose that D is bounded and that $z_0 = 0$. If D is not bounded at the outset, we can map D conformally on a bounded domain with the aid of at most two auxiliary transformations. If the closure of D is the whole plane and if a and b are boundary points of D, then either determination of

$$(17.1.3) \qquad\qquad \left\{\frac{z - a}{z - b}\right\}^{\frac{1}{2}} \equiv z_1$$

is single-valued in D and maps D conformally on a simply-connected domain D_1. If $z_1 = c$ is in D_1, then a full neighborhood of $z_1 = -c$ is exterior to D_1, and the transformation

$$(17.1.4) \qquad\qquad \frac{1}{z_1 + c} \equiv z_2$$

maps D_1 on a bounded simply-connected domain D_2. If $\bar{D}$ is not the whole plane, only the second transformation is needed. Finally, the translation $z_3 = z_2 + d$ maps D_2 on a domain D_3 containing the origin such that $z_3 = 0$ is the image of the point z_0 in the original domain D.

We may suppose that these preliminary simplifications have been made. Thus D shall be a bounded, simply-connected domain containing the origin, and the mapping function is to be normalized by the initial conditions

$$(17.1.5) \qquad\qquad f(0) = 0, \quad f'(0) > 0.$$

We now consider the class $\mathscr{F}$ of functions $f(z)$ having the following properties:

   (i) $f(z)$ *is holomorphic in* D,
   (ii) $f(z)$ *is univalent in* D, *that is,* $f(z_1) \neq f(z_2)$ *if* $z_1 \neq z_2$,
   (iii) $f(z)$ *satisfies* (17.1.5), *and*
   (iv) $|f(z)| < 1$ *in* D.

$\mathscr{F}$ is not vacuous for it contains functions of the form kz provided that k is a sufficiently small positive number. For $\mathscr{F}$ we now pose the following maximum problem:

Find that element of $\mathscr{F}$ for which $f'(0)$ has the largest possible value.

We observe first that the set of positive numbers defined by $\{f'(0) \,|\, f \in \mathscr{F}\}$

is bounded. In fact, if D contains the circle $|z| = \rho$, then Cauchy's estimates combined with condition (iv) show that

$$f'(0) < \frac{1}{\rho}.$$

Let us denote the supremum of the set $\{f'(0)\}$ by μ. We have to show that there is an element $g(z) \in \mathcal{F}$ such that $g'(0) = \mu$. Later it will be verified that $g(z)$ is indeed the desired mapping function. Condition (iv) makes $\mathcal{F}$ into a normal family. We can then find a sequence $\{f_n(z)\}$ of elements of $\mathcal{F}$ such that

$$\lim_{n \to \infty} f_n'(0) = \mu,$$

and

$$\lim_{n \to \infty} f_n(z) \equiv g(z)$$

exists uniformly on compact subsets of D. It is clear that $g'(0) = \mu$ and that $g(z)$ satisfies conditions (i), (iii), and (iv). For $g(z)$ clearly is not a constant, so the maximum principle shows that the inequality in (iv) holds also for $g(z)$. That $g(z)$ is also univalent in D follows from Theorem 14.3.4, the theorem of Hurwitz, for if there are two distinct points z_1 and z_2 in D where $g(z)$ takes on the same value a, then there are neighborhoods of z_1 and z_2 in D in each of which $f_n(z)$ must assume the same value a for all large values of n, and this contradicts the assumption that $f_n(z)$ is univalent. Thus $g(z) \in \mathcal{F}$.

We come now to the mapping properties of $g(z)$. It is clear that

$$w = g(z)$$

maps D conformally on a simply-connected domain D_0 which is a subset of the disk $|w| < 1$. If D_0 is a proper subset of the disk, then there must exist a point w_0 with $|w_0| < 1$ which is a boundary point of D_0. We shall show that this leads to a contradiction by constructing an element $h(z)$ of $\mathcal{F}$ with $h'(0) > \mu$. For this purpose we use Koebe's transformation. Suppose that

$$w_0 = ae^{i\alpha},$$

and define

(17.1.6)

$$
\begin{cases}
g_1(z) = e^{-i\alpha}g(z), \\[2mm]
p(z) = \dfrac{a - g_1(z)}{1 - ag_1(z)}, \\[2mm]
q(z) = [p(z)]^{\frac{1}{2}}, \quad q(0) = +a^{\frac{1}{2}}, \\[2mm]
g_2(z) = \dfrac{a^{\frac{1}{2}} - q(z)}{1 - a^{\frac{1}{2}}q(z)}, \\[2mm]
h(z) = e^{i\alpha}g_2(z).
\end{cases}
$$

All these functions are holomorphic in D. This is trivial for $g_1(z)$, and since $a < 1$, $p(z)$ is also holomorphic in D. Since $p(z) \neq 0$ in D, it follows that $q(z)$ is also holomorphic there. Now the mapping $g_1(z) \to p(z)$ is defined by a Möbius transformation which leaves the unit disk invariant. It follows that $p(z)$ maps D on a simply-connected domain D_1 in the unit disk. Likewise, $q(z)$ is holomorphic in D and maps D on a simply-connected domain D_2 in the unit disk. Since $a < 1$, we see that $g_2(z)$ is holomorphic in D and, further, that the same is true for $h(z)$ and that $h(z)$ maps D on a domain D_3 in the unit disk. Now $h(z) \in \mathscr{F}$ for it clearly satisfies conditions (i), (ii), and (iv), and $h(0) = 0$. To verify (iii) we compute $h'(z)$ by the chain rule, obtaining

$$h'(z) = e^{i\alpha} \frac{1 - a}{[1 - a^{\frac{1}{2}}q(z)]^2} \cdot \frac{1}{2} [p(z)]^{-\frac{1}{2}} \cdot \frac{1 - a^2}{[1 - ag_1(z)]^2} e^{-i\alpha} g'(z),$$

whence

$$h'(0) = \frac{1 + a}{2a^{\frac{1}{2}}} \mu > \mu.$$

Thus the assumption that D_0 is a proper subset of the unit disk leads to a contradiction.

Hence $w = g(z)$ maps D conformally onto the unit disk in the w-plane, and the initial conditions (17.1.5) are satisfied. It remains to prove uniqueness of the mapping function. This follows from Theorem 15.1.2. For if $w = G(z)$ maps D conformally on the unit disk and if $G(z)$ satisfies the same initial conditions (17.1.5), then the composed function

$$G[\check{g}(w)],$$

where $\check{g}(w)$ is the inverse of $g(z)$, maps the unit disk of the w-plane onto itself. It vanishes for $w = 0$ and has a positive derivative for $w = 0$. By Theorem 15.1.2 such a function must coincide with w, whence it follows that

$$G(z) \equiv g(z),$$

and the mapping theorem is proved.

DEFINITION 17.1.1.   *The number $r_i \equiv 1/\mu$ is called the interior mapping radius of D with respect to the origin.*

The function

(17.1.7) $$w = r_i g(z) = z + c_2 z^2 + c_3 z^3 + \cdots$$

maps D conformally on the disk

$$| w | < r_i$$

in such a manner that $z = 0$ goes into $w = 0$ and the transformation coincides with the identity mapping at the origin. We can get this mapping function directly as the solution of the following minimax problem:

Let $\mathscr{F}_0$ be the class of all functions $f(z)$ holomorphic and univalent in the domain D and such that

$$f(0) = 0, \quad f'(0) = 1.$$

Find that element of $\mathscr{F}_0$ for which

(17.1.8) $$\max_{z \in D} | f(z) |$$

is as small as possible.

The solution is unique and is given by (17.1.7).

EXERCISE 17.1

1. Let D be a simply-connected domain interior to the unit circle and containing the origin, which is at the distance a from the boundary of D. Replace $g(z)$ by z in (17.1.6), take $\alpha = 0$, and find the corresponding function $h(z)$. Show that $w = h(z)$ maps D on a domain D_1, interior to the unit circle and containing the origin, whose distance from the boundary of D_1 is at least equal to a. (*Hint:* Follow the mapping of the disk $|z| < a$. This property of the transformations (17.1.6) is basic for Koebe's proof.)

17.2. The kernel function. Another approach to the conformal mapping problem is furnished by the L_2-theory of holomorphic functions. The relevant aspects of this theory have been developed in numerous papers by Stefan Bergman starting with his Berlin dissertation of 1921.

Let D be a domain having at least two boundary points. We restrict ourselves to the simply-connected case, but D need not be bounded. As usual we denote by $H[D]$ the set of functions $f(z)$ holomorphic in D and by $L_2H[D]$ the subset of functions which are also quadratically integrable over D. If $f \in L_2H[D]$, we set

(17.2.1)
$$\| f \| = \left\{ \iint_D |f(z)|^2 \, dx \, dy \right\}^{\frac{1}{2}},$$

and this is called the L_2-norm of f.[1] Normally the integral must be taken in the sense of Lebesgue. We also introduce the inner product

(17.2.2)
$$(f, g) = \iint_D f(z)\overline{g(z)} \, dx \, dy,$$

in terms of which we have

(17.2.3)
$$\| f \|^2 = (f, f)$$

and Schwarz's inequality

(17.2.4)
$$| (f, g) | \leq \| f \| \, \| g \|.$$

The inner product has the following noteworthy properties:

(17.2.5)
$$(C_1 f_1 + C_2 f_2, g) = C_1(f_1, g) + C_2(f_2, g),$$
$$(f, C_1 g_1 + C_2 g_2) = \overline{C}_1(f, g_1) + \overline{C}_2(f, g_2),$$
$$(g, f) = \overline{(f, g)}.$$

---

[1] For the terminology and properties of abstract spaces used below, the reader should consult Section 4.7. There is an excellent account of inner-product spaces and Hilbert spaces, including the Gram-Schmidt orthogonalization process, to be found in Angus E. Taylor, *An Introduction to Functional Analysis* (John Wiley & Sons, Inc., New York, 1958), pp. 106–121.

We shall need a relation between the L_2-metric in $L_2H[D]$ and the sup-metric of $H[D]$. This is furnished by

LEMMA 17.2.1. *If* $f(z) \in L_2H[D]$ *and if* $z_0 \in D$ *at the distance* r_0 *from the boundary of* D, *then*

$$(17.2.6) \qquad |f(z_0)| \leq [\pi r_0^2]^{-\frac{1}{2}} \|f\|.$$

Proof. Compare the proof of Theorem 15.2.6. We have

$$\|f\|^2 > \int_0^{2\pi} \int_0^{r_0} \left| \sum_{n=0}^{\infty} \frac{f^{(n)}(z_0)}{n!} r^n e^{ni\theta} \right|^2 r\, dr\, d\theta$$

$$= \pi \sum_{n=0}^{\infty} \frac{|f^{(n)}(z_0)|^2}{(n!)^2} \frac{r_0^{2n+2}}{n+1} \geq \pi r_0^2 |f(z_0)|^2,$$

which is the desired inequality.

The set $L_2H[D]$ is a linear vector space. The only property that requires a proof is the fact that the sum of two elements of the set is also an element of the set. This follows from formulas (17.2.3) and (17.2.4). In fact,

$$(f+g, f+g) = (f,f) + (f,g) + (g,f) + (g,g)$$

$$\leq (f,f) + 2\,|(f,g)| + (g,g)$$

$$\leq \|f\|^2 + 2\,\|f\|\,\|g\| + \|g\|^2$$

$$= [\|f\| + \|g\|]^2.$$

This implies that

$$(17.2.7) \qquad \|f+g\| \leq \|f\| + \|g\|,$$

so that (17.2.1) really has the properties of a norm.

We now define the distance between the elements f and g to be $\|f-g\|$.

THEOREM 17.2.1. *The space* $L_2H[D]$ *is complete in the metric defined by* (17.2.1).

Proof. Suppose that the sequence $\{f_n\} \subset L_2H[D]$ is a Cauchy sequence, that is, for every given $\varepsilon > 0$ there is an integer N_ε such that

$$\|f_m - f_n\| \leq \varepsilon \quad \text{for} \quad m, n > N_\varepsilon.$$

Let D_δ be a (measurable) subset of D whose distance from the boundary of D equals δ. For $z \in D_\delta$, Lemma 17.2.1 gives

$$|f_m(z) - f_n(z)| \leq (\pi \delta^2)^{-\frac{1}{2}} \|f_m - f_n\|,$$

whence it follows that the sequence $\{f_n(z)\}$ converges everywhere in D and uniformly with respect to z in D_δ. Its limit function $f(z)$ then belongs to $H[D]$. On the other hand, the set $\{\|f_n\|\}$ is bounded, and from the inequality

$$\iint_{D_\delta} |f_n(z)|^2\, dx\, dy < \iint_D |f_n(z)|^2\, dx\, dy \leq M^2$$

we obtain, by uniform convergence,

$$\iint_{D_\delta} |f(z)|^2 \, dx \, dy \leq M^2.$$

Since this holds for every D_δ, we infer that $f \in L_2 H[D]$. Similarly, from

$$\iint_{D_\delta} |f_m(z) - f_n(z)|^2 \, dx \, dy \leq \iint_D |f_m(z) - f_n(z)|^2 \, dx \, dy \leq \varepsilon^2$$

we conclude that

$$\iint_{D_\delta} |f(z) - f_n(z)|^2 \, dx \, dy \leq \varepsilon^2,$$

by the uniform convergence of $f_m(z)$ on D_δ, and, hence, ultimately

$$\|f - f_n\| \leq \varepsilon.$$

This shows that the space $L_2 H[D]$ is complete in its metric.

It follows that $L_2 H[D]$ is a (B)-space. It is also a Hilbert space since the metric is defined by an inner product. Now in any Hilbert space we may find an orthonormal system S, that is, a set of elements $\{f_n\}$ such that

(17.2.8) $(f_m, f_n) = \delta_{mn},$

where the Kronecker δ equals 1 if $m = n$ and 0 otherwise. We shall see later how such a system can be found in $L_2 H[D]$. For the time being let us explore the properties of such a system, supposed to exist, the underlying Hilbert space $\mathscr{H}$ being arbitrary.

A set of n elements $g_1, g_2, \cdots, g_n$ of a linear space X spans a linear manifold M_n, namely, the set of all elements of the form

$$c_1 g_1 + c_2 g_2 + \cdots + c_n g_n,$$

where the c's are arbitrary (complex) numbers. M_n obviously contains the zero element of X obtained by setting $c_1 = c_2 = \cdots = c_n = 0$. We recall that the elements $\{g_k\}$ are said to be *linearly independent* if

(17.2.9) $c_1 g_1 + c_2 g_2 + \cdots + c_n g_n = 0$

implies that

(17.2.10) $c_1 = c_2 = \cdots = c_n = 0.$

They are *linearly dependent* if (17.2.9) can hold for constants c_k not all 0. The space X is of *dimension* n if it contains a set of n linearly independent elements and if every set of $(n + 1)$ elements is linearly dependent. It is of infinite dimension if for every n there exists a set of n linearly independent elements.

LEMMA 17.2.2.   *The elements of an orthonormal set $S = \{f_n\}$ in a Hilbert space $\mathscr{H}$ are linearly independent.*

Proof.   For

$$\sum_{j=1}^{n} c_j f_j = 0$$

implies

$$\left(\sum_{j=1}^{n} c_j f_j, f_k \right) = 0, \quad k = 1, 2, \cdots, n,$$

whence

$$\sum_{j=1}^{n} c_j (f_j, f_k) = 0 \quad \text{and} \quad c_k = 0.$$

Thus the manifold spanned by S has as its dimension the number of elements in S. The interesting case is that in which S spans all of $\mathcal{H}$. Suppose to start with that S is an infinite set, let f be any element of $\mathcal{H}$, and form the infinite series

(17.2.11) $$\sum_{k=1}^{\infty} (f, f_k) f_k.$$

This is the Fourier series of the element f in the system S, and (f, f_k) is the kth Fourier coefficient. We shall show that the series converges and defines an element of $\mathcal{H}$.

THEOREM 17.2.2.   *For each n and each choice of complex numbers $c_1, c_2, \cdots, c_n$ we have*

(17.2.12) $$\left\| f - \sum_{k=1}^{n} (f, f_k) f_k \right\| \le \left\| f - \sum_{k=1}^{n} c_k f_k \right\|.$$

Further,

(17.2.13) $$\sum_{k=1}^{\infty} |(f, f_k)|^2 \le (f, f),$$

and the series (17.2.11) converges to an element g of $\mathcal{H}$ such that for all n

(17.2.14) $$(f - g, f_n) = 0.$$

Proof.   We have

$$\left(f - \sum_{j=1}^{n} c_j f_j, f - \sum_{k=1}^{n} c_k f_k \right)$$

$$= (f, f) - \sum_{j=1}^{n} c_j (f_j, f) - \sum_{k=1}^{n} \overline{c_k} (f, f_k) + \sum_{j=1}^{n} \sum_{k=1}^{n} c_j \overline{c_k} (f_j, f_k)$$

$$= (f, f) - \sum_{j=1}^{n} c_j \overline{(f, f_j)} - \sum_{k=1}^{n} \overline{c_k} (f, f_k) + \sum_{k=1}^{n} |c_k|^2$$

$$= (f, f) - \sum_{k=1}^{n} |(f, f_k)|^2 + \sum_{k=1}^{n} |c_k - (f, f_k)|^2$$

$$\ge (f, f) - \sum_{k=1}^{n} |(f, f_k)|^2$$

$$= \left(f - \sum_{j=1}^{n} (f, f_j) f_j, f - \sum_{k=1}^{n} (f, f_k) f_k \right).$$

This proves (17.2.12). Since the last member of this chain of relations is non-negative, we see that

$$\sum_{k=1}^{n} | (f, f_k) |^2 \leq (f, f)$$

for every n, and this implies (17.2.13). Further, since the series in (17.2.13) converges, for any $\varepsilon > 0$ there exists an N_ε such that

$$\left(\sum_{k=m}^{n} (f, f_k) f_k, \sum_{k=m}^{n} (f, f_k) f_k \right) = \sum_{k=m}^{n} | (f, f_k) |^2 < \varepsilon$$

for $n > m > N_\varepsilon$. This asserts that the partial sums of the series (17.2.11) form a Cauchy sequence in $\mathscr{H}$, and since $\mathscr{H}$ is a complete space, it follows that the series converges to an element g of $\mathscr{H}$. Finally, if $p > n$ we have

$$\left(\sum_{k=1}^{p} (f, f_k) f_k, f_n \right) = (f, f_n).$$

Since the limit of the left member, as $p \to \infty$, is (g, f_n), we have also proved (17.2.14).

Formula (17.2.12) asserts that the nth partial sum of the Fourier series of f gives the best approximation of f in the manifold M_n spanned by the first n elements of S. Formula (17.2.13) is known as *Bessel's inequality* after the German astronomer Friedrich Wilhelm Bessel (1784–1846), who also gave his name to the Bessel functions.

The most interesting case is that in which the orthonormal system S has the property that every Fourier series converges to the function. S is then said to be *closed*, or *complete*. This property can be expressed in four different ways, all equivalent.

THEOREM 17.2.3.   *If S is an orthonormal system in $\mathscr{H}$, the following statements are equivalent*:

(i) *If $(f, f_n) = 0$ for all n, then $f = 0$.*
(ii) *For each $f \in \mathscr{H}$ we have*

(17.2.15)
$$f = \sum_{n=1}^{\infty} (f, f_n) f_n.$$

(iii) *Linear combinations of the f_n's are dense in $\mathscr{H}$.*
(iv) *For each $f \in \mathscr{H}$ we have*

(17.2.16)
$$\sum_{n=1}^{\infty} | (f, f_n) |^2 = (f, f).$$

Proof.   We shall prove the implications

$$(i) \Rightarrow (ii) \Rightarrow (iii) \Rightarrow (iv) \Rightarrow (i),$$

and this will show the equivalence.

Suppose that (i) holds. Then (17.2.14) shows that for each $f \in \mathscr{H}$ we have $f = g$, which is (ii).

Suppose that (ii) holds. Then the partial sums of (17.2.15) converge to f, that is, (iii) holds since f is arbitrary in $\mathcal{H}$.

If (iii) holds, then for any $f \in \mathcal{H}$ and any $\varepsilon > 0$, there exist an integer n and constants $c_1, c_2, \cdots, c_n$ such that

$$\left\| f - \sum_{k=1}^{n} c_k f_k \right\| < \varepsilon.$$

By the proof of (17.2.12) this implies that

$$(f, f) - \sum_{k=1}^{n} |(f, f_k)|^2 < \varepsilon^2,$$

and this implies (iv), which is known as *Parseval's identity*.

Finally, if (iv) holds, then clearly $(f, f_n) = 0$ for all n implies that $(f, f) = 0$, so that $\| f \| = 0$ and $f = 0$, which is (i).

Let us now return to the space $L_2 H[D]$ and suppose that

$$S: \quad \{\omega_n(z)\}$$

is a complete orthonormal system in this space. Owing to the special properties of the space $L_2 H[D]$, the system S is endowed with many important properties for which there is no analogue in the general case.

THEOREM 17.2.4.   *The Fourier series*

$$(17.2.17) \qquad\qquad \sum_{n=1}^{\infty} (f, \omega_n) \omega_n(z)$$

converges pointwise to $f(z)$ in D, uniformly with respect to z in any domain D_δ.

Proof.   For $z \in D_\delta$ Lemma 17.2.1 shows that

$$\left| f(z) - \sum_{k=1}^{n} (f, \omega_k) \omega_k(z) \right| \leq (\pi \delta^2)^{-\frac{1}{2}} \left\| f - \sum_{k=1}^{n} (f, \omega_k) \omega_k \right\|.$$

By the preceding theorem the right member tends to zero as $n \to \infty$, and this proves the assertion.

THEOREM 17.2.5.   *If $t \in D$ and the distance of t from the boundary of D equals r, then*

$$(17.2.18) \qquad\qquad \sum_{k=1}^{\infty} |\omega_k(t)|^2 \leq (\pi r^2)^{-1}.$$

Proof.   We get the inequality from a study of the following extremal problem:

Let M_n be the linear manifold spanned by the functions $\omega_1(z), \omega_2(z), \cdots, \omega_n(z)$. If $f(z) \in M_n$ and if

$$f(t) = 1$$

where t is a fixed point in D, find the least value of $\| f \|$.

By assumption

$$f(z) = \sum_{k=1}^{n} c_k \omega_k(z),$$

so that

(17.2.19)
$$\| f \|^2 = \sum_{k=1}^{n} | c_k |^2$$

and

(17.2.20)
$$\sum_{k=1}^{n} c_k \omega_k(t) = 1.$$

The question is to find the minimum of the right-hand side of (17.2.19) when (17.2.20) holds. We apply Cauchy's inequality to (17.2.20). Now in

(17.2.21)
$$\left| \sum_{k=1}^{n} a_k b_k \right|^2 \leq \sum_{k=1}^{n} | a_k |^2 \sum_{k=1}^{n} | b_k |^2$$

we have equality if and only if there is a number λ such that

$$b_k = \lambda \overline{a_k}, \quad k = 1, 2, \cdots, n.$$

Thus we have

$$1 = \left| \sum_{k=1}^{n} c_k \omega_k(t) \right|^2 \leq \sum_{k=1}^{n} | c_k |^2 \sum_{k=1}^{n} | \omega_k(t) |^2$$

$$= \| f \|^2 \sum_{k=1}^{n} | \omega_k(t) |^2$$

with equality if and only if for each k

(17.2.22)
$$c_k = \lambda \overline{\omega_k(t)}.$$

Substituting back in (17.2.20) we see that

$$\lambda = \lambda_n \equiv \left\{ \sum_{k=1}^{n} | \omega_k(t) |^2 \right\}^{-1}.$$

Actually we have the solution of the minimum problem right here, but all we need for our present purposes is an inequality which follows when we apply Lemma 17.2.1 to the function $f(z)$ determined by (17.2.22). With this choice of c_k, $k = 1, 2, \cdots, n$, we have

(17.2.23)
$$\| f \| = (\lambda_n)^{\frac{1}{2}}.$$

We substitute this value of $\| f \|$ in (17.2.6), take $z_0 = t$, and obtain

$$1 \leq (\pi r^2)^{-\frac{1}{2}} (\lambda_n)^{\frac{1}{2}}$$

or

$$\sum_{k=1}^{n} | \omega_k(t) |^2 \leq (\pi r^2)^{-1}$$

for all n; this implies (17.2.18).

COROLLARY.   *The Fourier series (17.2.17) converges absolutely throughout D.*

Proof.   By Cauchy's inequality

$$\left\{ \sum_{k=1}^{\infty} |(f, \omega_k)| \, |\omega_k(z)| \right\}^2 \leq \sum_{k=1}^{\infty} |(f, \omega_k)|^2 \cdot \sum_{k=1}^{\infty} |\omega_k(z)|^2$$

$$\leq \| f \|^2 (\pi r^2)^{-1},$$

if r is the distance of z from the boundary of D.

THEOREM 17.2.6.   *The set of elements of the space $L_2H[D]$ coincides with the set of series*

$$(17.2.24) \qquad \sum_{k=1}^{\infty} c_k \omega_k(z) \quad with \quad \sum_{k=1}^{\infty} |c_k|^2 < \infty.$$

Proof.   We know that every element of $L_2H[D]$ is of this form. Conversely, such a series converges absolutely in D and uniformly with respect to z in any domain D_δ, as shown by Cauchy's inequality. Its sum then is an element $f(z)$ of $H[D]$. But it is also an element of $L_2H[D]$, for the partial sums of the series form a Cauchy sequence in $L_2H[D]$ since

$$\left\| \sum_{k=m}^{n} c_k \omega_k(z) \right\|^2 = \sum_{k=m}^{n} |c_k|^2 \to 0 \quad \text{as } m \to \infty.$$

There is then an element of $L_2H[D]$ which has the given series as its Fourier series in the system S, and this element must be $f(z)$.

This result is the Riesz-Fischer theorem for the system S in $L_2H[D]$. It is named after F. Riesz and Ernst Fischer (1875–1954), who proved the original theorem for ordinary Fourier series in the space $L_2(-\pi, \pi)$ in 1907.

We shall now introduce Bergman's kernel function

$$(17.2.25) \qquad K(z, t) \equiv \sum_{n=1}^{\infty} \omega_n(z)\overline{\omega_n(t)}, \quad z, t \in D.$$

By our preceding estimates the series is absolutely convergent in $D \times D$ and uniformly convergent in $D_\delta \times D_\delta$. Further,

$$(17.2.26) \qquad |K(z, t)|^2 \leq K(z, z)K(t, t).$$

For each fixed t, $K(z, t)$ is an element of $L_2H[D]$ whose norm is

$$(17.2.27) \qquad \| K(\cdot, t) \| = [K(t, t)]^{\frac{1}{2}}.$$

Similarly the L_2-norm of $K(z, \cdot)$ is $[K(z, z)]^{\frac{1}{2}}$.

THEOREM 17.2.7.   *For any $f(z) \in L_2H[D]$ and any $z \in D$,*

$$(17.2.28) \qquad f(z) = \int\int_D f(t)K(z, t) \, du \, dv, \quad t = u + iv.$$

REMARK.   This shows why $K(z, t)$ is often called a *reproducing kernel*.

Proof.   Schwarz's inequality shows that the right member exists. The equality is clearly valid if $f(z)$ is one of the functions $\omega_k(z)$ or a linear combination of a finite number of such functions. If

$$f_n(z) = \sum_{k=1}^{n} (f, \omega_k)\omega_k(z),$$

Schwarz's inequality gives

$$\left| \int\!\!\int_D [f(t) - f_n(t)]K(z, t)\, du\, dv \right|^2 \leq \| f - f_n \|^2 K(z, z),$$

and this tends to zero as $n \to \infty$. Hence

$$\int\!\!\int_D f(t)K(z, t)\, du\, dv = \lim_{n \to \infty} \int\!\!\int_D f_n(t)K(z, t)\, du\, dv$$

$$= \lim_{n \to \infty} f_n(z) = f(z)$$

as asserted. We note the inequality

(17.2.29) $$|f(z)| \leq \| f \| [K(z, z)]^{\frac{1}{2}}.$$

We come now to the extremal property of the kernel function which is the key to the applications to conformal mapping.

THEOREM 17.2.8.   *If* $t \in D$, $f(z) \in L_2H[D]$,

(17.2.30) $$f(t) = 1, \quad \| f \| = \min,$$

then

(17.2.31) $$f(z) = g(z, t) \equiv \frac{K(z, t)}{K(t, t)}.$$

Proof.   This is the analogue of the extremal problem in Theorem 17.2.5 with the difference that $f(z)$ is no longer restricted to belong to one of the finite subspaces M_n. We proceed in the same manner as above. By assumption $f(z)$ is of the form

$$f(z) = \sum_{k=1}^{\infty} c_k\omega_k(z),$$

so that the extremal problem calls for a sequence $\{c_n\}$, satisfying

$$\sum_{n=1}^{\infty} c_n\omega_n(t) = 1,$$

which minimizes the Hermitian form

$$\sum_{n=1}^{\infty} |c_n|^2.$$

The choice

$$c_n = \frac{\overline{\omega_n(t)}}{K(t, t)}$$

gives $f(z) = g(z, t)$ and

$$\| g \| = [K(t, t)]^{-1}.$$

It remains to show that this choice really gives the minimum value to the norm.

For this purpose we set
$$f(z) = g(z, t) + h(z, t)$$
where
$$h(z, t) = \sum_{n=1}^{\infty} \eta_n \omega_n(z), \quad \| h \|^2 = \sum_{n=1}^{\infty} | \eta_n |^2, \quad \sum_{n=1}^{\infty} \eta_n \omega_n(t) = 0.$$
This gives
$$\| f \|^2 = \sum_{n=1}^{\infty} | [K(t, t)]^{-1} \overline{\omega_n(t)} + \eta_n |^2$$
$$= [K(t, t)]^{-1} \left\{ 1 + \sum_{n=1}^{\infty} \eta_n \omega_n(t) + \sum_{n=1}^{\infty} \overline{\eta_n} \; \overline{\omega_n(t)} \right\} + \sum_{n=1}^{\infty} | \eta_n |^2$$
$$= [K(t, t)]^{-1} + \sum_{n=1}^{\infty} | \eta_n |^2 = \| g \|^2 + \| h \|^2.$$
This shows that
$$\min \| f \| = \| g \|$$
as asserted.

COROLLARY. $K(z, t)$ *depends only upon the domain D and not upon the particular orthonormal system used to define the kernel.*

We come now to the application of the kernel function to conformal mapping.

THEOREM 17.2.9.   *The function*

(17.2.32)
$$F(z, t) \equiv [K(t, t)]^{-1} \int_t^z K(s, t) \, ds$$

maps D conformally on a disk $| w | < r_i(t)$ *in such a manner that*

(17.2.33)
$$F(t, t) = 0, \quad \frac{\partial}{\partial z} F(z, t) \Big|_{z=t} = 1.$$

Here $r_i(t)$ *is the interior mapping radius of D with respect to the point t and*

(17.2.34)
$$r_i(t) = [\pi K(t, t)]^{-\frac{1}{2}}.$$

PRELIMINARY REMARKS.   This theorem shows that the mapping function has still another extremal property:

In the subclass of $H[D]$ *made up of functions whose derivative is in* $L_2 H[D]$ *and such that*
$$f(t) = 0, \quad f'(t) = 1,$$
the function which maps D on the disk $| w | < r_i(t)$ *minimizes the integral*

(17.2.35)
$$\iint_D | f'(z) |^2 \, dx \, dy.$$

In other words, *this function minimizes the area of the map.* Compare Sections 4.6 and 14.4. Now it is obvious that the function $F(z, t)$ minimizes the area in view of the result of the preceding theorem, but it is far from obvious that the minimum area is circular. To give a direct proof of this fact is fairly difficult and such a proof seems to exist only for the case in which D is bounded by an analytic curve. If we are willing, however, to use the already proved existence of the mapping function in question, it is not difficult to prove that it coincides with the function $F(z, t)$. For this purpose we need two lemmas.

LEMMA 17.2.3.   *A complete orthonormal system for the disk Δ_R: $|z| < R$ is given by*

(17.2.36) $$\varphi_n(z) = \left(\frac{n}{\pi}\right)^{\frac{1}{2}} R^{-n} z^{n-1}, \quad n = 1, 2, 3, \cdots.$$

Proof.   The verification of the orthonormal property is left to the reader. To prove completeness we note that if

$$f(z) = \sum_{k=0}^{\infty} a_k z^k$$

is an element of $L_2 H[\Delta_R]$, then

$$\int\int_{|z|<R} f(z) \bar{z}^{n-1} \, dx \, dy = \frac{\pi}{n} R^{2n} a_{n-1}.$$

Thus, if all these integrals are zero, so is $f(z)$. By Theorem 17.2.2 this implies completeness of the system.

LEMMA 17.2.4.   *If $f(z, t)$ maps D conformally on the disk Δ_R: $|w| < R$ in such a manner that*

$$f(t, t) = 0, \quad f'(t, t) = 1,$$

then a complete orthonormal system for D is given by

(17.2.37) $$\omega_n(z) \equiv \left(\frac{n}{\pi}\right)^{\frac{1}{2}} R^{-n} [f(z, t)]^{n-1} f'(z, t), \quad n = 1, 2, 3, \cdots,$$

where the prime indicates differentiation with respect to z.

Proof.   In the integral

$$\int\int_D \omega_m(z) \overline{\omega_n(z)} \, dx \, dy$$

we introduce the explicit expression for $\omega_n(z)$ furnished by (17.2.37) and set $f(z, t) = w$. This maps D onto Δ_R and, evaluating the Jacobian of the transformation, we see that

$$f'(z, t) \, dx \, dy = du \, dv, \quad w = u + iv,$$

so that the integral reduces to a normalization factor times

$$\int\int_D w^m \bar{w}^n \, du \, dv,$$

which is 0 if $m \neq n$. Thus the system is orthonormal by virtue of Lemma 17.2.3. Similarly, if $f(z) \in L_2 H[D]$ and if $\check{z}(w)$ is the inverse of $f(z, t)$, then

$$[\check{z}(w)] \check{z}'(w) \in L_2 H[\Delta_R].$$

Hence, if $f(z)$ is orthogonal to every $\omega_n(z)$, then $f\,[\check{z}(w)]\check{z}'(w)$ is orthogonal to every w^{n-1}. Such a function is identically zero and therefore $f(z)$ must have the same property. Thus the system is complete.

Proof of Theorem 17.2.9.   The kernel function is independent of the particular complete orthogonal system that is used. It takes a particularly simple form in the system $\{\omega_n(z)\}$ of Lemma 17.2.4, since

$$\omega_1(t) = \pi^{-\frac{1}{2}}R^{-1}, \quad \omega_n(t) = 0, n > 1.$$

It follows that

$$K(z, t) = \pi^{-1}R^{-2}f'(z, t), \quad K(t, t) = \pi^{-1}R^{-2},$$

whence we get

$$F(z, t) = f(z, t), \quad R = r_i(t) = [\pi K(t, t)]^{-\frac{1}{2}}$$

as asserted.

We note the inequality

(17.2.38)
$$\left| \int_t^z K(s, t)\, ds \right| \leq \left[\frac{1}{\pi} K(t, t) \right]^{\frac{1}{2}}.$$

We obtained above a complete orthonormal system for $L_2 H[D]$ with the aid of the conformal mapping theorem. It is of some interest to note that we can construct such a system directly, using merely the variational methods which are natural in the L_2-theory.

THEOREM 17.2.10.   *For each* n, $n = 0, 1, 2, \cdots$, *let* F_n *be the subset of* $L_2 H[D]$ *made up of functions* $f(z)$ *satisfying the conditions:*

(17.2.39) $f^{(k)}(t) = 0, k = 0, 1, \cdots, n-1, \quad f^{(n)}(t) = 1.$

Then F_n *contains a unique element* $g_n(z) = g_n(z, t)$ *whose norm is a minimum, and the functions* $\{g_n(z, t)\}$ *form a complete orthogonal system.*

Proof. F_n is not vacuous, for if D is bounded, as we may assume without restricting the generality, any polynomial of the form

$$\frac{1}{n!}(z - t)^n + \sum_{k=n+1}^{n+p} c_k(z - t)^k$$

is in F_n. It is clear that $\|f\|$ has a non-negative infimum for f in F_n. If this is denoted by A_n, we can find a Cauchy sequence $\{f_{n,k}\}$ in $L_2 H[D] \cap F_n$ such that $\lim_{k\to\infty} \|f_{n,k}\| = A_n$. Since

$$|f_{n,j}(z) - f_{n,k}(z)| \leq (\pi\delta^2)^{-\frac{1}{2}} \|f_{n,j} - f_{n,k}\|$$

for $z \in D_\delta$, we see that the sequence $\{f_{n,k}(z)\}$ converges to a limit $g_n(z)$, uniformly in D_δ. Here $g_n(z) \in L_2 H[D]$, $\|g_n\| = A_n$, and the uniform convergence shows that in a neighborhood of $z = t$ we have

$$g_n(z) = \frac{1}{n!}(z - t)^n + \cdots$$

so that $g_n(z) \not\equiv 0, g_n(z) \in F_n$ and $A_n > 0$. Thus $g_n(z)$ is a solution of the extremal problem.

In order to show that the solution is unique we set

$$f(z) = g_n(z) + h(z),$$

where $h(z) \in L_2H[D]$ and

$$h(t) = h'(t) = \cdots = h^{(n)}(t) = 0.$$

Hence

$$\|f\|^2 = (f, f) = (g_n, g_n) + (g_n, h) + (h, g_n) + (h, h)$$
$$= A_n^2 + \|h\|^2 + 2\Re[(h, g_n)].$$

We claim that

(17.2.40) $(h, g_n) = 0$

for every $h(z)$ satisfying the above conditions. Suppose that this is not the case, and let $\eta(z)$ be such a function with $(\eta, g_n) \neq 0$. Then for

$$c = -(g_n, \eta)/(\eta, \eta)$$

we find that

$$\|g_n + c\eta\|^2 = A_n^2 - |(\eta, g_n)|^2/(\eta, \eta) < A_n^2,$$

which is a contradiction. It follows that (17.2.40) holds and

$$\|f\|^2 = \|g_n\|^2 + \|h\|^2 > \|g_n\|^2$$

for every admissible $h(z)$, not identically 0. This shows that the extremal problem has a unique solution.

It also shows that the functions $g_n(z) = g_n(z, t)$ form an orthogonal system, for if n is fixed and $p > n$ then $g_p(z, t)$ is an admissible function $h(z)$ in (17.2.40), that is,

$$(g_n, g_p) = 0, \quad p > n.$$

We now normalize the orthogonal system and set

$$\omega_n(z) = [(g_{n-1}, g_{n-1})]^{-\frac{1}{2}} g_{n-1}(z), \quad n = 1, 2, 3, \cdots.$$

We shall show that this system is complete.

Let $f(z)$ be an arbitrary element of $L_2H[D]$ and form its Fourier series

$$g(z) = \sum_{k=1}^{\infty} (f, \omega_k)\omega_k(z) \equiv \sum_{k=1}^{\infty} a_k\omega_k(z).$$

This is also an element of $L_2H[D]$ and we aim to show that $g(z) \equiv f(z)$. To this end we introduce a sequence of functions $\{f_n(z)\}$ such that

$$f_n(z) = \sum_{k=1}^{n} b_k^n \omega_k(z)$$

and

(17.2.41) $f_n^{(j)}(t) = f^{(j)}(t), \quad j = 0, 1, 2, \cdots, n - 1.$

It is possible to satisfy these conditions, for the system of equations defining the coefficients b_k^n has a triangular matrix and the elements in the main diagonal are different from 0 so we can determine $b_1{}^n, b_2{}^n, \cdots, b_n{}^n$ successively. Since the function $f(z) - f_n(z)$ vanishes for $z = t$ together with its derivatives of order $\leqq n - 1$, formula (17.2.40) shows that

$$(f - f_n, \omega_k) = 0, \quad k = 1, 2, \cdots, n.$$

Hence

$$b_k{}^n = a_k,$$

and

$$f_n(z) = \sum_{k=1}^{n} a_k \omega_k(z)$$

is simply the nth partial sum of the series $g(z)$. We now return to (17.2.41) and note that $f_n{}^{(j)}(z)$ converges uniformly to $g^{(j)}(z)$ when $n \to \infty$. It follows that

$$f^{(j)}(t) = g^{(j)}(t)$$

for all j. By the identity theorem $g(z) \equiv f(z)$, and the theorem is proved.

EXERCISE 17.2

1. If $g_1, g_2, \cdots, g_n$ are n elements of $\mathscr{H}$, show that they are linearly independent if and only if $\det | (g_j, g_k) | \neq 0$. [This determinant is known as the Gramian after the Danish actuary and mathematician Jörgen Pedersen Gram (1850–1916).]

2. If the elements g_k are linearly independent, an orthonormal system $f_1, f_2, \cdots, f_n$ can be constructed by the following orthogonalization process. Set

$$f_k = c_{k1} g_1 + c_{k2} g_2 + \cdots + c_{kk} g_k, \quad k = 1, 2, \cdots, n,$$

and determine the coefficients c_{kj} so that $(f_k, f_l) = \delta_{kl}$. This is the Gram-Schmidt process.

3. Find the kernel function for the disk $|w| < R$.

4. Verify that for the annulus $0 < r < |z| < R$ the following functions form a complete orthonormal system:

$$\omega_{2n-1}(z) = z^{n-1} \left(\frac{n}{\pi}\right)^{\frac{1}{2}} (R^{2n} - r^{2n})^{-\frac{1}{2}}, \quad n = 1, 2, 3, \cdots,$$

$$\omega_2(z) = z^{-1} [2\pi \log (R/r)]^{-\frac{1}{2}},$$

$$\omega_{2n}(z) = z^{-n} \left\{ \frac{\pi}{n-1} [r^{-2n+2} - R^{-2n+2}] \right\}^{-\frac{1}{2}}, \quad n = 2, 3, \cdots.$$

The corresponding kernel function is expressible in terms of the Weierstrass $\wp$-function.

5. What conditions should the coefficients of the Laurent series

$$f(z) = \sum_{n=-\infty}^{\infty} c_n z^n$$

satisfy in order that $f(z)$ belong to the class $L_2 H[D]$ corresponding to the annulus?

6. If $z_1, z_2, \cdots, z_n$ are n distinct points in D, prove that the Hermitian form

$$\sum_{j=1}^{n} \sum_{k=1}^{n} t_j \bar{t}_k K(z_j, z_k) \geq 0.$$

7. Prove that the reproducing property (17.2.28) characterizes $K(z, t)$ uniquely.

17.3. Fekete polynomials and the exterior mapping problem. The mapping of the complement of a bounded continuum E on the exterior of a circle can be studied with the aid of the Fekete polynomials $F_n(z; E)$ defined by (16.2.12). This gives another existence proof for mapping functions and brings out important properties of the latter and of the transfinite diameter.

THEOREM 17.3.1. *Let E be a bounded continuum whose complement G is simply-connected. Let $F_n(z) \equiv F_n(z; E)$ be the nth Fekete polynomial associated with E. Define*

$$(17.3.1) \qquad f_n(z) \equiv [F_n(z)]^{\frac{1}{n}} = z \prod_{k=1}^{n}\left(1 - \frac{z_{n,k}}{z}\right)^{\frac{1}{n}},$$

where the root has its principal value 1 at $z = \infty$. Then

$$(17.3.2) \qquad \lim_{n \to \infty} f_n(z) \equiv F(z) \equiv z\left[1 + \sum_{k=1}^{\infty} A_k z^{-k}\right]$$

exists everywhere in G, and $F(z)$ maps G conformally on

$$(17.3.3) \qquad D_0: \quad |w| > \rho(E),$$

where $\rho(E)$ is the transfinite diameter of E.

Proof. We refer to Section 16.2 and Theorem 16.5.6 for the properties of $F_n(z; E)$. We recall that the zeros $\{z_{n,k}\}$ of $F_n(z; E)$ are located on the boundary of E and maximize the nth Vandermonde determinant. Further, if

$$K_n(E) = \max_{z \in E} |F_n(z; E)|,$$

then

$$\lim_{n \to \infty} [K_n(E)]^{\frac{1}{n}} = \rho(E).$$

Finally, (16.5.17) asserts that if $\varepsilon > 0$ is given, then there exist a $\delta = \delta(\varepsilon)$ and an $N = N_\varepsilon$ such that any point z whose distance from E exceeds ε lies outside the lemniscate

$$(17.3.4) \qquad |F_n(z; E)| = [\rho(E) + \delta]^n, \quad n > N_\varepsilon.$$

For the proof of the mapping theorem we note first that each function $f_n(z)$ is single-valued in G since all its branch points are in E. Let $\delta > 0$ be fixed and consider the domain

$$D_\delta: \quad [w \,|\, |w| > \rho(E) + \delta].$$

We shall show that for $n > N(\delta)$ the functions $f_n(z)$ assume every value w_0 in D_δ once and only once in G.

Let w_0 be such a value with

$$| w_0 | = R \geq \rho(E) + \delta,$$

and consider the lemniscate

$$L: \quad | F_n(z; E) | = R^n.$$

If $\delta = \delta(\varepsilon)$ in the notation above, we may choose $N(\delta) = N_\varepsilon$ and then from (17.3.4) we conclude that L lies in G for $n > N_\varepsilon$. L is a simple closed curve which surrounds E, and if we describe L once in the positive sense then $\arg F_n(z)$ increases by $2n\pi$ since $F_n(z)$ has n zeros in E. Moreover, $\arg F_n(z)$ is steadily increasing as we describe L for otherwise there would be some complex number W_0 with $| W_0 | = R^n$ such that the equation

$$F_n(z) = W_0{}^n$$

would have more than n roots on L, and this is absurd. It follows that $\arg f_n(z)$ is also steadily increasing on L, and it increases by 2π when we describe L. It follows that $f_n(z)$ takes on the value w_0 once on L and, evidently, only once. Since $| f_n(z) | > R$ outside L and $< R$ inside L, we conclude that the equation

$$f_n(z) = w_0$$

has one and only one root in G if $n > N_\varepsilon$.

On the other hand, a value w_0 with $| w_0 | < \rho(E) - \delta$ can be assumed by $f_n(z)$ in G for at most a finite number of values of n. This follows also from (17.3.4). As a consequence we see that $\{f_n(z)\}$ is a normal family in G. We can then find a subsequence which converges to a limit function in G, uniformly with respect to z on compact subsets. Let this limit function be

$$F(z) = z\left[1 + \sum_{k=1}^{\infty} A_k z^{-k} \right],$$

where the series converges for z outside the smallest circular disk with center at $z = 0$ which contains E. We shall show that $F(z)$ is univalent in G and takes on every value in D_0 once and only once in G. That $| F(z) | \geq \rho(E)$ in G follows from the fact that

$$| f_n(z) | > \rho(E) - \delta \quad \text{for} \quad n > N_\varepsilon.$$

Since $F(z)$ is not a constant, we cannot have $| F(z) | = \rho(E)$ anywhere in G, so $| F(z) | > \rho(E)$ is necessary. No value w_0 with $| w_0 | > \rho(E)$ can be omitted by $F(z)$ in G, for each $f_n(z)$ with $n > N(w_0)$ assumes the value w_0 at some point $z_n(w_0)$ in G and, as we have seen, the set $\{z_n(w_0)\}$ is bounded away from E and from infinity. Hence, any limit point of the set is bounded away from E and infinity. There is necessarily such a limit point z_0 and at it $F(z_0) = w_0$.

Moreover, since the value w_0 is assumed only once by each $f_n(z)$, the same must be true for $F(z)$ by Theorem 14.3.4. This shows that $F(z)$ maps G conformally on the annulus (17.3.3).

Now if we do not have

$$\lim_{n \to \infty} f_n(z) = F(z),$$

there must be a subsequence which converges to a limit function $G(z) \neq F(z)$, and this function has exactly the same mapping properties as $F(z)$. If $\check{G}(w)$ is the inverse of $G(z)$, the composed function

$$F[\check{G}(w)]$$

maps the annulus (17.3.3) conformally onto itself, leaving invariant the point at infinity as well as the magnification there. Any such function must reduce to the identity so that $G(z) \equiv F(z)$. This completes the proof of the theorem.

Several remarks should be appended to the theorem. If G, the infinite component of the complement of a continuum E, is mapped conformally on the exterior of a circle with center at $w = 0$ by a function

$$F(z) = z + a_0 + a_1 z^{-1} + a_2 z^{-2} + \cdots,$$

then the radius of the circle is called the *exterior mapping radius* of E and is denoted by $r_e(E)$. Thus we have the

COROLLARY.   *The exterior mapping radius of a continuum E equals its transfinite diameter.*

The limit property (17.3.2) is not restricted to Fekete polynomials. In 1919 Georg Faber surmised that the Chebichev polynomials would have this property, and three years later he proved a limit theorem for a fairly general class of polynomials. Both the Chebichev and the Fekete polynomials of E would seem to satisfy his conditions. Franciszek Leja, who since 1933 has studied in great detail the transfinite diameter and related questions, has also proved several limit theorems including Theorem 17.3.1 (even under more general assumptions on the set E). The theory of orthogonal polynomials on a "scroc" leads to similar limit theorems. Thus, a study of the asymptotic representation of the polynomial $P_n(z)$ in terms of the exterior mapping function $F(z)$ led G. Szegö in 1921 to the formula

$$(17.3.5) \qquad \lim_{n \to \infty} [P_n(z)]^{\frac{1}{n}} = \lim_{n \to \infty} \frac{P_{n+1}(z)}{P_n(z)} = F(z).$$

There are interesting connections between the mapping function $F(z)$ and the equilibrium potential $U(z; \nu)$ studied in Section 16.4. We start by showing how the coefficients A_k of the mapping function can be expressed in terms of the moments of the equilibrium distribution $\nu(S)$. Our point of departure is the expansion

(17.3.6) $\qquad f(z; \zeta, \alpha, n) \equiv \prod_{j=1}^{n} \left(1 - \frac{\zeta_j}{z}\right)^{\alpha} \equiv 1 + \sum_{k=1}^{\infty} A_k(\zeta, \alpha, n) z^{-k},$

where the series converges for $|z| > \max |\zeta_j|$. The series reduces to that of $f_n(z)$ for

$$\zeta_j = z_{n,j}, \quad \alpha = \frac{1}{n}.$$

Now

$$\left(1 - \frac{\zeta}{z}\right)^{\alpha} = \exp\left\{\alpha \log\left(1 - \frac{\zeta}{z}\right)\right\} = \exp\left\{-\alpha \sum_{m=1}^{\infty} \frac{1}{m}\left(\frac{\zeta}{z}\right)^m\right\}.$$

Hence, if we set

(17.3.7) $\qquad \sum_{j=1}^{n} (\zeta_j)^m \equiv S_{n,m},$

then we have

$$f(z; \zeta, \alpha, n) = \exp\left\{-\alpha \sum_{m=1}^{\infty} \frac{1}{m} S_{n,m} z^{-m}\right\}$$

$$= \prod_{m=1}^{\infty} \exp\left\{-\frac{\alpha}{m} S_{n,m} z^{-m}\right\}$$

$$= \prod_{m=1}^{\infty} \left\{\sum_{p=0}^{\infty} \frac{(-\alpha)^p}{p!} \left(\frac{S_{n,m}}{m}\right)^p z^{-mp}\right\}.$$

This shows that

(17.3.8) $\quad A_k(\zeta, \alpha, n) = \sum \frac{(-\alpha)^{p_1+p_2+\cdots+p_k}}{p_1! \, p_2! \cdots p_k!} \left(\frac{S_{n,1}}{1}\right)^{p_1} \left(\frac{S_{n,2}}{2}\right)^{p_2} \cdots \left(\frac{S_{n,k}}{k}\right)^{p_k}$

where the summation extends over the non-negative integers $p_1, p_2, \cdots, p_k$ such that

(17.3.9) $\qquad p_1 + 2p_2 + \cdots + kp_k = k.$

In particular, we find that the coefficient $A_{n,k}$ of z^{-k+1} in the series for $f_n(z)$ may be written

(17.3.10) $\quad A_{n,k} = \sum \frac{(-1)^{p_1+p_2+\cdots+p_k}}{p_1! \, p_2! \cdots p_k!} \left(\frac{M_{n,1}}{1}\right)^{p_1} \left(\frac{M_{n,2}}{2}\right)^{p_2} \cdots \left(\frac{M_{n,k}}{k}\right)^{p_k}$

where

(17.3.11) $\qquad M_{n,m} = \frac{1}{n} \sum_{k=1}^{n} (z_{n,k})^m = \int_E t^m \, d\mu_n(t),$

and $\mu_n(S)$ is the point distribution associated with the nth Fekete polynomial (see the proof of Theorem 16.4.4). Thus $\{M_{n,m}\}$ is the sequence of moments of

$\mu_n(S)$. Using (16.4.16) we see that $M_{n,m}$ tends to a limit, M_m say, when $n \to \infty$, and that

$$(17.3.12) \qquad M_m = \int_E t^m \, d\nu(t),$$

that is, M_m is the mth moment of the equilibrium distribution on E. It is clear that $\lim\limits_{n\to\infty} A_{n,k}$ exists and equals A_k, so that A_k is expressible in terms of the M_m's in the same manner as $A_{n,k}$ is expressible in terms of the moments $M_{n,m}$. This leads to the following theorem, the first part of which has already been proved:

THEOREM 17.3.2. *If E is a bounded continuum, $\nu(S)$ is its equilibrium distribution, and $F(z)$, defined by (17.3.1) and (17.3.2), is the exterior mapping function which maps $G = \mathbf{C}(E)$ conformally on $|w| > r_e(E)$, then the coefficients A_k of $F(z)$ are expressed in terms of the moments M_m of $\nu(S)$ by the formula*

$$(17.3.13) \qquad A_k = \sum \frac{(-1)^{p_1 + p_2 + \cdots + p_k}}{p_1!\, p_2! \cdots p_k!} \left(\frac{M_1}{1}\right)^{p_1} \left(\frac{M_2}{2}\right)^{p_2} \cdots \left(\frac{M_k}{k}\right)^{p_k}.$$

Conversely,
(17.3.14)
$$M_k = k \sum (-1)^{p_1 + p_2 + \cdots + p_k} \frac{(p_1 + p_2 + \cdots + p_k - 1)!}{p_1!\, p_2! \cdots p_k!} A_1^{p_1} A_2^{p_2} \cdots A_k^{p_k},$$

where in both cases the summation extends over the non-negative integers p_j which satisfy (17.3.9). Further, for $|z| > R$,

$$(17.3.15) \qquad \frac{F'(z)}{F(z)} = \sum_{k=0}^{\infty} M_k z^{-k-1}.$$

Finally, if $U(z;\nu)$ is the equilibrium potential of E and $z \in G$,

$$(17.3.16) \qquad \log \left| F(z) \right| = -U(z;\nu).$$

Proof. Formula (17.3.13) has already been proved. The associated formulas (17.3.13) and (17.3.14) are familiar from another context, namely, the theory of symmetric functions of the roots of an algebraic equation. The first formula expresses the kth elementary symmetric function of the roots in terms of their power sums, and the second expresses the power sums in terms of the elementary functions. Actually the discussion given below is patterned on the proofs commonly used in the theory of symmetric functions. Both formulas are consequences of what is known as the *formulas of Newton*[1]

$$(17.3.17) \quad M_n + A_1 M_{n-1} + A_2 M_{n-2} + \cdots + A_{n-1} M_1 + n A_n = 0,$$

which are valid for $n = 1, 2, 3, \cdots$.

---

[1] These relations occur in Newton's *Arithmetica Universalis*, published in 1707. The explicit formulas were given by Edward Waring (1734–98) in his *Meditationes Algebraicae* of 1770. Waring became famous for several conjectures in number theory which remained unproved until the first quarter of this century.

We shall verify (17.3.15), from which (17.3.17) easily follows, but we shall leave (17.3.14) to the reader since we have no direct use for these relations. We have

$$\log \frac{f_n(z)}{z} = \frac{1}{n} \sum_{j=1}^{n} \log \left(1 - \frac{z_{n,j}}{z} \right)$$

$$= - \sum_{m=1}^{\infty} \frac{1}{m} \left[\frac{1}{n} \sum_{j=1}^{n} (z_{n,j})^m \right] z^{-m} = - \sum_{m=1}^{\infty} \frac{1}{m} M_{n,m} z^{-m},$$

whence

(17.3.18) $$\log \frac{F(z)}{z} = - \sum_{m=1}^{\infty} \frac{1}{m} M_m z^{-m}, \quad |z| > R.$$

Term-by-term differentiation of this series gives (17.3.15), and from the identity

$$- \sum_{k=1}^{\infty} k A_k z^{-k-1} = \sum_{m=0}^{\infty} M_m z^{-m} \sum_{k=0}^{\infty} A_k z^{-k}, \quad A_0 = 1,$$

we get (17.3.17) by equating coefficients of like powers. Finally, we note that the right member of (17.3.18) equals

$$\int_E \log \left(1 - \frac{t}{z} \right) d\nu(t),$$

where the logarithm has its principal value. This gives

(17.3.19) $$\log F(z) = \int_E \log (t - z) \, d\nu(t),$$

where principal values correspond to each other. Taking real parts we obtain (17.3.16). This completes the proof.

A remark concerning the numerical coefficients in (17.3.13) and (17.3.14) is in order. In the first formula $k!$ is the common denominator. The numerators are integers whose sum is 0 and whose absolute values add up to $k!$. It follows that the sum of the coefficients is 0 if $k > 1$, and the sum of their absolute values is 1. Since

(17.3.20) $$|M_k| \leq R^k, \quad R = \max_{z \in E} |z|,$$

we get also

(17.3.21) $$|A_k| \leq R^k.$$

In the case of (17.3.14) on the other hand, all coefficients are integers which add up to -1 and whose absolute values have the sum $2^k - 1$.

There are other relations involving the moments M_k which give alternate expressions for them and which throw further light on the equilibrium distribution. By (17.3.15) we have

(17.3.22) $$M_k = \frac{1}{2\pi i} \int_{C_r} \frac{F'(z)}{F(z)} z^k \, dz,$$

the integral being taken in the positive sense along a level curve

$$C_r: \quad |F(z)| = r > r_e(E).$$

Here we introduce the inverse function

(17.3.23)
$$z = \check{F}(w) = w\left[1 + \sum_{n=1}^{\infty} \alpha_n w^{-n}\right],$$

where the series converges for

$$|w| > r_e(E).$$

This gives

(17.3.24)
$$M_k = \frac{1}{2\pi i} \int_{|w|=r} [\check{F}(w)]^k \frac{dw}{w}.$$

Thus M_k is the constant term in the Laurent expansion of $[\check{F}(w)]^k$, which we obtain by taking the kth power of the expansion (17.3.23). This leads to expressions of the form

(17.3.25)
$$M_k = \Sigma\, N_{p_1, p_2, \cdots, p_k} \alpha_1^{p_1} \alpha_2^{p_2} \cdots \alpha_k^{p_k},$$

where the N's are integers and the summation extends over the non-negative integers p_j which satisfy (17.3.9). Conversely, we may of course express the α's in terms of the M's. We find that

$$\alpha_1 = M_1, \quad \alpha_2 = \tfrac{1}{2}M_2 - \tfrac{1}{2}M_1^2, \quad \alpha_3 = \tfrac{1}{3}M_3 - M_1 M_2 + \tfrac{2}{3}M_1^3, \quad \cdots,$$

and obtain a general formula of the type

(17.3.26)
$$\alpha_k = \Sigma\, R_{p_1, p_2, \cdots, p_k} M_1^{p_1} M_2^{p_2} \cdots M_k^{p_k}$$

where the R's are rational numbers whose sum is 0 if $k > 1$ and the p's satisfy (17.3.9). We shall not elaborate this point any further.

A more interesting observation is perhaps the following: In the integral (17.3.24) we have $w = re^{i\theta}$, that is,

(17.3.27)
$$M_k = \int_{C_r} z^k\, d\left(\frac{\theta}{2\pi}\right).$$

This means that we can also regard M_k as the kth moment of a mass distribution ν_r on C_r. If an arc of C_r is mapped on the circular arc

$$\theta_1 \leq \arg w \leq \theta_2, \quad |w| = r$$

by $F(z)$, then the ν_r-measure of the arc on C_r equals

$$\frac{\theta_2 - \theta_1}{2\pi},$$

and this holds for all $r > r_e(E)$.

In the case in which E is bounded by a simple closed rectifiable oriented curve C, we can let $r \to r_e(E)$ in these formulas for, as we shall see in Section

17.5, in this case $F(z)$ and its inverse are continuous and one-to-one in the closed regions $G \cup C$ and $| w | \geq r_e(E)$ respectively. This gives

THEOREM 17.3.3.   *Let the boundary of E be a simple closed rectifiable curve C. Let Γ be an arc of C and let $w = F(z)$ be the exterior mapping function of $\mathbf{C}(E)$. If $F(z)$ maps Γ on an arc of $| w | = r_e(E)$ whose angular measure is $\alpha\,(= angle subtended at the origin)$ and if $\nu(S)$ is the equilibrium distribution on C, then*

$$(17.3.28) \qquad\qquad \nu(\Gamma) = \frac{\alpha}{2\pi}.$$

EXERCISE 17.3

1. Prove that
$$A_1 = -\alpha_1 = M_1.$$

The point $z = M_1$ is known as the *conformal mass center* of G (K. Löwner and P. Frank, 1919). Show that it is the center of gravity of the equilibrium distribution.

2. If E is bounded by a simple closed rectifiable curve C and if C_r is the level line $| F(z) | = r > r_e(E)$ with the mass distribution ν_r as in (17.3.27), show that ν and ν_r have the same center of gravity.

3. If E is symmetric with respect to $z = z_0$, show that $M_1 = z_0$.

4. Suppose that E is bounded by a "scroc" C which is symmetric with respect to (i) $z = 0$, (ii) the real axis, (iii) the imaginary axis. Prove that the equilibrium distribution $\nu(S)$ has corresponding symmetry properties.

5. Prove that in case (i) of Problem 4 all moments and all coefficients A_k of odd order are 0. Prove that in case (ii) all moments and coefficients are real, whereas in case (iii) moments of even order are real and those of odd order are purely imaginary and the same holds for the coefficients. What happens if C has two of these symmetry properties? if it has all three?

6. If we place the mass 1 at $z = 1$, the corresponding moments M_k are all equal to 1 and (17.3.15) defines a (trivial) mapping function

$$w = z - 1.$$

Use this observation to prove that the sum of the numerical coefficients in (17.3.13) is 0 if $k > 1$ and apply the same argument in (17.3.26).

17.4. Univalent functions.   In the preceding sections we have encountered classes of univalent functions which correspond to problems of conformal mapping. We shall now examine these functions in more detail.

There are two classes of primary interest. We denote by $\mathscr{S}$ the functions

$$(17.4.1) \qquad\qquad f(z) = z + a_2 z^2 + a_3 z^3 + \cdots$$

which are holomorphic and univalent in the unit disk K. Such functions map K in a one-to-one and conformal manner onto domains

$$D = f[K]$$

which contain the origin. D has interior mapping radius 1 with respect to the origin.

We denote by $\mathcal{U}$ the functions

(17.4.2) $$F(z) = z + \alpha_0 + \alpha_1 z^{-1} + \alpha_2 z^{-2} + \cdots$$

which are holomorphic, except for the pole at infinity, and univalent in the domain Y: $|z| > 1$, that is, the northern hemisphere. Such functions map Y in a one-to-one and conformal manner onto domains G which contain the point at infinity

$$G = F[Y].$$

G has exterior mapping radius 1 or, equivalently, its complement E has transfinite diameter 1.

These classes are obviously related. Thus

(17.4.3) $$f(z) \in \mathcal{S} \quad \text{implies that} \quad \left[f\left(\frac{1}{z} \right) \right]^{-1} \in \mathcal{U},$$

and every function $F(z)$ in $\mathcal{U}$ is obtainable in this manner.

These classes $\mathcal{S}$ and $\mathcal{U}$ are not linear and they are not left invariant by any of the regular operations of algebra or analysis. It should be noted, however, that

(17.4.4) $$f(z) \in \mathcal{S} \quad \text{implies that} \quad [f(z^n)]^{\frac{1}{n}} \in \mathcal{S}$$

for any positive integer n. Here we can replace f and $\mathcal{S}$ by F and $\mathcal{U}$ respectively, provided that $F(z) \neq 0$ for $|z| > 1$.

The coefficients of the functions belonging to $\mathcal{S}$ and to $\mathcal{U}$ are subject to severe restrictions. We start by proving the following result, known as the *area theorem*, which is due to T. H. Gronwall (1914).

THEOREM 17.4.1. *If $F(z) \in \mathcal{U}$, then*

(17.4.5) $$\sum_{n=1}^{\infty} n \, |\alpha_n|^2 \leq 1.$$

Proof. Let $r > 1$ and consider the image C_r of the circle $|z| = r$ under the transformation $w = F(z)$. Since C_r is a simple closed analytic curve surrounding E, it bounds a region whose area is given by the classical formula

$$A(r) = \frac{1}{2} \int_0^{2\pi} \left[U \frac{dV}{d\theta} - V \frac{dU}{d\theta} \right] d\theta$$

where $z = re^{i\theta}$, $F(z) = U + iV$. The integral is the imaginary part of

(17.4.6) $$\frac{1}{2} \int_{|z|=r} \overline{F(z)} F'(z) \, dz.$$

Here we substitute (17.4.2) and obtain

$$\frac{1}{2} \int_{|z|=r} \left[\bar{z} + \sum_{m=0}^{\infty} \bar{\alpha}_m \bar{z}^{-m} \right] \left[1 - \sum_{n=1}^{\infty} n \alpha_n z^{-n-1} \right] dz.$$

Since

$$\int_{|z|=r} \bar{z}^{-m} z^{-n-1} \, dz = \begin{cases} 0, & m \neq n, \\ 2\pi i r^{-2m}, & n = m, \end{cases}$$

the imaginary part of (17.4.6) reduces to

$$\pi \left\{ r^2 - \sum_{n=1}^{\infty} n \mid \alpha_n \mid^2 r^{-2n} \right\} \geqq 0.$$

This holds for every $r > 1$, hence also in the limit when $r \to 1$, whence (17.4.5) results.

From the geometric interpretation of the inequality, it follows that equality holds in (17.4.5) if and only if E is of (two-dimensional) measure 0.

COROLLARY. $\mid \alpha_1 \mid \leqq 1$ and equality holds if and only if E is a line segment of length 4.

The first assertion is obvious. If $\mid \alpha_1 \mid = 1$, then all other coefficients are 0 so that

$$F(z) = z + \alpha_0 + e^{i\beta} z^{-1}$$

where β is real. This function maps Y on the whole w-plane slit along the line segment $[-2e^{\frac{1}{2}i\beta} + \alpha_0, 2e^{\frac{1}{2}i\beta} + \alpha_0]$.

THEOREM 17.4.2.   For $\mid z \mid > 1$ we have

$$(17.4.7) \qquad\qquad \mid F'(z) \mid \leqq \frac{\mid z \mid^2}{\mid z \mid^2 - 1},$$

with equality holding for a particular value $z = z_0$ if and only if

$$(17.4.8) \qquad\qquad F(z) = z + \alpha_0 - \frac{\mid z_0 \mid^2 - 1}{z_0(\bar{z}_0 z - 1)}.$$

Proof.   Cauchy's inequality gives

$$\mid F'(z) - 1 \mid = \left| \sum_{n=1}^{\infty} n \alpha_n z^{-n-1} \right|$$

$$\leqq \left\{ \sum_{n=1}^{\infty} n \mid \alpha_n \mid^2 \right\}^{\frac{1}{2}} \left\{ \sum_{n=1}^{\infty} n \mid z \mid^{-2n-2} \right\}^{\frac{1}{2}}$$

$$\leqq (\mid z \mid^2 - 1)^{-1},$$

and this implies (17.4.7). Equality cannot hold for all values of z but may possibly hold for a particular value z_0. For this to be the case, we must have equality in both places above. For equality to hold in the first step, a number μ must exist such that

$$\alpha_n = \mu \bar{z}_0^{-n-1}, \quad n = 1, 2, 3, \cdots,$$

and the second step requires that

$$|\mu| \left\{ \sum_{n=1}^{\infty} n |z_0|^{-2n-2} \right\}^{\frac{1}{2}} = 1$$

or

$$|\mu| = |z_0|^2 - 1.$$

A simple calculation shows that if $\mu = -|\mu|$, then the mapping function is given by (17.4.8) and has the desired property.

We list next some observations which bear on the restrictions imposed on E by the assumption that $\rho(E) = 1$. The first is a statement concerning the conformal mass center of G.

THEOREM 17.4.3. *If G does not contain the origin, then*

(17.4.9)
$$|\alpha_0| \leq 2,$$

where equality holds if and only if E is the line segment $[0, 4e^{i\beta}]$.

Proof. The assumption means that $F(z) \neq 0$ in Y, so that $[F(z^2)]^{\frac{1}{2}}$ also belongs to $\mathscr{U}$. Now

$$[F(z^2)]^{\frac{1}{2}} = \left[z^2 + \alpha_0 + \sum_{n=1}^{\infty} \alpha_n z^{-2n} \right]^{\frac{1}{2}}$$

$$= z \left[1 + \alpha_0 z^{-2} + \sum_{n=1}^{\infty} \alpha_n z^{-2n-2} \right]^{\frac{1}{2}}$$

$$= z + \tfrac{1}{2}\alpha_0 z^{-1} + \tfrac{1}{2}(\alpha_1 - \tfrac{1}{4}\alpha_0^2) z^{-3} + \cdots.$$

The Corollary of Theorem 17.4.1 gives

$$|\tfrac{1}{2}\alpha_0| \leq 1 \quad \text{or} \quad |\alpha_0| \leq 2$$

with equality holding if and only if

$$[F(z^2)]^{\frac{1}{2}} = z + e^{i\beta} z^{-1}$$

with β real, that is,

(17.4.10)
$$F(z) = z + 2e^{i\beta} + e^{2i\beta} z^{-1}.$$

This function maps Y on the whole w-plane slit along the line segment from 0 to $4e^{i\beta}$, as asserted.

COROLLARY. *The distance from $z = \alpha_0$ to any point of E is at most 2 where equality holds only for the mapping functions which differ from (17.4.10) by a constant.*

Proof. It is clear that the maximum distance is reached on the boundary of E. But if w_0 is a boundary point of E, then $F(z) - w_0$ satisfies the conditions of Theorem 17.4.3, so that $| w_0 - \alpha_0 | \leq 2$ and equality can hold if and only if $F(z) - w_0$ is of the form (17.4.10).

We turn now to the class $\mathscr{S}$ of functions which are holomorphic and univalent in the unit disk K. Here the outstanding coefficient problem is to verify Bieberbach's conjecture of 1916 that

(17.4.11) $| a_n | \leq n$ if $f(z) \in \mathscr{S}$.

The functions

(17.4.12) $f(z) = \dfrac{z}{(1 + e^{i\beta}z)^2}$,

where β is real, belong to $\mathscr{S}$, and here we have $| a_n | = n$ for every n. Thus, if (17.4.11) is true for all n, no better estimate is possible. The conjecture has been verified for $n = 2, 3$, and 4; it has led to a vast amount of fascinating mathematics including important variational methods for the study of extremal problems, but the conjecture itself is still unproved. The case $n = 2$ was proved by Bieberbach; it has a number of very important consequences.

THEOREM 17.4.4.   *If $f(z) \in \mathscr{S}$, then*

(17.4.13) $| a_2 | \leq 2$

with equality holding only for the functions (17.4.12).

   Proof.   We apply Theorem 17.4.3 to the function

(17.4.14) $\left\{ f\left(\dfrac{1}{z}\right) \right\}^{-1} = z - a_2 + (a_2{}^2 - a_3)z^{-1} + \cdots$,

which does not vanish for $1 < | z |$. It follows that $| a_2 | \leq 2$ and equality holds if and only if the right-hand side of (17.4.14) reduces to

$$z + 2e^{i\beta} + e^{2i\beta}z^{-1} = z[1 + e^{i\beta}z^{-1}]^2,$$

which implies (17.4.12).

   This extremal function maps K on the whole w-plane slit radially from $w = \frac{1}{4}e^{-i\beta}$ to infinity. Thus a boundary point of $f(K)$ has the distance $\frac{1}{4}$ from the origin. P. Koebe in 1907 proved that there is a positive number C such that for any $f(z) \in \mathscr{S}$ the distance of the boundary of $f(K)$ from $w = 0$ is at least C. That the "Koebe constant" has the value $\frac{1}{4}$ was proved by Bieberbach in 1916.

THEOREM 17.4.5.   *We have*

(17.4.15) $\bigcap_{f \in \mathscr{S}} f[K] = [w \, | \, | w | < \tfrac{1}{4}].$

   Proof.   Let $f(z) \in \mathscr{S}$, let w_0 be a boundary point of $f[K]$, and form

(17.4.16) $g(z) = \dfrac{f(z)}{1 - w_0{}^{-1}f(z)}$.

This function belongs to $\mathscr{S}$, for the denominator is different from 0 in K and $g(z_1) = g(z_2)$ if and only if $f(z_1) = f(z_2)$. Now

$$g(z) = z + (a_2 + w_0^{-1})z^2 + \cdots,$$

and by the preceding theorem

$$| a_2 + w_0^{-1} | \leqq 2,$$

whence

(17.4.17) $$| w_0 |^{-1} \leqq 2 + | a_2 |.$$

Since $| a_2 | \leqq 2$ we get

$$| w_0 |^{-1} \leqq 4, \quad | w_0 | \geqq \tfrac{1}{4},$$

that is, every boundary point of $f[K]$ has a distance from the origin which is at least equal to $\tfrac{1}{4}$. On the other hand, we have already seen that every point of the circle $| w | = \tfrac{1}{4}$ is a boundary point of one of the domains $f[K]$ corresponding to the extremal functions (17.4.12). This completes the proof.

If $f(z)$ is not one of the extremal functions, then formula (17.4.17) gives a more favorable lower bound for the distance of $\partial f[K]$ from the origin.

We come now to the *distortion theorem* of Koebe.

THEOREM 17.4.6.   *If $f(z) \in \mathscr{S}$, then*

(17.4.18) $$\frac{1-r}{(1+r)^3} \leqq | f'(z) | \leqq \frac{1+r}{(1-r)^3} \quad if \quad | z | \leqq r.$$

These are best possible bounds: equality is reached for the functions (17.4.12).

The existence of bounds for $| f'(z) |$ valid for the class $\mathscr{S}$ was proved by Koebe in 1907, and the best bounds were given by Gronwall in 1916. We base the proof on

LEMMA 17.4.1.   *If $f(z) \in \mathscr{S}$ and $| z | < 1$,*

(17.4.19) $$\left| \tfrac{1}{2}(1 - | z |^2)\frac{f''(z)}{f'(z)} - \bar{z} \right| \leqq 2.$$

Proof.   We prove (17.4.19) with z replaced by z_0. The Möbius transformation

$$z \to \frac{z + z_0}{1 + \bar{z}_0 z}$$

maps K onto itself, taking 0 into z_0. It follows that for any choice of constants C_1 and C_2

$$g(z) = C_1 + C_2 f\!\left(\frac{z + z_0}{1 + \bar{z}_0 z}\right)$$

is univalent in K. Here we determine C_1 and C_2 so that $g(0) = 0, g'(0) = 1$, that is, so that $g(z) \in \mathscr{S}$. These conditions give

$$C_1 = -\frac{f(z_0)}{f'(z_0)[1 - |z_0|^2]}, \quad C_2 = \frac{1}{f'(z_0)[1 - |z_0|^2]}.$$

We have

$$g'(z) = \frac{1}{f'(z_0)} \frac{1}{(1 + \bar{z}_0 z)^2} f'\left(\frac{z + z_0}{1 + \bar{z}_0 z}\right),$$

$$g''(z) = -\frac{1}{f'(z_0)} \frac{2\bar{z}_0}{(1 + \bar{z}_0 z)^3} f'\left(\frac{z + z_0}{1 + \bar{z}_0 z}\right)$$

$$+ \frac{1}{f'(z_0)} \frac{1 - |z_0|^2}{(1 + \bar{z}_0 z)^4} f''\left(\frac{z + z_0}{1 + \bar{z}_0 z}\right).$$

It follows that

$$g(z) = z + \left\{\tfrac{1}{2}[1 - |z_0|^2]\frac{f''(z_0)}{f'(z_0)} - \bar{z}_0\right\} z^2 + \cdots.$$

Since $g(z) \in \mathscr{S}$, the coefficient of z^2 is in absolute value ≤ 2, and this is the assertion of the lemma.

Proof of Theorem 17.4.6.   By the lemma we have for $|t| < 1$

$$\frac{f''(t)}{f'(t)} - \frac{2\bar{t}}{1 - |t|^2} = 4\frac{\eta(t)}{1 - |t|^2}, \quad |\eta(t)| \leq 1.$$

This expression we integrate from 0 to z along a straight line in K, obtaining

(17.4.20) $$\log f'(z) + \log (1 - |z|^2) = 4\int_0^z \frac{\eta(t)\, dt}{1 - |t|^2},$$

where the right member does not exceed

$$4\int_0^r \frac{ds}{1 - s^2} = 2\log\frac{1 + r}{1 - r}$$

in absolute value. From a consideration of the real parts we get the inequality

$$2\log\frac{1 - r}{1 + r} \leq \log[(1 - |z|^2)|f'(z)|] \leq 2\log\frac{1 + r}{1 - r}.$$

This holds for $|z| \leq r < 1$. It is clear that $\log|f'(z)|$ assumes its maximum and minimum values on the boundary of the disk $|z| \leq r$. This leads to the inequality (17.4.18). If $f(z)$ is given by (17.4.12),

$$f'(z) = \frac{1 - \omega z}{(1 + \omega z)^3}, \quad \omega = e^{i\beta},$$

and the absolute value of this function reaches the upper bound in (17.4.18) for $z = -\bar{\omega}r$ and the lower bound for $z = \bar{\omega}r$.

COROLLARY.   *We have*

(17.4.21) $|\arg f'(z)| \leq 2 \log \dfrac{1 + r}{1 - r}$ *for* $|z| \leq r.$

For this is what we get by equating imaginary parts in (17.4.20). This *rotation theorem*, discovered by Bieberbach in 1919, is not the best result available. The best possible bounds were found by G. M. Golusin (1906–1952) in 1936 and are

(17.4.22) $|\arg f'(z)| \leq \begin{cases} 4 \arc \sin r, & 0 < r \leq 2^{-\frac{1}{2}}, \\ \pi + \log \dfrac{r^2}{1 - r^2}, & 2^{-\frac{1}{2}} < r < 1. \end{cases}$

Here the argument has the value 0 at $z = 0$ and the arc sine has its principal value.

THEOREM 17.4.7.   *If* $f(z) \in \mathscr{S}$, *then*

(17.4.23) $\dfrac{r}{(1 + r)^2} \leq |f(z)| \leq \dfrac{r}{(1 - r)^2},$ $|z| \leq r.$

Proof.   To get the upper bound for $|f(z)|$ we integrate the upper bound for $|f'(t)|$, taken from (17.4.18), along the line segment from 0 to z. Since $f(0) = 0$, we have

$$|f(z)| \leq \int_0^z |f'(t)| \, |dt| \leq \int_0^r \frac{1 + s}{(1 - s)^3} \, ds = \frac{r}{(1 - r)^2}.$$

To verify the lower bound, we note that

$$|f(z)| = \int_0^z |f'(t)| \, |dt|,$$

provided we integrate from 0 to z along the path L whose image is the line segment $[0, f(z)]$. On L we have

$$|dt| \geq d\,|t|, \quad |f'(t)| \geq \frac{1 - |t|}{(1 + |t|)^3},$$

and the inequality follows. Again (17.4.12) provides extremal functions for both inequalities.

LEMMA 17.4.2.   *Let* $F(z)$ *be holomorphic in* $|z| \leq R$ *and denote by* $J(r)$ *the area of the image of the disk* $|z| \leq r$, $0 \leq r \leq R$, *under the mapping* $w = F(z)$. *Then*

(17.4.24) $4 \displaystyle\int_0^R J(r) \frac{dr}{r} = \int_0^{2\pi} |F(Re^{i\theta})|^2 \, d\theta - 2\pi \, |F(0)|^2.$

Proof. If

$$F(z) = \sum_{n=0}^{\infty} a_n z^n,$$

then

$$F(0) = a_0, \quad \int_0^{2\pi} |F(Re^{i\theta})|^2 \, d\theta = 2\pi \sum_{n=0}^{\infty} |a_n|^2 R^{2n},$$

$$J(r) = \pi \sum_{n=1}^{\infty} n \, |a_n|^2 r^{2n},$$

and (17.4.24) follows.

This lemma can be used to derive an inequality for the L_1-mean of $f(z) \in \mathscr{S}$.

THEOREM 17.4.8.   *If $f(z) \in \mathscr{S}$, then*

(17.4.25) $$\frac{1}{2\pi} \int_0^{2\pi} |f(re^{i\theta})| \, d\theta \leq \frac{r}{1-r}, \quad 0 < r < 1.$$

Proof.   Since $f(z) \in \mathscr{S}$, the same is true for $F(z) = [f(z^2)]^{\frac{1}{2}}$. We apply the lemma to $F(z)$. Here $F(0) = 0$ and

$$\int_0^{2\pi} |F(Re^{i\theta})|^2 \, d\theta = \int_0^{2\pi} |f(R^2 e^{2i\theta})| \, d\theta = \int_0^{2\pi} |f(R^2 e^{i\vartheta})| \, d\vartheta.$$

On the other hand, (17.4.23) shows that

$$|f(z^2)|^{\frac{1}{2}} \leq \frac{r}{1-r^2}, \quad |z| \leq r,$$

and since $F(z)$ is univalent

$$J(r) = J(r; F) \leq \pi \left(\frac{r}{1-r^2} \right)^2.$$

It follows that

$$\int_0^R J(r) \frac{dr}{r} \leq \pi \int_0^R \frac{r \, dr}{(1-r^2)^2} = \frac{1}{2} \pi \frac{R^2}{1-R^2}.$$

Hence, by Lemma 17.4.2,

$$\frac{1}{2\pi} \int_0^{2\pi} |f(R^2 e^{i\vartheta})| \, d\vartheta \leq \frac{R^2}{1-R^2},$$

which is equivalent to (17.4.25).

The inequality is of the right order of magnitude: for the functions (17.4.12) the mean value equals

(17.4.26) $$\frac{r}{1-r^2}.$$

Theorem 17.4.8 throws some light on Bieberbach's conjecture (17.4.11). We have the following

COROLLARY.　*If $f(z) \in \mathscr{S}$, then for all n*

(17.4.27)
$$|a_n| < en.$$

Proof.　For any r with $0 < r < 1$ we have

$$a_n = r^{-n} \frac{1}{2\pi} \int_0^{2\pi} f(re^{i\theta}) e^{-ni\theta} \, d\theta,$$

and by (17.4.25) this gives

$$|a_n| \leqq \frac{r^{1-n}}{1-r}.$$

The minimum value of the right member is reached for

$$r = \frac{n-1}{n}$$

and equals

$$\left(1 + \frac{1}{n-1}\right)^{n-1} n < en$$

as asserted.

This inequality was discovered by J. E. Littlewood in 1923. Sharper estimates are known; thus G. M. Golusin replaced the factor e by $\frac{3}{4}e$ in 1948, and three years later I. E. Bazilyevich proved the estimate

$$|a_n| \leqq \tfrac{1}{2}en + 1.51.$$

Bieberbach's conjecture has been verified for several special classes of univalent functions, among which are those for which $f(K)$ is symmetric with respect to the real axis or starlike with respect to the origin. More stringent bounds are known if $f(K)$ is bounded or convex.

We shall derive some of these results and we start with the case in which $f(K)$ is symmetric with respect to the real axis. This assumption clearly implies and is implied by the coefficients a_n being real. The corresponding result is due to Jean Dieudonné (1931).

THEOREM 17.4.9.　*If $f(z) \in \mathscr{S}$ and has real coefficients, then*

(17.4.28)
$$|a_n| \leqq n.$$

Proof.　We refer to formula (14.2.4), which expresses the coefficients of a power series as Fourier coefficients of its imaginary part. We set $\Im[f(z)] = V(z)$ and note that the sign of $V(z)$ is the same as that of $\Im[z]$. We have

$$a_n r^n = \frac{i}{\pi} \int_{-\pi}^{\pi} V(re^{i\theta}) e^{-ni\theta} \, d\theta, \quad n = 1, 2, 3, \cdots,$$

and, taking real parts on both sides,

$$a_n r^n = \frac{1}{\pi} \int_{-\pi}^{\pi} V(re^{i\theta}) \sin n\theta \, d\theta = \frac{2}{\pi} \int_0^{\pi} V(re^{i\theta}) \sin n\theta \, d\theta$$

by the parity properties of the integrand. The identity

$$e^{ni\theta} - e^{-ni\theta} = [e^{i\theta} - e^{-i\theta}] \sum_{k=0}^{n-1} e^{(n-1-2k)i\theta}$$

shows that

$$|\sin n\theta| \leq n |\sin \theta|.$$

Hence

$$|a_n| r^n \leq \frac{2}{\pi} \int_0^\pi V(re^{i\theta}) |\sin n\theta| \, d\theta$$

$$\leq n \frac{2}{\pi} \int_0^\pi V(re^{i\theta}) \sin \theta \, d\theta = na_1 = n.$$

This, in the limit $r \to 1$, gives the desired result.

For the starlike case we need preliminary considerations which are of some independent interest.

THEOREM 17.4.10.　*Suppose that* (i) $g(z)$ *is holomorphic in the unit disk* K, (ii) $g(0) = 1$, (iii) $\Re[g(z)] > 0$ *in* K, *and* (iv) $g(z) = \sum_{n=0}^\infty c_n z^n$, *then*

(17.4.29)　　　$|c_n| \leq 2, n \geq 1, \quad \text{and} \quad |g(z)| \leq \frac{1+r}{1-r}, |z| \leq r.$

Proof.　We get from (14.2.3)

$$c_n r^n = \frac{1}{\pi} \int_{-\pi}^\pi U(re^{i\theta}) e^{-ni\theta} \, d\theta, \quad U(z) = \Re[g(z)],$$

whence

$$|c_n| r^n \leq \frac{1}{\pi} \int_{-\pi}^\pi U(re^{i\theta}) \, d\theta = c_0 + \bar{c}_0 = 2.$$

Passing to the limit with r we get the inequality for $|c_n|$, from which the inequality for $g(z)$ follows. Both inequalities are sharp since

$$g(z) = \frac{1+z}{1-z} = 1 + 2 \sum_{n=1}^\infty z^n$$

satisfies the conditions of the theorem.

Suppose now that $f(z) \in \mathscr{S}$ and $f(K)$ is starlike with respect to the origin. Consider also the restricted range $f_r(K)$: the set of values assumed by $f(z)$ in the disk $|z| \leq r$ where $0 < r < 1$. It is clear that $f(K)$ is starlike if each of the restricted ranges $f_r(K)$ has this property. It is an easy matter to find a condition that will ensure that each $f_r(K)$ is starlike with respect to the origin, and it may be shown that this condition is also necessary in order that $f(K)$ be starlike. We shall prove the sufficiency but omit the necessity of the condition.

THEOREM 17.4.11.   *A (necessary and) sufficient condition that $f(K)$ be*
starlike with respect to the origin when $f(z) \in \mathscr{S}$ is that for each z with $|z| < 1$
we have

(17.4.30)
$$\Re\left[z\frac{f'(z)}{f(z)} \right] > 0.$$

Proof.   Since $f(z) \in \mathscr{S}$ by assumption, we can define

$$\arg f(z) \equiv \Theta(r, \theta), \quad z = re^{i\theta},$$

uniquely by analytic continuation and the initial condition

$$\Theta(0, \theta) = \theta.$$

This function $\Theta(r, \theta)$ is differentiable and

(17.4.31)
$$\frac{\partial}{\partial\theta}\,\Theta(r,\theta) = \frac{\partial}{\partial\theta}\,\Im[\log f(z)] = \Im\left[\frac{f'(z)}{f(z)}\frac{\partial z}{\partial\theta}\right]$$
$$= \Im\left[iz\frac{f'(z)}{f(z)}\right] = \Re\left[z\frac{f'(z)}{f(z)}\right].$$

By assumption this expression is positive, that is, $\Theta(r, \theta)$ is an increasing
function of θ for fixed values of r. Hence, when z describes the circle $|z| = r$
once in positive sense, then $w = f(z)$ traces a curve C_r once in positive sense.

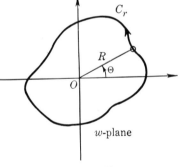

Figure 18

See Figure 18. Moreover, this curve has an equation in polar coordinates of the
form

$$R = F(\Theta)$$

where $F(\Theta)$ is a single-valued analytic function of Θ of period 2π. Every point
of such a curve is "visible" from the origin, and this implies that the set of all
points w which lie on the curves C_s with $0 \le s \le r < 1$ is starlike with respect
to the origin. But this is the restricted range $f_r(K)$, so the theorem is proved.

Combining the last two theorems we get

THEOREM 17.4.12.   *If $f(z) \in \mathscr{S}$ and $f(K)$ is starlike with respect to the origin, then for all n*

(17.4.32) $$|a_n| \leq n.$$

Proof. We shall prove that condition (17.4.30) implies (17.4.32). Suppose that

(17.4.33) $$z \frac{f'(z)}{f(z)} = 1 + \sum_{n=1}^{\infty} c_n z^n.$$

Then by Theorem 17.4.10 we have $|c_n| \leq 2$ for all n. We multiply both sides of (17.4.33) by $f(z)$, substitute the power series for $f(z)$, and compare coefficients. In this way we get the recurrence formulas

$$a_2 = c_1,$$
$$2a_3 = c_2 + a_2 c_1,$$
$$\cdots \cdots \cdots \cdots \cdots \cdots \cdots \cdots \cdots \cdots$$
$$(n-1)a_n = c_{n-1} + a_2 c_{n-2} + \cdots + a_{n-1} c_1,$$
$$\cdots \cdots \cdots \cdots \cdots \cdots \cdots \cdots \cdots \cdots$$

We have clearly $|a_2| \leq 2$, and (17.4.32) follows by induction on n.

The estimates (17.4.28) and (17.4.32) are evidently both sharp since they are reached for

$$f(z) = \frac{z}{(1-z)^2},$$

which belongs to both classes.

Finally we note the following theorem, the proof of which is left to the reader.

THEOREM 17.4.13.   *If $f(z) \in \mathscr{S}$ and $f(K)$ is convex, then*

(17.4.34) $$|a_n| \leq 1.$$

The estimate is sharp, for $f(z) = (1-z)^{-1}$ maps K onto the half-plane $\Re(w) > \frac{1}{2}$, which is a convex set.

EXERCISE 17.4

1. Show that the function

$$w = z(1 + z^{-3})^{\frac{2}{3}}$$

belongs to the class $\mathscr{U}$ and that it maps Y on the w-plane cut along the three line segments $[0, \omega^k]$ where $\omega = \exp(\frac{2}{3}\pi i)$ and $k = 0, 1, 2$. [This is an extremal function for the problem of majorizing the coefficient α_2 in the class $\mathscr{U}$. In the present case $|\alpha_2| = \frac{2}{3}$.]

2. If ν is real, prove that the function

$$z(1 - z)^{-\nu}$$

belongs to $\mathscr{S}$ if and only if $0 \leq \nu \leq 2$, and that for ν in this range $f(K)$ is starlike with respect to the origin. (*Hint:* Discuss the derivative.)

3. If $f(z) \in \mathscr{S}$ and $f(K)$ is convex, prove that

$$\Re\left[z\frac{f''(z)}{f'(z)}\right] > -1, \quad z \in K.$$

4. Under the same assumptions show that the mapping defined by $zf'(z)$ is starlike with respect to the origin.

5. Prove Theorem 17.4.13.

6. If C_r is the image of the circle $|z| = r$ under the mapping $w = f(z)$, prove that the curvature of C_r at the point $w = f(z)$ is given by

$$\frac{1}{\rho} = \frac{1}{|zf'(z)|} \Re\left[1 + z\frac{f''(z)}{f'(z)}\right].$$

7. Take $f(z) = z(1 - z)^{-2}$ and prove that the curvature is negative at the point $w = f(-r)$ if $r > 2 - \sqrt{3}$.

8. If $f(z) \in \mathscr{S}$, prove that $f_r(K)$ is convex provided $0 < r \leq 2 - \sqrt{3}$. In view of Problem 7 the constant is sharp. (The German geometer Eduard Study (1862–1930) in 1913 introduced the notion of *radius of convexity* (German *Rundungsschranke*) for a power series. In the class $\mathscr{S}$ every element has a radius of convexity which is $\geq 2 - \sqrt{3}$. This bound was discovered by R. Nevanlinna in 1920.)

9. Verify (17.4.4).

17.5. The boundary problem. We return now to conformal mapping proper and plan to pay some attention to the problem of the correspondence between the boundaries. We shall work with the inverse mapping function, that is, the function, suitably normalized, which maps the open unit disk K onto the simply-connected domain D. If we assign

$$f(0) = w_0 \in D, \quad f'(0) > 0,$$

the function $f(z)$ is uniquely defined, as we saw in Section 17.1. We shall impose various conditions on D as we go along. Without restricting the generality we may suppose that the area of D is finite. Here area means the two-dimensional Lebesgue measure of D.

If the mapping function is

(17.5.1)
$$f(z) = \sum_{n=0}^{\infty} a_n z^n,$$

the area condition implies and is implied by the convergence of the series

(17.5.2)
$$\sum_{n=1}^{\infty} n \, | \, a_n \, |^2.$$

This condition, in conjunction with the Riesz-Fischer theorem (compare Theorem 17.2.6), ensures that

(17.5.3)
$$\sum_{n=0}^{\infty} a_n e^{ni\theta}$$

is the Fourier series of a function in $L_2(0, 2\pi)$. We denote this function by $f(e^{i\theta})$ or $F(\theta)$. The first symbol suggests that we are dealing with the boundary values of $f(z)$ on the unit circle. We shall try to find out in what sense this is the case. The first bit of evidence is furnished by

LEMMA 17.5.1. $F(\theta)$ is the limit in the mean of order 2 of $f(re^{i\theta})$, that is,

(17.5.4)
$$\lim_{r \to 1} \int_0^{2\pi} | \, f(re^{i\theta}) - F(\theta) \, |^2 \, d\theta = 0.$$

Proof.   Since the value of the integral equals

$$2\pi \sum_{n=1}^{\infty} | \, a_n \, |^2 (1 - r^{2n}),$$

the conclusion is immediate.

It is natural to assume that we have also pointwise convergence of $f(re^{i\theta})$ to $F(\theta)$, at least almost everywhere. This is indeed the case and was proved in 1906 by Pierre Fatou (1878–1929) in his Paris dissertation. Fatou worked with bounded holomorphic functions, not necessarily univalent, but the result is true under more general boundedness conditions. In order to prove this theorem we have to introduce the Poisson integral and we also have to recall some properties of the Lebesgue integral.

We suppose that $f(z)$ is holomorphic in K, $f \in H[K]$ in our usual notation, not necessarily univalent. The coefficients $\{a_n\}$ must satisfy the condition

(17.5.5)
$$\sum_{n=0}^{\infty} | \, a_n \, |^2 < \infty,$$

which is weaker than (17.5.2). Of course, (17.5.5) is the precise condition that will ensure that (17.5.3) defines a function $F(\theta)$ in $L_2(0, 2\pi)$. This class of analytic functions is often denoted by H_2, the "H" referring to G. H. Hardy (1877–1947).

If $0 \leq |z| = r < a < R < 1$, then

$$f(z) = \frac{R}{2\pi} \int_0^{2\pi} \frac{f(Re^{i\varphi})e^{i\varphi}}{Re^{i\varphi} - z} \, d\varphi.$$

Here we can let R tend to 1, for the integral with R replaced by 1 exists and the difference of the two integrals equals

$$\frac{z}{2\pi}(1 - R) \int_0^{2\pi} \frac{f(Re^{i\varphi})e^{i\varphi}}{(Re^{i\varphi} - z)(e^{i\varphi} - z)} \, d\varphi + \frac{1}{2\pi} \int_0^{2\pi} \frac{f(Re^{i\varphi}) - F(\varphi)}{e^{i\varphi} - z} e^{i\varphi} \, d\varphi.$$

The absolute value of the first term does not exceed

$$(2\pi)^{-\frac{1}{2}}(1 - R)(R - a)^{-1}(1 - a)^{-1} \left\{ \sum_{n=0}^{\infty} |a_n|^2 \right\}^{\frac{1}{2}},$$

while the second term is dominated by

$$(2\pi)^{-\frac{1}{2}}(1 - a)^{-1} \left\{ \int_0^{2\pi} |f(Re^{i\varphi}) - F(\varphi)|^2 \, d\varphi \right\}^{\frac{1}{2}},$$

and both expressions tend to zero as $R \to 1$. It follows that

$$(17.5.6) \qquad f(z) = \frac{1}{2\pi} \int_0^{2\pi} \frac{F(\varphi) \, d\varphi}{1 - e^{-i\varphi}z}.$$

Thus every function in H_2 is representable as a Cauchy integral in terms of its own boundary values (in the L_2-sense). For our purposes the representation in terms of a Poisson integral offers advantages since the Poisson kernel is real and positive. To obtain this representation we note that

$$\int_0^{2\pi} F(\varphi)e^{ni\varphi} \, d\varphi = 0, \quad n = 1, 2, 3, \cdots.$$

Hence

$$0 = \frac{1}{2\pi} \int_0^{2\pi} \frac{F(\varphi)e^{i\varphi}\bar{z} \, d\varphi}{1 - e^{i\varphi}\bar{z}},$$

and if we add this expression to (17.5.6) and simplify we get

$$(17.5.7) \qquad f(re^{i\theta}) = \frac{1}{2\pi} \int_0^{2\pi} \frac{(1 - r^2)F(\varphi) \, d\varphi}{1 - 2r\cos(\theta - \varphi) + r^2}.$$

This is Poisson's formula (Siméon Denis Poisson, 1781–1840, produced a faulty convergence proof for Fourier series based on such formulas in 1820).

We set

$$(17.5.8) \qquad P(t; r) = \frac{1}{2\pi} \frac{1 - r^2}{1 - 2r\cos t + r^2},$$

so that (17.5.7) becomes

$$(17.5.9) \qquad f(re^{i\theta}) = \int_0^{2\pi} F(\varphi)P(\theta - \varphi; r) \, d\varphi.$$

The Poisson kernel $P(t; r)$ has a number of noteworthy properties, among which we list the following:

(17.5.10)

    (i) $P(t; r) > 0, 0 \leqq r < 1$, and all t;

    (ii) $\displaystyle\int_0^{2\pi} P(t; r)\, dt = 1$ for all r;

    (iii) $\displaystyle P(t; r) < \frac{\pi(1 - r)}{\pi^2(1 - r)^2 + 4t^2}$, $|t| \leqq \tfrac{1}{2}\pi$; and

    (iv) $\displaystyle\lim_{r \to 1} P(t; r) = 0, t \not\equiv 0 \pmod{2\pi}$, the convergence being uniform with respect to t in $0 < \delta \leqq t \leqq 2\pi - \delta$.

Here (ii) follows from (17.5.9) upon setting $f(z) \equiv 1$. The other relations follow from

$$0 < 1 - r^2 < 2(1 - r),$$
$$1 - 2r \cos t + r^2 = (1 - r)^2 + 4 \sin^2 \tfrac{1}{2}t$$
$$> (1 - r)^2 + \frac{4}{\pi^2} t^2 > 0.$$

For the discussion of the Poisson integral we need some properties of the Lebesgue integral. The reader is probably familiar with the basic fact that if $g(t) \in L(a, b)$, then

$$(17.5.11) \qquad \frac{d}{dt} \int_0^t g(s)\, ds = g(t)$$

for almost all t.[1] A fairly direct consequence of this proposition is the following:

LEMMA 17.5.2.   *Let* $g(t) \in L(a, b)$. *For almost all* t_0 *in* $[a, b]$ *the limit*

$$(17.5.12) \qquad \lim_{h \to 0} \frac{1}{h} \int_{t_0}^{t_0 + h} |g(s) - g(t_0)|\, ds$$

exists and is equal to zero, h being restricted to be such that $t_0 + h$ is also in $[a, b]$.

Proof.   We can obviously restrict ourselves to the case in which $g(t)$ is real-valued. Let $\{r_n\}$ be an ordering of the rationals and set

$$G_n(t) = \int_a^t |g(s) - r_n|\, ds.$$

Then there exists a set $N_n \subset [a, b]$ of measure zero such that if t is not in N_n,

---

then $G_n(t)$ has a derivative which is finite and equal to $\mid g(t) - r_n \mid$. The set $N = \cup N_n$ also has measure zero. Now let t_0 be any point in $[a, b] \ominus N$. There is a rational number r_n such that $\mid g(t_0) - r_n \mid < \frac{1}{2}\varepsilon$ where $\varepsilon > 0$ is given. Then t_0 is not in N_n, so the function $G_n(t)$ has at t_0 a derivative equal to $\mid g(t_0) - r_n \mid$. It follows that

$$0 \leq \frac{1}{h} \int_{t_0}^{t_0+h} \mid g(s) - g(t_0) \mid ds$$

$$\leq \frac{1}{h} \int_{t_0}^{t_0+h} \mid g(s) - r_n \mid ds + \frac{1}{h} \int_{t_0}^{t_0+h} \mid g(t_0) - r_n \mid ds$$

$$= \frac{1}{h} [G_n(t_0 + h) - G_n(t_0)] + \mid g(t_0) - r_n \mid,$$

where h may be positive or negative. As $h \to 0$ the limit of the first term in the last member is

$$G_n'(t_0) = \mid g(t_0) - r_n \mid < \frac{1}{2}\varepsilon.$$

It follows that the second member of the inequality has a superior limit not exceeding ε, and since ε is arbitrary the lemma is proved.

We can now state and prove the main limit theorem for the Poisson integral.

THEOREM 17.5.1.   *If $g(t) \in L(-\pi, \pi)$ and if*

(17.5.13) $$g(t; r) = \int_{-\pi}^{\pi} g(s + t)P(s; r) \, ds,$$

then

(17.5.14) $$\lim_{r \to 1} g(t; r) = g(t)$$

for almost all values of t. In particular, the limit relation holds at all points of continuity of $g(t)$ and uniformly with respect to t in any closed interval of continuity.

Proof. Let us denote by E the set of points t_0 in $(-\pi, \pi]$ for which the limit in (17.5.12) exists and equals zero. E is often referred to as the Lebesgue set of the function $g(t)$. We shall prove that (17.5.14) holds for $t = t_0 \in E$. We set

$$G(t; t_0) = \int_0^t \mid g(t_0 + s) - g(t_0) \mid ds.$$

Suppose that $t_0 \in E$ and let $\varepsilon > 0$ be given. Then there is a $\delta = \delta(t_0, \varepsilon)$ so small that

$$\mid G(t; t_0) \mid \leq \varepsilon \mid t \mid \quad \text{if} \quad \mid t \mid \leq \delta.$$

We have

$$g(t_0; r) - g(t_0) = \int_{-\pi}^{\pi} [g(t_0 + s) - g(t_0)]P(s; r) \, ds$$

by (17.5.10), part (i). We divide the interval of integration into three parts $(-\pi, -\delta), (-\delta, \delta), (\delta, \pi)$ and denote the corresponding integrals by I_1, I_2, and I_3 respectively.

In the first interval $P(s; r)$ has its biggest value at the right endpoint so that

$$| I_1 | \leq P(\delta; r) \int_{-\pi}^{-\delta} | g(t_0 + s) - g(t_0) | \, ds,$$

and a similar estimate holds for I_3. This gives

$$| I_1 | + | I_3 | < P(\delta; r)[\| g(\cdot) \| + 2\pi | g(t_0) |],$$

where the first term in the brackets is the L_1-norm of $g(\cdot)$. This expression tends to zero when $r \to 1$. The estimates hold uniformly in any interval of continuity or, more generally, in any closed interval in which $| g(t) |$ is bounded.

To handle I_2 we use (17.5.10), (iii) followed by an integration by parts. This leads to simpler expressions than a direct integration by parts involving $P(s; r)$ itself. Thus the first estimate is

$$| I_2 | \leq \rho \int_{-\delta}^{\delta} \frac{| g(t_0 + s) - g(t_0) |}{\rho^2 + 4s^2} \, ds, \quad \rho = \pi(1 - r);$$

and integration by parts then gives

$$\left[\frac{\rho G(t; t_0)}{\rho^2 + 4t^2} \right]_{-\delta}^{\delta} + 8\rho \int_{-\delta}^{\delta} \frac{G(s; t_0)s \, ds}{(\rho^2 + 4s^2)^2} .$$

By the choice of δ this expression does not exceed

$$\frac{2\rho\delta\varepsilon}{\rho^2 + 4\delta^2} + 8\rho\varepsilon \int_{-\infty}^{\infty} \frac{s^2 \, ds}{(\rho^2 + 4s^2)^2} \leq \tfrac{1}{2}(1 + \pi)\varepsilon,$$

which is an upper bound for $| I_2 |$ that can be made arbitrarily small by choosing ε small. It follows that (17.5.14) holds for $t \in E$.

In any closed interval of continuity we can choose δ independently of t_0, so the estimates hold uniformly in such an interval. This implies uniform convergence in the interval and completes the proof of the theorem.

From this theorem we obtain the theorem of Fatou as an immediate

COROLLARY. *If* $f(z) \in H_2$, *then* $\lim_{r \to 1} f(re^{i\theta})$ *exists and equals* $F(\theta)$ *for every* θ *in* E, *the Lebesgue set of* $F(\theta)$. *In particular, the limit exists at every point of continuity and uniformly with respect to* θ *in any closed interval of continuity.*

The assumption that $f(z) \in H_2$ holds, in particular, if condition (17.5.2) is satisfied, that is, if the area of $D = f[K]$ is finite. But in this case we can prove a stronger result with the aid of Fejér's lemma (Lemma 11.5.2), namely,

THEOREM 17.5.2.   *If $f(z)$ maps K conformally onto a domain D of finite area, then the series (17.5.3) converges to $F(\theta)$ everywhere in E. Moreover, the convergence is uniform in any closed interval of continuity.*

We shall now study the relations between the set

(17.5.15) $$B = [F(\theta) \mid \theta \in E]$$

and ∂D, the boundary of D. We start by proving

LEMMA 17.5.3.   *If $\{z_n\} \subset K$ and if $\lim\limits_{n \to \infty} |z_n| = 1$, then any limit point of the sequence $\{f(z_n)\}$ belongs to ∂D. In particular, $B \subset \partial D$, and $B = \partial D$ if $f(z)$ is continuous in $\bar{K}$.*

Proof.   Suppose that the first assertion is false and that there is a point $w_0 \in D$ and a sequence $\{z_n\} \subset K$ with $\lim\limits_{n \to \infty} |z_n| = 1$ such that $\lim f(z_n) = w_0$. Suppose $f(z_0) = w_0$. Since the mapping is one-to-one, we can take a small neighborhood $N(w_0)$ of w_0 and find its inverse image, which is a small neighborhood $N(z_0)$ of z_0. Since ultimately $f(z_n) \in N(w_0)$, the points z_n must ultimately belong to $N(z_0)$. This is absurd since $|z_n| \to 1$. It follows that every limit point belongs to ∂D and, in particular, $B \subset \partial D$.

Suppose now that $f(z)$ is continuous in $\bar{K}$. This implies and is implied by the series (17.5.3) being the Fourier series of a continuous function namely $F(\theta)$. Suppose that $\beta \in \partial D$. Then there exists a sequence $\{z_n\}$ with $\lim |z_n| = 1$ such that $f(z_n) \to \beta$. A subsequence of $\{z_n\}$ must converge to a point z_0 on ∂K. By the continuity of $f(z)$ we have $f(z_0) = \beta$, and, thus, there exists a $\theta_0 = \arg z_0$ with $f(e^{i\theta_0}) = \beta$. Thus, $\partial D \subset B$ and, hence, $B = \partial D$. This completes the proof.

Conversely, $f(z)$ is continuous in K if the boundary of D is sufficiently simple. It suffices that ∂D is the union of a finite number of Jordan arcs. We shall discuss only the simplest case, that in which the boundary is a simple closed curve. We need some preliminary results for the proof.

LEMMA 17.5.4.   *If $F(z)$ is holomorphic and bounded in K and if $F(z)$ tends to a constant limit when z approaches an arc of the unit circle, then $F(z)$ is a constant.*

Proof.   We may assume that the constant limit is 0 and that the arc has a length $> 2\pi/n$. Let $\omega = \exp(2\pi i/n)$ and form

$$G(z) \equiv F(z)F(\omega z)F(\omega^2 z) \cdots F(\omega^{n-1} z).$$

This function is holomorphic and bounded in K and tends uniformly to the limit 0 as $|z_n| \to 1$. By the principle of the maximum, $G(z) \equiv 0$ and, hence, $F(z) \equiv 0$ as asserted.

LEMMA 17.5.5. *If*

$$\sum_{n=0}^{\infty} c_n e^{ni\theta}$$

is the Fourier series of a bounded function $G(\theta)$, *then* $G(\theta)$ *cannot have a simple discontinuity anywhere. In particular, if* $G(\theta)$ *is of bounded variation, then* $G(\theta)$ *is continuous.*

Proof. Suppose that this is not true and that, for instance, $\Re[G(\theta)]$ has a simple discontinuity at a point which we may assume to be $\theta = 0$. We shall show that this implies proper divergence of the Fourier series of the imaginary part at $\theta = 0$. This is impossible in view of the boundedness of $G(\theta)$. The behavior of the series at $\theta = 0$ is typified by the series

$$i \log (1 - e^{i\theta}) = -i \sum_{n=1}^{\infty} \frac{1}{n} e^{ni\theta} \equiv A(\theta) + iB(\theta).$$

Here

$$A(\theta) = \tfrac{1}{2} \operatorname{sgn} \theta \cdot (\pi - |\theta|), \quad -\pi < \theta < \pi,$$

has the saltus π at $\theta = 0$. Set $G(\theta) = g_1(\theta) + ig_2(\theta)$ and suppose that $g_1(\theta)$ has the saltus δ at $\theta = 0$. Then

$$g_1(\theta) - \frac{\delta}{\pi} A(\theta) \equiv h_1(\theta)$$

is continuous at $\theta = 0$. Set

$$g_2(\theta) - \frac{\delta}{\pi} B(\theta) \equiv h_2(\theta).$$

Now if

$$h_1(\theta) \sim \sum_{-\infty}^{\infty} \alpha_n e^{ni\theta}, \quad \text{then} \quad h_2(\theta) \sim -i \sum_{-\infty}^{\infty} \alpha_n \operatorname{sgn} n \; e^{ni\theta}.$$

The nth partial sum of the second series is expressible in terms of $h_1(t)$ and the conjugate Dirichlet kernel

(17.5.16) $$S_n[h_2; \theta] = -\frac{1}{\pi} \int_{-\pi}^{\pi} h_1(t) \tilde{D}_n(t - \theta) \, dt,$$

where

$$\tilde{D}_n(v) = \sum_{k=1}^{n} \sin kv = [\cos \tfrac{1}{2}v - \cos (n + \tfrac{1}{2})v][2 \sin \tfrac{1}{2}v]^{-1}.$$

We note that

(17.5.17) $$|\tilde{D}_n(v)| \leq \min\left[n, \frac{2\pi}{|v|}\right].$$

We have

$$S_n[g_2; 0] = S_n\left[g_2 - \frac{\delta}{\pi} B; 0\right] + \frac{\delta}{\pi} S_n[B; 0].$$

We desire to prove that

(17.5.18) $$\lim_{n \to \infty} \frac{S_n[g_2; 0]}{\log n} = -\frac{\delta}{\pi}.$$

Since

$$S_n[B; 0] = -\sum_{k=1}^{n} \frac{1}{k} = -\log n + O(1),$$

it suffices to show that

$$S_n\left[g_2 - \frac{\delta}{\pi} B; 0\right] = o(\log n).$$

To this end we use (17.5.16). Here $h_1(t) = g_1(t) - \frac{\delta}{\pi} A(t)$ is bounded and continuous at $t = 0$. Without restricting the generality we may suppose that $h_1(0) = 0$. Using (17.5.17) we then get

$$|S_n[h_2; 0]| \leq \frac{1}{\pi}\left\{\int_{-\pi}^{-\pi/n} + \int_{-\pi/n}^{\pi/n} + \int_{\pi/n}^{\pi}\right\} |h_1(t)| \, |\tilde{D}_n(t)| \, dt$$

$$\leq \frac{n}{\pi}\int_{-\pi/n}^{\pi/n} |h_1(t)| \, dt + 2\left\{\int_{-\pi}^{-\pi/n} + \int_{\pi/n}^{\pi}\right\} |h_1(t)| \frac{dt}{t}$$

$$= o(1) + o(\log n) = o(\log n)$$

as asserted. This implies (17.5.18) and contradicts the boundedness of $G(\theta)$. Hence $G(\theta)$ cannot have any simple discontinuities.

Since a function of bounded variation can have simple discontinuities only, we conclude that if $G(\theta)$ is of bounded variation, then $G(\theta)$ is continuous. Actually $G(\theta)$ is absolutely continuous in this case.

We can now state and prove

THEOREM 17.5.3. *Let D be the interior of a simple closed curve C and let $f(z)$ be a function which maps K conformally onto D. Then $f(z)$ is continuous in $\overline{K}$, the series (17.5.1) converges uniformly in $\overline{K}$, and the correspondence between $\overline{K}$ and $\overline{D}$ is one-to-one and bicontinuous.*

Proof. We may assume that $f(0) = 0, f'(0) > 0$. Let

$$C: \quad w = b(t),$$

where $b(t)$ is a continuous function of the real variable t of period 2π such that $\arg b(t)$ increases by 2π when t goes from 0 to 2π. Since C is a Jordan curve, $b(t_1) = b(t_2)$ implies $t_1 = t_2$ if $0 \leq t_2 - t_1 < 2\pi$.

Now consider the radial images

$$C(\theta): \quad w = f(re^{i\theta}), \quad 0 \leq r < 1, \theta \text{ fixed}.$$

These arcs emanate from the origin, where $C(\theta)$ is tangent to $\arg w = \theta$. If $\theta \in E$, then $C(\theta)$ has a definite endpoint $F(\theta)$ on C. No two arcs $C(\theta_1)$ and $C(\theta_2)$ can have a point in common except for the origin and, possibly, also the endpoint on C. The existence of such a common point would violate the univalency of $f(z)$.

But the assumption that $C(\theta_1)$ and $C(\theta_2)$ have the same endpoint w_0 on C is also untenable. For if this were true, then $C(\theta_1)$ and $C(\theta_2)$ would together form the boundary of a curvilinear lens having only one point w_0 in common with C. All the arcs $C(\theta)$ with $\theta_1 < \theta < \theta_2$ are confined to this lens. By Lemma 17.5.3 these arcs must have limit points on C, and since w_0 is the only point available, we conclude that

$$\lim_{r \to 1} f(re^{i\theta}) = w_0, \quad \theta_1 \le \theta \le \theta_2.$$

By Lemma 17.5.4 this implies that $f(z) \equiv w_0$, which is clearly absurd. Hence, if θ_1 and θ_2 are distinct elements of E, then $F(\theta_1) \ne F(\theta_2)$.

For every $\theta \in E$ there is a definite value of t, $t = t(\theta)$, such that

$$(17.5.19) \qquad F(\theta) = b(t(\theta)),$$

and we have just seen that $t(\theta_1) \ne t(\theta_2)$ if $0 < \theta_2 - \theta_1 < 2\pi$.

The arcs $C(\theta)$ form an ordered family in the following sense: The curve

$$w = f(re^{i\theta}), \quad 0 \le \theta < 2\pi, r \text{ fixed}, 0 < r < 1,$$

intersects each curve $C(\theta)$ once and only once; if θ increases from 0 to 2π, then $C(\theta_1)$ is encountered before $C(\theta_2)$ if $\theta_1 < \theta_2$, and $C(\theta_3)$ lies between $C(\theta_1)$ and $C(\theta_2)$ if $\theta_1 < \theta_3 < \theta_2$. This is independent of r. Hence the ordering must also extend to the boundary; that is, $F(\theta_1)$ precedes $F(\theta_2)$ on C, and $F(\theta_3)$ lies between $F(\theta_1)$ and $F(\theta_2)$ if all three values of θ are in E. This means that the function $t(\theta)$ of (17.5.19) is strictly increasing.

Next we prove that the set B is dense on C. If this were not the case, then there would exist an arc of C without any points $F(\theta)$, and this would imply that the increasing function $t(\theta)$ has a simple discontinuity. But then the bounded periodic function

$$b(t(\theta)) = F(\theta) \sim \sum_{n=0}^{\infty} a_n e^{ni\theta}$$

would also have a simple discontinuity, and this is excluded by Lemma 17.5.5. Hence $t(\theta)$ is continuous and B is dense on C.

We are now practically through. We have to prove that $f(z)$ can be defined as a continuous function on the unit circle. We have already $f(e^{i\theta}) = F(\theta)$ in E, that is, for almost all θ. Suppose θ_0 is not in E and take a sequence of nested intervals $[\theta_{n1}, \theta_{n2}]$ such that $\theta_{n1}, \theta_{n2} \in E$, $\theta_{n1} < \theta_0 < \theta_{n2}$, and $\theta_{n2} - \theta_{n1} \to 0$. But then there exists a corresponding sequence of intervals $[t_{n1}, t_{n2}]$ such that $t_{n1} = t(\theta_{n1})$, $t_{n2} = t(\theta_{n2})$. Since $t(\theta)$ is continuous on E, the nested t-intervals converge to a unique point t_0 which is then the value of $f(e^{i\theta_0}) = F(\theta_0)$. Thus $F(\theta)$ is actually continuous in $[0, 2\pi]$. This makes $f(z)$ continuous in $\bar{K}$ and induces uniform convergence of (17.5.1) in $\bar{K}$. We have seen that the correspondence is one-to-one also on the boundaries. Finally, since $f(z)$ is continuous and one-to-one in the closed set $\bar{K}$, its inverse is also continuous. Thus, the mapping is bicontinuous. This completes the proof.

There are many other important properties of the mapping function which are worth mentioning. We list some here without proofs; for the proofs we refer to Zygmund's treatise cited in the Collateral Reading.

THEOREM 17.5.4.   *If the boundary C of D is rectifiable and of length L, then* (i) *the series* (17.5.1) *for the mapping function converges absolutely on* $|z| = 1$ *and*

$$(17.5.20) \qquad\qquad \sum_{n=0}^{\infty} |a_n| \leq \tfrac{1}{2}L;$$

(ii) *the function* $F(\theta) = f(e^{i\theta})$ *is absolutely continuous; and* (iii) *sets of measure zero on* $|z| = 1$ *correspond to sets of measure zero on* C, *and conversely.*

THEOREM 17.5.5.   *If the boundary curve C has a tangent at a point $w_0 \in C$, then the mapping $w = f(z)$ is angle preserving at the point z_0 which corresponds to w_0.*

EXERCISE 17.5

1. If $g(t) \in L_2(0, 2\pi)$, prove that

$$\lim_{h \to 0} \int_0^{2\pi} |g(t + h) - g(t)|^2 \, dt = 0$$

(continuity in the mean).

2. If the real-valued function $g(t) \in L_2(0, 2\pi)$ and $g(t; r)$ is defined by (17.5.13), show that the harmonic conjugate of $g(t; r)$ is given by

$$h(t; r) = -\int_{-\pi}^{\pi} g(s + t) \, Q(s; r) \, ds$$

where the conjugate Poisson kernel is

$$Q(s; r) = \frac{1}{\pi} \frac{r \sin s}{1 - 2r \cos s + r^2}.$$

3. Does $h(t; r)$ approach a limit in the L_2-sense as $r \to 1$?

4. Verify (17.5.16) and (17.5.17).

5. Suppose that ∂D consists of a Jordan curve C and a Jordan arc A where one endpoint of A lies on C. Prove an analogue of Theorem 17.5.3 by an adaptation of the proof. Note that A has two sides which correspond to different arcs of the unit circle.

6. Verify that Theorem 17.5.3 holds also for the exterior mapping function. Compare Theorem 17.3.3.

17.6. Special mappings. In earlier parts of this treatise we have encountered a number of functions which enable us to solve special mapping problems. We refer in the first place to Chapters 3 and 6, where we encountered fractional linear functions, roots and powers, the function $(z^2 + 1)/(2z)$, and the exponential and trigonometric functions and their inverses. These functions are very useful in conformal mappings involving circular disks, half-planes, sectors, strips and half-strips, and regions bounded by arcs of conics. We shall give some examples involving composite transformations.

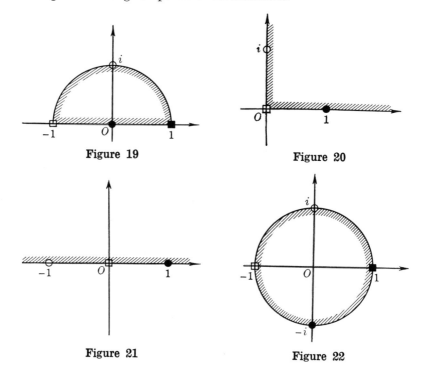

Figure 19

Figure 20

Figure 21

Figure 22

Figures 19–22 show various stages in the mapping of a semicircular disk onto a full disk. The three mapping functions are

$$(17.6.1) \qquad w_1 = \frac{1+z}{1-z}, \quad w_2 = w_1^2, \quad w_3 = \frac{w_2 - i}{w_2 + i}.$$

Here w_1 maps the upper half of K onto the first quadrant, which is taken into the upper half-plane by w_2, and, finally, w_3 maps the half-plane onto the disk K. Thus we find that

$$(17.6.2) \qquad w = \frac{(z+1)^2 - i(z-1)^2}{(z+1)^2 + i(z-1)^2}$$

maps the upper half of the unit disk into the whole disk leaving the points $1, -1,$ and i invariant.

If we replace the exponent 2 in (17.6.2) by some other positive integer n, we get a function which maps a crescent, formed by two circles meeting at an angle π/n at $z = \pm 1$, onto K.

Figure 23

The "double horn" D shown in Figure 23 is defined by the inequality

$$\left| \frac{z}{a} - 1 \right| > 1 > \left| \frac{z}{b} - 1 \right|.$$

The inversion

$$w_1 = \frac{1}{z}$$

maps D onto the strip

$$\frac{1}{2b} < \Re(w_1) < \frac{1}{2a} \, ,$$

and the strip is mapped onto the upper half-plane by the exponential function

$$w_2 = \exp\left\{ \frac{2ab}{b-a} \, \pi i \left(w_1 - \frac{1}{2b} \right) \right\}.$$

Finally

(17.6.3) $$w = i \tan \left\{ \frac{ab}{b-a} \, \pi \left(\frac{1}{z} - \frac{1}{2b} \right) - \frac{\pi}{4} \right\}$$

maps D onto K.

The function

$$w = \frac{1}{2}\left(z + \frac{1}{z} \right)$$

gives a correspondence between curvilinear quadrilaterals bounded in the z-plane by rays and concentric circles and in the w-plane by arcs of confocal conics. We can map

$$R_1 < |z| < R_2, \quad \alpha < \arg z < \beta$$

on a rectangle by means of a logarithm, but to map the rectangle on a half-plane we need elliptic functions. Compare Figures 8 and 9, Section 12.5. If $\omega_1 > 0$, $-i\omega_3 > 0$, the Weierstrass $\wp$-function

$$\wp(z; \omega_1, \omega_3)$$

maps the rectangle with vertices at $z = 0$, ω_1, $\omega_1 + \omega_3$, ω_3 (that is, one quarter of the period parallelogram) onto the upper half of the w-plane.

This is a special case of the mapping of the interior of a polygon on a half-plane. Normally the mapping function is not single-valued and it is easier to find explicit expressions for the inverse mapping function which takes the half-plane into the polygonal region. The correspondence is given by the so-called Schwarz-Christoffel formulas [Elwin Bruno Christoffel (1829–1900)]. Formula (17.6.4) below was known to Schwarz in 1864, and was published by Christoffel in 1867 and by Schwarz in 1866 and 1869.

Let there be given a simple closed polygon $A_1, A_2, \cdots, A_n, A_1$, where the numbering of the vertices agrees with the positive orientation, and let $\alpha_1\pi$, $\alpha_2\pi, \cdots, \alpha_n\pi$ be the corresponding interior angles. The α's satisfy the inequalities

$$0 < \alpha_j < 2, \quad \sum_{j=1}^{n} \alpha_j = 2(n - 2).$$

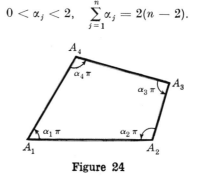

Figure 24

See Figure 24, where $n = 4$. Let $w = f(z)$ be a function holomorphic in the upper half-plane which maps $\Im(z) > 0$ conformally on the interior Π_i of the polygon Π. Then the points $A_1, A_2, \cdots, A_n$ correspond to points $a_1, a_2, \cdots, a_n$ on the real axis and

$$a_1 < a_2 < a_3 < \cdots < a_n.$$

We can assign three of these points, say the first three; this choice determines $f(z)$ uniquely, so that the remaining points $a_4, \cdots, a_n$ are uniquely determined functions of a_1, a_2, a_3 and the configuration Π. We note that $f(z)$ is continuous on the real axis by Theorem 17.5.3.

For the case of a rectangle, we know that $f(z)$ is an elliptic integral. Compare formula (12.5.21), where $a_1 = e_1$, $a_2 = e_2$, $a_3 = e_3$, and $a_4 = \infty$. This suggests that the general solution is of the form

$$(17.6.4) \quad f(z) = C_1 \int_{a_1}^{z} (t - a_1)^{\alpha_1 - 1}(t - a_2)^{\alpha_2 - 1} \cdots (t - a_n)^{\alpha_n - 1} \, dt + C_2$$

where C_1 and C_2 are constants determined by Π. A simple calculation shows that if this is true then

$$(17.6.5) \qquad g(z) \equiv \frac{f''(z)}{f'(z)} = \sum_{j=1}^{n} \frac{\alpha_j - 1}{z - a_j}$$

is a single-valued function of z whose only singularities are simple poles at the points $\{a_j\}$. Conversely, if we can show that the mapping function $f(z)$ satisfies (17.6.5), then it is necessarily of the form (17.6.4).

There are two things to be proved: (i) $g(z)$, the logarithmic derivative of $f'(z)$, exists in the whole plane as a single-valued function of z having no other singularities than the points $\{a_j\}$; (ii) each point a_j is a simple pole of $g(z)$ of residue $\alpha_j - 1$, and $g(z)$ vanishes at $z = \infty$.

The problem in the large can be solved with the aid of Schwarz's reflection principle, which was formulated by Schwarz, with this particular problem in mind, in his paper of 1869. The principle tells us that $f(z)$ is holomorphic on each of the line segments (a_j, a_{j+1}) which correspond to the sides (A_j, A_{j+1}). Moreover, we can cross any particular segment (a_j, a_{j+1}) and define $f(z)$ in the lower half-plane. The values taken on by $f(z)$ in the lower half-plane form the interior of the polygon Π_1 obtained by reflecting Π in the side (A_j, A_{j+1}). Using the reflection principle again, we see that we can cross each of the intervals (a_j, a_{j+1}) from below to get back into the upper half-plane and that the values which the new determination of $f(z)$ takes in $\Im(z) > 0$ form the interior of a polygon Π_2 obtained by reflecting Π_1 in one of the sides of Π_1. See Figure 25.

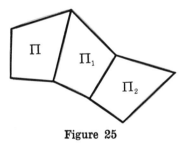

Figure 25

Let $f_1(z)$ be this new determination. Since Π and Π_2 are congruent, we see that

$$(17.6.6) \qquad f_1(z) = Af(z) + B$$

where A and B are constants and $|A| = 1$. This simply expresses that Π is carried into Π_2 by a translation and a rotation.

We can continue this process indefinitely, reflecting alternately the upper and the lower half-planes in the line segments (a_j, a_{j+1}). After an even number of reflections we are back with z at the starting point. Each reflection of a half-plane in the real axis corresponds to a reflection of a polygon in one of its sides. An even number of such reflections leads to a polygon which is congruent to Π, that is, to a determination of $f(z)$ which satisfies an equation of type (17.6.6). This implies that

$$\frac{f_1''(z)}{f_1'(z)} = \frac{f''(z)}{f'(z)} ;$$

that is, $g(z)$ is single-valued and its only singularities are the points $a_1, a_2, \cdots, a_n$ which correspond to the vertices. To this set we may have to add the point at infinity, if it is not already one of the points a_j. This finishes step (i).

We come now to the local problem, the nature of the singularities. Here the fact that A_j is a vertex of Π with interior angle $\alpha_j \pi$ is decisive. In the neighborhood of $z = a_j$ we must therefore have

$$f(z) = f(a_j) + (z - a_j)^{\alpha_j} h_j(z), \quad \Im(z) > 0,$$

where $h_j(z)$ is holomorphic and tends to a finite limit, $\neq 0$, when $z \to a_j$. This expresses the fact that a small semicircular disk $|z - a_j| < \rho$, $0 < \Im(z)$, is mapped on a small sector of opening $\alpha_j \pi$ at $w = A_j$. This gives

$$g(z) = \frac{\alpha_j - 1}{z - a_j} + \varphi_j(z), \quad 0 < \Im(z),$$

where $\varphi_j(z)$ is holomorphic and tends to a finite limit as $z \to a_j$. The limit is also finite for approach along the real axis. Now $g(z)$ is holomorphic in a punctured neighborhood of $z = a_j$, and the same is true for the first term on the right and, hence, also for $\varphi_j(z)$. Reflection in the real axis shows that $\varphi_j(z)$ must also be bounded in the lower half of the punctured neighborhood, and this implies that $z = a_j$ is a removable singularity of $\varphi_j(z)$. Thus

$$g(z) - \frac{\alpha_j - 1}{z - a_j} = \varphi_j(z)$$

is holomorphic in a full neighborhood of $z = a_j$.

It follows that

$$g(z) - \sum_{j=1}^{n} \frac{\alpha_j - 1}{z - a_j} \equiv \Phi(z)$$

is holomorphic in the finite plane. Here we have tacitly assumed that a_n is finite. The point $z = \infty$ is then mapped on an interior point of one of the sides so that $f(z)$ is holomorphic at infinity. This makes $g(z)$ holomorphic at $z = \infty$ and equal to 0 there. Hence $\Phi(z) \equiv 0$. It follows that (17.6.5) holds and hence also (17.6.4).

The actual determination of the remaining parameters in (17.6.4) is usually rather laborious unless the figure has a high degree of symmetry. We restate the result as

THEOREM 17.6.1. *Formula* (17.6.4) *defines the mapping of the upper half-plane onto a polygon of n sides and with interior angles* $\alpha_1 \pi, \alpha_2 \pi, \cdots, \alpha_n \pi$ *at the vertices which correspond to the points* $a_1, a_2, \cdots, a_n$ *respectively.*

Similar formulas are valid for the mapping function of the exterior of a polygon and for improper polygons having a vertex at infinity.

Another important case for which explicit formulas are available is that in which D is bounded by arcs of circles. We have encountered special instances of this mapping problem in the theory of modular functions, Section 13.6. For a systematic study of this important problem we need another concept introduced by Schwarz in 1873, namely, the expression known nowadays as the *Schwarzian derivative* or the *Schwarzian differential parameter*.

We saw above that the expression

$$\frac{d}{dz} \log w'$$

is invariant under the linear transformations

$$w \to Aw + B.$$

We are led to the Schwarzian derivative if we look for a differential invariant for the group of all fractional linear transformations.

If

$$W = \frac{Aw + B}{Cw + D},$$

then

$$W' = \frac{AD - BC}{(Cw + D)^2} w',$$

where the prime denotes differentiation with respect to z. By logarithmic differentiation this gives

$$\frac{W''}{W'} = \frac{w''}{w'} - 2\frac{Cw'}{Cw + D}.$$

Hence

$$\left(\frac{W''}{W'}\right)' = \left(\frac{w''}{w'}\right)' + 2\left(\frac{Cw'}{Cw + D}\right)^2 - 2\frac{Cw''}{Cw + D},$$

while

$$\left(\frac{W''}{W'}\right)^2 = \left(\frac{w''}{w'}\right)^2 + 4\left(\frac{Cw'}{Cw + D}\right)^2 - 4\frac{Cw''}{Cw + D}.$$

It follows that

$$\left(\frac{W''}{W'}\right)' - \frac{1}{2}\left(\frac{W''}{W'}\right)^2 = \left(\frac{w''}{w'}\right)' - \frac{1}{2}\left(\frac{w''}{w'}\right)^2.$$

Thus, the Schwarzian derivative defined by

(17.6.7) $\{w, z\} \equiv \left(\frac{w''}{w'}\right)' - \frac{1}{2}\left(\frac{w''}{w'}\right)^2 = \frac{w'''}{(w')^2} - \frac{3}{2}\frac{(w'')^2}{(w')^3}$

has the property of being invariant under linear fractional transformations acting on the first variable:

(17.6.8) $$\{W, z\} = \{w, z\}.$$

There is also a simple transformation involving the second variable. If

(17.6.9) $$Z = \frac{az + b}{cz + d}, \quad \text{then} \quad \{w, z\} = \left(\frac{dZ}{dz}\right)^2 \{w, Z\}.$$

A basic property of the Schwarzian derivative is given by

THEOREM 17.6.2.   *Let D be a simply-connected domain, $z_0 \in D$, let $p(z)$ be holomorphic in D, and let $w(z)$ be a solution of the third-order nonlinear differential equation*

(17.6.10) $$\{w, z\} = 2p(z)$$

which is holomorphic in some neighborhood of $z = z_0$. Then there exist two linearly independent solutions $u_1(z)$ and $u_2(z)$ of the second-order linear differential equation

(17.6.11) $$u'' + p(z)u = 0$$

such that

(17.6.12) $$w(z) = \frac{u_1(z)}{u_2(z)},$$

and this representation is valid in all of D. The representation is unique if we choose $u_2(z_0) = 1$. Conversely, every function of this form is a solution of (17.6.10).

Proof.   We note first that if $u(z)$ is a solution of (17.6.11), not identically zero, then $w(z)u(z)$ is also a solution provided that

$$uw'' + 2u'w' = 0$$

or

$$\frac{w''}{w'} = -2\frac{u'}{u}.$$

Differentiation gives

$$\left(\frac{w''}{w'}\right)' = -2\frac{u''}{u} + 2\left(\frac{u'}{u}\right)^2 = 2p(z) + \frac{1}{2}\left(\frac{w''}{w'}\right)^2,$$

that is, $w(z)$ must satisfy (17.6.10) and, if this is the case, then, conversely, $w(z)$ is the quotient of two solutions of (17.6.11). The latter must be linearly independent, for no constant can satisfy (17.6.10).

Suppose now that $w(z)$ is given as a holomorphic function of z satisfying (17.6.10) in some small neighborhood of $z = z_0$. The general existence theorem shows that $w(z)$ is uniquely defined by the initial values $w(z_0)$, $w'(z_0)$, $w''(z_0)$. Here we must have $w'(z_0) \neq 0$ since $\{w, z\}$ as function of the variables w', w'', w''' is not holomorphic at $w' = 0$. On the other hand, if $u_1(z)$ and $u_2(z)$ are any two linearly independent solutions of (17.6.11), then from the above considerations or direct verification we see that

$$\frac{u_1(z)}{u_2(z)}$$

is also a solution of (17.6.10). These solutions coincide if and only if their initial values do. Now knowing $w(z_0)$, $w'(z_0)$, and $w''(z_0)$ and choosing $u_2(z_0) = 1$, we find that

$$u_1(z_0) = w(z_0), \quad u_2{'}(z_0) = \frac{1}{2} \cdot \frac{w''(z_0)}{w'(z_0)}, \quad u_1{'}(z_0) = \frac{2w(z_0)w''(z_0) - [w'(z_0)]^2}{2w'(z_0)}$$

are uniquely determined. This means that $u_1(z)$ and $u_2(z)$ are uniquely determined. The discussion given in Section 10.7, especially in connection with formula (10.7.7), shows that $u_1(z)$ and $u_2(z)$ can be continued indefinitely in D. Since D is simply-connected, the theorem of monodromy (Theorem 10.3.1) shows that $u_1(z)$ and $u_2(z)$ are holomorphic in D. Their quotient is then a meromorphic function in D, and the law of permanence of functional equations shows that (17.6.12) is valid in all of D and is a solution of (17.6.10) in all of D. This completes the proof.

After these preliminaries we proceed to applications to conformal mapping.

THEOREM 17.6.3. *Given a simple closed curvilinear polygon* Π *whose sides are arcs of circles which form the angles* $\alpha_1\pi$, $\alpha_2\pi$, $\cdots$, $\alpha_n\pi$ *at the vertices* A_1, A_2, $\cdots$, A_n *where* $0 \leqq \alpha_j \leqq 2$. *Then there exist real numbers* a_1, a_2, $\cdots$, a_n, β_1, β_2, $\cdots$, β_n *such that*

$$a_1 < a_2 < \cdots < a_n, \quad \sum_{j=1}^{n} \beta_j = 0,$$

(17.6.13)

$$\sum_{j=1}^{n} [2\beta_j a_j + 1 - \alpha_j{}^2] = 0, \quad \sum_{j=1}^{n} [\beta_j a_j{}^2 + (1 - \alpha_j{}^2)a_j] = 0,$$

and the upper z *half-plane is mapped conformally onto the interior of* Π *by*

(17.6.14)
$$w = f(z) = \frac{u_1(z)}{u_2(z)},$$

where $u_1(z)$ *and* $u_2(z)$ *are two linearly independent solutions of the differential equation*

(17.6.15)
$$u'' + \left[\frac{1}{4} \sum_{j=1}^{n} \frac{1 - \alpha_j{}^2}{(z - a_j)^2} + \frac{1}{2} \sum_{j=1}^{n} \frac{\beta_j}{z - a_j} \right] u = 0.$$

The points a_j *correspond to the vertices* A_j.

REMARK. A linear fractional transformation T applied to w in (17.6.14) gives a function which maps the upper half-plane conformally on the interior of a polygon Π_T obtained by applying T to Π. The two polygons have the same angles.

Proof. We know a priori of the existence of a function $f(z)$ holomorphic in the upper half-plane which maps $\Im(z) > 0$ conformally on the interior of Π. We know also that if $w \to A_j$, then z has a real limit which we denote by a_j. We do not know the values of the a_j's. As a matter of fact, we can assign three of them arbitrarily, say a_1, a_2, a_3; the others are then uniquely determined. In order to prove that $f(z)$ is given by an expression of the form (17.6.14), it suffices to show that the Schwarzian derivative $\{f, z\}$ is holomorphic save for poles at the points a_j and that

$$(17.6.16) \qquad \{f, z\} = \frac{1}{2} \sum_{j=1}^{n} \frac{1 - \alpha_j^2}{(z - a_j)^2} + \sum_{j=1}^{n} \frac{\beta_j}{z - a_j},$$

where the β_j's satisfy (17.6.13).

Since $f(z)$ gives a conformal map of the upper half-plane, $f'(z) \neq 0$ for $\Im(z) > 0$, so that $\{f, z\}$ exists and is holomorphic for such values of z. By Theorem 17.5.3 the correspondence between z and w is also continuous and one-to-one on the boundaries. This means that $f(z)$ is continuous in each interval (a_j, a_{j+1}) and when z goes from a_j to a_{j+1}, then w goes from A_j to A_{j+1} along the circular arc. We can now invoke the extended form of the principle of symmetry formulated in the Corollary of Theorem 7.7.2. It states that $f(z)$ is holomorphic at each point of the open interval (a_j, a_{j+1}) and that $f(z)$ can be continued analytically across this line segment so as to be defined in the lower half-plane. Further, if $w = f(x + iy)$, $y > 0$, then $f(x - iy) = w^*$ where w^* is obtained from w by a reflection ($=$ reciprocation) in the arc $A_j A_{j+1}$. Thus the values of $f(x - iy)$ make up the interior of a polygon Π_1 adjacent to Π and "symmetric" to it with respect to the common arc. See Figure 26, where $n = 3$

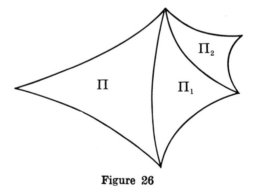

Figure 26

and each $\alpha_j = \frac{1}{6}$. The correspondence between the lower half-plane and the interior of Π_1 is one-to-one and conformal. Hence $f'(z) \neq 0$ in each of the n lower half-planes obtained by reflection in the n sides of Π. It is also seen that $f'(z) \neq 0$ on the line segments (a_j, a_{j+1}) since otherwise the correspondence could not be one-to-one. This makes $\{f, z\}$ holomorphic on the line segments as well as in the lower half-plane.

Actually $\{f, z\}$ is real on (a_j, a_{j+1}). This is seen as follows: We can map (A_j, A_{j+1}) on a segment of the real axis by a fractional linear transformation which takes $f(z)$ into

$$(17.6.17) \qquad F(z) = \frac{Af(z) + B}{Cf(z) + D}.$$

Since $F(z)$ is real on (a_j, a_{j+1}), the same holds for its derivatives. Hence $\{F, z\}$ is real and, since $\{F, z\} = \{f, z\}$, the same is true for $\{f, z\}$. Thus $\{f, z\}$ is real on the real axis except possibly at the points $\{a_j\}$.

We can get back to the upper half-plane by a second reflection in the real axis. This means reflection of an adjacent polygon Π_1 in one of its sides, which we may assume is not the side which it has in common with Π. The resulting secondary polygon Π_2 is obtainable from the original polygon Π by a fractional linear transformation which leaves $\{f, z\}$ invariant. It follows that $\{f, z\}$ is actually single-valued and exists in the whole plane except for the singular points $\{a_j\}$.

It remains to discuss the singularities. We assume $\alpha_j \neq 0, 1, 2$. We use again the fact that $\{f, z\}$ is invariant under Möbius transformations. We can then choose the coefficients in (17.6.17) in such a manner that a particular vertex A_j goes into the origin and the adjacent curvilinear sides become straight line segments one of which lies along the positive real axis. For small values of $|z - a_j|$ we then have

$$F(z) = (z - a_j)^{\alpha_j} H_j(z),$$

where $H_j(z)$ is holomorphic and tends to a limit $\neq 0$ when $z \to a_j$. Further, $H_j(z)$ is real for small real values of $z - a_j$. A simple calculation then gives

$$\{f, z\} = \{F, z\} = \left(\frac{F''}{F'}\right)' - \frac{1}{2}\left(\frac{F''}{F'}\right)^2$$

$$= \frac{1}{2}\frac{1 - \alpha_j^2}{(z - a_j)^2} + \frac{\beta_j}{z - a_j} + h_j(z),$$

where β_j is real and $h_j(z)$ is holomorphic at $z = a_j$.

We can now form

$$\Delta(z) \equiv \{f, z\} - \frac{1}{2}\sum_{j=1}^{n}\frac{1 - \alpha_j^2}{(z - a_j)^2} - \sum_{j=1}^{n}\frac{\beta_j}{z - a_j},$$

which is seen to be holomorphic in the finite part of the plane. We have, of course, assumed tacitly that each a_j is finite. Then $z = \infty$ is mapped into an interior point of one of the sides of Π. This requires that $f(z)$ be holomorphic at infinity, and an elementary calculation shows that this requires that $z^4\{f, z\}$ be holomorphic at infinity. This shows that $\Delta(z)$ is also holomorphic at $z = \infty$. Since it vanishes there, we must have $\Delta(z) \equiv 0$. This verifies (17.6.16). Finally, relations (17.6.13) express the fact that $z^4\{f, z\}$ is holomorphic at $z = \infty$. This completes the proof.

The solution depends upon a number of parameters, and to determine them is usually very difficult unless $n = 3$ or the polygon is regular. In either case the solution of the mapping problem leads to hypergeometric functions.

In the case $n = 3$ we can dispose of a_1, a_2, a_3 as we please. The remaining three parameters are then uniquely determined. We have three linear equations in the three unknowns β_1, β_2, β_3, and the determinant is

$$2(a_1 - a_2)(a_2 - a_3)(a_3 - a_1) \neq 0$$

so we can express β_1, β_2, β_3 in terms of a_1, a_2, a_3 and α_1, α_2, α_3. The resulting expression for the Schwarzian derivative is

$$(17.6.18) \qquad \frac{1}{2} \prod_{j=1}^{3} (z - a_j)^{-1} \sum_{j=1}^{3} (1 - \alpha_j^2) \frac{(a_j - a_k)(a_j - a_l)}{z - a_j}.$$

We shall place the singular points at $z = 0, 1$, and ∞. This we can do by setting $a_1 = 0$, $a_2 = 1$, and letting $a_3 \to \infty$ in (17.6.18). We obtain a result which may be written

$$(17.6.19) \quad \{w, z\} = \frac{1}{2} \frac{1 - \alpha_1^2}{z^2} + \frac{1}{2} \frac{1 - \alpha_2^2}{(z - 1)^2} + \frac{1}{2} \frac{\alpha_1^2 + \alpha_2^2 - \alpha_3^2 - 1}{z(z - 1)} \equiv 2p(z).$$

The corresponding second-order equation

$$(17.6.20) \qquad u'' + p(z)u = 0$$

has its singularities at the points $z = 0, 1$, and ∞, and all three are regular singular points. Further, the difference of the roots of the corresponding indicial equation is α_1 at 0, α_2 at 1, and α_3 at ∞. The canonical equation with three regular singular points at 0, 1, ∞ is the hypergeometric one

$$(17.6.21) \qquad z(1 - z)y'' + [c - (a + b + 1)z]y' - aby = 0.$$

Actually we can transform (17.6.20) into a hypergeometric equation with

$$(17.6.22) \quad a = \tfrac{1}{2}(1 - \alpha_1 - \alpha_2 + \alpha_3), \quad b = \tfrac{1}{2}(1 - \alpha_1 - \alpha_2 - \alpha_3), \quad c = 1 - \alpha_1$$

by setting

$$u(z) = z^\rho (1 - z)^\sigma y(z)$$

where

$$\rho = \tfrac{1}{2}(1 - \alpha_1), \quad \sigma = \tfrac{1}{2}(1 - \alpha_2).$$

It is clear that if $y_1(z)$ and $y_2(z)$ are solutions of (17.6.21) corresponding to the solutions $u_1(z)$ and $u_2(z)$ of (17.6.20), then

$$\frac{u_1(z)}{u_2(z)} = \frac{y_1(z)}{y_2(z)},$$

and if $u_1(z)$ and $u_2(z)$ form a fundamental system, that is, are linearly independent, then $y_1(z)$ and $y_2(z)$ also form a fundamental system.

It follows that if $y_1(z)$ and $y_2(z)$ form a fundamental system for (17.6.21), then their quotient maps the upper z half-plane onto the interior of a curvilinear triangle with the angles $\alpha_1\pi$, $\alpha_2\pi$, $\alpha_3\pi$. Different fundamental systems lead to different triangles, but all triangles in the set have the same angles and we can pass from one triangle to another by a Möbius transformation. The canonical fundamental system at the origin

$$(17.6.23) \qquad \begin{aligned} y_{01}(z) &= z^{1-c}F(a - c + 1, b - c + 1, 2 - c; z), \\ y_{02}(z) &= F(a, b, c; z) \end{aligned}$$

is particularly interesting. Here we assume

$$0 < a, \quad 0 < b, \quad 0 < c < 1, \quad 0 < a - c + 1, \quad 0 < b - c + 1.$$

The corresponding triangle OPQ has the vertex O at the origin, P lies on the positive real axis, OP and OQ are straight line segments, and PQ is a circular arc. The conditions on the coefficients are satisfied if, for instance, $\alpha_1 + \alpha_2 + \alpha_3 < 1$. The points P and Q are expressible in terms of Gamma functions.

The representation of the mapping functions as the quotient of $y_{01}(z)$ and $y_{02}(z)$ is in terms of power series convergent for $|z| < 1$. It is possible to obtain the analytic continuation of these power series in the whole plane cut along the real axis from $z = +1$ to $z = +\infty$ in terms of definite integrals. In fact we have the representation

$$(17.6.24) \quad F(a, b, c; z) = \frac{\Gamma(c)}{\Gamma(a)\Gamma(c - a)} \int_0^1 t^{a-1}(1 - t)^{c-a-1}(1 - zt)^{-b}\, dt$$

valid in the cut plane if

$$\Re(a) > 0, \quad \Re(c - a) > 0.$$

If these conditions are satisfied, and $|z| < 1$, then we can expand $(1 - zt)^{-b}$ in a binomial series, substitute in the integral, and integrate termwise. The resulting integrals are all of the form

$$\int_0^1 t^{\alpha-1}(1 - t)^{\beta-1}\, dt = \frac{\Gamma(\alpha)\Gamma(\beta)}{\Gamma(\alpha + \beta)}$$

for $\Re(\alpha) > 0$, $\Re(\beta) > 0$. Here $\alpha = a + n - 1$, $\beta = c - a$. The result is (17.6.24). On the other hand, the integral exists in the cut plane, so it gives the analytic continuation of the hypergeometric series outside the unit circle. It is obvious that we can get a similar representation for $y_{01}(z)$, namely,

$$\frac{\Gamma(2 - c)}{\Gamma(a - c + 1)\Gamma(1 - a)} z^{1-c} \int_0^1 t^{a-c}(1 - t)^{-a}(1 - zt)^{c-b-1}\, dt,$$

which requires

$$\Re(a) < 1, \quad \Re(c - a) < 1.$$

Thus both representations are valid if

$$0 < \Re(a) < 1, \quad 0 < \Re(c - a) < 1.$$

In particular, these conditions are satisfied if $\alpha_1 + \alpha_2 + \alpha_3 < 1$.

Formula (17.6.24) shows that

(17.6.25) $$F(a, b, c; 1) = \frac{\Gamma(c)\Gamma(c - a - b)}{\Gamma(c - a)\Gamma(c - b)}$$

provided the arguments of the Gamma functions have positive real parts. From this we obtain the point P

$$f(1) = \frac{\Gamma(2 - c)\Gamma(c - a)\Gamma(c - b)}{\Gamma(c)\Gamma(1 - a)\Gamma(1 - b)}.$$

The point Q can also be found, but this requires knowledge of the group of the hypergeometric equation or similar information, so we state the result without proof:

$$f(\infty) = e^{\pi i(1 - c)} \frac{\Gamma(b)\Gamma(c - a)\Gamma(2 - c)}{\Gamma(c)\Gamma(b - c + 1)\Gamma(1 - a)}.$$

The case of a "regular" curvilinear polygon with n sides also leads to elegant explicit formulas. We shall state the result and sketch the proof.

THEOREM 17.6.4.   *Given a polygon Π with center at the origin and vertices at the nth roots of unity whose sides are circular arcs of equal length which form the interior angle $\alpha\pi$ with each other where $0 \leq \alpha \leq 2$. The function $f(z)$ which maps the interior of the unit circle conformally onto the interior of Π with $f(0) = 0$ and $f'(0) > 0$ is*

(17.6.26)

$$f(z) = z \frac{\Gamma\left(1 - \frac{1}{n}\right)\Gamma\left(\frac{1}{2}(1 - \alpha) + \frac{1}{n}\right) F\left(\frac{1}{2}(1 - \alpha) + \frac{1}{n}, \frac{1}{2}(1 - \alpha), 1 - \frac{1}{n}; z^n\right)}{\Gamma\left(1 + \frac{1}{n}\right)\Gamma\left(\frac{1}{2}(1 - \alpha) - \frac{1}{n}\right) F\left(\frac{1}{2}(1 - \alpha) - \frac{1}{n}, \frac{1}{2}(1 - \alpha), 1 + \frac{1}{n}; z^n\right)}.$$

Proof.   It follows from the stated conditions that the roots of unity are fixed points under the mapping and that each ray $\arg z = 2k\pi i/n$ is mapped onto itself. This requires that $f(z)$ has a power series expansion of the form

$$f(z) = \sum_{p=0}^{\infty} a_p z^{np+1}, \quad |z| < 1,$$

so that

$$\{f, z\} = \sum_{p=1}^{\infty} b_p z^{np-2}.$$

Again, the proof of formula (17.6.16) applies with obvious modifications to the present situation and shows that

$$\{f, z\} = \tfrac{1}{2}(1 - \alpha^2) \sum_{k=1}^{n} \frac{1}{(z - \omega_k)^2} + \sum_{k=1}^{n} \frac{\beta_k}{z - \omega_k}, \quad \omega_k = e^{2k\pi i/n}.$$

Here

$$\sum_{k=1}^{n} \frac{1}{(z - \omega_k)^2} = -\frac{d}{dz} \sum_{k=1}^{n} \frac{1}{z - \omega_k} = -\frac{d}{dz} \frac{nz^{n-1}}{z^n - 1}$$

$$= n \left\{ \frac{nz^{2n-2}}{(z^n - 1)^2} - \frac{(n - 1)z^{n-2}}{z^n - 1} \right\}.$$

On the other hand, if

$$\sum_{k=1}^{n} \frac{\beta_k}{z - \omega_k} = \sum_{p=1}^{\infty} c_p z^{np - 2}, \quad |z| < 1,$$

then the sum must be of the form

$$\frac{C z^{n-2}}{z^n - 1},$$

for the sum is a rational function of z which vanishes at ∞ and whose denominator is $z^n - 1$. If the series is to have the required form, $C z^{n-2}$ is the only possible term in the numerator. Here we determine C by the condition that $z^4\{f, z\}$ is holomorphic at infinity. This gives

$$C = -\tfrac{1}{2}n(1 - \alpha^2),$$

so that

(17.6.27) $$\{f, z\} = \tfrac{1}{2}n^2(1 - \alpha^2) \frac{z^{n-2}}{(z^n - 1)^2}.$$

The associated second-order differential equation

(17.6.28) $$u'' + \tfrac{1}{4}n^2(1 - \alpha^2) \frac{z^{n-2}}{(z^n - 1)^2} u = 0$$

may be reduced to a hypergeometric equation by the introduction of new variables

$$t = z^n, \quad y = (z^n - 1)^{-\frac{1}{2}(1 - \alpha)} u.$$

A lengthy but elementary calculation gives

(17.6.29)

$$t(1 - t)y'' + \left[\left(1 - \frac{1}{n}\right) - \left(2 - \alpha - \frac{1}{n}\right)t \right] y' - \frac{1}{2}(1 - \alpha) \left[\frac{1}{2}(1 - \alpha) - \frac{1}{n} \right] y = 0,$$

that is, a hypergeometric equation with

$$a = \frac{1}{2}(1 - \alpha) - \frac{1}{n}, \quad b = \frac{1}{2}(1 - \alpha), \quad c = 1 - \frac{1}{n}.$$

Here the fundamental system at the origin is

(17.6.30)

$$y_{01}(t) = F\left(\frac{1}{2}(1 - \alpha) - \frac{1}{n}, \frac{1}{2}(1 - \alpha), 1 - \frac{1}{n}; t\right),$$

$$y_{02}(t) = t^{\frac{1}{n}} F\left(\frac{1}{2}(1 - \alpha) + \frac{1}{n}, \frac{1}{2}(1 - \alpha), 1 + \frac{1}{n}; t\right),$$

and by formula (17.6.25)

$$y_{01}(1) = \frac{\Gamma(\alpha)\Gamma\left(1 - \dfrac{1}{n}\right)}{\Gamma\left(\dfrac{1}{2}(1 + \alpha)\right)\Gamma\left(\dfrac{1}{2}(1 + \alpha) - \dfrac{1}{n}\right)},$$

$$y_{02}(1) = \frac{\Gamma(\alpha)\Gamma\left(1 + \dfrac{1}{n}\right)}{\Gamma\left(\dfrac{1}{2}(1 + \alpha)\right)\Gamma\left(\dfrac{1}{2}(1 + \alpha) + \dfrac{1}{n}\right)}.$$

The required mapping function is now

(17.6.31)
$$f(z) = \frac{y_{01}(1)}{y_{02}(1)} \frac{y_{02}(z^n)}{y_{01}(z^n)},$$

and this is formula (17.6.26). This completes the discussion.

EXERCISE 17.6

1. On what domain is the unit disk mapped by the function

$$w = \log \frac{1}{1 - z} ?$$

2. Same question for $w = $ arc tan z.

3. Show that the function

$$w = \frac{\pi \text{ arc tan } z}{4 \text{ arc tan } z + \pi}$$

maps the unit disk on the domain

$$\Re(w) < \tfrac{1}{4}\pi, \quad |w - \tfrac{3}{8}\pi| > \tfrac{1}{8}\pi.$$

4. Show that

$$w = \tan^2 \left(\tfrac{1}{4}\pi\sqrt{z}\right)$$

maps the interior of the parabola with focus at $z = 0$ and vertex at $z = 1$ conformally on the unit disk K (Schwarz, 1864).

5. Show that

$$w = \int_0^z [s(1 - s^2)]^{-\frac{1}{2}} ds$$

maps the upper half-plane $\Im(z) > 0$ conformally onto the interior of a square (Schwarz, 1869). Where are the vertices?

6. Show that the function

$$w = \int_0^z (1 - s^n)^{-\frac{2}{n}} ds$$

maps the unit disk in the z-plane onto the interior of a regular polygon of n sides and with center at the origin (Schwarz, 1864). Show that the radius of the circumscribed circle is

$$\frac{\Gamma\left(1 - \frac{2}{n}\right)\Gamma\left(\frac{1}{n}\right)}{\Gamma\left(1 - \frac{1}{n}\right)}.$$

7. Verify (17.6.8).

8. Verify (17.6.9).

9. If $f(z)$ is holomorphic at infinity, show that $z^4\{f, z\}$ is also holomorphic there.

10. Given a sphere

$$X = R \sin \theta \cos \varphi, \quad Y = R \sin \theta \sin \varphi, \quad Z = R \cos \varphi,$$

where $0 \leq \theta \leq \pi$, $0 \leq \varphi < 2\pi$, and the so-called Mercator's projection $x = \varphi$, $y = \log \tan \frac{1}{2}\theta$, prove that the sphere is mapped conformally onto a plane strip.

17.7. The theorem of Bloch. The present section is devoted to a brief discussion of a special case of the theorem of Bloch and its application to Picard's theorem. Bloch's theorem may be regarded as a generalization of that of Koebe, Theorem 17.4.5, when the assumption that the mapping be one-to-one is dropped. We start with

LEMMA 17.7.1. *Let $f(z)$ be holomorphic in the disk D_R: $|z| \leq R$, and denote by $f[D_R]$ the set of values assumed by $f(z)$ in D_R. Let $f(0) = 0$, $|f'(0)| = a > 0$ and let $|f'(z)| \leq M$ in D_R. Then*

$$(17.7.1) \qquad \left[w \mid |w| < \frac{a^2 R}{4M}\right] \subset f[D_R].$$

Proof. We have

$$|f(z)| \leq RM \quad \text{for} \quad |z| \leq R.$$

Suppose that $w = c$ is not in $f[D_R]$. Then $c \neq 0$, and $1 - \dfrac{1}{c} f(z)$ is holomorphic and $\neq 0$ in D_R. Set

$$h(z) = \left\{ 1 - \frac{1}{c} f(z) \right\}^{\frac{1}{2}} \equiv \sum_{n=0}^{\infty} a_n z^n.$$

Here

$$a_0 = 1, \quad a_1 = -\frac{f'(0)}{2c}, \quad \cdots,$$

and by the quadratic mean value theorem (Problem 6, Exercise 8.2 on page 208 of Volume I)

$$\sum_{n=0}^{\infty} |a_n|^2 R^{2n} = \frac{1}{2\pi} \int_{-\pi}^{\pi} |h(Re^{i\theta})|^2 \, d\theta.$$

Since

$$|h(Re^{i\theta})|^2 \leq 1 + \frac{RM}{|c|},$$

we see that

$$1 + \frac{|f'(0)|^2}{4|c|^2} R^2 \leq 1 + \frac{RM}{|c|},$$

whence

$$|c| \geq \frac{R}{4M} |f'(0)|^2 = \frac{a^2 R}{4M},$$

and this implies (17.7.1).

We shall now state and prove a weak form of Bloch's theorem.

THEOREM 17.7.1.   *Let $f(z)$ be holomorphic in the closed unit disk and let $f[\overline{K}]$ denote its range. Let $f'(0) = 1$. There exists an absolute constant $L > 0$ such that $f(\overline{K})$ contains a disk of radius $\geq L$.*

REMARKS.   (1) Actually Bloch proved the following stronger statement: *There exists an absolute constant $B > 0$ and a disk of radius $\geq B$ which is the one-to-one image of a subdomain of K under the mapping $w = f(z)$.*

(2) For the applications that we have in mind the weaker statement suffices. The proof given below is due to Landau (1926), and it gives $L \geq \frac{1}{16}$.

(3) B is known as Bloch's constant, L as Landau's. Their exact values are unknown. Some bounds are given below.

Proof of Theorem 17.7.1.   Set

$$M(s) = \max_{|z| \leq s} |f'(z)|, \quad 0 \leq s \leq 1.$$

The continuous function $sM(1 - s) \equiv C(s)$ is 0 for $s = 0$ and 1 for $s = 1$. Let r be the least value of s for which $C(s) = 1$. We have $0 < r \leq 1$.

Let a be such that

$$|a| \leq 1 - r, \quad |f'(a)| = M(1 - r)$$

and consider the function

$$g(z) = f(z + a) - f(a),$$

which is holomorphic for $|z| \leq r$. We have

$$g(0) = 0, \quad |g'(0)| = |f'(a)| = M(1 - r) = \frac{1}{r},$$

and for $|z| \leq \frac{1}{2}r$

$$|g'(z)| = |f'(z + a)| \leq M(1 - \frac{1}{2}r) < \frac{2}{r}$$

by the choice of r. Here we apply Lemma 17.7.1 to $g(z)$ with $R = \frac{1}{2}r$, $a = 1/r$, $M \leq 2/r$. It follows that the range of $g(z)$ covers a disk with center at $w = 0$ whose radius is at least $\frac{1}{16}$. But this means that $f(\overline{K})$ contains a disk with center at $w = f(a)$ whose radius is at least $\frac{1}{16}$. Hence L exists and $L \geq \frac{1}{16}$.

It is known that

(17.7.2) $0.43 \cdots \leq B \leq 0.47 \cdots < 0.5 \leq L \leq 0.56 \cdots$.

The upper bounds are related to the mapping functions of Theorem 17.6.4 for $n = 3$ with α equal to 0, $\frac{1}{6}$, or $\frac{1}{3}$. We denote the corresponding triangles by $\triangle_\alpha$ where $\triangle_0$ and $\triangle_{1/6}$ are curvilinear but $\triangle_{1/3}$ is an ordinary equilateral triangle. Let $f_\alpha(z)$ be the corresponding mapping function and form

$$F_1(z) = f_{1/3}[\check{f}_{1/6}(z)], \quad F_2(z) = f_{1/3}[\check{f}_0(z)],$$

where $\check{f}_\alpha(z)$ is the inverse of $f_\alpha(z)$. $F_1(z)$ maps the triangle $\triangle_{1/6}$ conformally onto $\triangle_{1/3}$ and $F_2(z)$ maps $\triangle_0$ conformally onto $\triangle_{1/3}$. Both mapping functions may be continued analytically in their complete domains of existence with the aid of the symmetry principle of Schwarz.

The boundary circles of $\triangle_{1/6}$ are orthogonal to the circle

$$C: |z| = (1 + 3^{\frac{1}{2}})^{\frac{1}{2}}.$$

Successive reflections in the sides of $\triangle_{1/6}$ and the resulting triangles lead to a net of curvilinear triangles which fills the interior of C without overlapping. All these triangles have the same angles, and their sides are arcs of circles orthogonal to C. In the w-plane we get a corresponding net of equilateral triangles. Since six curvilinear triangles which have a vertex in common form a full neighborhood of this vertex, we see that the corresponding equilateral triangles over the w-plane have a vertex in common and cover a neighborhood of this vertex twice. Thus all the vertices in the w-plane are simple branch points of a Riemann surface over the w-plane which is the image of $|z| < R$ under the mapping $w = F_1(z)$. Since the radius of the circumscribed circle of $\triangle_{1/3}$ is 1, we see that the Riemann surface contains smooth disks of radius 1 which are in one-to-one correspondence with subdomains of $|z| < R$, and these are the largest disks possible. Although the function $F_1(z)$ does not belong to the class to which Theorem 17.7.1 applies,

$$f(z) \equiv \frac{F_1(Rz)}{RF_1'(0)}$$

is such a function since it is holomorphic in K and $f'(0) = 1$. The largest smooth disk now has a radius equal to

$$\frac{f_{1/6}'(0)}{f_{1/3}'(0)} \frac{1}{R} = \frac{\Gamma(\tfrac{1}{3})\Gamma(\tfrac{11}{12})}{(1 + 3^{1/2})^{1/2}\Gamma(\tfrac{1}{4})} = 0.47 \cdots.$$

This upper bound for B was found by L. Ahlfors and H. Grunsky in 1937. Their conjecture that this is the true value of B has been neither proved nor disproved.

For L we consider the function $F_2(z)$ instead. It may be extended to the interior of K by the symmetry principle. The boundary of $\triangle_0$ is orthogonal to the unit circle, and all triangles obtained by the reflection process have the same property. All these triangles have their vertices on the unit circle and each vertex belongs to infinitely many triangles. The image of $|z| < 1$ is now a Riemann surface over the w-plane built up of equilateral triangles. But in this case each vertex in the w-plane carries a branch point of infinite order, and such a point does not belong to $F_2[K]$. Thus the largest disk which belongs to the surface has the radius 1. To get a function whose derivative is 1 at the origin we divide by $F_2'(0)$ and find that

$$L \leq \frac{1}{F_2'(0)} = \frac{f_0'(0)}{f_{1/3}'(0)} = \frac{\Gamma(\tfrac{2}{3})}{\Gamma(\tfrac{5}{3})\Gamma(\tfrac{1}{3})} = \frac{3}{2\Gamma(\tfrac{1}{3})} = 0.56 \cdots.$$

Picard's theorem, Theorem 14.5.1, is a simple consequence of Theorem 17.7.1. We follow Landau and base the proof on the following lemma:

LEMMA 17.7.2. *Let D be either a closed disk $|z| \leq R$, $R > 0$ or the finite plane. Let $F(z)$ be holomorphic in D and let*

$$F(z) \neq 0, \quad F(z) \neq 1, \quad z \in D.$$

Then there exists a function $f(z)$ holomorphic in D such that

(17.7.3) $$F(z) = -\exp\{\pi i \cosh[2f(z)]\},$$

$f(0)$ is uniquely determined by $F(0)$, and the range of $f(z)$ does not cover any disk of radius 1.

Proof. We can determine successively functions $h(z)$, $u(z)$, $v(z)$, and $f(z)$, all holomorphic in D, which satisfy the conditions listed below. In each case it is necessary to make a definite choice among various possibilities, but it suffices to make this choice, say, at $z = 0$; the function in question is then uniquely defined in D. We start by choosing an $h(z)$ such that

$$F(z) = \exp\,[2\pi i h(z)], \quad 0 \leqq \Re[h(0)] < 1.$$

Such a function exists since $F(z) \neq 0$. Next we choose $u(z)$ so that

$$h(z) = u^2(z), \quad 0 \leqq \Re[u(0)].$$

This is possible since $F(z) \neq 1$ implies $h(z) \neq 0$. Then we choose $v(z)$ so that

$$h(z) = 1 + v^2(z), \quad 0 \leqq \Re[v(0)].$$

Since $F(z) \neq 1$, we have $h(z) \neq 1$, so that $v(z)$ exists. Since $u^2(z) - v^2(z) = 1$, we have $u(z) \neq v(z)$, and there exists a uniquely defined $f(z)$ such that

$$u(z) - v(z) = e^{f(z)}, \quad -\pi < \Im[f(0)] \leqq \pi.$$

We have then

$$u(z) + v(z) = [u(z) - v(z)]^{-1} = e^{-f(z)},$$

$$u(z) = \tfrac{1}{2}[e^{f(z)} + e^{-f(z)}] = \cosh\,[f(z)],$$

$$2\pi i h(z) = 2\pi i u^2(z) = \pi i \cosh\,[2f(z)] + \pi i,$$

and finally

$$F(z) = -\exp\,\{\pi i \cosh\,[2f(z)]\}$$

as asserted. It is clear that $f(0)$ depends only upon $F(0)$.

Next we note that $f[D]$ omits all the points of the form

(17.7.4) $$\pm \log\,(\sqrt{m} + \sqrt{m-1}) + \tfrac{1}{2}n\pi i,$$

where m is a positive integer and n an integer. Suppose that this were not true and that we could find a $z_0 \in D$ such that $f(z_0)$ has a value belonging to the set (17.7.4). Since $\cosh 2u$ has the period πi, we see that

$$\cosh\,[2f(z_0)] = \tfrac{1}{2}\{[\sqrt{m} + \sqrt{m-1}]^2 + [\sqrt{m} - \sqrt{m-1}]^2\} = 2m - 1,$$

so that

$$F(z_0) = -\exp\,[(2m-1)\pi i] = +1,$$

and this contradicts the hypotheses. It follows that $f[D]$ omits all points of this set, which we denote by E.

The points of E are spaced in such a manner that they form the vertices of a rectangular net. The height of each rectangle is $\tfrac{1}{2}\pi < \sqrt{3}$. The width is steadily decreasing as we recede from the imaginary axis, and its maximal value is $\log\,(1 + \sqrt{2}) < 1$. It follows that any disk of radius 1 must contain points of E. This completes the proof.

We can now prove Picard's theorem in a few lines.

THEOREM 17.7.2. *An entire function which omits the values 0 and 1 is a constant.*

Proof. Let $F(z)$ be the entire function, let D be the finite plane, and let

$f(z)$ be the function given by Lemma 17.7.2. Then $f(z)$ is also entire and its range $f[D]$ omits all the points of the set E. If $F(z)$ is not a constant, $f(z)$ has the same property. Take a point $z = a$ where $f'(a) \neq 0$ and form the entire function

$$Lf\left(\frac{z}{Lf'(a)} + a\right) = Lf(a) + z + \cdots,$$

where L is Landau's constant. The range of this function omits all points $L \cdot E$ obtained by multiplying the points of E by L. The upper bound for the radius of a disk contained in the range of this function is $< L$, and this contradicts Theorem 17.7.1. It follows that $F(z)$ must be a constant.

The same type of argument applies to Landau's theorem:

THEOREM 17.7.3. *Given a power series*

$$F(z) = \alpha + z + \cdots.$$

There exists a $g(\alpha) > 0$ such that in the disk $|z| \leq g(\alpha)$ the function $F(z)$ lacks at least one of the properties (i) *$F(z)$ is holomorphic,* (ii) *$F(z) \neq 0$, and* (iii) *$F(z) \neq 1$.*

Proof. Let $R > 0$ be such that $F(z)$ has all three properties in the disk $D: |z| \leq R$. Form the corresponding function $f(z)$. It is holomorphic in the same disk, $f(0)$ depends only upon α, and $f[D]$ omits the set E. Further

$$F'(z) = -2\pi i \sinh [2f(z)]f'(z)F(z),$$

and for $z = 0$

$$1 = -2\pi \alpha i \sinh [2f(0)]f'(0).$$

This shows that $f(0) \neq 0$, $f'(0) \neq 0$, and that $f'(0)$ depends only upon α. The function

$$\frac{f(Rz)}{Rf'(0)} = a_0 + z + \cdots$$

is holomorphic in K and its range covers no open disk of radius $[R|f'(0)|]^{-1}$. It follows that this quantity must exceed L, so that

(17.7.5)
$$R < \frac{L}{|f'(0)|} \equiv g(\alpha).$$

This completes the proof.

EXERCISE 17.7

1. Use Landau's method to prove Theorem 15.4.6, Schottky's theorem.

2. There is a sharp $\frac{1}{16}$-theorem. Prove that if $F(z) \in H(K)$, if $F(0) = 0$, $F'(0) = 1$, if $F(z) \neq 0$ for $0 < z < 1$, and if $|\alpha| < \frac{1}{16}$, then the equation $F(z) = \alpha$ has a solution in K. Here $F(z) = \frac{1}{16}\nu\left(\frac{1}{\pi i} \log z\right)$ is the extremal function where $\nu(w)$ is the inverse modular function (Z. Nehari).

COLLATERAL READING

As general references are recommended

CARATHÉODORY, C. *Conformal Representation.* Cambridge Tracts in Mathematics and Mathematical Physics, No. 28. Cambridge University Press, London, 1932.

GOLUSIN, G. M. *Geometrische Funktionentheorie.* Deutscher Verlag der Wissenschaften, Berlin, 1957.

NEHARI, Z. *Conformal Mapping.* McGraw-Hill Book Company, Inc., New York, 1952.

PÓLYA, G., and SZEGÖ, G. *Aufgaben und Lehrsätze aus der Analysis*, Vols. I, II, especially pp. 13–29 of Volume II. Springer-Verlag, Berlin, 1925.

In connection with Section 17.2 see

BERGMAN, S. *The Kernel Function and Conformal Mapping.* Mathematical Surveys, Vol. V. American Mathematical Society, New York, 1950.

Some references for Section 17.3 are:

FABER, G. "Über nach Polynomen fortschreitende Reihen," *Sitzungsberichte der mathematisch-physikalischen Klasse der Bayerischen Akademie der Wissenschaften zu München*, 1922, pp. 157–178.

LEJA, F. "Sur une suite de polynomes et la représentation conforme d'un domain plan quelconque sur le cercle," *Annales de la Société Polonaise de Mathématique*, Vol. 14 (1936), pp. 116–134.

SZEGÖ, G. *Orthogonal Polynomials*, Chapter XI. Colloquium Publications, Vol. 23. American Mathematical Society, New York, 1939.

In connection with Section 17.4 we list

JENKINS, J. A. *Univalent Functions and Conformal Mapping.* Ergebnisse der Mathematik, New Series, No. 18. Springer-Verlag, Berlin, 1958.

SCHIFFER, M. "Extremum Problems and Variational Methods in Conformal Mapping," in *Proceedings of the International Congress of Mathematicians 1958*, pp. 211–231. Cambridge University Press, London, 1960.

For various aspects of the boundary problem, see Carathéodory's Tract, cited above, and

PRIWALOW, I. I. *Randeigenschaften analytischer Funktionen.* Deutscher Verlag der Wissenschaften, Berlin, 1956.

ZYGMUND, A. *Trigonometric Series*, Second Edition, Vol. I, Chapter VII. Cambridge University Press, New York, 1959.

A detailed discussion of the polygonal mapping functions is to be found in the treatises of Golusin and Nehari. For the properties of hypergeometric functions needed in this discussion, see

WHITTAKER, E. T., and WATSON, G. N. *A Course of Modern Analysis*, Fourth Edition, Chapter 14. Cambridge University Press, New York, 1952.

The best collection of special mappings known to the author is

KOBER, H. *Dictionary of Conformal Representations*. Dover Publications, Inc., New York, 1957.

A brief return to the point of departure is often instructive. With this in mind the reader might find reading some of Schwarz's papers on conformal mapping an interesting experience. See

SCHWARZ, H. A. *Gesammelte Mathematische Abhandlungen*, Vol. II. Verlag von Julius Springer, Berlin, 1890. Chelsea Publishing Company, New York, 1972.

For Landau's method in Section 17.7, see

LANDAU, E. *Darstellung und Begründung einiger neuerer Ergebnisse der Funktionentheorie*, Second Edition, Chapter 7. Springer-Verlag, Berlin, 1929. Reprinted in: *Das Kontinuum*. Chelsea Publishing Company, New York, 1960, 1973.

18

MAJORIZATION

18.1. The Phragmén-Lindelöf principle. In the present chapter we shall study some principles of majorization for which the prototype is given by results discovered by Edvard Phragmén (1863–1937) in 1904 and generalized by him and E. Lindelöf in 1908.

To elucidate Phragmén's discovery, let us consider the function

$$(18.1.1) \qquad f(z) = \exp(\gamma z^{\alpha}), \quad \gamma > 0, \alpha \geq \tfrac{1}{2}, |\arg z| < \pi,$$

where $z^{\alpha} > 0$ when $z > 0$. The absolute value of this function equals 1 when

$$\arg z = (2k + 1)\frac{\pi}{2\alpha}, \quad |k| \leq \alpha - \tfrac{1}{2}.$$

These rays define sectors

$$S_k: \quad (2k - 1)\frac{\pi}{2\alpha} < \theta < (2k + 1)\frac{\pi}{2\alpha},$$

and $f(z)$ is bounded in the sectors S_k for which k is an *odd* integer. Each sector has the opening π/α. Since

$$(18.1.2) \qquad \log M(r;f) = \gamma r^{\alpha},$$

we see that it is possible for a function satisfying such a growth condition to be bounded on two rays making an angle of π/α with each other without being bounded in the sector formed by the rays. Phragmén observed that this is a best possible result in the following sense:

THEOREM 18.1.1. *If in a sector S of opening π/α a function $f(z)$ is holomorphic, if $f(z)$ is bounded on the rays bounding S, and if in the interior of S we have*

$$(18.1.3) \qquad \log M(r;f) = o(r^{\alpha}),$$

then $f(z)$ is bounded in S.

We shall not prove this theorem here. It will appear as a special case of the next theorem, and sharper results will be proved in Section 18.4. Let us remark in passing that the result can be considered as an example of the vague "principle of moderation" mentioned in Section 8.1. In the present case we see that a function which is holomorphic in a sector of opening ω and which is bounded on the boundary of the sector either is bounded in the closed sector or else

$$(18.1.4) \qquad \limsup_{r \to \infty} r^{-\pi/\omega} \log M(r;f) > 0.$$

We proceed to a statement and proof of the principle of Phragmén-Lindelöf.

THEOREM 18.1.2.   *Suppose that $f(z)$ is holomorphic in the simply-connected domain D whose boundary ∂D is the union of two disjoint sets E_0 and E_1. Suppose that*

(i) *there exists a constant $C > 0$ such that for every $\varepsilon > 0$ we have*

$$|f(z)| \leqq C + \varepsilon,$$

provided that $z \in D$ and its distance from E_0 is sufficiently small;

(ii) *there exists a function $\omega(z)$, holomorphic in D where it is different from 0 and its absolute value does not exceed 1, such that for any $\varepsilon > 0$, $\eta > 0$ we have*

$$|f(z)[\omega(z)]^\eta| \leqq C + \varepsilon,$$

provided that $z \in D$ and its distance from E_1 is sufficiently small.

Then, for all z in D we have

$$(18.1.5) \qquad\qquad |f(z)| \leqq C.$$

If the sign of equality holds anywhere in D, then it holds everywhere and $f(z)$ is a constant of absolute value C.

Proof.   If E_1 is void, we obtain the classical principle of the maximum as formulated in Theorem 7.9.3. If E_1 is not void, we choose a fixed $\eta > 0$ and consider the function

$$F(z) \equiv f(z)[\omega(z)]^\eta,$$

which is holomorphic in D and satisfies the conditions of Theorem 7.9.3. Since $\omega(z) \neq 0$ in D, and D is simply-connected, we can choose any one of the admissible determinations of $[\omega(z_0)]^\eta$ at some point $z = z_0$ in D, and define $[\omega(z)]^\eta$ by analytic continuation. The resulting power is single-valued in D and, hence, holomorphic. It follows that $F(z)$ is indeed holomorphic in D and that Theorem 7.9.3 applies, so that

$$|F(z)| \leqq C.$$

This gives

$$|f(z)| \leqq C \, |\omega(z)|^{-\eta},$$

and, since this holds for every $\eta > 0$, we see that (18.1.5) must hold.

The assumptions made above may be modified considerably without affecting the validity of the conclusion. Thus, the theorem remains valid if

(i) D is not simply-connected,

(ii) $f(z)$ and $\omega(z)$ are analytic in D and locally holomorphic, their absolute values being single-valued in D.

For the maximum principle applies to single-valued harmonic functions, and $\log|F(z)|$ is by assumption single-valued and harmonic in D provided we omit small neighborhoods of any zeros of $f(z)$, where $\log|F(z)|$ becomes negatively infinite. It follows that

$$\log|F(z)| \leqq \log C \quad \text{and} \quad \log|f(z)| \leqq \log C - \eta \log|\omega(z)|.$$

Since η is arbitrary, (18.1.5) holds. The case in which equality holds at a point of D is covered by the Corollary of Theorem 7.9.2 given on page 207 of Volume I.

We may also suppose that

$$\partial D = \bigcup_{n=1}^{\infty} E_n,$$

and that for each set E_n with $n \geq 1$ there is a corresponding function $\omega_n(z)$ such that

$$|f(z)[\omega_n(z)]^{\eta}| \leq C + \varepsilon$$

provided the distance of z from E_n is small. This case, however, may be reduced to the preceding one by defining a function

(18.1.6) $$\omega(z) = \prod_{n=1}^{\infty} [\omega_n(z)]^{\alpha_n}$$

where the α_n's are positive numbers so chosen that the product converges absolutely in D, uniformly with respect to z on compact subsets. Such a choice is always possible and the resulting function $\omega(z)$ is holomorphic in D as well as $\neq 0$ and it does not exceed 1 in absolute value. Since for each n we have

$$|f(z)[\omega(z)]^{\eta}| \leq |f(z)[\omega_n(z)]^{\alpha_n\eta}| \leq C + \varepsilon,$$

if z is sufficiently near to E_n, it is seen that the previous argument applies.

In the preceding discussion we have tacitly assumed D to be bounded. The extension to unbounded domains is immediate.

Generalizations of the Phragmén-Lindelöf principle in a different direction were given by F. and R. Nevanlinna in 1922. These involve the Green's functions of D and of subdomains of D. The extension is listed without proof in the next section.

We shall now give some applications of Theorem 18.1.2. Suppose that $f(z)$ is holomorphic in a bounded domain D and, for $n = 1, 2, 3, \cdots$, let E_n consist of a single point $z_n \in \partial D$. Suppose that for each fixed $\varepsilon > 0$ and each fixed n

(18.1.7) $$\lim (z - z_n)^{\varepsilon} f(z) = 0$$

as $z \to z_n$ in D, uniformly with respect to z. On the rest of the boundary we suppose that condition (i) holds. Then

$$|f(z)| \leq C$$

everywhere in D. For if $z = a$ is a point exterior to D, we may take

$$\omega_n(z) = \frac{z - z_n}{M_n(z - a)},$$

where

$$M_n = \sup_{z \in D} \left| \frac{z - z_n}{z - a} \right|.$$

This function $\omega_n(z)$ is holomorphic in D, $\neq 0$, and ≤ 1 in absolute value in D. Thus the assumptions of the extended Theorem 18.1.2 are satisfied.

Next we prove Phragmén's Theorem 18.1.1. We assume S to be the sector

$$S: \quad | \arg z | \leq \frac{\pi}{2\alpha} .$$

Here E_0 is the set of points $z = r \exp (\pm \pi i/2\alpha)$ with $0 \leq r < \infty$, and E_1 is the point at infinity. Let us first suppose that (18.1.3) holds in the stronger form

$$(18.1.8) \qquad \lim_{r \to \infty} \sup r^{-\beta} \log M(r; f) = 0$$

for some fixed β, $\beta < \alpha$, and let $\beta < \gamma < \alpha$. We now take

$$\omega(z) = \exp (-z^\gamma).$$

Then in S

$$| \omega(z) | \leq \exp (-\rho r^\gamma), \quad \rho = \cos \frac{\pi \gamma}{2\alpha} > 0,$$

so that

$$| f(z)[\omega(z)]^\eta | \leq M(\delta) \exp (\delta r^\beta - \eta \rho r^\gamma),$$

where δ is an arbitrarily small fixed positive number and the inequality is true for $| z | = r > r(\delta)$. Since $\beta < \gamma$, the limit when $r \to \infty$ is 0 for any fixed $\eta > 0$, and this holds uniformly in z. It follows that condition (ii) holds for this choice of $\omega(z)$. Hence the theorem is true if condition (18.1.8) is valid.

The general case is reduced to this particular one by a simple device. For every fixed $\delta > 0$ we have

$$| f(z) | \leq M(\delta) \exp (\delta r^\alpha) \quad \text{if} \quad r > r(\delta).$$

We now take

$$\omega(z) = \exp (-z^\alpha)$$

and set

$$F_\eta(z) = f(z)[\omega(z)]^\eta.$$

This function is holomorphic in S, on the finite boundary of which its absolute value has the same bound C as $f(z)$. For $z \in S$ and $| z | > r(\delta)$ we have

$$| F_\eta(z) | \leq M(\delta) \exp \{[\delta - \eta \cos (\alpha\theta)]r^\alpha\}.$$

If $\delta < \eta$, as we may assume, we see that for $\theta = 0$, that is, on the positive real axis, $\lim F_\eta(z) = 0$ as $z \to \infty$. It follows that $| F_\eta(z) |$ has a finite positive maximum C_1 on the positive real axis. If $C_2 = \max (C, C_1)$, we conclude that $| F_\eta(z) | \leq C_2$ on the three rays $\arg z = k\pi/(2\alpha)$, where $k = 0, \pm 1$. The real axis bisects the sector S into two subsectors S_+ and S_-, each of opening $\pi/(2\alpha)$. Since $F_\eta(z)$ is bounded on the finite boundary of S_+ and $\log M(r; F_\eta) = o(r^\alpha)$ in S_+, we can apply the particular case of the theorem proved above to the function $F_\eta(z)$ in the sector S_+ and conclude that $| F_\eta(z) | \leq C_2$ in S_+. The same argument applies to S_-, so that $| F_\eta(z) | \leq C_2$ in all of S. Here we must have $C_2 = C$ since otherwise we would obtain a contradiction of the classical principle of the maximum. Hence $| F_\eta(z) | \leq C$ everywhere in S and for every $\eta > 0$. It follows that $| f(z) | \leq C$ in S, and the theorem is proved.

This result has an important bearing on the theory of entire functions and gives an extension of Liouville's theorem.

THEOREM 18.1.3. *Suppose that $f(z)$ is an entire function of finite order which does not exceed $\rho > 0$. Suppose $f(z)$ is bounded on a set of rays, $\arg z = \theta_j$, $j = 1, 2, \cdots, n$, such that the angles between consecutive rays are less than π/ρ. Then $f(z)$ is a constant.*

For the rays determine n sectors which together exhaust the plane, and, by Theorem 18.1.1, $f(z)$ is bounded in each sector and, hence, in the whole plane.

COROLLARY. *An entire function of order $\rho < \frac{1}{2}$ cannot be bounded on any ray. If $\rho = \frac{1}{2}$ and $\tau(f) < \infty$, there is at most one ray on which $f(z)$ is bounded.*

For if $\rho < \frac{1}{2}$ and $f(z)$ is bounded on a ray, then the two sides of the ray determine a sector of opening $2\pi < \pi/\rho$. In the case $\rho = \frac{1}{2}$ the functions

$$z^{-\frac{1}{2}} \sin z^{\frac{1}{2}} \quad \text{and} \quad z^{-\frac{1}{2}}(1 - \cos z^{\frac{1}{2}})$$

show that a ray of boundedness may exist.

By conformal mapping of a sector we can obtain results analogous to Theorem 18.1.1. The following result for a half-strip is of sufficient interest to quote:

THEOREM 18.1.4. *If the domain D is contained in a strip of width π/α, if $f(z)$ is holomorphic in D and satisfies condition (i) of Theorem 18.1.2 at all finite points of ∂D, and if, for each $\varepsilon > 0$, uniformly with respect to z,*

(18.1.9) $$\lim \exp\left(-\varepsilon e^{\alpha|z|}\right) |f(z)| = 0$$

as $z \to \infty$ in D, then $f(z)$ is bounded in D, $|f(z)| \leq C$.

Since a logarithmic transformation takes sectors into strips, the theorem is a consequence of Theorem 18.1.1.

EXERCISE 18.1

1. Verify Theorem 18.1.4.

2. Suppose that $f(z)$ is holomorphic in $\Re(z) > 0$ and satisfies condition (i) of Theorem 18.1.2 at all points of the imaginary axis except at $z = 0$, where instead we have, for every $\varepsilon > 0$,

$$\lim_{z \to \infty} \exp\left(-\frac{\varepsilon}{|z|}\right) f(z) = 0,$$

uniformly in z. Prove that $|f(z)| \leq C$ in $\Re(z) > 0$.

3. Verify that for any fixed $\delta > 0$ the function $\exp(\delta/z)$ satisfies condition (i) on the imaginary axis for $z \neq 0$. Since this function is not bounded in the right half-plane, the condition of the preceding problem is the best of its kind.

4. Suppose that $f(z)$ is holomorphic in $\Re(z) > -\delta$, $\delta > 0$, and that

$$\lim_{y \to +\infty} f(iy) = \alpha, \quad \lim_{y \to +\infty} f(-iy) = \beta$$

exist and are finite. If

$$M(r; f) = \sup |f(re^{i\theta})|, \quad |\theta| \leq \tfrac{1}{2}\pi,$$

and if

$$\log M(r; f) = o(r) \quad \text{as} \quad r \to \infty,$$

prove that $\beta = \alpha$ and that

$$\lim_{r \to \infty} f(re^{i\theta}) = \alpha, \quad |\theta| \leq \tfrac{1}{2}\pi,$$

uniformly with respect to θ. (*Hint:* Use Theorems 18.1.1 and 15.4.4.)

5. If $\beta \neq \alpha$ in the preceding problem prove that

$$\limsup_{r \to \infty} r^{-1} \log M(r; f) > 0.$$

6. Prove the assertions concerning the function $\omega(z)$ of (18.1.6).

18.2. Dirichlet's problem; Lindelöf's principle. In this section and later sections we shall have to make use of the existence of certain classes of harmonic functions and of their properties. In order to put the discussion on a firm basis we have to include a discussion of Dirichlet's problem, which has been mentioned several times in earlier parts of this treatise.

The *first boundary problem of potential theory*, or *Dirichlet's problem*, is the following:

Given a domain D bounded by a finite number of disjoint Jordan arcs Γ and a measurable function $u(\zeta)$ on Γ, to find a solution of Laplace's equation

$$\Delta U \equiv \frac{\partial^2 U}{\partial x^2} + \frac{\partial^2 U}{\partial y^2} = 0$$

in D which approaches $u(\zeta)$ when z tends to ζ.

We know the solution of this problem for the particular case of a circular disk $|z| < \rho$. Let

$$(18.2.1) \qquad P(r, \theta; \rho, \varphi) = \frac{1}{2\pi} \frac{\rho^2 - r^2}{\rho^2 + r^2 - 2\rho r \cos(\theta - \varphi)}$$

be the Poisson kernel. If $u(\rho e^{i\varphi}) \in L(-\pi, \pi)$, then

$$(18.2.2) \qquad U(re^{i\theta}) \equiv \int_{-\pi}^{\pi} u(\rho e^{i\varphi}) P(r, \theta; \rho, \varphi) \, d\varphi$$

defines a harmonic function in $|z| < \rho$ which by Theorem 17.5.1 tends to $u(\rho e^{i\theta})$ when $r \to \rho$ for almost all values of θ. Similarly, for the upper half-plane, if $u(x) \in L(-\infty, \infty)$, then

$$(18.2.3) \qquad U(x, y) \equiv \frac{y}{\pi} \int_{-\infty}^{\infty} \frac{u(s) \, ds}{(x - s)^2 + y^2}$$

gives the solution of Dirichlet's problem.

Naturally, explicit elementary formulas can be expected for the solution only in a very few cases, but we shall prove the existence of a unique solution under fairly mild restrictions on D and on $u(\zeta)$. The solution will be expressed in terms of the Green's function of D.

THEOREM 18.2.1.   *Let D be bounded by a finite number of rectifiable Jordan curves Γ. Let $g(z, z_0; D)$ be the Green's function of D and let $-h(z, z_0; D)$ be a conjugate harmonic function of g. Let $u(\zeta)$ be given as a continuous function of ζ on Γ. Then*

$$(18.2.4) \qquad U(z) = \frac{1}{2\pi} \int_\Gamma u(\zeta)\, d_\zeta h(\zeta, z; D)$$

exists for $z \in D$. $U(z)$ is harmonic in D and for each $\zeta \in \Gamma$

$$(18.2.5) \qquad \lim_{z \to \zeta} U(z) = u(\zeta).$$

Proof.   Without restricting the generality we may assume that D is bounded. To start with we shall impose quite severe restrictions on Γ and on $u(\zeta)$. These will be removed later. We assume that the components of Γ are disjoint *analytic* Jordan curves $\Gamma_0, \Gamma_1, \cdots, \Gamma_n$, where each of the curves $\Gamma_1, \cdots, \Gamma_n$ lies in the interior of Γ_0 and Γ_j lies in the exterior of Γ_k if $j, k = 1, \cdots, n$, $j \neq k$. D then is interior to Γ_0 and exterior to $\Gamma_1, \cdots, \Gamma_n$. Each curve is oriented in such a way that D lies to the left of the curve in the positive orientation.

The assumption that the curves are analytic has the following sense: There are functions $f_j(s)$, $j = 0, 1, \cdots, n$, periodic and of period 1, which are holomorphic in s in a narrow strip $-\alpha < \Im(s) < \alpha$ such that

$$(18.2.6) \qquad \Gamma_j: \quad \zeta = f_j(s), \quad 0 \leq s \leq 1, \quad j = 0, 1, \cdots, n.$$

We shall assume also that $f_j{}'(s) \neq 0$ when s is real. This assumption is normally made in speaking of analytic curves.

We shall further assume that the boundary values $u(\zeta)$ are *analytic functions of the local parameter s*, that is, for each j the function $u[f_j(s)]$ is periodic and holomorphic in some strip which we may take to be $-\alpha < \Im(s) < \alpha$. It follows, in particular, that $u(\zeta)$ has a continuous partial derivative in the direction of the interior normal to Γ.

The existence of the Green's function $g(z, z_0; D)$ follows from the discussion in Section 16.5. We recall that $g(z, z_0; D) = 0$ on Γ, $g(z, z_0; D) > 0$ in D, and

$$g(z, z_0; D) + \log |z - z_0|$$

is harmonic in D. We can then find a conjugate harmonic function $-h(z, z_0; D)$ such that

$$(18.2.7) \qquad F(z) \equiv \exp(-g + ih)$$

is locally holomorphic in D and vanishes at $z = z_0$. We have $|F(z)| < 1$ in D

and $| F(z) | \to 1$ as $z \to \zeta \in \Gamma$. If D is multiply-connected, $F(z)$ is normally not single-valued in D, but $| F(z) |$ has this property. Further, if z describes a closed circuit in D, $F(z)$ is multiplied by a constant of absolute value 1. $F(z)$ is single-valued and hence holomorphic in D if $\Gamma = \Gamma_0$ so that D is simply-connected. In the latter case $F(z)$ maps D conformally onto the unit disk in such a manner that $z = z_0$ corresponds to $w = 0$.

The assumptions of analyticity for Γ and for $u(\zeta)$ imply further properties of $F(z)$ which we shall now exploit. Suppose that $\zeta_0 \in \Gamma_j$ and that $f_j(s_0) = \zeta_0$. Since $f_j'(s_0) \neq 0$, there is a small circular neighborhood of $s = s_0$ which is mapped conformally on a neighborhood of $z = \zeta_0$. Let Δ be the intersection of the latter with D and let Σ be the inverse image of Δ in the s-plane. Then Σ is located in the upper half-plane and part of its boundary is an interval σ of the real axis, a subinterval of $(0, 1)$. For $s \in \Sigma$ we form the composite function

$$F_j(s) \equiv F[f_j(s)],$$

which is holomorphic in Σ and whose limiting values on σ have the absolute value 1. By Schwarz's principle of symmetry, Theorem 7.7.2, $F_j(s)$ can be continued analytically across σ into the lower half-plane and is well defined in Σ^*, the conjugate of Σ. This means that $F(z)$ can be continued analytically across an arc of Γ_j containing the point ζ_0. But ζ_0 is an arbitrary point on Γ. It follows that $F(z)$ is analytic everywhere on Γ, and this implies that $g(z, z_0; D)$ and $h(z, z_0; D)$ have continuous derivatives everywhere on Γ.

We conclude from this discussion that the right member of (18.2.4) exists. We shall now prove that if the boundary problem has a solution for the given data, then the solution is given by (18.2.4). To this end we use Green's formula.

Let $\varepsilon > 0$ be a small number, less than the distance of z_0 from Γ, and let D_ε be that part of D in which $| z - z_0 | > \varepsilon$. Let Γ_ε be the boundary of D_ε. If U and V are two functions having continuous first- and second-order partials with respect to x and y in D_ε, then Green's formula gives

$$(18.2.8) \qquad \iint_{D_\varepsilon} [U \, \Delta V - V \, \Delta U] \, dx \, dy = - \int_{\Gamma_\varepsilon} \left[U \frac{\partial V}{\partial n} - V \frac{\partial U}{\partial n} \right] ds,$$

where we integrate on the left over D_ε and on the right over the rectifiable curve Γ_ε. The normal derivative is taken in the direction of the *interior* normal.

Suppose now that $U(z)$ is a solution of the given boundary problem and apply (18.2.8) with

$$U = U(z), \quad V = g(z, z_0; D).$$

Since $\Delta U = \Delta V = 0$ and $g = 0$ on Γ, the formula reduces to

$$(18.2.9) \qquad \int_\Gamma U \frac{\partial g}{\partial n} \, ds + \int_{\gamma_\varepsilon} U \frac{\partial g}{\partial n} \, ds - \int_{\gamma_\varepsilon} g \frac{\partial U}{\partial n} \, ds = 0$$

where γ_ε: $|z - z_0| = \varepsilon$. The first integral exists since the integrand is a holomorphic function of the local parameter s. On γ_ε we have

$$g(z, z_0; D) = -\log \varepsilon + g_0(z)$$

where $g_0(z)$ is harmonic. Further,

$$\frac{\partial}{\partial n} g(z, z_0; D) = -\frac{1}{\varepsilon} + g_1(z).$$

This shows that the second integral in (18.2.9) equals

$$-2\pi U(z_0) + O(\varepsilon),$$

while the third integral is

$$O\left(\varepsilon \log \frac{1}{\varepsilon}\right).$$

Hence, letting $\varepsilon \to 0$, we get

$$(18.2.10) \qquad U(z_0) = \frac{1}{2\pi} \int_\Gamma u(\zeta) \frac{\partial}{\partial n} g(\zeta, z_0; D) \, ds.$$

Here the kernel $\dfrac{\partial}{\partial n} g(\zeta, z_0; D)$ is non-negative since $g = 0$ on Γ and $g > 0$ in D.

Since $g - ih$ is locally holomorphic on Γ, the Cauchy-Riemann equations hold. A simple calculation shows that in terms of the variables n and s we have

$$(18.2.11) \qquad \frac{\partial g}{\partial n} = \frac{\partial h}{\partial s},$$

so that (18.2.10) becomes

$$U(z_0) = \frac{1}{2\pi} \int_\Gamma u(\zeta) \, d_\zeta h(\zeta, z_0; D),$$

and this is (18.2.4).

As a consequence of this formula we obtain

LEMMA 18.2.1. *The functions $g(z, z_0; D)$ and $h(z, z_0; D)$ are harmonic functions of z_0 as well as of z as long as $z_0 \neq z$. Moreover*

$$(18.2.12) \qquad g(z, z_0; D) = g(z_0, z; D), \quad h(z, z_0; D) = h(z_0, z; D).$$

Proof. We know that (18.2.4) gives the solution of Dirichlet's problem provided the latter exists. There is one case in which we have a priori information and that is when

$$u(\zeta) = \log | \zeta - z |$$

where z is a fixed point in D. We have then

$$U(z_0) = g(z_0, z; D) + \log | z_0 - z |,$$

so that

$$(18.2.13) \quad g(z_0, z; D) + \log | z_0 - z | = \frac{1}{2\pi} \int_\Gamma \log | \zeta - z | \, d_\zeta h(\zeta, z_0; D).$$

For a fixed z_0 the right member is a harmonic function of z since $\log |\zeta - z|$ has this property. This shows that $g(z_0, z; D)$ is harmonic in z except for $z = z_0$, where it becomes infinite as $-\log |z - z_0|$. To prove the symmetry property we consider the difference

$$\delta(z, z_0) \equiv g(z, z_0; D) - g(z_0, z; D).$$

This is a harmonic function of both variables everywhere in $D \times D$ including the locus $z = z_0$. If $z \to \zeta \in \Gamma$ we have

$$\limsup \delta(z, z_0) = -\limsup g(z_0, z; D) \leqq 0.$$

This implies that $\delta(z, z_0) \leqq 0$ for every $z_0 \in D$. On the other hand, if $z_0 \to \zeta \in \Gamma$, then

$$\limsup \delta(z, z_0) = \limsup g(z, z_0; D) \geqq 0,$$

so that $\delta(z, z_0) \geqq 0$ everywhere in D. This shows that $\delta(z, z_0) \equiv 0$ and proves the first formula under (18.2.12).

The symmetry property of $h(z, z_0; D)$ follows from that of $g(z, z_0; D)$, and hence it follows also that $h(z, z_0; D)$ is harmonic in both arguments except for $z = z_0$.

We now return to the proof of Theorem 18.2.1. Formula (18.2.4) = (18.2.10) defines the unique solution of Dirichlet's problem for the given data provided the problem has a solution. Now there is a fairly general case in which the existence of a solution is known in advance. Suppose that $V(z)$ is harmonic in $D \cup \Gamma$ and take $u(\zeta) = V(\zeta)$ on Γ. Then $V(z)$ is the unique solution, that is, for such a function we have

$$(18.2.14) \qquad V(z) = \frac{1}{2\pi} \int_\Gamma V(\zeta) \frac{\partial}{\partial n} g(\zeta, z; D) \, ds.$$

In particular, for $V(z) \equiv 1$

$$(18.2.15) \qquad \frac{1}{2\pi} \int_\Gamma \frac{\partial}{\partial n} g(\zeta, z; D) \, ds \equiv 1.$$

Equivalently we may write

$$(18.2.16) \qquad V(z) = \frac{1}{2\pi} \int_\Gamma V(\zeta) \, d_\zeta h(\zeta, z; D),$$

$$(18.2.17) \qquad 1 = \frac{1}{2\pi} \int_\Gamma d_\zeta h(\zeta, z; D).$$

Formula (18.2.4) makes sense also under the more general assumptions of Theorem 18.2.1. We complete the proof of the theorem in two steps: (i) we prove (18.2.16) for functions harmonic in D and continuous in $D \cup \Gamma$ where Γ satisfies the conditions of the theorem; (ii) we form (18.2.4) and prove directly that it gives the solution of Dirichlet's problem. Here we use (18.2.16) in the convergence proof.

Let $\Gamma_0, \Gamma_1, \cdots, \Gamma_n$ be rectifiable Jordan curves. The Green's function $g(z, z_0; D)$ is still well defined, and its conjugate $-h(z, z_0; D)$ is defined up to an additive constant. D is of course the domain bounded by the curves Γ_j. Let λ be a small positive constant and consider the curve

$$\Gamma_\lambda: \quad g(z, z_0; D) = \lambda.$$

If λ is sufficiently small, this curve consists of $n + 1$ ovals $\Gamma_{\lambda 0}, \Gamma_{\lambda 1}, \cdots, \Gamma_{\lambda n}$ which approximate the curves $\Gamma_0, \Gamma_1, \cdots, \Gamma_n$ respectively and bound a domain D_λ contained in D and containing $z = z_0$. The curves Γ_λ are analytic. We have

$$g(z, z_0; D_\lambda) = g(z, z_0; D) - \lambda, \quad h(z, z_0; D_\lambda) = h(z, z_0; D).$$

The function $h(z, z_0; D)$ increases when z describes Γ_λ in the positive sense. Since (18.2.17) holds with Γ replaced by Γ_λ, we see that the increase of $h(z, z_0; D)$ when z describes $\Gamma_{\lambda j}$ is $\leq 2\pi$. Moreover, for each j the increase is independent of λ provided λ is sufficiently small.

If now $V(z)$ is harmonic in D and continuous in $\bar{D}$, then $V(z)$ is analytic in terms of the local parameter on $\Gamma_{\lambda j}$. We can use

$$s = h(\zeta, z_0; D), \quad \zeta \in \Gamma_{\lambda j},$$

to parametrize all curves $\Gamma_{\lambda j}$ for a fixed j since the right-hand side is increasing and independent of λ.

Formula (18.2.16) now is valid for $V(z)$ when $z \in D_\lambda$ and Γ is replaced by Γ_λ so that

$$V(z) = \frac{1}{2\pi} \int_{\Gamma_\lambda} V(\zeta)\, d_\zeta h(\zeta, z; D).$$

This integral is the sum of $n + 1$ integrals extended over $\Gamma_{\lambda 0}, \Gamma_{\lambda 1}, \cdots, \Gamma_{\lambda n}$. Using $h(\zeta, z; D)$ as parameter, we can write

$$(18.2.18) \qquad V(z) = \sum_{j=0}^{n} \frac{1}{2\pi} \int_0^{\alpha_j} V[\zeta_{\lambda j}(s)]\, ds.$$

Here $\zeta_{\lambda j}(s)$ denotes the point on $\Gamma_{\lambda j}$ where $g = \lambda$, $h = s$, and α_j is the increment of $h(\zeta, z; D)$ on $\Gamma_{\lambda j}$. There is no restriction in taking 0 as lower limit of integration instead of integrating over an interval of length α_j. Now $V[\zeta_{\lambda j}(s)]$ is uniformly continuous in (λ, s) in the region under consideration. Thus

$$\lim_{\lambda \to 0} V[\zeta_{\lambda j}(s)] \equiv V[\zeta_j(s)]$$

exists uniformly in s for each j. Hence

$$V(z) = \sum_{j=0}^{n} \frac{1}{2\pi} \int_0^{\alpha_j} V[\zeta_j(s)]\, ds = \frac{1}{2\pi} \int_\Gamma V(\zeta)\, d_\zeta h(\zeta, z; D),$$

that is, (18.2.16) holds under the weaker assumptions: Γ is made up of rectifiable arcs and $V(z)$ is harmonic in D and continuous in $\bar{D}$. This is the first step.

Suppose now that $u(\zeta)$ is a continuous function defined on Γ. Then formula (18.2.4) defines a function in D

$$U(z) \equiv \frac{1}{2\pi} \int_\Gamma u(\zeta)\, d_\zeta h(\zeta, z; D).$$

The integral exists in the Riemann-Stieltjes sense since $u(\zeta)$ is continuous and $h(\zeta, z; D)$ is continuous and increasing on each curve Γ_j. The integral can be approximated by a sum of the form

$$U_m(z) = \frac{1}{2\pi} \sum_{k=1}^m u(\sigma_k)[h(\zeta_k, z; D) - h(\zeta_{k-1}, z; D)].$$

Each such sum is evidently harmonic in D. Since the total variation of $h(\zeta, z; D)$ equals 2π independently of z, it is an easy matter to show that $U_m(z)$ converges uniformly to $U(z)$ in D. A uniformly convergent sequence of harmonic functions converges to a harmonic function, so $U(z)$ is harmonic in D.

Next it should be proved that

(18.2.19) $$\lim_{z \to \zeta} U(z) = u(\zeta)$$

if ζ is any point on Γ. Here we need (18.2.17) and the following lemma:

LEMMA 18.2.2. *Let $\Gamma = \alpha \cup \beta$ where α and β are arcs and $\alpha \cap \beta = \emptyset$. If z approaches a point of α, then*

(18.2.20) $$\lim \int_\beta d_\zeta h(\zeta, z; D) = 0.$$

Proof. We restrict ourselves to the case in which α is a single arc which may or may not contain its endpoints. Suppose that $z \to \zeta_0 \in \alpha$. We choose a function $V(z)$ harmonic in the closure of D such that $V(\zeta) - V(\zeta_0) > 0$ for $\zeta \in \Gamma$, $\zeta \neq \zeta_0$. We use formulas (18.2.16) and (18.2.17) to obtain

$$V(z) - V(\zeta_0) = \frac{1}{2\pi} \int_\Gamma [V(\zeta) - V(\zeta_0)]\, d_\zeta h(\zeta, z; D)$$

$$= \frac{1}{2\pi} \int_\alpha + \frac{1}{2\pi} \int_\beta.$$

Let $\inf_{\zeta \in \beta} [V(\zeta) - V(\zeta_0)] = \mu > 0$. We have then

$$V(z) - V(\zeta_0) > \frac{1}{2\pi} \int_\beta \geq \frac{\mu}{2\pi} \int_\beta d_\zeta h(\zeta, z; D).$$

As $z \to \zeta_0$, $V(z) \to V(\zeta_0)$. Since μ is fixed positive, (18.2.20) follows.

We can now complete the proof as follows: We have

$$U(z) - u(\zeta_0) = \frac{1}{2\pi} \int_\Gamma [u(\zeta) - u(\zeta_0)]\, d_\zeta h(\zeta, z; D) = \frac{1}{2\pi} \int_\alpha + \frac{1}{2\pi} \int_\beta \equiv I_1 + I_2.$$

Here we choose the arc α, containing ζ_0, so small that for a given $\varepsilon > 0$ and $\zeta \in \alpha$ we have $|\, u(\zeta) - u(\zeta_0)\, | \leqq \varepsilon$. This gives

$$| I_1 | \leqq \frac{\varepsilon}{2\pi} \int_\alpha d_\zeta h(\zeta, z; D) \leqq \varepsilon.$$

If $|\, u(\zeta)\, | \leqq M$, then

$$| I_2 | \leqq M \frac{1}{\pi} \int_\beta d_\zeta h(\zeta, z; D),$$

and this tends to zero when $z \to \zeta_0$ by Lemma 18.2.2. It follows that

$$\limsup_{z \to \zeta_0} |\, U(z) - u(\zeta_0)\, | \leqq \varepsilon.$$

Since ε is arbitrary, we conclude that (18.2.19) holds everywhere on the boundary. Hence (18.2.4) gives the solution of the given boundary value problem.

It should be observed that (18.2.4) remains valid under much more general assumptions on Γ and on $u(\zeta)$. In the following we shall restrict ourselves to rectifiable boundary curves, but we shall allow $u(\zeta)$ to have a finite number of discontinuities of the first kind. The convergence proof given above obviously applies at all points of continuity of such a function.

The remainder of this section and the two following sections are devoted to the application of potential theoretic concepts to the problem of majorization. We start with the form given to the Phragmén-Lindelöf principle by the brothers Nevanlinna in 1922.

THEOREM 18.2.2.   *Let D be a finitely-connected domain and f (z) be a function holomorphic in D. Let $[\Gamma_\delta \,|\, 0 < \delta \leqq 1]$ be a family of rectifiable curves in D such that Γ_δ bounds a domain D_δ. Suppose that $D_{\delta_2} \subset D_{\delta_1}$ if $\delta_1 < \delta_2$ and that for each $z \in D$ there is a $\delta(z)$ such that $z \in D$ for $\delta < \delta(z)$. Let $g(z, z_0; D_\delta)$ be the Green's function of D_δ and let $-h(z, z_0; D_\delta)$ be one of its conjugates.*

Suppose that for each $\varepsilon > 0$ there exists a positive measurable function $C_\varepsilon(z)$ such that

(i) $|f(z)| \leqq C_\varepsilon(z)$ *for every* $z \in D$;

(ii) *there is at least one point* $z_0 \in D$ *such that*

$$\liminf_{\delta \to 0} \int_{\Gamma_\delta} \log^+ C_\varepsilon(z) \, d_z h(z, z_0; D_\delta) < \varepsilon.$$

This implies that everywhere in D

$$|f(z)| \leqq 1.$$

For the proof we refer to the original article.

We proceed to what is known as *Lindelöf's principle* (1908).

THEOREM 18.2.3.   *Let D be a domain in the z-plane, G one in the w-plane, each bounded by a finite number of rectifiable Jordan curves. Let $w = f(z)$ be a function which is holomorphic in D and whose values in D belong to G. Then for any choice of a pair of points z and z_0 in D, $z \neq z_0$, we have*

$$(18.2.21) \qquad\qquad g[f(z), f(z_0); G] \geq g(z, z_0; D).$$

If the sign of equality holds for a single pair (z, z_0), then it holds for all.

Proof.   We may assume $z_0 \neq \infty$. The two functions

$$g[f(z), f(z_0); G] + \log |f(z) - f(z_0)| \quad \text{and} \quad g(z, z_0; D) + \log |z - z_0|$$

are harmonic in D by the definition of the Green's function. Hence their difference

$$g[f(z), f(z_0); G] - g(z, z_0; D) + \log \left| \frac{f(z) - f(z_0)}{z - z_0} \right|$$

is also harmonic in D. It follows that

$$\delta(z, z_0) \equiv g[f(z), f(z_0); G] - g(z, z_0; D)$$

is harmonic in D except possibly at the zeros of $f(z) - f(z_0)$, where it may approach $+\infty$. When z tends to a point of ∂D, the function $\delta(z, z_0)$ can approach no limit less than 0. It follows that $\delta(z, z_0) \geq 0$ in D for all (z, z_0). The principle of the minimum shows that either $\delta(z, z_0) > 0$ for all (z, z_0) or $\delta(z, z_0) \equiv 0$. This completes the proof.

The case in which D and G are simply-connected is worthy of special consideration. Let the two auxiliary functions $F(u)$ and $H(u)$ map $|u| < 1$ conformally on D and G respectively in such a manner that

$$F(0) = z_0, \quad F'(0) > 0, \quad G(0) = f(z_0), \quad G'(0) > 0.$$

Let D_r and G_r be the images of $|u| \leq r$, $0 < r < 1$, under these mappings. Then Lindelöf's principle implies the

COROLLARY.   *The values taken on by $f(z)$ in D_r belong to G_r, and if one of these values lies on ∂G_r, then $w = f(z)$ maps D conformally onto G.*

As an example of the use of this principle, let us consider the class $\mathscr{P}$ of functions $f(z)$ holomorphic in the unit disk, whose real parts are positive. We suppose that $f(0) > 0$. Here D is the unit disk, G is the right half-plane, and we take $z_0 = 0$, $w_0 = f(0)$. From (18.2.21) we get

$$(18.2.22) \qquad\qquad \left| \frac{f(z) - f(0)}{f(z) + f(0)} \right| \leq |z|.$$

For the Green's functions involved see Problems 1 and 2 of the Exercise below.

This gives

$$f(z) = f(0) \frac{1+t}{1-t},$$

where $|t| \leq r$ if $|z| \leq r$, and, hence, the double inequality

$$(18.2.23) \qquad f(0) \frac{1-r}{1+r} \leq |f(z)| \leq f(0) \frac{1+r}{1-r}, \quad f(z) \in \mathscr{P}, |z| \leq r.$$

The function

$$f(z) = \frac{1+z}{1-z} \in \mathscr{P}$$

and it maps the unit disk on the right half-plane. Here we have equality between the first and second members of (18.2.23) when $z = -r$ and between the second and third members when $z = r$.

EXERCISE 18.2

1. Show that the Green's function for the unit disk is

$$\log \left| \frac{1 - z\bar{z}_0}{z - z_0} \right|.$$

2. Show that the Green's function for the right half-plane is

$$\log \left| \frac{z + \bar{z}_0}{z - z_0} \right|.$$

3. What inequalities does (18.2.22) imply for the real and the imaginary parts of a function in $\mathscr{P}$?

4. Apply Lindelöf's principle to the case in which $D = [z \mid |z| < 1]$ and $G = [w \mid -1 < \Re(w) < 1]$, $z_0 = w_0 = 0$. Obtain inequalities for the real and the imaginary parts of functions of this class and determine the extremal functions for which equality holds.

5. Prove the existence of a function $V(\zeta)$ with the properties required in the proof of Lemma 18.2.2.

6. Determine a function harmonic in the unit circle which is 1 on an arc of length α on the unit circle and 0 on the complementary arc.

7. Let D be the upper half-plane and let α be an interval of the real axis. Let $z \in D$ and let φ be the angle under which α is seen from z. Show that φ/π is harmonic in D and takes the value 1 on α and the value 0 on the remainder of the real axis.

8. Let D be the domain $[z \mid |z| < 1, 0 < \Im(z)]$ and let α be the semicircular boundary. If φ is the angle under which the diameter $(-1, 1)$ is seen from the point z in D, show that $2(\pi - \varphi)/\pi$ is harmonic in D and takes the value 1 on the semicircle and the value 0 on the diameter.

18.3. Harmonic measure. The last three problems in the preceding Exercise are examples of a special class of harmonic functions known as *harmonic measures*.

DEFINITION 18.3.1.   *Let D be a bounded domain bounded by a finite number of rectifiable Jordan curves Γ. Let $\Gamma = \alpha \cup \beta$, Int $(\alpha) \cap$ Int $(\beta) = \emptyset$, where α and β are finite sets of Jordan arcs. The function $\omega(z, \alpha; D)$ which is harmonic in D and assumes the value 1 on α and the value 0 on β is called the harmonic measure of α with respect to D, evaluated at the point z.*

It follows from the definition that

(18.3.1) $\omega(z, \alpha; D) + \omega(z, \beta; D) \equiv 1, \quad z \in D,$

for the left side is a harmonic function which equals 1 everywhere on Γ and, hence, it is identically 1 in D. This extends to the case in which

$$\Gamma = \alpha_1 \cup \alpha_2 \cup \cdots \cup \alpha_n, \quad \alpha_j \cap \alpha_k = \emptyset,$$

where we have

(18.3.2) $\sum_{k=1}^{n} \omega(z, \alpha_k; D) \equiv 1, \quad z \in D.$

Among the important properties of harmonic measures should be noted the *principle of extension of the domain*, observed in 1921 by Torsten Carleman (1892–1948).

THEOREM 18.3.1.   *Let $D \subset D_1$, $D \subset D_2$ be three domains bounded by a finite number of rectifiable Jordan arcs. Let $\partial D = \alpha \cup \beta$, $\alpha \cap \beta = \emptyset$. Let $\beta \subset D_1$, $\alpha \subset \partial D_1$ and let $\alpha \subset D_2$, $\beta \subset \partial D_2$. Then for $z \in D$*

(18.3.3) $\omega(z, \alpha; D) \leq \omega(z, \alpha; D_1), \quad \omega(z, \beta; D_2) \leq \omega(z, \beta; D).$

Proof.   We consider the difference

$$\delta(z) = \omega(z, \alpha; D_1) - \omega(z, \alpha; D),$$

which is harmonic in D. On α both measures equal 1 so that $\delta(z) = 0$. On β we have $\omega(z, \alpha; D_1) \geq 0$ while $\omega(z, \alpha; D) = 0$ so that $\delta(z) \geq 0$. Since $\delta(z) \geq 0$ everywhere on ∂D, it follows that $\delta(z) \geq 0$ in D. The second inequality is proved in the same manner.

The theorem asserts that the harmonic measure $\omega(z, \alpha; D)$ increases if we extend the domain across its β-boundary and decreases for extensions across the α-boundary. The value of the principle lies in the fact that it enables us to replace a given domain by a simpler one for which it is easier to compute or to estimate harmonic measures.

Next we observe that harmonic measures may be used to represent the solution of Dirichlet's problem. To this end we consider a suitable parametrization of the boundary Γ which we take as in Theorem 18.2.1. Define a function $f(s)$ in the interval $(0, n + 1)$ such that

$$\Gamma_j: \quad \zeta = f(s), \quad j \leq s \leq j + 1, \quad j = 0, 1, \cdots, n.$$

Denote by $\alpha(s)$ the "arc" ($=$ collection of at most $n + 1$ distinct arcs) on Γ which corresponds to the interval $(0, s)$ and define the harmonic measure

$$(18.3.4) \qquad \omega(z, s) \equiv \omega(z, \alpha(s); D), \quad z \in D.$$

For a fixed z the function $\omega(z, s)$ is continuous in s and never decreasing; it goes from 0 to 1 when s goes from 0 to $n + 1$. We now form

$$(18.3.5) \qquad U(z) \equiv \frac{1}{2\pi} \int_{\Gamma} u[f(s)] \, d_s \omega(z, s), \quad z \in D,$$

where $u(\zeta)$ is the given boundary function, which we assume to be continuous except for a finite number of discontinuities of the first kind. $U(z)$ is a bounded function whose values lie between the greatest lower and the least upper bounds of $u(\zeta)$. Since $U(z)$ is the uniform limit of a sequence of Riemann-Stieltjes sums, each of which is harmonic in D, it follows that $U(z)$ is harmonic in D.

If $z \to f(s_0)$, then

$$(18.3.6) \qquad \lim \omega(z, s) = \begin{cases} 0, & 0 \leq s < s_0, \\ 1, & s_0 < s \leq n + 1. \end{cases}$$

From this we conclude that

$$\lim U(z) = u[f(s_0)] = u(\zeta_0)$$

at all points of continuity of $u(\zeta)$. It follows that (18.3.5) represents the solution of Dirichlet's problem for the given data.

A beautiful application of harmonic measures is furnished by the n *constants theorem* discovered by the brothers Nevanlinna and by A. Ostrowski in 1922.

THEOREM 18.3.2.   *Suppose that $f(z)$ is holomorphic in a domain D whose boundary is the union of n distinct rectifiable arcs $\alpha_1, \cdots, \alpha_n$. Suppose that for each j there is a constant M_j such that if z approaches any point of α_j then the limits of $f(z)$ do not exceed M_j in absolute value. Then for each $z \in D$*

$$(18.3.7) \qquad \log |f(z)| \leq \sum_{j=1}^{n} \omega(z, \alpha_j; D) \log M_j.$$

Proof.   Let $\delta(z)$ denote the difference between the right and the left members of (18.3.7). Then $\delta(z)$ is harmonic everywhere in D except at the zeros of $f(z)$, where $\delta(z) \to +\infty$. On Γ we have $\delta(z) \geq 0$ everywhere. From this it follows that $\delta(z) \geq 0$ in D, and (18.3.7) is proved.

THEOREM 18.3.3.   *Under the assumptions of the preceding theorem, let C be a compact subset of D. Then there exist n positive constants $\lambda_1, \lambda_2, \cdots, \lambda_n$ such that for $z \in C$ we have*

$$(18.3.8) \qquad |f(z)| \leq \prod_{j=1}^{n} M_j^{\lambda_j}, \quad \sum_{j=1}^{n} \lambda_j = 1.$$

Proof.   Suppose that the numbering is such that M_n is the largest of the constants M_j. Using (18.3.2) and (18.3.7) we get

$$\log |f(z)| \leq \log M_n + \sum_{j=1}^{n-1} \log \frac{M_j}{M_n} \omega(z, \alpha_j; D).$$

Set

$$\lambda_j = \min \omega(z, \alpha_j; D), \quad z \in C, \quad j = 1, 2, \cdots, n-1.$$

Then $0 < \lambda_j < 1$, and

$$1 \equiv \omega(z, \alpha_n; D) + \sum_{j=1}^{n-1} \omega(z, \alpha_j; D) \geq \omega(z, \alpha_n; D) + \sum_{j=1}^{n-1} \lambda_j$$

shows that for $z \in C$ we have

$$0 < \omega(z, \alpha_n; D) \leq 1 - \sum_{j=1}^{n-1} \lambda_j \equiv \lambda_n.$$

This gives

$$\log |f(z)| \leq \log M_n + \sum_{j=1}^{n-1} \lambda_j \log \frac{M_j}{M_n} = \sum_{j=1}^{n} \lambda_j \log M_j,$$

which is equivalent to (18.3.8).

The case $n = 2$ leads to a number of interesting applications, some of which are listed below. We start with the so-called *three circles theorem* of J. Hadamard.

THEOREM 18.3.4. *Let $f(z)$ be holomorphic in $|z| < R$ and set $M(r) = M(r; f) = \max\limits_{\theta} |f(re^{i\theta})|$. If $0 < r_1 \leq r \leq r_2 < R$, then*

$$(18.3.9) \quad \log M(r) \leq \frac{\log r_2 - \log r}{\log r_2 - \log r_1} \log M(r_1) + \frac{\log r - \log r_1}{\log r_2 - \log r_1} \log M(r_2).$$

Proof. We shall give two proofs of this theorem, the first of which is based on the two constants theorem.

I. We take $D = [z \mid r_1 < |z| < r_2]$ and let α be the circle $|z| = r_1$, β the circle $|z| = r_2$. Then

$$\omega(z, \alpha; D) = \frac{\log r_2 - \log r}{\log r_2 - \log r_1}, \quad \omega(z, \beta; D) = \frac{\log r - \log r_1}{\log r_2 - \log r_1}.$$

We take $n = 2$, $M_1 = M(r_1)$, $M_2 = M(r_2)$, and substitute in (18.3.7). The result is (18.3.9).

II. Let σ be a real number to be disposed of later. The function $z^\sigma f(z)$ has a single-valued absolute value in the annulus under consideration, so that the principle of the maximum applies and gives

$$r^\sigma M(r) \leq \min [r_1{}^\sigma M(r_1), r_2{}^\sigma M(r_2)].$$

If we take

$$\sigma = \frac{\log M(r_1) - \log M(r_2)}{\log r_2 - \log r_1},$$

then the two expressions in the brackets become equal and the resulting estimate reduces to (18.3.9).

An equivalent form of (18.3.9) is the determinant condition

$$(18.3.10) \quad \begin{vmatrix} \log M(r) & \log r & 1 \\ \log M(r_1) & \log r_1 & 1 \\ \log M(r_2) & \log r_2 & 1 \end{vmatrix} \geq 0.$$

We can also connect the theorem with the theory of *convex functions*.

A function $g(x)$ is said to be convex in the interval $[a, b]$ if for each x, x_1, x_2 with $a \leq x_1 \leq x \leq x_2 \leq b$ we have

$$(18.3.11) \qquad g(x) \leq \frac{x_2 - x}{x_2 - x_1} g(x_1) + \frac{x - x_1}{x_2 - x_1} g(x_2).$$

A comparison with (18.3.9) leads to the

COROLLARY. $\log M(r)$ is a convex function of $\log r$.

Such convexity properties are also valid for certain mean values of $f(z)$.

In Section 15.4 we proved several results due to Lindelöf, who obtained them by potential theoretic considerations. We shall state and prove a weaker form of Theorem 15.4.4 using the two constants theorem.

THEOREM 18.3.5. *If $f(z)$ is holomorphic and bounded in the upper half-plane, if $f(z)$ is continuous at all finite points of the real axis, and if $f(x) \to a$ when $x \to +\infty$, then*

$$(18.3.12) \qquad \lim_{z \to \infty} f(z) = a$$

uniformly in any sector $0 \leq \arg z \leq \pi - \delta$, $0 < \delta$.

Proof. Without restricting the generality we may assume that $|f(z)| \leq 1$ for $0 < \Im(z)$ and that $a = 0$. Given an $\varepsilon > 0$, we can then find an A such that

$$|f(x)| \leq \varepsilon, \quad A \leq x < \infty.$$

We use the two constants theorem with D equal to the upper half-plane and with α as the line segment (A, ∞), β as the complement, $M_1 = \varepsilon$, and $M_2 = 1$. This gives

$$\log |f(z)| \leq \omega(z, \alpha; D) \log \varepsilon.$$

Here

$$(18.3.13) \qquad \omega(z, \alpha; D) = \frac{\varphi}{\pi} = 1 - \frac{1}{\pi} \arg(z - A).$$

This is clearly a harmonic function in the upper half-plane which approaches 1 if z tends to a point on (A, ∞) and 0 on the complementary segment. We note that φ is the angle under which (A, ∞) is seen from the point z.

In the sector

$$0 \leq \arg(z - A) \leq \pi(1 - \delta)$$

we have

$$\delta \leq \omega(z, \alpha; D) \leq 1$$

and, hence,

$$\log |f(z)| \leq \delta \log \varepsilon, \quad |f(z)| \leq \varepsilon^\delta.$$

Here δ is a fixed positive number and ε is at our disposal. If $\eta > 0$ is given, we can choose ε so small that $\varepsilon^\delta \leq \eta$, that is, $|f(z)| \leq \eta$ in the corresponding sector $0 \leqq \arg (z - A) \leqq \pi(1 - \delta)$, where, of course, A depends upon the choice of ε. This proves the theorem.

The following is an amplification of Theorem 15.4.4. See also Problem 4 of Exercise 18.1.

THEOREM 18.3.6.   *Suppose that $f(z)$ is holomorphic in the upper half-plane and continuous on the real axis. Suppose that*

$$(18.3.14) \qquad \lim_{x \to +\infty} f(x) = a, \quad \lim_{x \to -\infty} f(x) = b.$$

If

$$\log M(r;f) = o(r), \quad M(r;f) = \max_\theta |f(re^{i\theta})|,$$

then $a = b$, and

$$\lim_{z \to \infty} f(z) = a$$

holds uniformly in $0 \leqq \arg z \leqq \pi$.

Proof.   By Theorem 18.1.1, $f(z)$ is bounded in $\Im(z) > 0$. We then apply Theorem 18.3.5 to

$$f(z) - a \quad \text{in} \quad 0 \leqq \arg z \leqq \tfrac{3}{4}\pi,$$

and to

$$f(z) - b \quad \text{in} \quad \tfrac{1}{4}\pi \leqq \arg z \leqq \pi.$$

These functions tend uniformly to zero in the sectors indicated. Since the sectors overlap, we must have $a = b$ and we conclude that $f(z) \to a$ uniformly in the union of the sectors.

As a last application we give the sharper form of Theorem 18.1.1 which was discovered by the brothers Nevanlinna in 1922.

THEOREM 18.3.7.   *Suppose that $f(z)$ is holomorphic in the upper half-plane and that its limiting values on the real axis are at most 1 in absolute value. Set*

$$(18.3.15) \qquad \lambda = \liminf_{r \to \infty} \frac{1}{r} \log M(r;f).$$

If $\lambda = 0$, then $|f(z)| \leqq 1$ in the upper half-plane. If $0 < \lambda < \infty$, then

$$(18.3.16) \qquad |f(x + iy)| \leqq \exp\left[\frac{4\lambda}{\pi} y\right].$$

REMARK.   It will be shown in the next section that $\lambda \geqq 0$. An equivalent form of (18.3.16) is that

$$(18.3.17) \qquad f(z) = F(z) \exp\left[-\frac{4\lambda}{\pi} iz\right], \quad \Im(z) > 0,$$

where $|F(z)| \leqq 1$.

Proof.   By assumption there exists a sequence $\{r_n\}$, $r_n \to \infty$, such that

$$\lim_{n \to \infty} \frac{1}{r_n} \log M(r_n; f) = \lambda.$$

We apply the two constants theorem to the semicircular disk

$$[z \mid |z| < r, 0 < \Im(z)].$$

Further, α is the semicircle, β is the diameter, $M_1 = M(r)$, $M_2 = 1$.
Then

$$\log |f(z)| \leq \omega(z, \alpha; D) \log M(r).$$

According to Problem 8 of Exercise 18.2 the harmonic measure is

$$\omega(z, \alpha; D) = \frac{2}{\pi} (\pi - \varphi),$$

where φ is the angle under which the diameter $(-r, r)$ is seen from the point z.
According to Figure 27, $\pi - \varphi$ is the sum of the angles θ and ψ, so that

$$\omega(z, \alpha; D) = \frac{2}{\pi} \left[\text{arc tan} \frac{y}{r + x} + \text{arc tan} \frac{y}{r - x} \right].$$

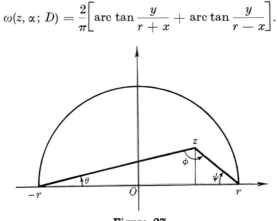

Figure 27

Here x and y are fixed and we shall let r become infinite. Expanding the arc
tangents we get

$$\omega(z, \alpha; D) = \frac{4y}{\pi r} \left\{ 1 + O\left(\frac{1}{r^2}\right) \right\},$$

so that

$$\log |f(z)| \leq \frac{4y}{\pi r} \log M(r; f) \left\{ 1 + O\left(\frac{1}{r^2}\right) \right\}$$

Here we let $r = r_n \to \infty$ and obtain

$$\log |f(z)| \leq \frac{4\lambda}{\pi} y,$$

which is (18.3.17).

EXERCISE 18.3

1. Prove that a uniformly convergent sequence of harmonic functions converges to a harmonic limit. (*Hint:* Use a representation of harmonic functions such as Poisson's integral or (18.2.4).)

2. Prove that $\omega(z, \alpha(s); D)$ is continuous and never decreasing. Verify (18.3.6) and that $\lim U(z) = u(\zeta)$ at points of continuity of $u(\zeta)$.

3. Prove that a function is convex in $[a, b]$ if $g''(x) \geqq 0$. Give an example of a convex function which fails to have a second derivative at some points.

4. If $f(z) = \sum_{n=0}^{\infty} a_n z^n$, $|z| < R$, and if

$$M_2(r;f) = \left\{ 2\pi \sum_{n=0}^{\infty} |a_n|^2 r^{2n} \right\}^{\frac{1}{2}},$$

verify that $\log M_2(r;f)$ is a convex function of $\log r$. (*Hint:* Set $r = e^s$ and compute the second derivative. The latter is the quotient of two power series in e^{2s} with non-negative coefficients.)

5. For every sector of opening $\delta\pi$, $0 < \delta < 2$, there are entire functions which are bounded in the sector and in no larger sector. Show that the Mittag-Leffler function $E_\alpha(z)$ is bounded in $\frac{1}{2}\alpha\pi < \arg z < 2\pi - \frac{1}{2}\alpha\pi$.

18.4. The Nevanlinna-Ahlfors-Heins theorems. Let us return to the point of departure of this chapter, namely, the behavior of a holomorphic function for approach to infinity in a sector. It is sufficient to consider the case of the right half-plane since we can always map a sector conformally onto a half-plane. Theorems 18.1.1 and 18.3.6 give us the following alternatives:

If $f(z)$ is holomorphic in the right half-plane and bounded on the imaginary axis, then either $f(z)$ is bounded in the right half-plane or

$$(18.4.1) \qquad\qquad \liminf_{r \to \infty} r^{-1} \log M(r;f) > 0.$$

Maurice Heins proved in 1946 that in the second case $r^{-1} \log M(r;f)$ tends to a definite limit which may be $+\infty$. This result will be proved below.

Instead of the function $M(r;f)$ we may use other growth-measuring functions. Thus, in 1922 the Nevanlinna brothers introduced the function

$$(18.4.2) \qquad\qquad m(r;f) = \int_{-\frac{1}{2}\pi}^{\frac{1}{2}\pi} \log^+ |f(re^{i\theta})| \cos\theta \, d\theta,$$

and they proved that either $f(z)$ is bounded in $\Re(z) > 0$ or else

$$(18.4.3) \qquad\qquad \liminf_{r \to \infty} r^{-1} m(r;f) > 0.$$

In 1937 L. V. Ahlfors proved that $r^{-1}m(r;f)$ is a nondecreasing function of r, so that the ratio has a finite or infinite limit. Heins also gave a simple proof of Ahlfors' theorem; he proved that the two limits involved are cofinite and he gave the relation holding between them. Following Heins we shall present these results below. The basis of the discussion is a lemma which is also of independent interest.

LEMMA 18.4.1.   *Suppose that $F(z)$ is holomorphic in $\Re(z) > 0$ and that $|F(z)| \leq 1$. Let $\mu(r, \varepsilon)$ be the Lebesgue measure of the set*

(18.4.4) $$E(r, \varepsilon) = [\theta \mid \log |F(re^{i\theta})| < -\varepsilon r, |\theta| < \tfrac{1}{2}\pi].$$

If

(18.4.5) $$\limsup_{r \to \infty} \mu(r, \varepsilon) > 0,$$

then there exists a positive constant κ such that

(18.4.6) $$\log |F(x + iy)| < -\kappa x, \quad 0 < x.$$

Proof.   For $|\theta| < \tfrac{1}{2}\pi$, let $\chi(\theta; r, \varepsilon)$ be the characteristic function of the set $E(r, \varepsilon)$. We extend this function for all θ by the following conventions:

(18.4.7) $$\begin{cases} \chi(\tfrac{1}{2}\pi; r, \varepsilon) = \chi(-\tfrac{1}{2}\pi; r, \varepsilon) = 0, \\ \chi(\theta; r, \varepsilon) = -\chi(\pi - \theta; r, \varepsilon) \quad \text{for} \quad \tfrac{1}{2}\pi < \theta < \tfrac{3}{2}\pi, \\ \chi(\theta + 2\pi; r, \varepsilon) = \chi(\theta; r, \varepsilon). \end{cases}$$

We shall use the Poisson kernel $P(r, \theta; \rho, \varphi)$, for the properties of which we refer to Section 17.5. We shall also need

(18.4.8) $$2\pi P(r, \theta; \rho, \varphi) = 1 + \frac{2r}{\rho} \cos(\varphi - \theta) + O\left(\frac{r}{\rho}\right)^2,$$

valid for small values of r/ρ, and the following simple properties of $\chi(\theta; r, \varepsilon)$:

(18.4.9) $$\int_{-\pi}^{\pi} \chi(\varphi; \rho, \varepsilon)\, d\varphi = \int_{-\pi}^{\pi} \chi(\varphi; \rho, \varepsilon) \sin \varphi\, d\varphi = 0.$$

We define

(18.4.10) $$G(r, \theta; \rho) \equiv \int_{-\pi}^{\pi} \chi(\varphi; \rho, \varepsilon) P(r, \theta; \rho, \varphi)\, d\varphi,$$

and our aim is to establish the basic inequality

(18.4.11) $$\log |F(re^{i\theta})| \leq -\varepsilon\rho\, G(r, \theta; \rho), \quad |\theta| < \tfrac{1}{2}\pi.$$

Now $G(r, \theta; \rho)$ is for fixed ρ and $r < \rho$ a harmonic function of (r, θ) since the Poisson kernel has this property. We have

$$G(r, \pm\tfrac{1}{2}\pi; \rho) = 0,$$

for by (18.4.7) and the symmetry properties of the Poisson kernel we have

$$\chi(\varphi; \rho, \varepsilon) P(r, \pm\tfrac{1}{2}\pi; \rho, \varphi) = -\chi(\pi - \varphi; \rho, \varepsilon) P(r, \pm\tfrac{1}{2}\pi; \rho, \pi - \varphi).$$

For $|\theta| < \tfrac{1}{2}\pi$ we have

(18.4.12) $$-\varepsilon\rho \lim_{r \to \rho} G(r, \theta; \rho) = -\varepsilon\rho\, \chi(\theta; \rho, \varepsilon) \geq \log |F(\rho e^{i\theta})|.$$

By Theorem 17.5.1 the limit exists and has the stated value at all points of continuity of $\chi(\varphi; \rho, \varepsilon)$. Since $E(r, \varepsilon)$ consists of a finite or countable number of intervals, this means that the exceptional set where the limit relation does not hold consists of the endpoints of these intervals. At such a point the second member of (18.4.12) becomes $-\frac{1}{2}\varepsilon\rho$ instead. The inequality follows from the definition of $E(r, \varepsilon)$ together with the fact that $|F(z)| \leq 1$. It is valid for all values of θ, $|\theta| < \frac{1}{2}\pi$. Hence

$$-\varepsilon\rho\, G(r, \theta; \rho) - \log|F(re^{i\theta})| \geq 0$$

on the boundary of the semicircular disk $|x + iy| < \rho$, $0 < x$. The difference is a harmonic function of (r, θ) except at the zeros of $F(z)$, if any, where it becomes positively infinite. It follows that the inequality holds also in the interior of the disk, that is, (18.4.11) holds as asserted.

If r/ρ is small, (18.4.8) and (18.4.9) give

$$G(r, \theta; \rho) = \frac{1}{2\pi}\int_{-\pi}^{\pi}\chi(\varphi; \rho, \varepsilon)\, d\varphi + \frac{r}{\rho\pi}\int_{-\pi}^{\pi}\chi(\varphi; \rho, \varepsilon)\cos(\varphi - \theta)\, d\varphi + O\left(\frac{r}{\rho}\right)^2$$

$$= \frac{r}{\pi\rho}\cos\theta\int_{-\pi}^{\pi}\chi(\varphi; \rho, \varepsilon)\cos\varphi\, d\varphi + O\left(\frac{r}{\rho}\right)^2.$$

In order to get an upper bound for $\log|F(re^{i\theta})|$ it suffices to get a lower bound for the integral in the last member which equals twice

$$(18.4.13) \qquad\qquad \int_{-\frac{1}{2}\pi}^{\frac{1}{2}\pi}\chi(\varphi; \rho, \varepsilon)\cos\varphi\, d\varphi.$$

Since χ is 0 or 1, we get the least possible value for the integral if the set $E(r, \varepsilon)$ reduces to two intervals of equal length, one at each end of the interval of integration where $\cos\varphi$ is small. It follows that (18.4.13) cannot be less than

$$(18.4.14) \qquad\qquad 2\{1 - \cos[\tfrac{1}{2}\mu(\rho, \varepsilon)]\} \equiv \tfrac{1}{4}\gamma(\rho, \varepsilon).$$

Combining these estimates we finally get

$$(18.4.15) \qquad\qquad \log|F(re^{i\theta})| \leq -\frac{\varepsilon}{2\pi}\gamma(\rho, \varepsilon)\, r\cos\theta + O\left(\frac{r^2}{\rho}\right).$$

The lemma is an immediate consequence of this inequality since (18.4.5) implies that $\lim\sup \gamma(\rho, \varepsilon) > 0$.

We can now state and prove the main theorem of Heins.

THEOREM 18.4.1. *Suppose that* $f(z)$ *is holomorphic in* $\Re(z) > 0$ *and* $\lim\sup_{x \to 0}|f(x + iy)| \leq 1$ *for all* y. *Set*

$$(18.4.16) \qquad \liminf_{r \to \infty} r^{-1}M(r; f) = \lambda, \quad \limsup_{r \to \infty} r^{-1}M(r; f) = \mu.$$

Then $\lambda \geq 0$ *and* $\lambda = \mu$. *If* $0 < \lambda < \infty$, *then either*

$$(18.4.17) \qquad\qquad \log M(r; f) < \lambda r$$

for all r *or*

$$(18.4.18) \qquad\qquad f(z) = Ce^{\lambda z}, \quad |C| = 1.$$

Proof.   We exclude two trivial possibilities: (i) $f(z) \equiv 0$ and (ii) $\lambda = +\infty$. We separate two cases.

Case I. $0 < \lambda < \infty$. In this case it is immediate from (18.3.17) that

$$\mu \leq \frac{4}{\pi}\lambda.$$

In the present case

(18.4.19) $$f(z) = F(z) \exp\left(\frac{4\lambda}{\pi} z\right),$$

where $|F(z)| \leq 1$. Let κ be the largest non-negative number for which

(18.4.20) $$\log |F(x + iy)| \leq -\kappa x$$

holds for all $x > 0$. Such a number κ exists since $F(z) \not\equiv 0$. Set

$$F(z) = e^{-\kappa z}G(z).$$

Here $G(z)$ is holomorphic in $x > 0$, its absolute value does not exceed 1, and $G(z)$ cannot satisfy an inequality of type (18.4.20) with a strictly positive κ. From the definition of κ it follows that

$$\mu \leq \frac{4\lambda}{\pi} - \kappa.$$

If now $\lambda < \mu$, then we have

$$\lambda < \frac{4\lambda}{\pi} - \kappa.$$

Moreover, from the definition of λ it follows that there exists a sequence $\{r_k\}$, $r_k \to \infty$, such that

$$\log M(r_k; f) < (\lambda + \varepsilon)r_k,$$

ε fixed positive. Further, if $\lambda < \mu$ then there exist positive numbers γ and δ such that the inequality

$$\log |G(r_k e^{i\theta})| \leq -\gamma r_k$$

holds for all k and all θ in the interval $(-\delta, \delta)$. We can clearly take

$$-\gamma = \lambda + \varepsilon - \left(\frac{4\lambda}{\pi} - \kappa\right)\cos\delta,$$

which is negative if δ is sufficiently small. It follows that the function $G(z)$ satisfies the assumptions of Lemma 18.4.1 with $\limsup \mu(r, \gamma) \geq 2\delta$. The lemma then asserts that there is a positive κ_0 such that

$$\log |G(x + iy)| \leq -\kappa_0 x$$

for all $x > 0$. Since this contradicts the definition of $G(z)$, it follows that the hypothesis $\lambda < \mu$ is untenable.

Hence $\lambda = \mu$ and we have

$$\lambda = \frac{4\lambda}{\pi} - \kappa,$$

so that

(18.4.21) $f(z) = e^{\lambda z} F(z), \quad |F(z)| \leq 1, 0 < \Re(z).$

It follows that

$$\log |f(x + iy)| \leq \lambda x.$$

If there exists a $z_0 = x_0 + iy_0$ with $0 < x_0$ for which we have equality, then $|F(z_0)| = 1$ and we must have $|F(z)| = 1$ for all z or $F(z) = C$, a constant of absolute value 1.

Case II. $-\infty < \lambda < 0$. We have $f(z) \not\equiv 0$ by assumption and we want to prove that this case cannot arise. Let κ denote the largest positive number such that

(18.4.22) $\log |f(x + iy)| \leq -\kappa x$

for all $x > 0$. The existence of such a number follows from (18.3.17). We have now

$$f(z) = e^{-\kappa z} G(z)$$

where $G(z)$ is holomorphic for $\Re(z) > 0$ and is of modulus 1 at most. Further, $G(z)$ cannot satisfy an inequality of type (18.4.22) for any positive κ. The proof then proceeds as in Case I. We have

$$G(z) = e^{\kappa z} f(z),$$

and there is a sequence $\{r_k\}$, $r_k \to \infty$, such that

$$\log |G(r_k e^{i\theta})| \leq [\kappa \cos \theta + \lambda + \varepsilon] r_k.$$

Since $\lambda < 0$ we can find θ-intervals of fixed positive length in which

$$\kappa \cos \theta + \lambda + \varepsilon \leq -\eta,$$

where η is a suitably chosen positive number. This implies that condition (18.4.5) is satisfied for $\varepsilon = \eta$, and we get the same contradiction as before. It follows that Case II cannot arise, and the theorem is proved.

Heins used the same method to prove the theorem of Ahlfors.

THEOREM 18.4.2.   *Under the same assumptions on $f(z)$ as in the preceding theorem, the ratio*

(18.4.23) $r^{-1} m(r; f)$

is a never decreasing function of r.

Proof.   We define a function $L(\rho, \varphi)$ for $\rho > 0$ and all real φ by

(18.4.24)
$$
\begin{aligned}
L(\rho, \varphi) &\equiv \log^+ |f(\rho e^{i\varphi})|, \quad |\varphi| < \tfrac{1}{2}\pi, \\
L(\rho, \pm\tfrac{1}{2}\pi) &= 0, \\
L(\rho, \varphi) &= -L(\rho, \pi - \varphi), \quad \tfrac{1}{2}\pi < \varphi < \tfrac{3}{2}\pi, \\
L(\rho, \varphi + 2\pi) &= L(\rho, \varphi).
\end{aligned}
$$

Since $\limsup |f(x + iy)| \leq 1$, it follows that $L(\rho, \varphi)$ is a continuous function of φ for all φ.

For $r < \rho$ and $|\theta| < \tfrac{1}{2}\pi$ we have the basic inequality

$$
\log^+ |f(re^{i\theta})| \leq \int_{-\pi}^{\pi} L(\rho, \varphi) P(r, \theta; \rho, \varphi)\, d\varphi
$$

(18.4.25)

$$
= \int_{-\frac{1}{2}\pi}^{\frac{1}{2}\pi} \log^+ |f(\rho e^{i\varphi})| \, [P(r, \theta; \rho, \varphi) - P(r, \pi - \theta; \rho, \varphi)]\, d\varphi,
$$

which is proved by the same type of argument as was used for (18.4.11). We leave the details to the reader.

Next we note that

$$
\int_{-\frac{1}{2}\pi}^{\frac{1}{2}\pi} \cos\theta \, [P(r, \theta; \rho, \varphi) - P(r, \pi - \theta; \rho, \varphi)]\, d\theta
$$

$$
= \int_{-\pi}^{\pi} \cos\theta \, P(r, \theta; \rho, \varphi)\, d\theta = \int_{-\pi}^{\pi} \cos\theta \, P(r, \varphi; \rho, \theta)\, d\theta = \frac{r}{\rho} \cos\varphi
$$

by the symmetry properties of the Poisson kernel.

If now we multiply the first and last members of (18.4.25) by $\cos\theta$ and integrate with respect to θ from $-\tfrac{1}{2}\pi$ to $\tfrac{1}{2}\pi$ and change the order of integration in the repeated integral, we obtain

$$
m(r; f) \leq \int_{-\frac{1}{2}\pi}^{\frac{1}{2}\pi} \log^+ |f(\rho e^{i\varphi})| \, \frac{r}{\rho} \cos\varphi \, d\varphi,
$$

that is,

$$
r^{-1} m(r; f) \leq \rho^{-1} m(\rho; f), \quad r < \rho,
$$

as asserted.

These two theorems assert that

$$
\lambda = \lim r^{-1} \log M(r; f)
$$

and

(18.4.26)
$$
\nu = \lim r^{-1} m(r; f)
$$

exist as finite numbers or $+\infty$. Heins proved the following relation between the two limits:

THEOREM 18.4.3.   *λ and ν are cofinite, and if λ is finite then*

(18.4.27)
$$
\nu = \tfrac{1}{2}\pi\lambda.
$$

Proof.   It is understood that $f(z) \not\equiv 0$. We have

$$m(r; f) = \int_{-\frac{1}{2}\pi}^{\frac{1}{2}\pi} \log^+ |f(re^{i\theta})| \cos \theta \, d\theta \leq 2 \log M(r; f),$$

so that

$$\nu \leq 2\lambda.$$

Hence, if $\lambda = 0$ so is ν, if λ is finite so is ν; and $\nu = \infty$ implies $\lambda = \infty$. We have also the following inequality:

(18.4.28) $$\lambda \leq \frac{2}{\pi} \nu.$$

We base the proof upon (18.4.8), which shows that for small values of r/ρ we have

$$P(r, \theta; \rho, \varphi) - P(r, \pi - \theta; \rho, \varphi) = \frac{2}{\pi} \frac{r}{\rho} \cos \theta \cos \varphi + O\left(\frac{r}{\rho}\right)^2.$$

It follows that

$$\log^+ |f(re^{i\theta})| \leq \frac{2}{\pi} r \cos \theta \frac{1}{\rho} \int_{-\frac{1}{2}\pi}^{\frac{1}{2}\pi} \log^+ |f(\rho e^{i\varphi})| \cos \varphi \, d\varphi + O\left(\frac{r}{\rho}\right)^2.$$

Here we pass to the limit with ρ and obtain

(18.4.29) $$\log^+ |f(re^{i\theta})| \leq \frac{2}{\pi} \nu x.$$

On the other hand, we know that

$$\log |f(re^{i\theta})| \leq \lambda x$$

and that λ is the largest constant for which such an estimate is true. From this it follows that (18.4.28) holds.

Our first conclusion is that $\nu = 0$ implies $\lambda = 0$ and that ν finite requires that λ be finite. To prove the theorem we have to be able to reverse the inequality. But this is immediate, for from

$$\log^+ |f(re^{i\theta})| \leq \lambda r \cos \theta$$

we get

$$m(r; f) = \int_{-\frac{1}{2}\pi}^{\frac{1}{2}\pi} \log^+ |f(re^{i\theta})| \cos \theta \, d\theta$$

$$\leq \lambda r \int_{-\frac{1}{2}\pi}^{\frac{1}{2}\pi} \cos^2 \theta \, d\theta = \tfrac{1}{2}\pi \lambda r;$$

so that

$$\nu \leq \tfrac{1}{2}\pi\lambda$$

and the theorem is proved.

EXERCISE 18.4

1. Prove (18.4.25).

2. Formulate analogues of the theorems of this section for a sector of opening $\alpha\pi$, $0 < \alpha < 2$.

*3. Let $f(z)$ be analytic in D: $-\frac{1}{2}\pi < y < \frac{1}{2}\pi$ and let $|f(x \pm \frac{1}{2}\pi i)| \leq 1$ for all x. Let Ω then be the set where $|f(z)| > 1$ (supposed non-void) and let $\Delta_x = [y \mid x + iy \in \Omega]$. Define

$$m(x) \equiv \int_{\Delta_x} \log |f(x + iy)| \cos y \, dy.$$

Show that $m(x)$ is twice differentiable except for values of x for which $\partial\Omega$ has a vertical tangent or forms an angle. Show that

$$m''(x) = m(x) + \left\{\left[\frac{\partial}{\partial x}\log|f|\frac{dy}{dx} - \frac{\partial}{\partial y}\log|f|\right]\cos y\right\}\Big|_1^2$$

where $\Big|_1^2$ means that we sum the values of the expression at the upper endpoints of the intervals which constitute Δ_x and subtract the values at the lower endpoints. The relation holds for all x where $m''(x)$ exists. (Ahlfors, 1937.)

*4. With the notation as in Problem 3, show that $m''(x) \geq m(x)$, except at certain isolated points, and show that for $a < x < b$

$$\begin{vmatrix} m(x) & e^x & e^{-x} \\ m(a) & e^a & e^{-a} \\ m(b) & e^b & e^{-b} \end{vmatrix} \geq 0.$$

18.5. Subordination. The principle of subordination may be traced back to Lindelöf's principle and, more precisely, to the Corollary of Theorem 18.2.3 for the case in which D is the unit disk. With a change of notation we formulate

DEFINITION 18.5.1. *Let $F(z)$ be holomorphic and univalent in the unit disk D: $[z \mid |z| < 1]$ and let $G = F(D)$ be the conformal image of D. Let $f(z)$ be holomorphic in D, but not necessarily univalent, and let the image set $f(D) \subset G$. Then $f(z)$ is said to be subordinate to $F(z)$ in D, and $F(z)$ is a univalent majorant of D, denoted by $f(z) \prec F(z)$.*

THEOREM 18.5.1. *If $f \prec F$ in D, then there exists a function $\omega(z)$, holomorphic in D and with $|\omega(z)| < 1$, such that*

(18.5.1) $$f(z) = F[\omega(z)], \quad z \in D.$$

If $f(0) = F(0)$, then $\omega(0) = 0$ and $|\omega(z)| \leq |z|$ in D.

Proof. The relation $f \prec F$ implies that every value taken on by $f(z)$ in D is also assumed by $F(z)$. Suppose that $w_0 \in f(D)$. Then there are points z_0 and z_1 in D such that

$$w_0 = F(z_0) = f(z_1).$$

Here z_0 is uniquely determined once w_0 or, equivalently, z_1 is given. It follows that z_0 is a single-valued function of z_1. We have

$$z_0 = \check{F}[f(z_1)] \equiv \omega(z_1),$$

so that

$$f(z_1) = F[\omega(z_1)]$$

as asserted. From its definition $\omega(z)$ is holomorphic in D and $|\omega(z)| < 1$.

If $f(0) = F(0)$, then

$$f(0) = F[\omega(0)] = F(0).$$

This implies $\omega(0) = 0$ since $F(z)$ is univalent. In this case Schwarz's lemma gives the stronger inequality $|\omega(z)| \leq |z|$. Here we can assert that if equality holds for a single z, then it holds for all and $\omega(z) = Cz$ where $|C| = 1$. This in turn implies that $f(D) = G$.

We note further

$$|f(z)| \leq \max |F(z)|, \quad |z| = r < 1$$

or

$$(18.5.2) \qquad\qquad M(r; f) \leq M(r; F),$$

which are immediate consequences of Theorem 18.5.1. Actually the Corollary of Theorem 18.2.3 gives a stronger result, which we restate as a theorem.

THEOREM 18.5.2. *Let $f \prec F$ in D. Let $f(0) = F(0)$. Let D_r be the disk $|z| < r$ and let $G_r = F(D_r)$; then $f(D_r) \subset G_r$ and is a proper subset of G_r unless $f(z) = F(Cz)$, $|C| = 1$.*

Since

$$f'(0) = F'(0)\,\omega'(0), \quad |\omega'(0)| \leq 1,$$

we have

$$(18.5.3) \qquad\qquad |f'(0)| \leq |F'(0)|.$$

Further relations between subordinate and majorant were discovered by Werner Rogosinski in 1943. We give some of his theorems.

THEOREM 18.5.3. *If $f \prec F$ in D and if $p > 0$, then*

$$(18.5.4) \qquad \int_{-\pi}^{\pi} |f(re^{i\theta})|^p \, d\theta \leq \int_{-\pi}^{\pi} |F(re^{i\theta})|^p \, d\theta, \quad r < 1.$$

Proof. We choose an R, $0 < R < 1$, in such a manner that $F(z)$ has no zero on the circle $|z| = R$. We then construct a function $B(z)$ such that (1) $B(z)$ has the same zeros with the same multiplicities in $|z| < R$ as $F(z)$ has, and

(2) $| B(z) | = 1$ on $| z | = R$. $B(z)$ is a finite Blaschke product where each factor defines a linear fractional transformation which maps the disk $| z | < R$ onto itself. In other words we choose

$$(18.5.5) \qquad B(z) = \left(\frac{z}{R}\right)^m \prod_{k=1}^{n} \frac{R(z - z_k)}{R^2 - \overline{z_k}z} .$$

It follows that

$$\left[\frac{F(z)}{B(z)}\right]^p \equiv g(z)$$

is holomorphic in $| z | < R$ for any choice of $p > 0$. We have merely to select one of the possible determinations of the power, say at $z = 0$; it is then uniquely determined in the disk.

If a function $g(z)$ is holomorphic in the disk $| z | \leq R$, then we can represent it by a Poisson integral in terms of its boundary values. For reasons which will become apparent in a moment, we write the integral in complex form:

$$(18.5.6) \qquad g(z) = \frac{1}{2\pi} \int_{-\pi}^{\pi} g(t)\Re\left(\frac{t + z}{t - z}\right) d\varphi, \quad | z | < R, t = Re^{i\varphi}.$$

Here we substitute for $g(z)$ the particular function defined above and in the result we replace z by $\omega(z)$. Since $F[\omega(z)] = f(z)$ we obtain

$$(18.5.7) \qquad \left\{\frac{f(z)}{B[\omega(z)]}\right\}^p = \frac{1}{2\pi} \int_{-\pi}^{\pi} \left\{\frac{F(t)}{B(t)}\right\}^p \Re\left\{\frac{t + \omega(z)}{t - \omega(z)}\right\} d\varphi.$$

Since the kernel is positive we have

$$\left|\frac{f(z)}{B[\omega(z)]}\right|^p \leq \frac{1}{2\pi} \int_{-\pi}^{\pi} \left|\frac{F(t)}{B(t)}\right|^p \Re\left[\frac{t + \omega(z)}{t - \omega(z)}\right] d\varphi.$$

Here $| B[\omega(z)] | < 1$ and $| B(t) | = 1$. It follows that

$$(18.5.8) \qquad |f(z)|^p \leq \frac{1}{2\pi} \int_{-\pi}^{\pi} | F(t) |^p \Re\left[\frac{t + \omega(z)}{t - \omega(z)}\right] d\varphi$$

where $z = re^{i\theta}$. We integrate with respect to θ from $-\pi$ to π and obtain

$$\int_{-\pi}^{\pi} |f(re^{i\theta})|^p d\theta \leq \frac{1}{2\pi} \int_{-\pi}^{\pi} | F(Re^{i\varphi}) |^p \int_{-\pi}^{\pi} \Re\left[\frac{t + \omega(z)}{t - \omega(z)}\right] d\theta \, d\varphi.$$

The integrand is a harmonic function of z, and by the mean value theorem for potential functions (Theorem 8.2.4)

$$\int_{-\pi}^{\pi} \Re\left[\frac{t + \omega(z)}{t - \omega(z)}\right] d\varphi = 1,$$

since $\omega(0) = 0$. This gives

$$\int_{-\pi}^{\pi} |f(re^{i\theta})|^p d\theta \leq \int_{-\pi}^{\pi} | F(Re^{i\theta}) |^p d\theta,$$

valid for every $r < R$. Since the left side is a continuous function of r, the inequality holds also for $r = R$. A continuity argument shows that the restriction $F(z) \neq 0$ for $|z| = R$ may be removed.

We can interpret this result in the language of functional analysis. In analogy with the notation used in Section 17.5 we say that a function $g(z)$ belongs to the class $H_p(D)$ if it is holomorphic in the unit disk D and if

$$(18.5.9) \qquad \lim_{r} \sup \left\{ \int_{-\pi}^{\pi} |g(re^{i\theta})|^p \, d\theta \right\}^{\frac{1}{p}} \equiv \| g \|_p$$

is finite. If $p \geq 1$, the class $H_p(D)$ becomes a (B)-space under the norm $\| g \|_p$. With this terminology we have the following

COROLLARY. *If $F(z)$ is univalent and $F \in H_p(D)$ and if $f(z) \prec F(z)$ in D, then $f \in H_p(D)$ and*

$$\| f \|_p \leq \| F \|_p.$$

Actually, in the proof of Theorem 18.5.3 we do not need the assumption that $F(z)$ is univalent. We use only the fact that there exists a function $\omega(z)$, holomorphic in D with $|\omega(z)| \leq |z|$, such that

$$(18.5.10) \qquad f(z) = F[\omega(z)].$$

It is customary to say, even in this case, that $f(z)$ is subordinate to $F(z)$ and to use the same notation $f \prec F$. The corollary is still valid.

Using the same method we can also prove that

$$(18.5.11) \qquad \int_{-\pi}^{\pi} \log^+ |f(re^{i\theta})| \, d\theta \leq \int_{-\pi}^{\pi} \log^+ |F(re^{i\theta})| \, d\theta$$

provided $f \prec F$, in the strict or in the wide sense.

We return to (18.5.4). If $p = 2$ and $F \in H_2(D)$, then the inequality asserts that

$$(18.5.12) \qquad \sum_{k=0}^{\infty} |a_k|^2 \leq \sum_{k=0}^{\infty} |A_k|^2$$

where

$$f(z) = \sum_{k=0}^{\infty} a_k z^k, \quad F(z) = \sum_{k=0}^{\infty} A_k z^k.$$

A much sharper result is contained in the following:

THEOREM 18.5.4. *If $f \prec F$ and $f(0) = F(0)$, then for each n*

$$(18.5.13) \qquad \sum_{k=0}^{n} |a_k|^2 = \sum_{k=0}^{n} |A_k|^2.$$

Proof. We use the following notation:

$$s_n(z) = \sum_{k=0}^{n} a_k z^k, \quad S_n(z) = \sum_{k=0}^{n} A_k z^k$$

and

$$(z) = s_n(z) + r_n(z), \quad F(z) = S_n(z) + R_n(z).$$

Since $f(z) = F[\omega(z)]$ we have

$$f(z) = S_n[\omega(z)] + R_n[\omega(z)],$$

where the second summand contains no power of z of degree $< n + 1$. It follows that

$$S_n[\omega(z)] = s_n(z) + T_n(z),$$

where again the second summand contains no power of degree $< n + 1$. Note that $\omega(0) = 0$. Now

$$S_n[\omega(z)] \prec S_n(z)$$

in the wide sense implies that

$$\int_{-\pi}^{\pi} |S_n(z)|^2 \, d\theta \geq \int_{-\pi}^{\pi} |S_n[\omega(z)]|^2 \, d\theta$$

$$= \int_{-\pi}^{\pi} |s_n(z)|^2 \, d\theta + \int_{-\pi}^{\pi} |T_n(z)|^2 \, d\theta$$

$$\geq \int_{-\pi}^{\pi} |s_n(z)|^2 \, d\theta,$$

where we have used the fact that $s_n(z)$ and $T_n(z)$ are orthogonal to each other. It follows that

$$\sum_{k=0}^{n} |A_k|^2 r^{2k} \geq \sum_{k=0}^{n} |a_k|^2 r^{2k}$$

for every $r < 1$ and hence also for $r = 1$, which was to be proved.

M. Schiffer obtained in 1937 sharp estimates for subordinate functions with the aid of corresponding estimates for univalent functions and of the principle of hyperbolic measure (see Sections 15.1 and 17.4).

THEOREM 18.5.5.　*Let $f(z)$ and $F(z)$ be holomorphic in the unit disk, with $f(0) = F(0) = 0$, $F'(0) = 1$. Let $F(z)$ be univalent and $f(z) \prec F(z)$. Then*

$$(18.5.14) \qquad |f(z)| \leq \frac{|z|}{1 - |z|^2}, \quad |f'(z)| \leq \frac{1 + |z|}{(1 - |z|)^3},$$

where equality is reached only for

$$(18.5.15) \qquad f(z) = \frac{z}{(1 - \eta z)^2}, \quad |\eta| = 1.$$

Proof.　We have $f(z) = F[\omega(z)]$, whence it follows that

$$|f'(z)| = |F'[\omega(z)]| \, |\omega'(z)|.$$

Since $F(w)$ is univalent, Theorem 17.4.6 gives the estimate

$$(18.5.16) \qquad |F'(w)| \leq \frac{1 + |w|}{(1 - |w|)^3}, \quad |w| < 1,$$

and here (18.5.15) is the extremal function for which equality is reached. On the other hand, Problem 8 of Exercise 15.1 applied to the function $w = \omega(z)$ shows that

$$\frac{|\omega'(z)|}{1 - |\omega(z)|^2} \leq \frac{1}{1 - |z|^2}.$$

It follows that

$$|f'(z)| \leq \left(\frac{1 + |\omega(z)|}{1 - |\omega(z)|}\right)^2 \frac{1}{1 - |z|^2}.$$

Next we note that $|\omega(z)| \leq |z|$ and that the function

$$\frac{1 + x}{1 - x}$$

is increasing in the interval $(0, 1)$. Hence

$$\frac{1 + |\omega(z)|}{1 - |\omega(z)|} \leq \frac{1 + |z|}{1 - |z|}.$$

Combining these inequalities we get the second inequality under (18.5.14). Integrating this we get the first inequality.

For a given z we can have equality in (18.5.14) if and only if we have equality in (18.5.16) and if, in addition, $|\omega(z)| = |z|$. This shows that $\omega(z) = \eta z$ with $|\eta| = 1$, and $F(z)$ is the extremal function (17.4.12) for which equality holds in (18.5.16). This completes the proof.

Under special assumptions on the domain $G = F(D)$ one may derive estimates of the coefficients of the corresponding subordinate functions. The following result is elegant and easy to prove:

THEOREM 18.5.6. *If $f(z)$ and $F(z)$ are holomorphic in D such that $f(0) = F(0) = 0$, $F'(0) = 1$; if $F(z)$ is univalent and $f(z) \prec F(z)$; if $G = F(D)$ is convex: then under all of these conditions $|a_n| \leq 1$ for all n.*

Proof. Let $\eta = \exp \dfrac{2\pi i}{m}$ and consider the function

$$\frac{1}{m}\left[f(z) + f(\eta z) + \cdots + f(\eta^{m-1}z)\right] \equiv g_m(z).$$

For each given $z \in D$ all the numbers $f(z), f(\eta z), \cdots, f(\eta^{m-1}z)$ lie in G and, since G is convex, the center of gravity of these m numbers is also in G, that is, $g_m(z) \in G$. This implies that $g_m(z) \prec F(z)$. Now

$$g_m(z) = \sum_{k=1}^{\infty} a_{km} z^{km} \quad \text{if} \quad f(z) = \sum_{n=1}^{\infty} a_n z^n.$$

It follows that

$$g_m(z^{1/m}) = \sum_{k=1}^{\infty} a_{km} z^k$$

is holomorphic in D and that its values also lie in G, that is, $g_m(z^{1/m}) \prec F(z)$.

We can now use Theorem 18.5.4 for $n = 1$. It follows that for each m

$$| a_m | \leq | F'(0) | = 1$$

as asserted.

EXERCISE 18.5

1. If $G = F(D) = D$, characterize the subordinate functions. Determine an $F(z)$ if it is known that $F'(0) = a + bi$.

2. Same question if G is the right half-plane.

3. G is the whole plane from which the line segments $(-\infty, -\alpha)$ and $(\alpha, +\infty)$ have been deleted, $\alpha > 0$. Prove that any subordinate function for which $f(0) = 0$ satisfies the inequality

$$| f'(0) | \leq 2\alpha.$$

4. G is the strip $-1 < \Re(z) < 1$. Prove that if $f(z)$ is subordinate and $f(0) = 0$, then

$$| f'(0) | \leq 4/\pi.$$

5. Under the assumptions of Theorem 18.5.4 show that

$$| a_2 | \leq \max \left(| A_1 |, | A_2 | \right).$$

6. If $F(z) \in H_2(D)$, show that

$$\sum_{k=0}^{n} | A_k |^2 = \int_{-\pi}^{\pi} \int_{-\pi}^{\pi} F(e^{is}) \overline{F(e^{it})} D_n(s - t) \, ds \, dt$$

where

$$D_n(u) = \frac{1}{2\pi} \frac{\sin (n + \tfrac{1}{2})u}{\sin \tfrac{1}{2}u}$$

is the nth Dirichlet kernel.

7. Obtain a similar representation for

$$\sum_{k=1}^{n} k | A_k |^2.$$

8. In Theorem 18.5.6 replace the assumption that G is convex by (i) G is star-shaped, (ii) G is symmetric to the real axis, and prove in both cases that $| a_n | \leq n$. Extremal functions are given by (18.5.15).

COLLATERAL READING

As general references for this chapter we list

Golusin, G. M. *Geometrische Funktionentheorie*, Chapter 8. Deutscher Verlag der Wissenschaften, Berlin, 1957.

Nevanlinna, R. *Eindeutige analytische Funktionen*, Second Edition, Chaps. II, III, IV. Springer-Verlag, Berlin, 1953.

The following papers, listed in chronological order, are basic for most of the chapter:

PHRAGMÉN, E. "Sur une Extension d'un Théorème Classique de la Théorie des Fonctions," *Acta Mathematica*, Vol. 28 (1904), pp. 351–368.

——— and LINDELÖF, E. "Sur une Extension d'un Principe Classique de l'Analyse et sur Quelques Propriétés des Fonctions Monogènes dans le Voisinage d'un Point Singulier," *Acta Mathematica*, Vol. 31 (1908), pp. 381–406.

LINDELÖF, E. "Memoire sur Certaines Inégalités dans la Théorie des Fonctions Monogènes et sur Quelques Propriétés Nouvelles de ces Fonctions dans le Voisinage d'un Point Singulier Essentiel," *Acta Societatis Scientiarum Fennicæ*, Vol. 35, No. 7 (1908), 35 pages.

NEVANLINNA, F. and R. "Über die Eigenschaften analytischer Funktionen in der Umgebung einer singulären Stelle oder Linie," *Acta Societatis Scientiarum Fennicæ*, Vol. 50, No. 5 (1922), 46 pages.

AHLFORS, L. V. "On Phragmén-Lindelöf's Principle," *Transactions of the American Mathematical Society*, Vol. 41 (1937), pp. 1–8.

HEINS, M. "On the Phragmén-Lindelöf Principle," *Transactions of the American Mathematical Society*, Vol. 60 (1946), pp. 238–244.

In connection with Section 18.2 consult

NEHARI, Z. *Conformal Mapping*, Chapter I and Section 3 of Chapter VII. McGraw-Hill Book Company, Inc., New York, 1952.

TSUJI, M. *Potential Theory in Modern Function Theory*. Maruzen, Tokyo, 1959.

For subordination see also

LITTLEWOOD, J. E. *Lectures on the Theory of Functions*, pp. 163–185. Oxford University Press, London, 1944.

ROGOSINSKI, W. W. "On the Coefficients of Subordinate Functions," *Proceedings of the London Mathematical Society*, Series 2, Vol. 48 (1943), pp. 48–82.

19

FUNCTIONS HOLOMORPHIC IN
A HALF-PLANE

19.1. The Hardy-Lebesgue classes. The functions holomorphic in a half-plane are in a one-to-one correspondence with the functions holomorphic in the unit disk. In view of this situation it would seem unnecessary to give special attention to the half-plane case. It is a fact, however, that many problems lead to functions which are holomorphic in a half-plane and to special representations of such functions which are not conveniently studied by mapping on the unit disk. This makes it desirable to develop the theory of such functions and their representations at some length.

We start with the Hardy-Lebesgue classes for the right half-plane.

DEFINITION 19.1.1. *Let p be fixed, $1 \leq p < \infty$. We say that $f(z) \in H_p(0)$ if* (i) $f(z)$ *is holomorphic in* $\Re(z) > 0$; *if* (ii) *for each fixed $x > 0$, $f(x + iy)$ as function of y belongs to $L_p(-\infty, \infty)$; and if* (iii) *the corresponding L_p-norm is a bounded function of x, that is, if*

$$(19.1.1) \qquad M_p(x; f) \equiv \left\{ \int_{-\infty}^{\infty} |f(x + iy)|^p \, dy \right\}^{\frac{1}{p}},$$

then

$$(19.1.2) \qquad \| f \|_p \equiv \sup_{x>0} M_p(x; f) < \infty.[1]$$

The notation indicates that we are going to consider $H_p(0)$ as a normed linear vector space. For the study of the properties of the elements of $H_p(0)$ we have to use two famous inequalities, that of Hölder (Otto Hölder, 1860–1937) and that of Minkowski. These are stated as separate lemmas and in a slightly more general setting than is actually needed for our purposes.

LEMMA 19.1.1. *If $f(t) \in L_\alpha(a, b)$ and $g(t) \in L_\beta(a, b)$ where*

$$(19.1.3) \qquad \frac{1}{\alpha} + \frac{1}{\beta} = 1,$$

then $f(t) g(t) \in L_1(a, b)$ and

$$(19.1.4) \qquad \int_a^b |f(t) g(t)| \, dt \leq \left\{ \int_a^b |f(t)|^\alpha \, dt \right\}^{\frac{1}{\alpha}} \left\{ \int_a^b |g(t)|^\beta \, dt \right\}^{\frac{1}{\beta}}.$$

---

[1] More generally, $f(z) \in H_p(a)$, a real, if $f(z + a) \in H_p(0)$.

LEMMA 19.1.2.   *If $f_1(t)$ and $f_2(t)$ belong to $L_p(a, b)$, so does $f_1(t) + f_2(t)$ and*

$$(19.1.5) \quad \left\{\int_a^b |f_1(t) + f_2(t)|^p \, dt\right\}^{\frac{1}{p}} \leq \left\{\int_a^b |f_1(t)|^p \, dt\right\}^{\frac{1}{p}} + \left\{\int_a^b |f_2(t)|^p \, dt\right\}^{\frac{1}{p}}.$$

Here (a, b) may be finite or infinite. We recall that $f(t) \in L_p(a, b)$ if $f(t)$ is measurable and if $|f(t)|^p$ is integrable in (a, b). Strictly speaking, the elements of $L_p(a, b)$ are not functions but equivalence classes of functions, where two functions belong to the same equivalence class if their difference is zero except in a set of measure zero. This subtle distinction is of little importance for the following.

Proof of Lemma 19.1.1.   Our point of departure is the elementary inequality

$$(19.1.6) \quad A^x B^{1-x} \leq xA + (1 - x)B, \quad 0 \leq x \leq 1, 0 < A, 0 < B,$$

where equality holds if and only if $A = B$. If

$$F(x; A, B) \equiv xA + (1 - x)B - A^x B^{1-x}$$

then

$$F(0; A, B) = F(1; A, B) = 0$$

and

$$F''(x; A, B) = -\log^2 \frac{B}{A} \cdot A^x B^{1-x} < 0$$

unless $A = B$, when the result is identically zero. It follows that $y = F(x; A, A) \equiv 0$, while for $A \neq B$ the curve $y = F(x; A, B)$ is concave downward and passes through the points $(0, 0)$ and $(1, 0)$. This means that

$$F(x; A, B) > 0, \quad 0 < x < 1, A \neq B,$$

and that (19.1.6) holds.

Taking

$$A = |f(t)|^\alpha, \quad B = |g(t)|^\beta, \quad x = \frac{1}{\alpha}, \quad 1 - x = \frac{1}{\beta},$$

we get

$$(19.1.7) \quad |f(t) g(t)| \leq \frac{1}{\alpha} |f(t)|^\alpha + \frac{1}{\beta} |g(t)|^\beta.$$

This shows that the measurable function $f(t) g(t)$ is dominated by the sum of two integrable functions and, hence, is also integrable by a famous theorem due to Lebesgue. With the usual notation for Lebesgue norms, this gives

$$(19.1.8) \quad \|fg\|_1 \leq \frac{1}{\alpha} \|f\|_\alpha^\alpha + \frac{1}{\beta} \|g\|_\beta^\beta.$$

Here we can replace $f(t)$ by $\lambda f(t)$ and $g(t)$ by $\lambda^{-1}g(t)$ where λ is any positive constant. This yields

$$\| fg \|_1 \leq \frac{1}{\alpha} \lambda^\alpha \| f \|_\alpha{}^\alpha + \frac{1}{\beta} \lambda^{-\beta} \| g \|_\beta{}^\beta,$$

and here the right-hand side takes on its minimum value for

$$\lambda = \| f \|_\alpha^{-\alpha/(\alpha+\beta)} \| g \|_\beta^{\beta/(\alpha+\beta)}.$$

For this choice of λ we get

(19.1.9) $$\| fg \|_1 \leq \| f \|_\alpha \| g \|_\beta,$$

which is Hölder's inequality. The proof presupposes $\| f \|_\alpha \neq 0$; however, (19.1.9) is trivially true if $\| f \|_\alpha = 0$.

Proof of Lemma 19.1.2.   It is readily seen that the sum of two functions in $L_p(a, b)$ is in the same space, for the sum is measurable and

$$| f_1(t) + f_2(t) |^p \leq 2^p \max [| f_1(t) |^p, | f_2(t) |^p],$$

which is an integrable function. Minkowski's inequality is trivial if $p = 1$. For $p > 1$ we have

$$\int_a^b | f_1(t) + f_2(t) |^p \, dt \leq \int_a^b | f_1(t) + f_2(t) |^{p-1} | f_1(t) | \, dt$$

$$+ \int_a^b | f_1(t) + f_2(t) |^{p-1} | f_2(t) | \, dt.$$

To each of the integrals on the right we apply Hölder's inequality with

$$f(t) = | f_1(t) + f_2(t) |^{p-1}, \quad g(t) = | f_k(t) |, \quad \alpha = \frac{p}{p-1},$$

and obtain

$$\| f_1 + f_2 \|_p{}^p \leq \| f_1 + f_2 \|_p^{p-1}[\| f_1 \|_p + \| f_2 \|_p].$$

Now (19.1.4) is clearly valid if $\| f_1 + f_2 \|_p = 0$, and if this number is > 0 instead, the desired inequality follows from the last display.

We have to use a number of properties of the spaces $L_p(-\infty, \infty)$. We list these as lemmas.

LEMMA 19.1.3.   *The space $L_p(a, b)$ is complete under the normed metric. Here $1 \leq p < \infty$ and $b - a \leq \infty$.*

For the case $p = 2$ the reader will find a simple direct proof in F. Riesz and B. Sz. Nagy, *Leçons d'Analyse Fonctionnelle* (Akadémiai Kiadó, Budapest, 1952), pp. 58–59. This proof carries over to the general case with trivial modifications.

LEMMA 19.1.4.   *If $p > 1$ and $a > 0$, then the linear combinations of the functions*

$$(19.1.10)\quad \omega_n(it; a) = \frac{(it - a)^{n-1}}{(it + a)^n}, \quad \omega_{-n}(it; a) = \frac{(it + a)^{n-1}}{(it - a)^n}, \quad n = 1, 2, 3, \cdots$$

are dense in $L_p(-\infty, \infty)$.

Proof.   The following proof is patterned after the argument used by M. Riesz in a similar situation involving Parseval's formula and the moment problem for an infinite interval. If $f(t) \in L_p(-\infty, \infty)$ and $\varepsilon > 0$ is given, we can find a continuous function $g(t)$ in the same class such that $\| f - g \|_p < \varepsilon$. It is always possible to modify this function at $t = 0$ and at $t = \pm\infty$ so that the even functions

$$h_1(t) = \tfrac{1}{2}(t^2 + a^2)[g(t) + g(-t)],$$

$$h_2(t) = \frac{1}{2t}(t^2 + a^2)[g(t) - g(-t)],$$

are continuous in the closed interval $[-\infty, \infty]$. The function

$$x = (t^2 + a^2)^{-1}$$

maps the interval $0 \le t \le \infty$ on the interval $0 \le x \le a^{-2}$ and defines a one-to-one correspondence between the classes $C[0, \infty]$ and $C[0, a^{-2}]$. By the theorem of Weierstrass, polynomials in x are dense in the space $C[0, a^{-2}]$. Hence, given an $\varepsilon > 0$, we can find an integer n and numbers α_k, β_k such that for all t

$$\left| h_1(t) - \sum_{k=0}^{n} \alpha_k(t^2 + a^2)^{-k} \right| < \varepsilon,$$

$$\left| h_2(t) - \sum_{k=0}^{n} \beta_k(t^2 + a^2)^{-k} \right| < \varepsilon.$$

It follows that

$$\left| \tfrac{1}{2}[g(t) + g(-t)] - \sum_{k=0}^{n} \alpha_k(t^2 + a^2)^{-k-1} \right| < \varepsilon(t^2 + a^2)^{-1},$$

$$\left| \tfrac{1}{2}[g(t) - g(-t)] - t\sum_{k=0}^{n} \beta_k(t^2 + a^2)^{-k-1} \right| < \varepsilon | t | (t^2 + a^2)^{-1},$$

and

$$\left| g(t) - \sum_{k=0}^{n} (\alpha_k + \beta_k t)(t^2 + a^2)^{-k-1} \right| < \varepsilon(1 + | t |)(t^2 + a^2)^{-1}.$$

Hence

$$\left\| g(t) - \sum_{k=0}^{n} (\alpha_k + \beta_k t)(t^2 + a^2)^{-k-1} \right\|_p < \varepsilon C(a, p)$$

where

$$C(a, p) = \left\{ 2\int_0^{\infty} \frac{(1 + t)^p}{(t^2 + a^2)^p}\, dt \right\}^{\frac{1}{p}}.$$

To complete the proof it is sufficient to prove the existence of numbers A_{jk} and B_{jk} such that

(19.1.11)
$$(t^2 + a^2)^{-k-1} = \sum_{j=-k-1}^{k+1} A_{jk}\omega_j(it; a),$$
$$k = 0, 1, 2, \cdots,$$
$$t(t^2 + a^2)^{-k-1} = \sum_{j=-k-1}^{k+1} B_{jk}\omega_j(it; a).$$

This is obvious for $k = 0$. A simple calculation shows that for $j > 1$ we have

(19.1.12)
$$\omega_j(it; a)\,\omega_1(it; a) = \frac{1}{2a}\,[\omega_j(it; a) - \omega_{j+1}(it; a)],$$
$$\omega_j(it; a)\,\omega_{-1}(it; a) = \frac{1}{2a}\,[\omega_{j-1}(it; a) - \omega_j(it; a)],$$

and these relations hold also for $j = 1$ provided that in the second formula the undefined symbol $\omega_0(it; a)$ is replaced by $\omega_{-1}(it; a)$. Since

(19.1.13)
$$\omega_{-j}(it; a) = -\overline{\omega_j(it; a)},$$

the formulas are also valid for negative values of j provided now the subscript 0 is replaced by $+1$ when $j = -1$. With the aid of these formulas one easily verifies the validity of (19.1.11) for arbitrary integers $k > 0$. It is also possible to establish recurrence relations from which the coefficients A_{jk} and B_{jk} may be computed explicitly.

Combining the results we see that for every $f(t) \in L_p(-\infty, \infty)$ and every $\varepsilon > 0$ there are numbers γ_k such that

$$\left\| f(t) - \sum_{k=-n-1}^{n+1} \gamma_k\omega_k(it; a) \right\|_p < \varepsilon[1 + C(a, p)],$$

and this is the assertion of the lemma.

A simple calculation gives

(19.1.14)
$$\int_{-\infty}^{\infty} \omega_j(it; a)\,\overline{\omega_k(it; a)}\, dt = \delta_{jk}\frac{\pi}{a},$$

so that the following functions form an orthonormal system:

$$\left(\frac{a}{\pi}\right)^{\frac{1}{2}}\omega_k(it; a), \quad k = \pm 1, \pm 2, \pm 3, \cdots.$$

If $f(t) \in L_p(-\infty, \infty)$, $1 < p < \infty$, and if

(19.1.15)
$$f_n = \frac{a}{\pi}\int_{-\infty}^{\infty} f(s)\,\overline{\omega_n(is; a)}\, ds, \quad n = \pm 1, \pm 2, \cdots,$$

then we refer to the numbers f_n as the Fourier coefficients and to the series

$$\sum_{-\infty}^{\infty} f_n\omega_n(it; a)$$

as the Fourier series of $f(t)$ in the system $\{\omega_n(it; a)\}$.

For $p = 2$ we have the usual Bessel inequality, but Lemma 19.1.4 implies mean convergence of the partial sums

$$(19.1.16) \qquad \lim_{n \to \infty} \left\| f(t) - \sum_{k=-n}^{n} f_k \omega_k(it; a) \right\|_2 = 0$$

and Parseval's identity

$$(19.1.17) \qquad \frac{\pi}{a} \sum_{-\infty}^{\infty} |f_k|^2 = \int_{-\infty}^{\infty} |f(s)|^2 \, ds.$$

Actually the extension of (19.1.16)

$$(19.1.18) \qquad \lim_{n \to \infty} \left\| f(t) - \sum_{k=-n}^{n} f_k \omega_k(it; a) \right\|_p = 0$$

holds for every p, $1 < p < \infty$ (generalization of theorem of M. Riesz for trigonometric Fourier series), but this result is quite deep and we shall not need it, so it will not be proved. The following lemma is a consequence of (19.1.18), but we shall give a direct proof later.

LEMMA 19.1.5.   If $f(t) \in L_p(-\infty, \infty)$, $1 < p < \infty$, and if $f_n = 0$ for all n, then $f(t) = 0$ almost everywhere.

For the discussion of $H_p(0)$ with $p > 1$ the Fourier series of the Cauchy kernel is very useful.

LEMMA 19.1.6.   We have

$$(19.1.19) \qquad \frac{1}{it - z} = 2a \sum_{n=1}^{\infty} \omega_{-n}(it; a) \frac{(z - a)^{n-1}}{(z + a)^n}, \qquad \Re(z) > 0,$$

$$(19.1.20) \qquad \frac{1}{it - z} = -2a \sum_{n=1}^{\infty} \omega_n(it; a) \frac{(z + a)^{n-1}}{(z - a)^n}, \qquad \Re(z) < 0.$$

The series converge absolutely and uniformly with respect to t, $-\infty < t < \infty$, and uniformly with respect to z on compact subsets of the half-planes in question.

Proof.   If $\Re(z) > 0$ and

$$w = \frac{z - a}{z + a},$$

then

$$\frac{z + a}{it - z} = \frac{2a}{it - a - (it + a)w}$$

has a unique expansion in terms of powers of w,

$$\frac{2a}{it - a} \sum_{n=0}^{\infty} \left(\frac{it + a}{it - a} \right)^n w^n,$$

convergent for $|w| < 1$, uniformly with respect to t and w, $-\infty < t < \infty$, and

$|w| \leq r < 1$. This is equivalent to the first expansion, and the second is proved in the same way. We can of course replace (19.1.19) by the series

$$(19.1.21) \qquad 2a \sum_{n=1}^{\infty} \omega_{-n}(it; a)\, \omega_n(z; a).$$

This is one of many bilinear formulas available for the representation of such expressions as the Cauchy kernel and the Laplace kernel e^{-zt}.

Our next lemma concerns the Lebesgue *modulus of continuity*.

LEMMA 19.1.7.   *If $g(t) \in L_p(-\infty, \infty)$, $1 \leq p < \infty$, then*

$$(19.1.22) \qquad \mu_p(s; g) \equiv \left\{ \int_{-\infty}^{\infty} |g(t+s) - g(t)|^p \, dt \right\}^{\frac{1}{p}}$$

is a continuous bounded function of s which is subadditive and has the value 0 for $s = 0$.

Proof.   By Minkowski's inequality

$$\mu_p(s_1 + s_2; g) = \left\{ \int_{-\infty}^{\infty} |g(t + s_1 + s_2) - g(t)|^p \, dt \right\}^{\frac{1}{p}}$$

$$\leq \left\{ \int_{-\infty}^{\infty} |g(t + s_1 + s_2) - g(t + s_2)|^p \, dt \right\}^{\frac{1}{p}}$$

$$+ \left\{ \int_{-\infty}^{\infty} |g(t + s_2) - g(t)|^p \, dt \right\}^{\frac{1}{p}},$$

so that

$$(19.1.23) \qquad \mu_p(s_1 + s_2; g) \leq \mu_p(s_1; g) + \mu_p(s_2; g),$$

which expresses the subadditive property.

Now it is clear that if $g(t)$ is already continuous then $\mu_p(s; g)$ is continuous and $\mu_p(0; g) = 0$. If g is not continuous, we can find a continuous function $h(t)$ such that $\| g - h \|_p < \varepsilon$. Since for all s

$$0 \leq \mu_p(s; g) \leq 2 \| g \|_p,$$

we conclude that $\mu_p(s; g - h) \leq 2\varepsilon$ and

$$0 \leq \mu_p(s; g) \leq \mu_p(s; h) + \mu_p(s; g - h) \leq \mu_p(s; h) + 2\varepsilon,$$

so that for small values of $|s|$ we have

$$0 \leq \mu_p(s; g) \leq 3\varepsilon.$$

This requires $\mu_p(s; g)$ to be continuous at $s = 0$ and to have the value 0 there. We have then by the subadditivity

$$\mu_p(s + \delta; g) \leq \mu_p(s; g) + \mu_p(\delta; g),$$
$$\mu_p(s; g) \leq \mu_p(s + \delta; g) + \mu_p(-\delta; g),$$

whence it follows that $\mu_p(s; g)$ is continuous everywhere.

We come now to theorems which concern the functions of $H_p(0)$.

THEOREM 19.1.1.   *If* $g(t) \in L_p(-\infty, \infty)$, $1 < p < \infty$, *and*

$$g(t) \sim \sum_{-\infty}^{\infty} g_n \omega_n(it; a),$$

then for $\Re(z) > 0$

(19.1.24) $$f_+(z) \equiv \frac{1}{2\pi} \int_{-\infty}^{\infty} \frac{g(t)\, dt}{z - it} = \sum_{n=1}^{\infty} g_n \frac{(z-a)^{n-1}}{(z+a)^n},$$

whereas for $\Re(z) < 0$

(19.1.25) $$f_-(z) \equiv \frac{1}{2\pi} \int_{-\infty}^{\infty} \frac{g(t)\, dt}{it - z} = \sum_{n=1}^{\infty} g_{-n} \frac{(z+a)^{n-1}}{(z-a)^n}.$$

The series are absolutely convergent in the domains indicated.
      If $f_-(z) \equiv 0$, *that is, if* $g_{-n} = 0$ *for* $n \geq 1$, *then* $f_+(z) \in H_p(0)$ *and*

(19.1.26) $$\lim_{x \to 0} \| f_+(x + iy) - g(y) \|_p = 0.$$

Proof.   The first two formulas are trivial consequences of Lemma 19.1.6 since the series (19.1.19) and (19.1.20) converge uniformly with respect to t in the infinite interval.

We have $f_-(z) \equiv 0$ for $\Re(z) < 0$ if and only if $g_{-n} = 0$ for every $n \geq 1$. If this condition is satisfied and $\Re(z) > 0$, we have

$$f_+(z) = f_+(z) + f_-(-\bar{z}).$$

Here we substitute the integral representations for the two functions on the right and obtain after simplification

(19.1.27) $$f_+(z) = \frac{x}{\pi} \int_{-\infty}^{\infty} \frac{g(t)\, dt}{(t-y)^2 + x^2},$$

that is, $f_+(z)$ is representable by a Poisson integral as well as by an integral of the Cauchy type. We have to show that all these integrals involve the boundary values of $f_+(z)$ in the sense of mean convergence. Here we use the elementary identity

$$\frac{x}{\pi} \int_{-\infty}^{\infty} \frac{dt}{(t-y)^2 + x^2} \equiv 1$$

to get

$$| f_+(x+iy) - g(y) | \leq \frac{x}{\pi} \int_{-\infty}^{\infty} \frac{| g(s+y) - g(y) |}{s^2 + x^2}\, ds.$$

It follows that

$$\int_{-\infty}^{\infty} | f_+(x+iy) - g(y) |^p\, dy \leq \left(\frac{x}{\pi}\right)^p \int_{-\infty}^{\infty} \left\{ \int_{-\infty}^{\infty} \frac{| g(s+y) - g(y) |}{s^2 + x^2}\, ds \right\}^p dy.$$

Next, we use Hölder's inequality, choosing the two factors as

$$\frac{| g(s+y) - g(y) |}{(s^2 + x^2)^{1/p}} \quad \text{and} \quad \frac{1}{(s^2 + x^2)^{1/q}}$$

respectively, where, as usual, $q = p/(p-1)$. This gives

$$\| f_+(x+iy) - g(y) \|_p \leq \frac{x}{\pi} \left\{ \int_{-\infty}^{\infty} \left(\int_{-\infty}^{\infty} \frac{ds}{s^2 + x^2} \right)^{\frac{p}{q}} \int_{-\infty}^{\infty} \frac{|g(s+y) - g(y)|^p}{s^2 + x^2} \, ds \, dy \right\}^{\frac{1}{p}}$$

$$= \left\{ \frac{x}{\pi} \int_{-\infty}^{\infty} \frac{ds}{s^2 + x^2} \int_{-\infty}^{\infty} |g(s+y) - g(y)|^p \, dy \right\}^{\frac{1}{p}},$$

and finally

$$\| f_+(x+iy) - g(y) \|_p \leq \left\{ \frac{x}{\pi} \int_{-\infty}^{\infty} \frac{\mu_p(s;g)}{s^2 + x^2} \, ds \right\}^{\frac{1}{p}}$$

in the notation of Lemma 19.1.7. Here $0 \leq \mu_p(s;g) \leq 2 \| g \|_p$, $\mu_p(s;g)$ is continuous and has the value 0 at $s = 0$. We want the limit of the right member as $x \to 0$. If $\varepsilon > 0$ is given, we can find a $\delta = \delta(\varepsilon)$ such that $\mu_p(s;g) \leq \varepsilon$ for $|s| \leq \delta$. We have then

$$\frac{x}{\pi} \int_{-\infty}^{\infty} \frac{\mu_p(s;g) \, ds}{s^2 + x^2} = \int_{-\infty}^{-\delta} + \int_{-\delta}^{\delta} + \int_{\delta}^{\infty}.$$

Since the integral of the kernel is identically 1, the second integral on the right does not exceed ε and each of the others is bounded by $(2/\pi) \| g \|_p$ arc cot (δ/x). It follows that the superior limit for $x \to 0$ does not exceed ε. Thus (19.1.26) holds as asserted. This completes the proof.

COROLLARY 1.   *A function $g(y)$ can furnish the boundary values on the imaginary axis, in the sense of mean convergence, of a function in $H_p(0)$, $1 < p < \infty$, if and only if (i) $g(y) \in L_p(-\infty, \infty)$ and (ii) $g_n = 0$ for $n \leq -1$.*

Proof.   The preceding discussion shows that the conditions are sufficient, and (i) is clearly also necessary. Suppose now that $g_{-k} \neq 0$ for some $k \geq 1$ and that (19.1.26) holds. We have then

$$g_{-k} = \frac{a}{\pi} \int_{-\infty}^{\infty} g(t) \overline{\omega_{-k}(it;a)} \, dt = \lim_{x \to 0} \frac{a}{\pi} \int_{-\infty}^{\infty} f_+(x+iy) \, \overline{\omega_{-k}(iy;a)} \, dy$$

since

$$\left| \int_{-\infty}^{\infty} [f_+(x+iy) - g(y)] \overline{\omega_{-k}(iy;a)} \, dy \right| \leq \| f_+(x+iy) - g(y) \|_p \| \omega_{-k} \|_q,$$

which tends to zero with x. On the other hand, by (19.1.24) and the Fubini theorem

$$\int_{-\infty}^{\infty} f_+(x+iy) \overline{\omega_{-k}(iy;a)} \, dy = -\frac{1}{2\pi} \int_{-\infty}^{\infty} \omega_k(iy;a) \int_{-\infty}^{\infty} \frac{g(t) \, dt}{x + iy - it} \, dy$$

$$= -\frac{1}{2\pi} \int_{-\infty}^{\infty} g(t) \int_{-\infty}^{\infty} \frac{\omega_k(iy;a) \, dy}{x + iy - it} \, dt.$$

Here the last integral with respect to y is 0 for $x > 0$ and k any positive integer. This follows either from (19.1.20), with y and t interchanged, or from direct computation by means of formula (9.1.10). It follows that $g_{-k} = 0$ for every $k \geq 1$. Thus condition (ii) is also necessary.

COROLLARY 2 [= LEMMA 19.1.5].   *If $g(t) \in L_p(-\infty, \infty)$, $1 < p < \infty$, and if $g_n = 0$ for all n, then $g(t) \sim 0$.*

Proof.   The assumption implies first that $f_-(z) = 0$ for $\Re(z) < 0$, which implies the validity of (19.1.27) and hence also of (19.1.26). Since $f_+(x + iy) \equiv 0$ for $x > 0$, this gives $\| g \|_p = 0$ and $g(t) \sim 0$ as asserted.

The discussion up to this point has shown that certain classes of Cauchy and Poisson integrals define functions which are in $H_p(0)$ with $p > 1$. We shall show conversely that every function of $H_p(0)$ has this form. This is done in several steps.

THEOREM 19.1.2.   *If $b > 0$ and $f(z) \in H_p(0)$, $1 \leq p < \infty$, then*

$$(19.1.28) \qquad f(z) = \frac{1}{2\pi} \int_{-\infty}^{\infty} \frac{f(b + it)\, dt}{z - b - it}, \quad \Re(z) > b,$$

$$(19.1.29) \qquad f(x + iy) = \frac{x - b}{\pi} \int_{-\infty}^{\infty} \frac{f(b + it)\, dt}{(x - b)^2 + (y - t)^2}, \quad x > b.$$

Proof.   We apply Cauchy's formula to the rectangle with vertices at the points

$$b \pm iY, \quad B \pm iY,$$

where $b < x < B$, $| y | < Y$. Thus, with the usual integrand,

$$2\pi i f(z) = \int_{b-iY}^{b+iY} + \int_{b+iY}^{B+iY} + \int_{B+iY}^{B-iY} + \int_{B-iY}^{b-iY}$$

$$\equiv J(b, Y) + J_2 - J(B, Y) + J_4.$$

Using a device due to R. E. A. C. Paley (1907–1933) and Norbert Wiener, we integrate with respect to Y between the limits T and $2T$ and divide by T where $T > 2 | y |$. Here

$$\frac{1}{T} \int_T^{2T} | J_2 |\, dY \leq \frac{1}{T} \int_T^{2T} \int_b^B \frac{| f(s + iY) |}{| z - s - iY |}\, ds\, dY.$$

If $p = 1$, we observe that the denominator of the integrand exceeds $\frac{1}{2}T$ and the right member is less than

$$\frac{2}{T^2}(B - b) \sup_s \int_T^{2T} | f(s + iY) |\, dY < \frac{2}{T^2}(B - b) \| f \|_1,$$

where the norm is defined by (19.1.2). This tends to zero as $T \to \infty$. If $p > 1$, we use Hölder's inequality to obtain a similar upper bound

$$C_p(B - b)T^{-1-\frac{1}{p}} \| f \|_p$$

which also tends to zero with $1/T$. The same estimates apply to J_4. Thus we have

$$2\pi i f(z) = \lim_{T \to \infty} \frac{1}{T} \int_T^{2T} [J(b, Y) - J(B, Y)] \, dY.$$

Now

(19.1.30)
$$\lim_{Y \to \infty} J(b, Y) = i \int_{-\infty}^{\infty} \frac{f(b + it) \, dt}{z - b - it}$$

exists as an absolutely convergent integral, and the same holds if we replace b by B. It follows that

$$2\pi i f(z) = J(b, \infty) - J(B, \infty)$$

or

$$f(z) = \frac{1}{2\pi} \int_{-\infty}^{\infty} \frac{f(b + it) \, dt}{z - b - it} - \frac{1}{2\pi} \int_{-\infty}^{\infty} \frac{f(B + it) \, dt}{z - B - it}.$$

Compare Theorem 5.1.10 for a similar conclusion involving sums instead of integrals.

Here we let $B \to \infty$. A simple estimate shows that for $|z| < \frac{1}{2}B$

$$|J(B, \infty)| < C(p)B^{-\frac{1}{p}} \|f\|_p.$$

Thus (19.1.28) is obtained by passing to the limit with B.

If $z_b{}^* = -x + 2b + iy$ is written instead of $z = x + iy$ in (19.1.28), then the result is 0 since $z_b{}^*$ lies outside of the rectangle used in deriving the formula. Hence, if we subtract this expression from (19.1.28) and simplify, we obtain Poisson's formula (19.1.29).

THEOREM 19.1.3.   *If $f(z) \in H_p(0)$, then*

(19.1.31)
$$|f(x + iy)| \leq C(p)x^{-\frac{1}{p}} \|f\|_p, \quad x > 0.$$

Furthermore, for each fixed $x > 0$, we have

(19.1.32)
$$\lim_{|y| \to \infty} f(x + iy) = 0.$$

Proof. If $p > 1$ we obtain (19.1.31) by applying Hölder's inequality to formula (19.1.29). This gives

$$|f(x + iy)| \leq \frac{|x - b|}{\pi} \left\{ \int_{-\infty}^{\infty} |f(b + it)|^p \, dt \right\}^{\frac{1}{p}} \left\{ \int_{-\infty}^{\infty} [(x - b)^2 + (y - t)^2]^{-q} \, dt \right\}^{\frac{1}{q}}$$

$$\leq \frac{1}{\pi} \|f\|_p (x - b)^{-\frac{1}{p}} \left\{ \int_{-\infty}^{\infty} (1 + s^2)^{-q} \, ds \right\}^{\frac{1}{q}}.$$

The integral can be expressed in terms of Gamma functions involving p. Since the estimate holds for all $b > 0$, we can let $b \to 0$ and obtain (19.1.31). The case $p = 1$ is immediate.

We prove the second inequality for $p = 1$. Here we assume that y is large and positive. Then for $x > b$

$$| f(x + iy) | \leqq \frac{1}{2\pi} \left\{ \int_{-\infty}^{\frac{1}{2}y} + \int_{\frac{1}{2}y}^{\infty} \right\} \frac{| f(b + it) | \, dt}{| x - b + i(y - t) |} .$$

Now $| x - b + i(y - t) |$ exceeds $\frac{1}{2}y$ in the first integral and $x - b$ in the second. Hence

$$| f(x + iy) | \leqq \frac{1}{\pi y} \| f \|_1 + \frac{1}{\pi(x - b)} \int_{\frac{1}{2}y}^{\infty} | f(b + it) | \, dt,$$

where both expressions on the right are arbitrarily small for large values of y. The same is true for large negative y, and if $p > 1$ we use Hölder's inequality instead.

THEOREM 19.1.4. $M_p(x; f)$ is a positive decreasing function of x and

(19.1.33) $$\lim_{x \to 0} M_p(x; f) = \| f \|_p.$$

Proof.   We use formula (19.1.29). If $p = 1$,

$$\int_{-\infty}^{\infty} | f(x + iy) | \, dy \leqq \frac{x - b}{\pi} \int_{-\infty}^{\infty} \frac{dy}{(x - b)^2 + (y - t)^2} \int_{-\infty}^{\infty} | f(b + it) | \, dt$$

$$= \int_{-\infty}^{\infty} | f(b + iy) | \, dy,$$

that is, for $0 < b < x$

$$M_1(x; f) \leqq M_1(b; f).$$

Since x and b are arbitrary, $0 < b < x < \infty$, the monotone character of $M_1(x; f)$ is obvious. If $p > 1$, we use Hölder's inequality and proceed as in the proof of (19.1.26). The inequality

$$M_p(x; f) \leqq M_p(b; f)$$

results. Formula (19.1.33) follows from the definition of $\| f \|_p$ and from the fact that $M_p(x; f)$ is monotone decreasing.

COROLLARY.   *If $f(z) \in H_p(0)$, $1 \leqq p < \infty$, if $f(z)$ can be represented in the form*

(19.1.34) $$f(z) = \frac{1}{2\pi} \int_{-\infty}^{\infty} \frac{g(t) \, dt}{z - it}, \quad \Re(z) > 0,$$

with a $g(t) \in L_p(-\infty, \infty)$, and if we define

(19.1.35) $$M_p(0; f) \equiv \| g \|_p,$$

then

(19.1.36) $$M_p(0; f) \geqq M_p(x; f)$$

for all $x > 0$ and

(19.1.37) $$\| f \|_p = M_p(0; f).$$

The proof is as above. That $M_p(x; f)$ is continuous for $x \geq 0$ will be proved below for the case $p > 1$.

LEMMA 19.1.8.   *If L is a linear bounded functional on $L_q(a, b)$, $1 < q < \infty$, then there exists an element $g(t)$ of $L_p(a, b)$ such that for each $h(t) \in L_q(a, b)$ we have*

$$(19.1.38) \qquad L[h] = \int_a^b g(t)\, \overline{h(t)}\, dt$$

and

$$(19.1.39) \qquad \| L \| \equiv \sup_{\|h\| = 1} | L[h] | = \| g \|_p.$$

For a proof and further explication consult pages 73–78 of the work by Riesz and Nagy cited on page 431.

We come now to the main theorem.

THEOREM 19.1.5.   *If $f(z) \in H_p(0)$, $1 < p < \infty$, and if*

$$(19.1.40) \qquad f(z) = \sum_{n=1}^{\infty} f_n(a)\, \omega_n(z; a), \quad \Re(z) > 0,$$

then

$$(19.1.41) \qquad \sum_{n=1}^{\infty} f_n(a)\, \omega_n(it; a)$$

is the Fourier series of a function $f(it) \in L_p(-\infty, \infty)$. We have

$$(19.1.42) \qquad f(z) = \frac{1}{2\pi} \int_{-\infty}^{\infty} \frac{f(it)\, dt}{z - it}, \quad \Re(z) > 0,$$

$$(19.1.43) \qquad \lim_{x \to 0} \| f(x + iy) - f(iy) \|_p = 0.$$

Finally, $M_p(x; f)$ is continuous for $0 \leq x$.

Proof.   Any function $f(z)$ which is holomorphic in the right half-plane admits of unique expansions of the form

$$(19.1.44) \qquad f(z) = \sum_{n=1}^{\infty} f_n(a, b)\, \omega_n(z - b; a),$$

where $b \geq 0$ and the series is absolutely convergent for $\Re(z) > b$. Further if $f(z) \in H_p(0)$, $1 < p < \infty$, and $b > 0$, then we know that the series

$$(19.1.45) \qquad \sum_{n=1}^{\infty} f_n(a, b)\, \omega_n(iy; a)$$

is the Fourier series of an element of $L_p(-\infty, \infty)$, namely, that of $f(b + iy)$. Our problem is to prove that this is also the case for $b = 0$. For this purpose we shall study the coefficients $f_n(a, b)$.

We have the representation

$$(19.1.46) \qquad f_n(a, b) = \frac{a}{\pi i} \int_C f\left(b + a\, \frac{1 + w}{1 - w}\right) \frac{dw}{(1 - w)w^n},$$

where the integral is taken over a small circle with center at $w = 0$. This is valid for any $f(z)$ holomorphic in $\Re(z) > 0$. It is clear from this representation that the coefficients $f_n(a, b)$ are continuous functions of b for $b \geq 0$ so that

$$(19.1.47) \qquad \lim_{b \to 0} f_n(a, b) = f_n(a, 0) \equiv f_n(a).$$

On the other hand, if $f(z) \in H_p(0)$, $1 < p < \infty$, then

$$(19.1.48) \qquad f_n(a, b) = \frac{a}{\pi} \int_{-\infty}^{\infty} f(b + it) \,\overline{\omega_n(it; a)} \, dt.$$

This suggests considering the linear bounded functional $L_b[h]$ defined on $L_q(-\infty, \infty)$ by

$$(19.1.49) \qquad L_b[h] \equiv \frac{a}{\pi} \int_{-\infty}^{\infty} f(b + it) \,\overline{h(t)} \, dt.$$

Its bound is

$$\frac{a}{\pi} \| f(b + it) \|_p \leq \frac{a}{\pi} \| f \|_p.$$

Thus

$$L_b[\omega_n] = \begin{cases} f_n(a, b), & n \geq 1, \\ 0, & n \leq -1. \end{cases}$$

Formula (19.1.47) now shows that

$$(19.1.50) \qquad \lim_{b \to 0} L_b[h] \equiv L[h]$$

exists for any h in $L_q(-\infty, \infty)$ which is a finite linear combination of the elements of the orthogonal system $\{\omega_n(it; a)\}$. Since such combinations are dense in $L_q(-\infty, \infty)$, we conclude that $L[h]$ can be defined as a linear functional everywhere in $L_q(-\infty, \infty)$. Moreover, $L[h]$ is clearly bounded with a bound not exceeding $a \| f \|_p/\pi$. Thus, by Lemma 19.1.8, a function $g(t) \in L_p(-\infty, \infty)$ exists such that

$$(19.1.51) \qquad L[h] = \frac{a}{\pi} \int_{-\infty}^{\infty} g(t) \,\overline{h(t)} \, dt.$$

Formulas (19.1.47) and (19.1.48) now show that the Fourier series of $g(t)$ is actually the series

$$\sum_{n=1}^{\infty} f_n(a) \,\omega_n(it; a),$$

and this justifies the notation $f(it)$ for $g(t)$. This is the main step in the proof, to show that this formal series is actually the Fourier series of a function in $L_p(-\infty, \infty)$.

We can now fall back on Theorem 19.1.1. We take $g(t) = f(it)$ in this theorem and note that $g_n = f_n(a)$ for $n > 0$ and that $g_n = 0$ for $n < 0$. This implies that $f_+(z) \equiv f(z)$, so that (19.1.42) holds, as well as (19.1.43).

If $f(z) \in H_p(0)$ and $h > 0$, then $f(z + h)$ and

$$\Delta_h(z) = f(z + h) - f(z)$$

also belong to $H_p(0)$. According to the Corollary of Theorem 19.1.4,

$$M_p(0; \Delta_h) \geq M_p(x; \Delta_h)$$

or

$$\| f(h + iy) - f(iy) \|_p \geq \| f(x + h + iy) - f(x + iy) \|_p.$$

We have just seen that the left member is small if h is small. From this we conclude that $M_p(x; f)$ is continuous for $0 \leq x$ since

$$M_p(x; f) - M_p(x + h; f) \leq \| f(x + h + iy) - f(x + iy) \|_p.$$

This completes the proof of the theorem.

It remains to say something about the structure of the class $H_p(0)$ for $1 < p < \infty$. It is clear that it is a normed linear vector space.

THEOREM 19.1.6. $H_p(0)$ *is complete under the norm* (19.1.2).

Proof.   This is a consequence of the corresponding fact for $L_p(-\infty, \infty)$. For if $\{f_n\}$ is a Cauchy sequence in $H_p(0)$, then

$$M_p(0; f_n - f_m) \geq M_p(x; f_n - f_m).$$

If $f_0(iy)$ is the limit in the mean of the sequence $\{f_n(iy)\}$, then

$$f_0(z) \equiv \frac{1}{2\pi} \int_{-\infty}^{\infty} \frac{f_0(it)\, dt}{z - it}$$

is in $H_p(0)$ and

$$M_p(0; f_n - f_0) \geq M_p(x; f_n - f_0).$$

This implies that $\{f_n(z)\}$ converges to $f_0(z)$ in the metric of $H_p(0)$.

$H_p(0)$ is not an algebra, but if $f(z)$ is the product of two functions in $H_p(0)$, then $f(z + h) \in H_p(0)$ for every $h > 0$. We have a similar situation with regard to differentiation: $f'(z)$ is usually not in $H_p(0)$, but $f'(z + h)$ belongs to $H_p(0)$ if $f(z)$ does.

EXERCISE 19.1

1. Verify (19.1.14).

2. When does a rational function of z belong to $H_p(0)$?

3. Show that if the roots have their principal determinations then

$$z^{-\frac{1}{2}} - (z + 1)^{-\frac{1}{2}}$$

belongs to $H_p(0)$ for $1 \leq p < 2$.

4. Verify that $H_p(0)$ is not an algebra.

5. If $f(z) \in H_p(0)$, show that $f^2(z)$ also has this property if and only if $f(iy) \in L_{2p}(-\infty, \infty)$.

6. Find sufficient conditions in order that $f'(z)$ shall belong to $H_p(0)$ when $f(z)$ does.

7. Verify the assertions concerning formulas (19.1.44) and (19.1.46).

8. Show that neither $\Gamma(z + 1)$ nor its reciprocal can belong to any class $H_p(0)$.

9. In the proof of the existence of $\lim L_b[h]$ we have made tacit use of a special case of the Banach-Steinhaus theorem: If $\{L_n\}$ is a sequence of linear bounded functionals on a (B)-space X such that $\| L_n \| \leq M$ and if $\lim L_n[x]$ exists for x in a set which is dense in X, then the limit exists for all x in X, and $\lim L_n[x] \equiv L[x]$ is a linear bounded functional in X of bound not exceeding M. Prove this for the case $X = L_q(a, b)$ which figures above.

10. Modify the system $\{\omega_n(it; a)\}$ by raising the exponent of the denominator from n to $n + 1$. Show that the new system has its linear combinations dense in $L_1(-\infty, \infty)$, but is not an orthogonal system.

***11.** Extend Theorem 19.1.5 to the case $p = 1$, using the functions of Problem 10 for purposes of approximation. Note that most of the formulas (19.1.40)–(19.1.51) make sense, but since the new functions are not orthogonal the characterization of $f(iy)$ is less direct. It is possible, however, to orthogonalize these functions, and the Fourier coefficients of $f(iy)$ are linear combinations of the $f_n(a)$ with multipliers independent of $f(z)$. It is also necessary to verify that $f(z)$ is representable by its Poisson integral in terms of $f(it)$ and to verify the analogue of (19.1.26).

12. If $f(z) \in H_p(0)$, $1 \leq p < \infty$, show that $\lim_{x \to 0} f(x + iy)$ exists pointwise for almost all y. (*Hint:* If

$$f(z) = f\left(\frac{1 - w}{1 + w}\right) = g(w),$$

then $g(w)$ is holomorphic in $| w | < 1$, $g(e^{i\theta}) \in L(-\pi, \pi)$, and Theorem 17.5.1 can be used.)

19.2. Bounded functions. We shall return to the class $H_p(0)$ later, but here we want to devote some attention to the omitted case $p = \infty$, the bounded functions. We denote by $B(0)$ the class of all functions which are holomorphic and bounded in the right half-plane. We set

(19.2.1) $\| f \| = \sup | f(z) |, \quad f(z) \in B(0),$

and note that $B(0)$ is a Banach algebra under this norm.

We have the analogue of the theorem of Fatou.

THEOREM 19.2.1.   *If $f(z) \in B(0)$, then*

(19.2.2) $\lim_{x \to 0} f(x + iy) \equiv f(iy)$

exists for almost all x and is a bounded measurable function of y.

The proof is similar to that of Problem 12 in the preceding Exercise.

It is clear that a function of $B(0)$ cannot always be expected to be representable as a Cauchy integral of its boundary values. However, the Poisson integral applies and we have

THEOREM 19.2.2.   *If $f(z) \in B(0)$, then*

$$(19.2.3) \qquad f(x + iy) = \frac{x}{\pi} \int_{-\infty}^{\infty} \frac{f(it)\, dt}{(t - y)^2 + x^2}, \quad x > 0.$$

Proof.   Let $0 < b < x < B < \infty$ and $|\, y\, | < Y$ and integrate along the perimeter of the rectangle R with vertices at the points

$$b \pm iY, \quad B \pm iY.$$

We have then

$$f(z) = \frac{1}{2\pi i} \int_R \frac{f(w)\, dw}{w - z}, \quad 0 = \frac{1}{2\pi i} \int_R \frac{f(w)\, dw}{w + \bar{z}},$$

whence, by subtraction,

$$f(z) = \frac{x}{\pi i} \int_R \frac{f(w)\, dw}{(w - z)(w + \bar{z})}.$$

Here we let $Y \to \infty$, $B \to \infty$, $b \to 0$ in this order and obtain (19.2.3) in the limit. For the first two steps we use merely the fact that $|\, f(w)\, |$ is bounded; for the third step we need (19.2.2) and the bounded convergence theorem of Lebesgue.

There are some properties of the Poisson integral, in addition to those mentioned in the preceding section, which will be used in the following. These are listed in

LEMMA 19.2.1.   *If $g(t)$ is a bounded measurable function, then*

$$(19.2.4) \qquad g(x, y) = \frac{x}{\pi} \int_{-\infty}^{\infty} \frac{g(t)\, dt}{(t - y)^2 + x^2}, \quad x > 0,$$

is a bounded harmonic function. If $g(t)$ is real, so is $g(x, y)$, and $g(x, y)$ lies between the essential infimum and the essential supremum of $g(t)$. Further,

$$(19.2.5) \qquad \lim_{x \to 0} g(x, y) = g(y)$$

for almost all y and, in particular, at all points of continuity of $g(t)$. The same limit relation holds if $g(t) \in L_p(-\infty, \infty)$ for some p, $1 \leq p < \infty$. Finally, if $g(t)$ is positive in some interval $(y_0 - \delta, y_0 + \delta)$ and tends to $+\infty$ as $t \to y_0$, then

$$(19.2.6) \qquad \lim_{x \to 0} g(x, y_0) = +\infty.$$

Proof.   If ess sup $|\, g(t)\, | = M$, then clearly

$$|\, g(x, y)\, | \leq M \frac{x}{\pi} \int_{-\infty}^{\infty} \frac{dt}{(t - y)^2 + x^2} \equiv M.$$

That $g(x, y)$ is harmonic in $x > 0$ follows from the fact that the Poisson kernel has this property.

It is clear that if $g(t) \geqq 0$ almost everywhere, then $g(x, y) \geqq 0$. Hence, if

$$\text{ess inf } g(t) = b, \quad \text{ess sup } g(t) = B,$$

then

$$g(t) - b \geqq 0, \quad B - g(t) \geqq 0$$

almost everywhere, so that

$$b \leqq g(x, y) \leqq B.$$

For the limit relation we note that

$$g(x, y) - g(y) = \frac{x}{\pi} \int_{-\infty}^{\infty} \frac{g(y + s) - g(y)}{s^2 + x^2} \, ds.$$

Suppose first that $g(y) \in C[-\infty, \infty]$ and set

$$(19.2.7) \qquad \mu(s; g) = \sup_{y} | g(y + s) - g(y) |.$$

This is the *modulus of continuity* of $g(t)$. It is a *subadditive continuous function of s and* $\mu(0; g) = 0$. These properties are proved as Lemma 19.1.7 was proved; we have merely to replace L_p-norms by C-norms. We have then

$$\sup_{y} | g(x, y) - g(y) | \leqq \frac{x}{\pi} \int_{-\infty}^{\infty} \frac{\mu(s; g) \, ds}{s^2 + x^2}.$$

This tends to zero with x by the argument used in proving formula (19.1.26). Thus, if $g(t) \in C[-\infty, \infty]$, then (19.2.5) holds uniformly in y.

Suppose now that $g(t)$ is continuous for $t = y_0$ and bounded for all t. Given an $\varepsilon > 0$ we can then find a $\delta = \delta(\varepsilon)$ such that $| g(y_0 + s) - g(y_0) | \leqq \varepsilon$ for $| s | \leqq \delta$. Hence

$$| g(x, y_0) - g(y_0) | \leqq \frac{x}{\pi} \left\{ \int_{-\infty}^{-\delta} + \int_{-\delta}^{\delta} + \int_{\delta}^{\infty} \right\} \frac{| g(y_0 + s) - g(y_0) |}{s^2 + x^2} \, ds$$

$$\leqq \frac{4}{\pi} M \text{ arc cot } \frac{\delta}{x} + \varepsilon,$$

and as $x \to 0$ the superior limit of the last member is ε so that (19.2.5) holds at every point of continuity.

The same conclusion holds in the so-called Lebesgue set of $g(t)$ to which almost every y_0 belongs by Lemma 17.5.2 provided $g(t)$ is integrable over every finite interval. If y_0 belongs to this set, we define

$$(19.2.8) \qquad G(s; y_0) = \int_{0}^{s} | g(y_0 + t) - g(y_0) | \, dt.$$

Given an $\varepsilon > 0$ we choose a $\delta = \delta(\varepsilon)$ such that

$$| G(s; y_0) | \leqq \varepsilon | s | \quad \text{for} \quad | s | \leqq \delta.$$

We have then

$$\frac{x}{\pi} \int_{-\delta}^{\delta} \frac{|\, g(y_0 + s) - g(y_0)\,|}{s^2 + x^2}\, ds = \frac{x}{\pi} \frac{G(s; y_0)}{s^2 + x^2}\Big|_{s=-\delta}^{s=\delta} + \frac{2x}{\pi} \int_{-\delta}^{\delta} \frac{sG(s; y_0)\, ds}{(s^2 + x^2)^2}$$

$$\leq \frac{1}{\pi} \frac{2\delta x}{\delta^2 + x^2}\, \varepsilon + \frac{4x}{\pi} \int_0^\delta \frac{s^2\, ds}{(s^2 + x^2)^2}\, \varepsilon$$

$$< \left\{ \frac{1}{\pi} + \frac{4}{\pi} \int_0^\infty \frac{t^2\, dt}{(t^2 + 1)^2} \right\} \varepsilon = \left(\frac{1}{\pi} + 1 \right) \varepsilon.$$

If we assume that $g(t)$ is bounded for all t, then the estimates for the intervals $(-\infty, -\delta)$ and (δ, ∞) are unchanged, and we find that (19.2.5) holds for y_0 in the Lebesgue set of $g(t)$, that is, for almost all y_0. The modifications necessary when $g(t) \in L_p(-\infty, \infty)$ may be left to the reader.

Finally, suppose that $g(t) \in L(-\infty, \infty)$ and is positive in $(y_0 - \delta, y_0 + \delta)$ and tends to $+\infty$ as $t \to y_0$. Suppose that A is a given large positive number. We can then choose ε so small that $g(t) \geq A$ for $|\, t - y_0\,| \leq \varepsilon$. This gives

$$g(x, y_0) \geq A \frac{x}{\pi} \int_{-\varepsilon}^{\varepsilon} \frac{ds}{s^2 + x^2} - \frac{x}{\pi} \left\{ \int_{-\infty}^{-\varepsilon} + \int_{\varepsilon}^{\infty} \right\} \frac{|\, g(y_0 + s)\,|\, ds}{s^2 + x^2}$$

$$\geq A \frac{2}{\pi} \arctan \frac{\varepsilon}{x} - \frac{x}{\pi \varepsilon^2} \int_{-\infty}^{\infty} |\, g(t)\,|\, dt,$$

so that

$$\liminf_{x \to 0} g(x, y_0) \geq A.$$

Since A is arbitrary, (19.2.6) must hold, and the lemma is proved.

THEOREM 19.2.3.   *If $f(z) \in B(0)$, if $y = y_0$ belongs to the Lebesgue set of $f(iy)$, and if $z \to iy_0$ along a path which is not tangent to the imaginary axis, then*

$$(19.2.9) \qquad \lim_{z \to iy_0} f(z) = f(iy_0).$$

Proof.   We have

$$f(x + iy) - f(iy_0) = \frac{x}{\pi} \int_{-\infty}^{\infty} [f(iv + iy) - f(iy_0)] \frac{dv}{v^2 + x^2}$$

$$= \frac{x}{\pi} \int_{-\infty}^{\infty} [f(iv + iy) - f(iv + iy_0)] \frac{dv}{v^2 + x^2}$$

$$+ \frac{x}{\pi} \int_{-\infty}^{\infty} [f(iv + iy_0) - f(iy_0)] \frac{dv}{v^2 + x^2}.$$

Since y_0 belongs to the Lebesgue set of $f(iy)$, the proof of the preceding lemma shows that the second integral in the last member tends to zero with x. On the other hand, for a fixed $\delta > 0$

$$\frac{x}{\pi} \left\{ \int_{-\infty}^{-\delta} + \int_{\delta}^{\infty} \right\} |\, f(iv + iy) - f(iv + iy_0)\,| \frac{dv}{v^2 + x^2} \leq \frac{4}{\pi} M \operatorname{arc\, cot} \frac{\delta}{x},$$

which tends to zero with x regardless of y and y_0. Here M is the essential supremum of $|f(iy)|$. We now set

$$\int_0^t |f(iv + iy) - f(iv + iy_0)| \, dv \equiv \mu(t; y, y_0).$$

For $t > 0$ this expression is dominated by

$$\int_0^t |f(iv + iy) - f(iy_0)| \, dv + \int_0^t |f(iv + iy_0) - f(iy_0)| \, dv$$

$$< \int_{-t-h}^{t+h} |f(iv + iy_0) - f(iy_0)| \, dv + F(t; y_0)$$

$$= F(t + h; y_0) - F(-t - h; y_0) + F(t; y_0),$$

where $h = |y - y_0|$ and $F(t; y_0)$ denotes the second integral in the first line. A similar estimate holds for $t < 0$.

The condition on the mode of approach of z to iy_0 implies the existence of a positive number K such that

$$|y - y_0| \leq Kx.$$

Hence, if $\varepsilon > 0$ is given, we can find a $\xi = \xi(\varepsilon)$ and a constant C such that for $0 \leq x \leq \xi$ and $|t| \leq \delta$ we have

$$|\mu(t; y, y_0)| \leq C(|t| + x)\varepsilon$$

by virtue of the above estimates and the properties of $F(t; y_0)$ in the Lebesgue set of $f(iy)$.

Thus, we have

$$\frac{x}{\pi} \int_{-\delta}^{\delta} |f(iv + iy) - f(iv + iy_0)| \frac{dv}{v^2 + x^2}$$

$$= \frac{1}{\pi} \frac{x}{\delta^2 + x^2} [\mu(\delta; y, y_0) - \mu(-\delta; y, y_0)] + \frac{2x}{\pi} \int_{-\delta}^{\delta} \frac{v\mu(v; y, y_0)}{(v^2 + x^2)^2} \, dv$$

$$\leq C \frac{\varepsilon}{\pi} \left\{ \frac{2x(\delta + x)}{\delta^2 + x^2} + 4x \int_0^{\delta} \frac{v(v + x)}{(v^2 + x^2)^2} \, dv \right\} < 3C\varepsilon.$$

Since ε is arbitrary we conclude that (19.2.9) holds.

If $f(z) \in B(0)$, then for $\Re(z) > 0$ we have

$$(19.2.10) \qquad f(z) = \sum_{n=1}^{\infty} f_n(a) \, \omega_n(z; a), \qquad \omega_n(z; a) = \frac{(z - a)^{n-1}}{(z + a)^n}.$$

Here the coefficients may be assumed to be given by (19.1.15) provided that $f(it)(it + a)^{-1} \in L(-\infty, \infty)$, and in any case by (19.1.46) with $b = 0$.

A function which is bounded in the right half-plane naturally omits "most" complex values, but even the values which are assumed are subject to severe limitations. The function cannot show too strong affinity to any particular value without running the risk of reducing identically to that value. We shall consider these aspects of the theory in the following and may restrict ourselves to the value 0.

There are actually four different problems to study. First, the function $f(z) \in B(0)$ may have zeros in the right half-plane, and if there are infinitely many zeros either the point at infinity or points on the imaginary axis are limit points of the set. It is also possible for $f(z)$ to tend to the limit 0 as z approaches points on the imaginary axis, it may tend to 0 as z tends to infinity in the right half-plane, and finally it may tend to 0 along the imaginary axis. Each case is governed by different rules which the function must obey unless it is identically 0. The distribution of the zeros requires the convergence of certain series and the approach to 0 is subject to various integrability conditions.

The limiting values on the imaginary axis are subject to the least stringent regulations. For the case of a function bounded and holomorphic in the unit disk, F. and M. Riesz proved in 1916 that 0 can be a limiting value on the circle at most in a set of measure 0. On the other hand, P. Fatou proved in his thesis of 1906 that for any closed set of measure 0 on the unit circle there exists a function bounded and holomorphic in the unit disk which is continuous in $|z| \leq 1$, which assumes the value 0 in the given set and nowhere else. These results carry over to the class $B(0)$, either by conformal mapping or by adaptation of the method of proof. We follow the second alternative. Thus we aim to prove the following theorem:

THEOREM 19.2.4.   *If $f(z) \in B(0)$, if $f(z) \not\equiv 0$, and if*

(19.2.11) $$\lim_{x \to 0} f(x + iy) = 0 \quad for \quad y \in N,$$

then the set N is of measure 0. Conversely, if N is a closed set of measure 0, then there exists an $f(z) \in B(0)$ such that $f(z) \not\equiv 0$, $f(z)$ is continuous in $\Re(z) \geq 0$, (19.2.11) holds, and $f(z) \neq 0$ for all other values of $z = x + iy$ with $x \geq 0$.

Proof.   Let $f(z)$ satisfy the conditions of the theorem and suppose that the measure of N is positive. Without restricting the generality we may assume that the set N is bounded. Let A be a given positive number and choose the number B so that

(19.2.12) $$A \int_N \frac{dt}{t^2 + 1} + B \int_{C[N]} \frac{dt}{t^2 + 1} = 0.$$

Define

$$g(t) = \begin{cases} A, & t \in N, \\ B, & t \in \mathbf{C}[N], \end{cases}$$

and

$$U(x, y) = \frac{x}{\pi} \int_{-\infty}^{\infty} \frac{g(t)\, dt}{(t - y)^2 + x^2}.$$

This is a function harmonic in the right half-plane which tends to the value A or the value B according as y belongs to N or to $\mathbf{C}[N]$ when $x \to 0$ and y does not belong to an exceptional set of measure 0. Further, condition (19.2.12) shows that $U(1, 0) = 0$. Let $V(x, y)$ be the conjugate harmonic function, normalized so that $V(1, 0) = 0$, and form

$$F(x + iy) = \exp\left[U(x, y) + iV(x, y)\right].$$

Here $F(z) \in B(0)$, $F(1) = 1$, and as $x \to 0$, $F(x + iy)$ tends to $\exp A$ or $\exp B$ according as y belongs to N or to $\mathbf{C}[N]$, a set of measure 0 being excepted.

The function $G(z) \equiv f(z)F(z)$ belongs to $B(0)$ and admits of a unique expansion of the form

$$G(z) = \sum_{n=0}^{\infty} c_n \left(\frac{z-1}{z+1}\right)^n,$$

where

$$c_0 = G(1) = f(1)F(1) = f(1).$$

On the other hand, $G(z)(z+1)^{-1} \in H_2(0)$ and has a unique expansion in terms of the orthogonal system $\omega_n(z; 1)$, the coefficients of which coincide with the numbers c_n. It follows, in particular, that

$$c_0 = \frac{1}{\pi} \int_{-\infty}^{\infty} f(it)F(it)(it + 1)^{-2}\, dt.$$

Here we can neglect the integral over the set N where $f(it) = 0$. Since $f(z) \in B(0)$ we have $|f(it)| \leq \|f\|$ everywhere. Further, $|F(it)| = \exp B$ almost everywhere in $\mathbf{C}[N]$. It follows that

$$|c_0| \leq \|f\|\, e^B\, \frac{1}{\pi} \int_{\mathbf{C}[N]} \frac{dt}{1 + t^2} < \|f\|\, e^B.$$

Since A is at our disposal, B is also, and we may choose B arbitrarily large negative. It follows that

$$c_0 = f(1) = 0.$$

We can now repeat the argument. Since $f(1) = 0$, we see that on the one hand

$$c_1 = 2f'(1)F(1) = 2f'(1),$$

and on the other

$$c_1 = \frac{1}{\pi} \int_{-\infty}^{\infty} f(it)F(it)\, \frac{it-1}{(it+1)^3}\, dt,$$

so that

$$|c_1| \leq \|f\|\, e^B$$

and, hence, $c_1 = 0$. In this manner one proves by induction first that the coefficients c_n are independent of $F(z)$, that is, of B, and secondly that they are all 0. This is possible if and only if $f(z) \equiv 0$, which is contrary to the assumption. It follows that the measure of N must be 0.

Suppose next that N is a given closed set of measure 0 and let us construct a function $f(z)$ with the desired properties. We first observe that $C[N]$ is the union of nonoverlapping open intervals (a_n, b_n). We may assume that there are infinitely many intervals since the finite case is trivial. For each n we define a function $P_n(t)$ in (a_n, b_n) which is positive and analytic in the interval, becomes infinite at the endpoints, and has a finite integral over (a_n, b_n). We may take

$$P_n(t) = (1 + t^2)^{-1}[(t - a_n)^{-\frac{1}{2}} + (b_n - t)^{-\frac{1}{2}}]$$

with obvious modifications if one of the endpoints is infinite. Let

$$J_n = \int_{a_n}^{b_n} P_n(t)\, dt$$

and choose positive constants c_n such that

$$\sum_{n=1}^{\infty} c_n J_n < \infty.$$

Next, since $(1 + t^2)^{-1} \in L(-\infty, \infty)$ we can choose positive numbers d_n which become infinite with n such that the series

$$\sum_{n=1}^{\infty} d_n \int_{a_n}^{b_n} (1 + t^2)^{-1}\, dt$$

is convergent. This is a consequence of a classical theorem on infinite series: If $u_n > 0$, $\Sigma\, u_n$ convergent, then the series

$$\sum_{n=1}^{\infty} R_{n-1}^{\alpha-1} u_n, \quad R_{n-1} = \sum_{k=n}^{\infty} u_k,$$

converges if $\alpha > 0$. Here the multipliers become infinite with n if $0 < \alpha < 1$.
We now define a function $P(t)$ by

$$P(t) = c_n P_n(t) + d_n(1 + t^2)^{-1}, \quad a_n < t < b_n,$$

where n runs through the positive integers. This function is positive on every interval (a_n, b_n), that is, for almost all t, and it is analytic in each interval. By the choice of c_n and d_n it belongs to $L(-\infty, \infty)$. Further, if $t \to t_0$, $t_0 \in N$, then $P(t) \to +\infty$. For $P(t) > d_n(1 + t^2)^{-1}$ in (a_n, b_n), and d_n becomes infinite with n. It follows that $P(t)$ is uniformly large in any small neighborhood of $t = t_0$ so that $P(t)$ must have the limiting value $+\infty$ at every point of N.
We now define

$$P(x, y) = \frac{x}{\pi} \int_{-\infty}^{\infty} \frac{P(t)\, dt}{(t - y)^2 + x^2}.$$

By Lemma 19.2.1 this is a positive harmonic function in the right half-plane, and

$$\lim_{x \to 0} P(x, y) = P(y)$$

for all y where, by definition, $P(y) = +\infty$ if $y \in N$.

There exists a conjugate harmonic function $Q(x, y)$ which is uniquely determined by the Cauchy-Riemann equations together with the initial condition

$$Q(1, 0) = 0.$$

This function $Q(x, y)$ is continuous, even analytic, on each of the intervals (a_n, b_n). Its values at the points of N are of no importance. We now set

$$(19.2.13) \qquad g(x + iy) = P(x, y) + iQ(x, y), \quad f(z) = \frac{1}{1 + g(z)}.$$

Then $g(z)$ is holomorphic in $\Re(z) > 0$ and $\Re[g(z)] > 0$, so that $f(z)$ is also holomorphic there, and $|f(z)| < 1$, so that $f(z) \in B(0)$. It is clear that $f(z) \not\equiv 0$. Finally, as $x \to 0$, $g(x + iy)$ tends to a finite limit everywhere in $\mathbf{C}[N]$ and to ∞ in N. Also, if $t_0 \in N$, then $|g(z)|$ is uniformly large for $\Re(z) > 0, |z - it_0| < \delta$. From this it then follows that $f(z)$ is continuous everywhere in $\Re(z) \geq 0$ and takes the value 0 in N and nowhere else. This completes the proof of Theorem 19.2.4.

This theorem does not exhaust the available information concerning the zeros of $f(z)$ on the imaginary axis. There are also limitations on what may be referred to as the order of the zeros. Thus, we shall prove that $\log |f(iy)|$ is integrable over every finite interval. Similar integrability conditions express that a bounded function $f(z)$ cannot approach its limits on the imaginary axis arbitrarily fast. To prove such results and to study related problems concerning zeros in the right half-plane and limiting values at infinity, we shall develop suitable machinery following the procedure of the brothers Nevanlinna (1922). The reader has encountered similar considerations in Chapter 18.

Consider a domain D in the complex plane and a function $f(z)$ meromorphic in D. Let Γ be the boundary of D. We shall assume that Γ is piecewise analytic. In the cases which occur in the following, Γ is made up of circular arcs and straight-line segments. Let

$$a_1, a_2, \cdots, a_m$$

be the zeros of $f(z)$ in D, and let

$$b_1, b_2, \cdots, b_n$$

be the poles, each repeated according to its multiplicity. We assume to start with that $f(z)$ has neither zeros nor poles on Γ. Let $\varepsilon > 0$ be a small number and delete a circular disk of radius ε about each of the zeros and poles of $f(z)$ in D. Let D_ε denote the remaining part of D and let its boundary be Γ_ε. We assume that D_ε is connected, as will be the case for sufficiently small values of ε. Finally, let $\lambda(z)$ be a real-valued function of z, continuous together with its first- and second-order partial derivatives with respect to x and y in D.

We now apply the formula of Green

(19.2.14) $$\iint_{D_\varepsilon} (U\Delta V - V\Delta U)\, dx\, dy = -\int_{\Gamma_\varepsilon} \left(U \frac{\partial V}{\partial n} - V \frac{\partial U}{\partial n} \right) ds$$

with

$$U(z) = \log |f(z)|, \quad V(z) = \lambda(z).$$

Here n is the interior normal. We have $\Delta U = 0$ in D_ε, so the left member reduces to

$$\iint_{D_\varepsilon} \log |f(z)|\, \Delta\lambda(z)\, dx\, dy.$$

Let γ be one of the small circles which form a part of $\Gamma_\varepsilon \ominus \Gamma$. Let c be its center where c is a zero or a pole of $f(z)$. On γ we have

$$U = k \log \varepsilon + O(1), \quad \frac{\partial U}{\partial n} = \frac{k}{\varepsilon} + O(1)$$

since the interior normal is in the direction of increasing values of ε. As in Section 18.2 we see that

$$\int_\gamma U \frac{\partial \lambda}{\partial n}\, ds \to 0, \quad \int_\gamma \lambda \frac{\partial U}{\partial n}\, ds \to 2\pi k \lambda(c)$$

for $\varepsilon \to 0$. In the limit we obtain the following basic formula:

$$\iint_D \log |f(z)|\, \Delta\lambda(z)\, dx\, dy = -\int_\Gamma \left\{ \log |f(\zeta)| \frac{\partial \lambda}{\partial n} - \lambda(\zeta) \frac{\partial}{\partial n} \log |f(\zeta)| \right\} ds$$

(19.2.15)
$$- 2\pi [\Sigma_1^n \lambda(b_k) - \Sigma_1^m \lambda(a_j)].$$

As our first application of this formula we consider the case in which D is a sector of an annulus

$$D: \quad 0 < a < |z| < b < \infty, \quad |\arg z| < \frac{\pi}{2\alpha}, \quad 1 < \alpha,$$

and choose

$$\lambda(z) = \frac{1}{\alpha} \cos(\alpha\theta) [r^\alpha - a^\alpha], \quad z = re^{i\theta}.$$

This function vanishes on three sides of the boundary and

$$\Delta\lambda(z) = \alpha \cos(\alpha\theta)\, a^\alpha r^{-2}.$$

To simplify the formulas we set

(19.2.16) $$m_\alpha(r; f) = \alpha \int_{-\pi/(2\alpha)}^{\pi/(2\alpha)} \log |f(re^{i\theta})| \cos(\alpha\theta)\, d\theta.$$

We have then

$$\iint_D \log |f(z)| \, \Delta\lambda(z) \, dx \, dy = a^\alpha \int_a^b m_\alpha(r;f) \frac{dr}{r},$$

$$\int_\Gamma \log |f(z)| \frac{\partial\lambda}{\partial n} \, ds = \frac{1}{\alpha} a^\alpha m_\alpha(a;f) - \frac{1}{\alpha} b^\alpha m_\alpha(b;f)$$

$$+ \int_a^b \log \{|f(re^{i\pi/(2\alpha)})| \, |f(re^{-i\pi/(2\alpha)})|\} [r^\alpha - a^\alpha] \frac{dr}{r},$$

$$\int_\Gamma \lambda(z) \frac{\partial}{\partial n} \log |f(z)| \, ds = -\frac{1}{\alpha} [b^\alpha - a^\alpha] \int_{-\pi/(2\alpha)}^{\pi/(2\alpha)} \frac{\partial}{\partial r} \log |f(be^{i\theta})| \cos(\alpha\theta) \, d\theta.$$

In the last integral we note that

$$b \frac{\partial}{\partial r} \log |f(be^{i\theta})| = \frac{\partial}{\partial\theta} \arg f(be^{i\theta}).$$

Using this and integrating by parts we obtain

$$-[b^\alpha - a^\alpha] b^{-1} \int_{-\pi/(2\alpha)}^{\pi/(2\alpha)} \arg f(be^{i\theta}) \sin(\alpha\theta) \, d\theta.$$

Finally we suppose that $f(z)$ is holomorphic in D for any choice of a, b, and α. We also assume that $f(z)$ is bounded in D, and we may assume that $|f(z)| \le 1$ without restricting the generality. Since there are no poles, the terms $\lambda(b_k)$ are missing and

$$\lambda(a_j) = \frac{1}{\alpha} \cos(\alpha\theta_j) [r_j^\alpha - a^\alpha], \quad a_j = r_j e^{i\theta_j}.$$

Combining this information and rearranging the terms in (19.2.15), we obtain the following result:

(19.2.17)
$$\frac{1}{\alpha} a^\alpha m_\alpha(a;f) + a^\alpha \int_a^b m_\alpha(r;f) \frac{dr}{r}$$

$$+ \int_a^b \log \{|f(re^{i\pi/(2\alpha)})| \, |f(re^{-i\pi/(2\alpha)})|\} [r^\alpha - a^\alpha] \frac{dr}{r}$$

$$- 2\pi \frac{1}{\alpha} \Sigma_j \cos(\alpha\theta_j) [r_j^\alpha - a^\alpha]$$

$$= \frac{1}{\alpha} b^\alpha m_\alpha(b;f) - \alpha(b^\alpha - a^\alpha) b^{-1} \int_{-\pi/(2\alpha)}^{\pi/(2\alpha)} \arg f(be^{i\theta}) \sin(\alpha\theta) \, d\theta.$$

In this formula we want to pass to the limit with α and with a. To validate the results we need Fatou's lemma.[1]

---

[1] See, e.g., L. M. Graves, *The Theory of Functions of Real Variables*, Second Edition (McGraw-Hill Book Company, Inc., New York, 1956), p. 194.

LEMMA 19.2.2.   *Let $g_\alpha(x)$ and $g_0(x)$ be integrable over the measurable set E and set $g(x) = \lim \inf g_\alpha(x)$. If $g_0(x) \leqq g_\alpha(x)$ almost everywhere, and if, moreover,* $\lim \inf \int_E g_\alpha(x)\,dx < \infty$, *then $g(x)$ is integrable over E, and*

$$\int_E g(x)\,dx \leqq \lim \inf \int_E g_\alpha(x)\,dx.$$

In formula (19.2.17) we now suppose that $b \leqq 1$ and that $f(be^{i\theta})$ and $f(be^{-i\theta})$ tend to finite limits different from 0 when $\theta \to \frac{1}{2}\pi$. This implies that

$$\log |f(be^{i\theta})| \quad \text{and} \quad \arg f(be^{i\theta})$$

are integrable over the interval $(-\frac{1}{2}\pi, \frac{1}{2}\pi)$. Hence the right member of (19.2.17) tends to a finite limit as $\alpha \to 1$. The same is obviously true if we let $a \to 0$. Moreover, the double limit exists and is independent of the order of the limiting processes. It follows that the right member is bounded for $1 < \alpha, 0 < a < 1$. Let $-M$ be a lower bound of the right member for such α and a.

Since all terms on the left are nonpositive, the following inequalities hold for $1 < \alpha, 0 < a < 1$:

$$-\frac{1}{\alpha} a^\alpha m_\alpha(a;f) < M,$$

$$-a^\alpha \int_a^b m_\alpha(r;f)\,\frac{dr}{r} < M,$$

(19.2.18)
$$-\int_a^b \log |f(re^{i\pi/(2\alpha)})|\,(r^\alpha - a^\alpha)\,\frac{dr}{r} < M,$$

$$-\int_a^b \log |f(re^{-i\pi/(2\alpha)})|\,(r^\alpha - a^\alpha)\,\frac{dr}{r} < M,$$

$$2\pi\frac{1}{\alpha}\Sigma_j \cos(\alpha\theta_j)\,(r_j^\alpha - a^\alpha) < M.$$

The last inequality shows that if $f(z)$ has infinitely many zeros in the semi-circular disk $|z| < 1, 0 < \Re(z)$, then the series

(19.2.19) $\Sigma_j \Re(z_j)$

extended over these zeros is convergent.

In the third inequality we fix $a > 0$ and let $\alpha \to 1$. Here

$$0 \leqq -\log |f(re^{i\pi/(2\alpha)})| \to -\log |f(ir)|$$

for almost all values of r by virtue of Theorem 19.2.3. The assumptions of Fatou's lemma are clearly satisfied, whence it follows that $\log |f(iy)|$ is integrable over the interval (a, b). The restrictions on a and b imposed during the proof are not essential, as we see by applying the result just obtained to the function $f(ic + z)$, where c is an arbitrary real number. We conclude that $\log |f(iy)|$ is integrable over every finite interval.

Fatou's lemma also applies to the integrals $m_\alpha(a; f)$ and shows the existence of $m(a; f) \equiv m_1(a; f)$ and of the inequality

$$(19.2.20) \qquad -rm(r; f) \leqq M, \quad 0 < r < 1.$$

This implies a limitation on the rate at which $f(z)$ can tend to zero as z tends to a point on the imaginary axis, in this case the origin. The example

$$f(z) = \exp\left(-\frac{1}{z}\right) \quad \text{with} \quad m(r; f) = -\frac{\pi}{2r}$$

shows that no better estimate than (19.2.20) can hold in general. The second inequality under (19.2.18) does not seem to yield any further information. We have consequently proved

THEOREM 19.2.5.   *If $f(z) \in B(0), f(z) \not\equiv 0$, then (i) $\log |f(iy)|$ is integrable over every finite interval, (ii) for every real c there is a finite positive constant $M(c)$ such that*

$$(19.2.21) \qquad -r\int_{-\pi/2}^{\pi/2} \log |f(ci + re^{i\theta})| \cos \theta \, d\theta < M(c), \quad 0 < r < 1,$$

and

$$(19.2.22) \qquad \Sigma_j \, \Re(z_j) < M(c)$$

where the summation extends over the zeros of $f(z)$ in $|z - ic| < 1, 0 < \Re(z)$.

In order to complete the discussion we now consider half an annulus

$$a < |z| < R, \quad |\arg z| < \frac{\pi}{2},$$

and take a slightly different function $\lambda(z)$, namely,

$$\lambda(z) = \cos \theta \left(\frac{1}{r} - \frac{1}{R}\right).$$

This function is positive in the semiannulus, vanishes on three sides of its boundary, and

$$\Delta\lambda(z) = \frac{\cos \theta}{Rr^2}.$$

We substitute this value of $\lambda(z)$ in formula (19.2.15) with $f(z)$ replaced by $f(z + \delta), \delta > 0$. In the result we pass to the limit with δ. This is permitted since all integrals involved exist by the preceding discussion. This may require avoiding values of a which form a set of measure 0. The result may be written

$$(19.2.23) \qquad \begin{aligned} &\frac{1}{R}\int_a^R m(r; f)\frac{dr}{r} + \int_a^R \log \{|f(iy)| \, |f(-iy)|\}\left(\frac{1}{y} - \frac{1}{R}\right)\frac{dy}{y} \\ &\quad + \frac{1}{R} m(R; f) - 2\pi\Sigma_j \cos \theta_j\left(\frac{1}{r_j} - \frac{1}{R}\right) = J(R), \end{aligned}$$

where

$$J(R) = \frac{1}{a} m(a; f) + \left(\frac{1}{a} - \frac{1}{R}\right)\int_{-\pi/2}^{\pi/2} \arg f(ae^{i\theta}) \sin \theta \, d\theta.$$

It is no restriction to assume that the last integral exists since this is the case for almost all values of a. It follows that $J(R)$ is bounded for $R \geq a$. The terms which occur on the left in (19.2.23) are all nonpositive by virtue of our assumption that $|f(z)| \leq 1$. As above we get inequalities for the separate terms which are valid for all values of R. A straightforward discussion of these inequalities leads to

THEOREM 19.2.6.   *If* $f(z) \in B(0), f(z) \not\equiv 0$, *then*

(i) $\displaystyle\int_{-\infty}^{\infty} \frac{\log |f(iy)|}{1 + y^2}\, dy$ *exists*,

(ii) $\dfrac{1}{r}\, m(r; f)$ *is bounded for* $1 \leq r$, *and*

(iii) *the series*

(19.2.24)
$$\sum_j \frac{\Re(a_j)}{1 + |a_j|^2}$$

extended over the zeros of $f(z)$ *in* $\Re(z) > 0$ *is convergent*.

In proving (i) it should be noted that for $a < y < \tfrac{1}{2}R$

$$\left(\frac{1}{y} - \frac{1}{R}\right)\frac{1}{y} > \frac{1}{2y^2} > \frac{1}{2}\frac{1}{1 + y^2},$$

and this inequality is also useful in discussing (iii).

The three conclusions of this theorem show what restrictions are imposed on the rate of convergence of $f(z)$ to the limit 0 along the imaginary axis and in the right half-plane as well as on the distribution of its zeros. This completes the study, proposed above, of the affinity that a function of $B(0)$ can show toward the value 0.

If $\{a_n\}$ is a given sequence of complex numbers with $\Re(a_n) > 0$ such that the series (19.2.24) converges, then we can construct a Blaschke product with these zeros. We take

(19.2.25)
$$B(z) = \prod \frac{z - a_j}{z + \overline{a_j}} \cdot \prod \frac{a_k - z\,\overline{a_k}}{\overline{a_k} + z\,a_k},$$

where the first product extends over the zeros with $|a_j| < 1$ and the second is restricted to those with $|a_k| \geq 1$. Each of these products is absolutely convergent for $\Re(z) > 0$ and defines a function in $B(0)$. We have

$$|B(z)| < 1, \quad 0 < \Re(z),$$

and

(19.2.26)
$$\lim_{x \to 0} |B(x + iy)| = 1$$

for almost all y. We shall not prove this relation. Further,

$$f(z) = B(z)E(z)$$

where $E(z)$ has no zeros in $\Re(z) > 0$. It may be shown that $E(z) \in B(0)$ and has the representation

$$(19.2.27) \qquad E(z) = \exp\left\{ -\frac{1}{\pi} \int_{-\infty}^{\infty} \frac{dq(t)}{z - it} + c \right\},$$

where $q(t)$ is a monotone increasing function of bounded variation in $[-\infty, \infty]$ and c is a constant.

The discussion given above of functions which are bounded, that is, bounded away from infinity, applies with obvious modifications to functions which are bounded away from zero. Such a function is the reciprocal of a function in $B(0)$. Theorems 19.2.5 and 19.2.6 are valid for such a function; we have merely to replace zeros by poles and the limit 0 by the limit ∞.

Finally the discussion extends also to functions which are quotients of bounded functions, *beschränktartig* in the terminology of R. Nevanlinna. For such functions the following theorem holds; we omit a proof except for the remark that the necessary character of the conditions follows from the preceding discussion.

THEOREM 19.2.7. *A function* $F(z)$, *meromorphic in* $\Re(z) > 0$, *is the quotient of two functions in* $B(0)$ *if and only if*

(i) $F(z)$ *has boundary values* $F(iy)$ *almost everywhere on the imaginary axis such that*

$$\int_{-\infty}^{\infty} |\log|F(iy)|| \frac{dy}{1 + y^2} < \infty,$$

(ii) *the mean value*

$$\frac{1}{\pi r} \int_{-\pi/2}^{\pi/2} |\log|F(re^{i\theta})|| \cos\theta \, d\theta$$

is bounded for $1 \leq r$,

(iii) *for every fixed real* c *there is an* $M(c)$ *such that*

$$r \int_{-\pi/2}^{\pi/2} |\log|F(ic + re^{i\theta})|| \cos\theta \, d\theta < M(c)$$

for $0 < r < 1$, *and*

(iv) *the series*

$$\sum \frac{\Re(a_j)}{1 + |a_j|^2} \quad and \quad \sum \frac{\Re(b_k)}{1 + |b_k|^2}$$

extended over the zeros and poles of $F(z)$ *are convergent.*

From (19.2.27) we conclude that

$$(19.2.28) \qquad F(z) = \frac{B_1(z)}{B_2(z)} \exp\left\{ -\frac{1}{\pi} \int_{-\infty}^{\infty} \frac{dq(t)}{z - it} + c \right\},$$

where $B_1(z)$ and $B_2(z)$ are Blaschke products formed with the zeros and poles of $F(z)$ respectively, and $q(t)$ is of bounded variation in $[-\infty, \infty]$.

EXERCISE 19.2

1. Is it possible to replace $|y - y_0| \leq Kx$ by a weaker condition in the proof of Theorem 19.2.3 so as to admit some tangential paths?

2. The point $y = 0$ does not belong to the Lebesgue set of the function $\exp(i/y)$ which, for $y \neq 0$, defines the boundary values of $\exp(-1/z)$. The latter tends to zero as $z \to 0$ along any path in the sector $|\arg z| \leq \frac{1}{2}\pi - \varepsilon$. Show that there are paths having first-order contact with the imaginary axis at the origin along which no limit exists.

3. Show that if $f(z) \in B(0)$ and $e^{|y|}|f(iy)|$ is bounded, then $f(z) \equiv 0$.

4. Verify the convergence of (19.2.25).

5. In (19.2.27) choose $q(t)$ to be a step function with jumps α_n at the points $t = \beta_n$ where $\alpha_n > 0$ and $\Sigma \alpha_n < \infty$. Show that $E(z)$ is continuous on the imaginary axis and 0 at the points $z = i\beta_n$ and nowhere else.

6. Prove the following theorem due to F. and R. Nevanlinna (1922): Let $f(z)$ be holomorphic in $\Re(z) \geq 0$ but not necessarily bounded. Let $A(t)$ be a non-negative function such that

$$T(r) = \int_a^r A(t)t^{-2}\, dt \to \infty \text{ with } r.$$

If n, p, q are constants and $\varepsilon(r) \to 0$, suppose that for $r \to \infty$

(i) $\frac{1}{2}\log\{|f(ir)||f(-ir)|\} < [p + \varepsilon(r)]A(r),$

(ii) $m(r;f) < [q + \varepsilon(r)]r\, T(r),$

(iii) $\sum_{a < |a_j| < r} \cos\theta_j > [n + \varepsilon(r)]A(r),$

where the summation extends over the zeros of $f(z)$. Then either

$$n\pi \leq p + q \quad \text{or} \quad f(z) \equiv 0.$$

(*Hint:* Formula (19.2.23) holds and gives the desired result.)

7. Show that the functions

$$\sin \pi z \quad \text{and} \quad \frac{1}{\Gamma(1-z)}$$

satisfy the conditions of the preceding theorem with $A(t) = t$. Determine n, p, q and show that equality holds in $n\pi \leq p + q$.

8. If $f(z)$ has zeros at the positive integers, under what assumptions on $f(iy)$ and on $m(r;f)$ does it follow that $f(z) \equiv 0$? Compare your result with Theorem 11.3.3.

19.3. Growth-measuring functions. In this section we shall study various means of measuring the rate of growth of analytic functions. We have already encountered and studied a number of such functions such as the maximum modulus, the p-means $M_p(r;f)$ on circles and $M_p(x;f)$ on vertical lines, and the various proximity functions of Sections 14.4, 18.4, and 19.2 among others. In Section 11.3 we introduced the *growth indicator* of Phragmén, Lindelöf, and Pólya. We shall give the omitted proofs of its properties here. We shall also discuss some functions which are needed in the next section, in particular the mu function of Lindelöf.

We recall the definition of the growth indicator. Given a function $g(z)$ holomorphic and of exponential type in $\Re(z) \geq 0$, we set

$$(19.3.1) \qquad h(\theta;g) = \limsup_{r \to \infty} \frac{1}{r} \log |g(re^{i\theta})|, \quad -\tfrac{1}{2}\pi \leq \theta \leq \tfrac{1}{2}\pi,$$

which we call the growth indicator of $g(z)$. Its basic properties are given in

THEOREM 19.3.1 [= LEMMA 11.3.1]. *Either we have $h(\theta;g) = -\infty$ for $-\tfrac{1}{2}\pi < \theta < \tfrac{1}{2}\pi$ or $h(\theta;g)$ is continuous in $(-\tfrac{1}{2}\pi, \tfrac{1}{2}\pi)$ and has finite right- and left-hand derivatives there. In the second case, $h(\theta;g)$ is the function of support of an unbounded closed convex set S (known as the indicator diagram of $g(w)$).*

Proof. That the first possibility may happen was shown in Section 11.3, formula (11.3.21) and the following discussion. Suppose now that $h(\theta;g)$ has finite values for $\theta = \theta_1$ and $\theta = \theta_2$ where

$$-\tfrac{1}{2}\pi \leq \theta_1 < \theta_2 \leq \tfrac{1}{2}\pi, \quad \theta_2 - \theta_1 < \pi,$$

and let us form a function

$$H_\varepsilon(\theta) = A_\varepsilon \cos \theta + B_\varepsilon \sin \theta$$

such that, for a given $\varepsilon > 0$, we have

$$H_\varepsilon(\theta_1) = h(\theta_1;g) + \varepsilon \equiv h_1, \quad H_\varepsilon(\theta_2) = h(\theta_2;g) + \varepsilon \equiv h_2.$$

We find that

$$A_\varepsilon = \frac{h_1 \sin \theta_2 - h_2 \sin \theta_1}{\sin(\theta_2 - \theta_1)}, \quad B_\varepsilon = \frac{h_2 \cos \theta_1 - h_1 \cos \theta_2}{\sin(\theta_2 - \theta_1)}.$$

We now form

$$g_\varepsilon(z) = \exp(-\overline{C_\varepsilon}z), \quad C_\varepsilon = A_\varepsilon + iB_\varepsilon,$$

and consider

$$F_\varepsilon(z) = f(z)g_\varepsilon(z), \quad \theta_1 \leq \arg z \leq \theta_2.$$

Since

$$|g_\varepsilon(z)| = \exp[-H_\varepsilon(\theta)r], \quad z = re^{i\theta},$$

we conclude that $F_\varepsilon(z)$ tends to zero when $z \to \infty$ along the sides of the sector. This implies that $F_\varepsilon(z)$ is bounded on the finite boundary of the sector whose opening is $\theta_2 - \theta_1 < \pi$. There exists a constant K such that in the interior of the sector

$$| F_\varepsilon(z) | \leq \exp (K \,|\, z \,|) = o[\exp (|\, z \,|^{\pi/(\theta_2 - \theta_1)})].$$

Hence, by Theorem 18.1.1, $F_\varepsilon(z)$ is bounded in the sector, that is,

$$h(\theta; g) \leq A_\varepsilon \cos \theta + B_\varepsilon \sin \theta, \quad \theta_1 \leq \theta \leq \theta_2.$$

Since this holds for every $\varepsilon > 0$, it holds also for $\varepsilon = 0$. The resulting inequality may be written

$$(19.3.2) \quad \sin (\theta_2 - \theta_1) h(\theta; g) \leq \sin (\theta_2 - \theta) h(\theta_1; g) + \sin (\theta - \theta_1) h(\theta_2; g),$$

which is valid for $\theta_1 \leq \theta \leq \theta_2$.

This inequality does not exclude the possibility that $h(\theta_0; g) = -\infty$ for some θ_0 with $\theta_1 < \theta_0 < \theta_2$. Suppose this to be the case and take a value θ_3 outside of $[\theta_1, \theta_2]$. To fix the ideas, suppose that $\theta_2 < \theta_3 < \frac{1}{2}\pi$. Take a large positive number M and choose an a such that $h(\theta_3; g) < a$. We replace $\theta_1, \theta_2,$ h_1, h_2 in the preceding discussion by $\theta_0, \theta_3, -M, a$ respectively and obtain

$$\sin (\theta_3 - \theta_0) h(\theta; g) \leq -M \sin (\theta_3 - \theta) + a \sin (\theta - \theta_0)$$

for any θ with $\theta_0 < \theta < \theta_3$. Since M is arbitrary, this means that

$$h(\theta; g) = -\infty, \quad \theta_0 \leq \theta < \theta_3.$$

In particular, this is true for $\theta = \theta_2$. This contradicts our original assumption that $h(\theta_2; g)$ is finite. Thus we see that either $h(\theta; g)$ equals $-\infty$ for all θ in $(-\frac{1}{2}\pi, \frac{1}{2}\pi)$ or else it is finite for all such values.

In the second case (19.3.2) holds for all $\theta, \theta_1, \theta_2$ with

$$-\tfrac{1}{2}\pi < \theta_1 < \theta < \theta_2 < \tfrac{1}{2}\pi.$$

We prove first that this inequality implies that $h(\theta; g)$ is convex in any interval where it is negative and that, in any case, it differs locally from a convex function by a *sinusoid*, that is, a function of the form

$$(19.3.3) \qquad\qquad A \cos \theta + B \sin \theta.$$

It is clear that such a function satisfies (19.3.2) with equality for all θ. Let us write

$$(19.3.4) \quad S[H] \equiv \sin (\theta_1 - \theta_2) H(\theta_3) + \sin (\theta_2 - \theta_3) H(\theta_1) + \sin (\theta_3 - \theta_1) H(\theta_2).$$

Then, as just observed,

$$(19.3.5) \qquad\qquad S[H] = 0$$

is satisfied by

$$H(\theta) = A \cos (\theta - \alpha)$$

for any choice of A and α. Actually this is the general solution of (19.3.5), but we shall not need this fact.

The inequality (19.3.2) now becomes

$$(19.3.6) \qquad S[h] \leq 0, \quad -\tfrac{1}{2}\pi < \theta_1 < \theta_2 < \theta_3 < \tfrac{1}{2}\pi,$$

and we see that if $h = h(\theta; g)$ satisfies this inequality, then so does

$$h_0 \equiv h(\theta; g) - A \cos(\theta - \alpha).$$

Repeated use will be made of this fact below.

Next we observe that if $h(\theta_1) < 0$ and $h(\theta_2) < 0$, then (19.3.2) implies that

$$h(\theta) < 0 \quad \text{for} \quad \theta_1 < \theta < \theta_2.$$

Moreover, this shows that the set

$$(19.3.7) \qquad N_h = [\theta \mid h(\theta) < 0]$$

is either an interval or void.

Suppose that we are in the first case. Then, since $(\sin\theta)/\theta$ is decreasing in $(0, \pi)$,

$$\frac{\sin\beta}{\sin\gamma} > \frac{\beta}{\gamma} \quad \text{for} \quad 0 < \beta < \gamma < \pi.$$

Since $h(\theta) < 0$ for $\theta_1 \leq \theta \leq \theta_2$, this enables us to conclude from (19.3.2) that

$$(19.3.8) \qquad (\theta_2 - \theta_1) h(\theta) < (\theta_2 - \theta) h(\theta_1) + (\theta - \theta_1) h(\theta_2)$$

for the values of $\theta, \theta_1, \theta_2$ under consideration. This familiar inequality says that $h(\theta)$ is *convex* in any interval where it is negative. In such an interval the graph of $h(\theta)$ always lies below the secant joining any two points of the graph. Moreover, the function has at most one negative minimum and no negative maximum.

We note that if $h(\theta) \leq K$ for all θ and if $h(\alpha) > 0$, then

$$h_\alpha(\theta) \equiv h(\theta) - [h(\alpha) + 2K] \cos(\theta - \alpha) < 0$$

for

$$\max(-\tfrac{1}{2}\pi, \alpha - \tfrac{1}{3}\pi) < \theta < \min(\tfrac{1}{2}\pi, \alpha + \tfrac{1}{3}\pi),$$

so that $h_\alpha(\theta)$ is convex in this interval.

Since the functions $h(\theta)$ and $h_\alpha(\theta)$ are bounded above, convexity in some open interval implies continuity in that interval. This shows that $h(\theta)$ is continuous in $(-\tfrac{1}{2}\pi, \tfrac{1}{2}\pi)$. Furthermore, it shows the existence of

$$\lim_{\delta \to 0} h(-\tfrac{1}{2}\pi + \delta) \equiv h(-\tfrac{1}{2}\pi + 0), \quad \lim_{\delta \to 0} h(\tfrac{1}{2}\pi - \delta) \equiv h(\tfrac{1}{2}\pi - 0),$$

where, however, either or both limits could possibly be $-\infty$. This possibility is excluded by (19.3.2), for we have

$$2 \cos\theta \, h(0) \leq h(\theta) + h(-\theta), \quad |\theta| < \tfrac{1}{2}\pi,$$

whence

$$(19.3.9) \qquad 0 \leqq h(\tfrac{1}{2}\pi - 0) + h(-\tfrac{1}{2}\pi + 0).$$

The same inequality is satisfied by $h(\tfrac{1}{2}\pi)$ and $h(-\tfrac{1}{2}\pi)$ if these numbers are defined.

Next we note that a continuous convex function has finite right- and left-hand derivatives at all points of an open interval of convexity. This follows from the fact that the curve lies below its secant. This implies that the slope of a secant $P_0 P$ through a fixed point P_0 on the curve is a monotone function of P and hence tends to a limit when P tends to P_0. We conclude that $h(\theta)$ has finite right- and left-hand derivatives for all θ with $-\tfrac{1}{2}\pi < \theta < \tfrac{1}{2}\pi$, a unique right-hand derivative at $\theta = -\tfrac{1}{2}\pi$, and a unique left-hand derivative at $\theta = \tfrac{1}{2}\pi$. Either or both of the latter may be infinite, as shown by the example (11.3.17) or by the simpler function

$$(19.3.10) \qquad \mu(\theta) = \cos\theta \log(2\cos\theta) + \theta\sin\theta.$$

It should be noted that if

$$-h(-\tfrac{1}{2}\pi + 0) = h(\tfrac{1}{2}\pi - 0) = b,$$

then neither of the following combinations is admissible:

$$h'(-\tfrac{1}{2}\pi + 0) < +\infty, \quad h'(\tfrac{1}{2}\pi - 0) = +\infty,$$
$$h'(-\tfrac{1}{2}\pi + 0) = -\infty, \quad h'(\tfrac{1}{2}\pi - 0) > -\infty.$$

This follows from the inequality

$$2\,\frac{\cos\theta}{\tfrac{1}{2}\pi - \theta}\,h(0) \le \frac{h(\theta) - b}{\tfrac{1}{2}\pi - \theta} + \frac{h(-\theta) + b}{\tfrac{1}{2}\pi - \theta},$$

which gives

$$2h(0) \ge h'(\tfrac{1}{2}\pi - 0) - h'(-\tfrac{1}{2}\pi + 0)$$

in the limit.

After this discussion of the properties of $h(\theta)$ we turn to the closed set of which $h(\theta)$ is the function of support (see the end of Section 2.3 for the terminology). We consider the half-planes

$$H_\varphi: \ \Re(ze^{-i\varphi}) \le h(\varphi), \quad -\tfrac{1}{2}\pi \le \varphi \le \tfrac{1}{2}\pi,$$

where at the endpoints we replace $h(\varphi)$ by its limits. Now define

$$S = \cap_\varphi H_\varphi.$$

Thus z belongs to S if and only if $\Re(ze^{-i\varphi}) \le h(\varphi)$ for all φ. For each φ there is at least one $z = z_\varphi$ in S such that

$$\Re(z_\varphi e^{-i\varphi}) = h(\varphi).$$

It is obvious from the definition that S is convex unless it is void, but it is not obvious that the second possibility must be excluded. In any case S is confined in a horizontal left half-strip

$$(19.3.11) \qquad \Re(z) \le h(0), \quad -h(-\tfrac{1}{2}\pi + 0) \le \Im(z) \le h(\tfrac{1}{2}\pi - 0).$$

By (19.3.9) it is possible to satisfy the last inequality.

In order to show that $S \neq \emptyset$, it is sufficient to show that $h(\theta)$ admits of a sinusoid as a lower bound. For if

$$\Re(z_0 e^{-i\theta}) = x_0 \cos\theta + y_0 \sin\theta \leq h(\theta), \quad -\tfrac{1}{2}\pi \leq \theta \leq \tfrac{1}{2}\pi,$$

then z_0 belongs to S and so does the line segment $y = y_0, x \leq x_0$. The condition is clearly necessary as well as sufficient.

If $h(\theta) \geq 0$ for all θ, then we can take $z_0 = 0$, and if $N_h = [\theta \mid h(\theta) < 0]$ is bounded away from $-\tfrac{1}{2}\pi$ and $+\tfrac{1}{2}\pi$, then we can take $z_0 = x_0 = m/\gamma$ where $m = \min h(\theta)$ and $\gamma = \min \cos\theta$ for $\theta \in N_h$. But if N_h extends to one or both of the endpoints, then the choice of z_0 is less obvious. Suppose first that

$$-h(-\tfrac{1}{2}\pi + 0) = h(\tfrac{1}{2}\pi - 0) = b \geq 0.$$

In this case we must have $y_0 = b$. Consider

(19.3.12) $\quad H(\theta) \equiv -A \cos(\theta + \delta) \quad$ where $\quad A \sin\delta = b, 0 < \delta < \tfrac{1}{2}\pi.$

Then the function

$$h_1(\theta) \equiv h(\theta) - H(\theta)$$

is 0 at the endpoints of the interval. If $h_1(\theta)$ were negative anywhere, then it would have to be negative everywhere by (19.3.2). But $h_1(0) = h(0) + A \cos\delta$ is clearly positive for A sufficiently large. For such values of A, $h(\theta) > H(\theta)$ for all θ, so that S again cannot be void. The same argument applies if $b < 0$ instead. Finally, suppose that

$$h(-\tfrac{1}{2}\pi + 0) < 0 < h(-\tfrac{1}{2}\pi + 0) + h(\tfrac{1}{2}\pi - 0)$$

and choose a number b such that

$$-h(-\tfrac{1}{2}\pi + 0) < b < h(\tfrac{1}{2}\pi - 0).$$

Define $H(\theta)$ by (19.3.12) and form $h_1(\theta)$ as above. This function $h_1(\theta)$ is positive at the two endpoints of the interval. There are two alternatives. If $h_1(\theta) > 0$ in the whole interval, then we are through since then $h(\theta) > H(\theta)$ for all θ. On the other hand, if $N_{h_1} = [\theta \mid h_1(\theta) < 0]$ is not void, then the interval N_{h_1} is bounded away from the endpoints of $(-\tfrac{1}{2}\pi, \tfrac{1}{2}\pi)$. In this case we know the existence of a sinusoid

$$H_1(\theta) \equiv x_0 \cos\theta$$

such that $h_1(\theta) > H_1(\theta)$, and we have then

$$h(\theta) > H(\theta) + H_1(\theta) = A \cos(\theta + \delta) + x_0 \cos\theta,$$

which is also a sinusoid. Thus, S is not vacuous. The remaining cases can be handled by the same method. Thus we see that S is never void, and this completes the proof of Theorem 19.3.1.

The discussion is easily extended to the case in which $g(z)$ is an entire function of exponential type and $h(\theta; g)$ is defined for all θ as a periodic function of period 2π. In this case $h(\theta)$ is continuous for all values of θ and (19.3.9) generalizes to

(19.3.13) $$h(\theta) + h(\theta + \pi) \geqq 0$$

for all θ. In the interval $[-\pi, \pi)$ there is room for at most one subinterval where $h(\theta) < 0$, and the length of this interval cannot exceed π. The discussion of the derivatives carries over without change. The set S is now bounded and convex.

We leave the growth indicator and pass over to growth properties along vertical lines. We shall start with the mu function introduced by E. Lindelöf in 1908 for a study of the Riemann zeta function.

Suppose that $f(z)$ is holomorphic and of *finite order* in the vertical half-strip

$$B\colon a < x < b, \quad 1 < y.$$

Here $b \leqq \infty$ and the order assumption means that to each pair of numbers $(a_1, b_1), a < a_1 < b_1 < b$, there correspond finite numbers σ and M such that

(19.3.14) $$|f(x + iy)| \leqq My^{\sigma}, \quad a_1 \leqq x \leqq b_1, 2 \leqq y.$$

DEFINITION 19.3.1.   *The mu function of $f(z)$ is*

(19.3.15) $$\mu(x; f) = \limsup_{y \to \infty} \frac{\log |f(x + iy)|}{\log y}, \quad a < x < b.$$

THEOREM 19.3.2. $\mu(x; f)$ *is either a continuous convex function of x in* (a, b) *or else identically* $-\infty$.

Proof.   We can imitate the proof of the preceding theorem; we cannot obtain the desired result by conformal mapping. We note first that the possibility $\mu(x; f) \equiv -\infty$ is not excluded a priori. It occurs, for instance, if

$$f(z) = [\Gamma (\log (-iz))]^{-1}, \quad \Re(z) > 0.$$

It is clear that $\mu(x; f) \equiv \mu(x)$ is well defined in (a, b) but may have the value $-\infty$. Suppose now that $\mu(x_1) = m_1$ and $\mu(x_2) = m_2$ are finite where $a < x_1 < x_2 < b$. Let $k_\varepsilon(x)$ be the linear function of x which takes the values $m_1 + \varepsilon$ and $m_2 + \varepsilon$ at $x = x_1$ and x_2 respectively where $\varepsilon > 0$ is given. We have

$$(x_2 - x_1)k_\varepsilon(x) = (m_2 + \varepsilon)(x - x_1) + (m_1 + \varepsilon)(x_2 - x).$$

We now form the function

$$A_\varepsilon(z) = \exp [-k_\varepsilon(z) \log (-iz)],$$

where the logarithm has its principal value in B. A simple computation shows that

$$k_\varepsilon(z) = k_\varepsilon(x) + iy \frac{m_2 - m_1}{x_2 - x_1},$$

$$\log (-iz) = \log y - i\frac{x}{y} + O\!\left(\frac{1}{y^2}\right),$$

whence it follows that

$$|A_\varepsilon(z)| \leqq \exp [-k_\varepsilon(x) \log y + O(1)].$$

We then consider the function

$$F_\varepsilon(z) \equiv A_\varepsilon(z) f(z)$$

in the half-strip

$$B_0: \ x_1 \le x \le x_2, \quad 2 \le y.$$

Here $F_\varepsilon(z)$ tends to zero along the vertical boundary of B_0, so it is bounded on ∂B_0. In the interior of B_0 we have

$$|\, F_\varepsilon(z) \,| \le M y^\sigma$$

for some finite M and σ. This means that condition (18.1.9) is amply satisfied, and Theorem 18.1.4 shows that $|\, F_\varepsilon(z) \,|$ is bounded in B_0. Hence

$$\mu(x) \le k_\varepsilon(x), \quad x_1 \le x \le x_2.$$

Since this holds for every $\varepsilon > 0$, it also holds for $\varepsilon = 0$. Hence

(19.3.16) $$(x_2 - x_1)\mu(x) \le (x_2 - x)\mu(x_1) + (x - x_1)\mu(x_2).$$

This expresses that $\mu(x)$ is convex. It is the analogue of (19.3.2) and evidently much simpler than the latter equation.

The proof that $\mu(x; f)$ is continuous proceeds as in the case of $h(\theta; g)$. The convexity implies the existence of left- and right-hand derivatives of $\mu(x)$ everywhere. This completes the proof.

T. Carleman proved in 1931 that if $\lambda(x)$ is any continuous convex function given in an interval (a, b), then there exists a function $f(z)$ holomorphic in the half-strip $B: a < x < b, 1 < y$, such that

(19.3.17) $$\mu(x; f) = \lambda(x).$$

His point of departure is the observation that for

(19.3.18) $$f(z) = \exp\left(-\tfrac{1}{2}\pi i \alpha z\right) \Gamma(\alpha(z - a) + \tfrac{1}{2}), \quad \mu(x; f) = \alpha(x - a).$$

Suppose now that

$$y = \alpha_n(x - a_n), \quad n = 1, 2, 3, \cdots,$$

is a system of tangents enveloping the curve

$$y = \lambda(x), \quad a < x < b.$$

Thus the points $(a_n, \lambda(a_n))$ are dense on the curve. Then, if the constants c_n tend to zero sufficiently rapidly, the series

(19.3.19) $$\sum_{n=1}^{\infty} c_n \exp\left(-\tfrac{1}{2}\pi i \alpha_n z\right) \Gamma(\alpha_n(z - a_n) + \tfrac{1}{2})$$

defines a holomorphic function in B which satisfies (19.3.17).

There are several other types of growth-measuring functions available when

$|f(x + iy)|$ grows faster on vertical lines than is permitted by (19.3.14). We mention two samples:

$$(19.3.20) \qquad \mu_a(x; f) = \limsup_{y \to \infty} \frac{\log |f(x + iy)|}{(\log y)^a}, \quad a > 1,$$

$$(19.3.21) \qquad M_\alpha(x; f) = \limsup_{y \to \infty} \frac{\log |f(x + iy)|}{y^\alpha}, \quad \alpha > 0.$$

Using the technique of Theorem 19.3.2 we can prove that these functions, when they are defined, either are identically $-\infty$ or are continuous and convex in the open interval of definition.

The mean value

$$(19.3.22) \qquad \mathfrak{M}(x; f) \equiv \limsup_{T \to \infty} \frac{1}{2T} \int_{-T}^{T} |f(x + iy)|^2 \, dy$$

was introduced by J. Hadamard (1899) for absolutely convergent Dirichlet series, in which case it is a continuous convex function of x. Later such mean values became basic in the theory of *almost periodic functions*. $\mathfrak{M}(x; f)$ can also be defined for the functions of $B(0)$.

A different type of growth-measuring function was introduced by T. Carleman in 1931. Let $f(z)$ be holomorphic and of finite order in a half-strip B and consider the integral

$$(19.3.23) \qquad J_p(x, \alpha) \equiv \int_{2}^{\infty} |f(x + iy)|^p y^{-\alpha} \, dy, \quad a < x < b,$$

where $p > 0$ and α may be complex. The integral has an abscissa of absolute convergence $\sigma = \sigma(x, p)$, and $J_p(x, \alpha)$ is a holomorphic function of α in the half-plane $\Re(\alpha) > \sigma(x, p)$. By a generalization of Theorem 5.7.1 to integrals (due to E. Landau), the point $\alpha = \sigma(x, p)$ is a singular point of $J_p(x, \alpha)$. Carleman proved that $\sigma(x, p)$ is a convex function of p and of x and gave important applications of these ideas.

These examples, together with those listed at the beginning of the section, should convince the reader of the existence of close relations between growth measure and convexity.

EXERCISE 19.3

1. Prove that $A \cos(\theta - \alpha)$ is the only differentiable solution of (19.3.5).

2. If the c's are distinct and the C's different from zero, determine $h(\theta; g)$ and the indicator diagram when

$$g(z) = C_1 e^{c_1 z} + C_2 e^{c_2 z} + \cdots + C_n e^{c_n z}.$$

3. Construct an entire function whose indicator diagram is the equilateral triangle with vertices at the cube roots of unity.

4. If S_1 and S_2 are convex closed sets, not necessarily bounded, with functions of support $h_1(\theta)$ and $h_2(\theta)$, both defined either for $|\theta| < \frac{1}{2}\pi$ or as periodic functions for all θ, show that the vector sum $S_1 + S_2$ is also convex and has a function of support $h(\theta)$ such that $h(\theta) \leq h_1(\theta) + h_2(\theta)$.

5. Find the function of support of (**a**) a point, (**b**) a line segment, (**c**) a circular disk.

6. Show that the unit disk $|w| \leq 1$ is mapped by $\log(1 - w) = z$ on the region $e^x \leq 2\cos y$. Show that the latter is convex and that its function of support is $\mu(\theta)$ defined by (19.3.10).

7. Verify (19.3.13).

8. Verify (19.3.18).

9. In the case of $\mu_a(x; f)$ of (19.3.20) the auxiliary function $A_\varepsilon(z)$ used in the discussion of $\mu(x; f)$ is replaced by $\exp\{-k_\varepsilon(z)[\log(-iz)]^a\}$. Verify that this works and that $\mu_a(x; f)$ has the stated properties.

10. If the Dirichlet series

$$\sum_{n=1}^{\infty} a_n e^{-\lambda_n z}, \quad 0 < \lambda_n < \lambda_{n+1} \to \infty,$$

converges absolutely for $\Re(z) > \sigma$, prove that for such values

$$\mathfrak{M}(x; f) = \sum_{n=1}^{\infty} |a_n|^2 e^{-2\lambda_n x},$$

the series being convergent, and verify the convexity.

11. Determine $\sigma(x; p)$ for the Carleman integral $J_p(x, \beta)$ if $f(z)$ is the function of (19.3.18).

19.4. Remarks on Laplace-Stieltjes integrals.

The theory of the Laplace-Stieltjes transform has grown into a discipline of its own. Under these circumstances it is impossible to do justice to this interesting field of research in a few pages. We restrict ourselves to the mentioning of a few basic facts and we emphasize those aspects of the theory which are connected with the subject matter of the preceding parts of this chapter. We refer to the treatises of G. Doetsch and D. V. Widder for further explications.

A Laplace-Stieltjes integral is of the form

$$(19.4.1) \qquad f(z) \equiv \int_0^\infty e^{-zt}\, dA(t),$$

where $A(t)$ is either continuous or of bounded variation in every finite interval $[0, \omega]$. In the latter case we write $V(t, A)$ for the total variation of $A(s)$ in the interval $[0, t]$. Further, we set

$$(19.4.2) \qquad f(z, \omega) \equiv \int_0^\omega e^{-zt}\, dA(t).$$

We say that the integral *converges* for $z = z_0 = x_0 + iy_0$ if

$$\lim_{\omega \to \infty} f(z_0, \omega)$$

exists as a finite number. Similarly, if

(19.4.3) $$F(x, \omega) = \int_0^\omega e^{-xt} \, dV(t, A),$$

then we say that (19.4.1) is *absolutely convergent* for $z = z_0$ if

$$\lim_{\omega \to \infty} F(x_0, \omega) < \infty.$$

THEOREM 19.4.1.   *If $f(z_0, \omega)$ is a bounded function of ω, then (19.4.1) converges for $\Re(z) > \Re(z_0)$. If $F(x_0, \omega)$ is bounded, then the integral is absolutely convergent for $\Re(z) \geq \Re(z_0)$.*

Proof.   We have

(19.4.4)
$$\begin{aligned}
f(z, \omega) &= \int_0^\omega e^{-(z-z_0)t} e^{-z_0 t} \, dA(t) = \int_0^\omega e^{-(z-z_0)t} \, d_t f(z_0, t) \\
&= e^{-(z-z_0)\omega} f(z_0, \omega) + (z - z_0) \int_0^\omega e^{-(z-z_0)t} f(z_0, t) \, dt,
\end{aligned}$$

from which the first assertion follows immediately. Further, if $x > x_0$, we have $F(x, \omega) < F(x_0, \omega)$, and the second assertion follows.

From this theorem we conclude the existence of an *abscissa of ordinary convergence* $\sigma_0[f]$ as well as an *abscissa of absolute convergence* $\sigma_a[f]$. These satisfy the inequality

(19.4.5) $$-\infty \leq \sigma_0[f] \leq \sigma_a[f] \leq \infty,$$

where any one of the logical possibilities may be realized. The integral converges for $\Re(z) > \sigma_0[f]$, diverges for $\Re(z) < \sigma_0[f]$, converges absolutely whenever $\Re(z) > \sigma_a[f]$, and fails to do so for any z with $\Re(z) < \sigma_a[f]$. For these abscissas we have the following expressions:

(19.4.6) $$\sigma_0[f] = \limsup_{\omega \to \infty} \frac{1}{\omega} \log | A(\infty) - A(\omega) |,$$

(19.4.7) $$\sigma_a[f] = \limsup_{\omega \to \infty} \frac{1}{\omega} \log [V(\infty, A) - V(\omega, A)].$$

Here $A(\infty)$ is to be replaced by 0 if $A(\infty)$ is actually ∞ or not well defined, and similarly for $V(\infty, A)$. For the case in which $A(\infty)$ is replaced by 0, formula (19.4.6) is proved with the aid of (19.4.4). If $\Re(z)$ exceeds the right member of (19.4.6), we set $z_0 = 0$ so that $f(z_0, \omega) = A(\omega)$, and formula (19.4.4) shows that $f(z, \omega)$ tends to a finite limit. Conversely, if $f(z_0, \omega)$ has a finite limit, then the same formula with $z = 0$ gives an estimate of $| A(\omega) |$ which shows that $\Re(z_0)$ cannot be less than the right member of (19.4.6), so that this expression actually gives $\sigma_0[f]$. The excluded case, $A(\infty)$ finite, is handled by a simple modification of (19.4.4), and (19.4.7) is proved in the same manner, replacing z, z_0, $A(t)$, $f(z, \omega)$ by x, x_0, $V(t, A)$, $F(x, \omega)$ respectively.

Formula (19.4.1) defines a holomorphic function in $\Re(z) > \sigma_0[f]$. We have also

$$(19.4.8) \qquad f(z) = z \int_0^\infty e^{-zt} A(t)\, dt, \quad \Re(z) > \max\{0, \sigma_0[f]\}.$$

From this we conclude that the Lindelöf mu function

$$(19.4.9) \qquad \mu(x; f) \leq 1.$$

This can be improved somewhat; we have actually

$$f(x + iy) = o(y),$$

but nothing better is true in general. If there exists a half-plane of absolute convergence, then $f(z)$ is bounded and $\mu(x; f) = 0$ for $x \geq \sigma_a[f] + \varepsilon$. If $A(t)$ is absolutely continuous so that the integral becomes a Laplace integral

$$(19.4.10) \qquad f(z) = \int_0^\infty e^{-zt} F(t)\, dt, \quad F(t) = A'(t),$$

then $f(z)$ tends to zero when $z \to \infty$ uniformly with respect to z in any sector $|\arg z| \leq \tfrac{1}{2}\pi - \varepsilon$, $0 < \varepsilon$. If the integral is absolutely convergent, we may allow z to tend to infinity in $\Re(z) \geq \sigma_a[f] + \varepsilon$.

We say that $f(z)$ is the Laplace-Stieltjes transform of $A(t)$, $f = \mathfrak{L}_1[A]$, if (19.4.1) holds, and the Laplace transform of $F(t)$, $f = \mathfrak{L}[F]$, if (19.4.10) holds.

A Laplace-Stieltjes transform is a function of very modest growth and as such the frequency of its values is subject to severe limitations. The earliest result of this nature is due to Matyáš Lerch (1860–1922), who proved in 1903 that a Laplace transform of a continuous function $F(t)$ which is 0 at the positive integers is identically 0 and that this implies $F(t) \equiv 0$. The first part of this theorem is of course only a special instance of later, much more general theorems (compare Theorem 11.3.3 and Problem 6, Exercise 19.2). But the fact that $f(z) \equiv 0$ implies $F(t) \equiv 0$ means that *there are no non-trivial representations of zero as a Laplace transform*. This generalizes to Laplace-Stieltjes transforms. To get a simpler wording of the results, we assume that $A(t)$ is normalized so that

$$(19.4.11) \qquad A(t) = \tfrac{1}{2}[A(t - 0) + A(t + 0)], \quad 0 < t, A(0) = 0.$$

This does not affect $\mathfrak{L}_1[A]$. We have then

THEOREM 19.4.2.   *If* $\mathfrak{L}_1[A](z) \equiv 0$, *then* $A(t) \equiv 0$.

This follows from the inversion formulas for the Laplace-Stieltjes integral, of which by this time there are legion. An inversion formula for $f = \mathfrak{L}_1(A)$ is an expression of the form $A = T(f)$ which enables us to compute $A(t)$ when $f(z)$ is known. It is not necessary to use all values of $f(z)$ in the half-plane of convergence. The complex inversion formulas use the values on a vertical line; other inversion formulas involve the values of $f(z)$ for large positive values of z; and finally there are formulas which, at least in the simplest cases, involve only the values of $f(z)$ and of its derivatives at a single point.

Formulas of the third type arise from a consideration of *bilinear expansions of the Laplace kernel* (compare formula (19.1.21) for the Cauchy kernel). Let $\{\varphi_n(t)\}$ be an orthonormal system of real-valued functions complete in $L_2(0, \infty)$. Set

$$(19.4.12) \qquad \psi_n(z) = \mathfrak{L}[\varphi_n] = \int_0^\infty e^{-zt} \varphi_n(t)\, dt, \quad \mathfrak{R}(z) > 0.$$

These functions are holomorphic in $\mathfrak{R}(z) > 0$ and belong to $H_2(0)$. In fact (by the Plancherel theorem)

$$\int_{-\infty}^\infty |\psi_n(x + iy)|^2\, dy = 2\pi \int_0^\infty e^{-2xt} |\varphi_n(t)|^2\, dt$$

$$\leq 2\pi \int_0^\infty |\varphi_n(t)|^2\, dt = 2\pi.$$

By Theorem 19.1.5 the corresponding boundary functions $\psi_n(iy)$ exist and are in $L_2(-\infty, \infty)$. Moreover, these functions form an orthogonal system which is not complete in $L_2(-\infty, \infty)$ but which becomes complete after the adjunction of the functions

$$\psi_{-n}(iy) = \psi_n(-iy).$$

We have now the bilinear formula

$$(19.4.13) \qquad e^{-zt} \sim \sum_{n=1}^\infty \psi_n(z)\varphi_n(t).$$

For a fixed z with $\mathfrak{R}(z) > 0$, this is the Fourier φ-series of e^{-zt} which is in $L_2(0, \infty)$ qua function of t. These considerations lead to

THEOREM 19.4.3.   *The class $H_2(0)$ coincides with the class of series*

$$(19.4.14) \qquad f(z) \sim \sum_{n=1}^\infty f_n \psi_n(z), \quad \sum_{n=1}^\infty |f_n|^2 < \infty.$$

Every such function is a Laplace transform of a function $F(t)$ in $L_2(0, \infty)$, $f = \mathfrak{L}[F]$ where

$$(19.4.15) \qquad F(t) \sim \sum_{n=1}^\infty f_n \varphi_n(t).$$

The last series may be regarded as an inversion formula for the Laplace integral.

The best-known special case is that in which

$$(19.4.16) \qquad \psi_n(z) = (2\pi)^{-\frac{1}{2}} \frac{(z - \frac{1}{2})^{n-1}}{(z + \frac{1}{2})^n} = (2\pi)^{-\frac{1}{2}} \omega_n(z, \tfrac{1}{2})$$

in the notation of (19.1.10). Here

$$(19.4.17) \qquad \varphi_n(t) = L_n(t) e^{-\frac{1}{2}t}, \quad L_n(t) = \frac{1}{n!} \frac{d^n}{dt^n} (e^{-t} t^n)$$

is the nth Laguerre polynomial of order 0, and

$$(19.4.18) \qquad f_n = (2\pi)^{\frac{1}{2}} \sum_{k=0}^{n-1} \binom{n-1}{k} \frac{1}{k!} f^{(k)}(\tfrac{1}{2}),$$

so that in this case the inversion formula involves only the values of $f(z)$ and its derivatives at $z = \frac{1}{2}$. We obtain this expression from (19.1.46) with $a = \frac{1}{2}, b = 0$.

In this special case very much more is known concerning the relation between the two Fourier series (19.4.14) and (19.4.15). Thus every function of $H_p(0)$, $1 < p \leq 2$, admits of an absolutely convergent expansion (19.4.14) and is the Laplace transform of a function $F(t) \in L_q(0, \infty)$ where $q = p/(p - 1)$. The series (19.4.15) is at least Abel summable to $F(t)$ in the Lebesgue set of $F(t)$.

For $p > 2$ the situation changes radically. Every function of $H_p(0)$ still admits of an absolutely convergent series (19.4.14), but the corresponding series (19.4.16) is now merely the derived series of a series representing a continuous function $A(t)$, normally not of bounded variation in any interval $[0, \omega]$. In this case we have

$$(19.4.19) \qquad f(z) = \int_0^\infty e^{-zt} \, dA(t) = z \int_0^\infty e^{-zt} A(t) \, dt, \quad \Re(z) > 0.$$

We leave the inversion formulas and turn to the ring properties of Laplace-Stieltjes transforms. Let

$$f(z) = \int_0^\infty e^{-zt} \, dA(t), \quad g(z) = \int_0^\infty e^{-zt} \, dB(t)$$

and suppose that $\Re(z) > \max \{0, \sigma_0[f], \sigma_0[g]\}$. To the two measures $A(t)$ and $B(t)$ corresponds a product measure

$$(19.4.20) \qquad C(t) = \int_0^t A(t - s) \, dB(s) = \int_0^t B(t - s) \, dA(s).$$

(Compare Section 16.3 for similar concepts.) Formally we have

$$(19.4.21) \qquad f(z)g(z) \equiv h(z) = \int_0^\infty e^{-zt} \, dC(t).$$

The situation is analogous to that holding for the Cauchy product of two series discussed in Section 5.2. If the integral in the last member converges, then it converges to $f(z)g(z)$. It is always $(C, 1)$-summable to this value, which in the present situation is equivalent to saying that in any case

$$h(z) = z \int_0^\infty e^{-zt} C(t) \, dt$$

with a convergent integral. Formula (19.4.21) is valid with a convergent integral if one of the factor integrals is absolutely convergent (analogue of Mertens' theorem), and it is absolutely convergent if both factors have this property (analogue of Theorem 5.2.2).

Let us denote by $\mathbf{M}$ the set of all Laplace-Stieltjes transforms with $\sigma_a[f] \leq 0$ and let $\mathbf{M}_0$ be the subset for which the integrals are still absolutely convergent on the imaginary axis. We have just seen that $\mathbf{M}$ and $\mathbf{M}_0$ are *algebras*. They are also *domains of integrity* since there can be no divisor of zero in either class. $\mathbf{M}_0$ can be made into a (B)-algebra, for instance, by taking $\| f \| = V[\infty, A]$.

We have seen that $H_p(0) \subset \mathbf{M}$ for $1 < p \leq 2$. It is not difficult to show also that $H_1(0) \subset \mathbf{M}$. On the other hand, the elements of $H_p(0)$ with $p > 2$ are normally not in $\mathbf{M}$.

If $f(z) \in \mathbf{M}_0$, then $\{f(n)\}$ is a moment sequence in the sense of Section 11.1, and all moment sequences are generated in this way. In particular, if $\Re(\alpha) > 0$, $\Re(a) > 0$, then

$$(19.4.22) \qquad (z + a)^{-\alpha} = \frac{1}{\Gamma(\alpha)} \int_0^\infty e^{-(z+a)t}\, t^{\alpha-1}\, dt$$

is an element of $\mathbf{M}_0$. According to Problem 4, Exercise 11.1, $[\log (z + a)]^{-\alpha} \in \mathbf{M}_0$ if $a > 1$, $\Re(\alpha) > 0$. It was shown by F. Hausdorff in 1921 that any function on the logarithmic scale

$$(19.4.23) \qquad (z + a_0)^{-\alpha_0}[\log (z + a_1)]^{-\alpha_1} \cdots [\log (z + a_p)]^{-\alpha_p} \in \mathbf{M}_0,$$

and a fortiori such a function also belongs to $\mathbf{M}$, provided that the first exponent $\alpha_k \neq 0$ has a positive real part and provided that

$$a_0 > 0, a_1 > 1, \cdots, a_k > e_{k-1} \quad \text{with} \quad e_k = e^{e_{k-1}}.$$

This shows that functions belong to $\mathbf{M}_0$ which are in no class $H_p(0)$ and which tend to zero arbitrarily slowly.

Toward the other end of the scale there figure functions like

$$(19.4.24) \qquad \exp \{-c[\log (z + b)]^a\} \quad \text{and} \quad \exp [-c(z + b)^\alpha]$$

which belong to $\mathbf{M}_0$. Here $0 < a$, $0 < b$, $0 < c$, $0 < \alpha \leq 1$. For the second function $\alpha = 1$ is exceptional. Here $A(t)$ is a step function. All the other functions belong to $H_2(-b)$ and hence also to $\mathbf{M}_0$.

It is far from clear what analytic functions belong to $\mathbf{M}$ or, more generally, what functions admit of a representation by a Laplace-Stieltjes integral convergent in some half-plane. Such a function must satisfy certain conditions related to regularity, rate of growth, and distribution of values. These are mostly of a negative character and exclude large classes of functions from further consideration. The inversion formulas give necessary and sufficient conditions for a function to be of the form $\mathfrak{L}_1[A]$ or $\mathfrak{L}[F]$ with A or F having specified properties, but the criteria so obtained are very difficult to apply to specific functions. There are also many sufficient conditions which are easier to use but which have only a fairly narrow range of applicability.

The value of the abscissa of convergence of $\mathfrak{L}_1[f]$ is obviously related to the analytic properties of $f(z)$ but not in any obvious manner. There is frequently no finite singular point on the line $x = \sigma_0[f]$, nor does the rate of growth change drastically to the left of the line of convergence.

All this makes it desirable to have methods of analytic continuation which are particularly adapted to Laplace-Stieltjes transforms. It is natural to think of the classical methods of summability in this connection. (C, α)-summability can be applied to a Laplace-Stieltjes integral; it gives a representation of $f(z)$ in a half-plane $\Re(z) > \sigma_\alpha[f]$. Here $\sigma_\alpha[f]$ is a never increasing function of α, and

$$\lim_{\alpha \to \infty} \sigma_\alpha[f] \equiv \gamma$$

is uniquely determined by analytic properties of $f(z)$. The half-plane $\Re(z) > \gamma$ is the largest half-plane in which $f(z)$ is both holomorphic and of finite order in the sense of (19.3.14). For any a such that $-a < \sigma_\alpha[f]$ we have

$$(19.4.25) \qquad f(z) = (z + a)^\alpha \int_0^\infty e^{-zt} \, dA_\alpha(t, a), \quad \Re(z) > \sigma_\alpha[f],$$

where $A_\alpha(t, a)$ is obtained from $A(t)$ by a process of fractional integration of order α. The outside factor on the right is the reciprocal of the function in (19.4.22). Thus in the half-plane $\Re(z) > \sigma_\alpha[f]$ the function $f(z)$ may be represented as the quotient of two Laplace-Stieltjes transforms where the denominator is given by (19.4.22). This fact suggests that instead of asking what functions are representable by convergent (absolutely convergent) Laplace-Stieltjes transforms we could ask:

What functions can be represented as the quotient of two convergent (absolutely convergent) Laplace-Stieltjes transforms in a given half-plane?

This question has a surprisingly simple answer:

THEOREM 19.4.4. *A necessary condition that $f(z)$ be the quotient of two elements of $\mathbf{M}_0$ is that $f(z)$ be the quotient of two elements of $B(0)$, that is, that the conditions of Theorem 19.2.7 hold. Conversely, if these conditions hold, then $f(z)$ is the quotient of two elements of $\mathbf{M}$.*

Proof. Since $\mathbf{M}_0 \subset B(0)$, the conditions are necessary. Conversely, if $f(z)$ is the quotient of two functions $f_1(z)$ and $f_2(z)$ of $B(0)$, then it is also the quotient of

$$f_1(z)(1 + z)^{-1} \quad \text{and} \quad f_2(z)(1 + z)^{-1},$$

which are in $H_2(0)$ and hence in $\mathbf{M}$.

From an algebraic point of view what we are doing is to embed the domain of integrity $\mathbf{M}$ or $\mathbf{M}_0$ in its quotient field and to characterize the elements of the latter by their analytic properties.

These concepts also throw some light on the summability problem. Suppose that $f(z)$ is a given Laplace-Stieltjes transform with $0 < \sigma_0[f] < \infty$ and let

$$g(z) = \int_0^\infty e^{-zt} \, dB(t), \quad \sigma_a[g] \leqq 0.$$

Then

$$f(z)g(z) = \int_0^\infty e^{-zt}\, dC(t), \quad \sigma_0[fg] \leqq \sigma_0[f].$$

If inequality holds we have

(19.4.26) $$f(z) = \frac{1}{g(z)} \int_0^\infty e^{-zt}\, dC(t), \quad \sigma_0[fg] < \Re(z).$$

This gives a method of summing $\mathfrak{L}_1[A]$ in the strip

$$\sigma_0[fg] < x \leqq \sigma_0[f].$$

The case $g(z) = z^{-\alpha}$ was considered above. The multiplier

(19.4.27) $$g(z) = \frac{\Gamma(z)\Gamma(\alpha)}{\Gamma(z+\alpha)} = \int_0^\infty e^{-zt}(1 - e^{-t})^{\alpha-1}\, dt, \quad 0 < \alpha,$$

has been used in the theory of Dirichlet series (the typical means of the second kind of M. Riesz). It is equivalent to $z^{-\alpha}$ in the right half-plane. More powerful multipliers are furnished by formula (19.4.24). Thus if b is sufficiently large we can take

$$g(z) = \exp\{-c[\log(z+b)]^a\}, \quad 1 < a,$$

for $\Re(z) > \gamma_a$, the largest half-plane in which the function $\mu_a(x; f)$ of (19.3.20) is $< c$. Similarly we can take

$$g(z) = \exp[-c(z+b)^\alpha], \quad 0 < \alpha < 1,$$

in the largest half-plane where $M_\alpha(x; f)$ of (19.3.21) stays $< c$. The case $\alpha = 1$ is useless.

With these indications concerning the *quotient method of summability*, we finish our discussion of the Laplace-Stieltjes transform.

EXERCISE 19.4

1. Prove Theorem 19.4.3.

2. If $f(z) \in B(0)$ prove that for every $\alpha > \frac{1}{2}$ there exists an $F_\alpha(t)$ such that $f(z) = (1+z)^\alpha \mathfrak{L}[F_\alpha]$, $0 < \Re(z)$.

3. If $f(z)$ is holomorphic and of finite order $\leqq \sigma$ for $\Re(z) > 0$, prove that a similar representation exists for $\alpha > \sigma + \frac{1}{2}$.

4. Verify (19.4.27).

5. The (C, α)-means of $\mathfrak{L}_1[A]$ are

$$\int_0^\omega \left(1 - \frac{s}{\omega}\right)^\alpha e^{-zs}\, dA(s).$$

Verify (19.4.25) for $a = 0$, $\Re(z) > \max(0, \sigma_\alpha[f])$.

COLLATERAL READING

The reader will find useful information in

BOAS, R. P., JR. *Entire Functions.* Academic Press, New York, 1954.

For most of the subject matter it is necessary to go back to the original papers. Here is a list of some papers to consult:

CARLEMAN, T. "Sur la Croissance de Certaines Classes de Fonctions Analytiques," *Matematisk Tidsskrift*, Series B, 1931, pp. 46–62.

FATOU, P. "Séries Trigonométriques et Séries de Taylor," *Acta Mathematica*, Vol. 30 (1906), pp. 335–400.

HARDY, G. H., INGHAM, A. E., and PÓLYA, G. "Theorems concerning Mean Values of Analytic Functions," *Proceedings of the Royal Society*, Series A, Vol. 113 (1927), pp. 542–569.

HILLE, E., and TAMARKIN, J. D. "A Theorem of Paley and Wiener," *Annals of Mathematics*, Series 2, Vol. 34 (1933), pp. 606–614.

———. "On the Absolute Integrability of Fourier Transforms," *Fundamenta Mathematicæ*, Vol. 25 (1935), pp. 329–352.

LINDELÖF, E. "Quelques Remarques sur la Croissance de la Fonction $\zeta(s)$," *Bulletin des Sciences Mathématiques*, Series 2, Vol. 32 (1908), pp. 341–356.

NEVANLINNA, F. and R. "Über die Eigenschaften analytischer Funktionen in der Umgebung einer singulären Stelle oder Linie," *Acta Societatis Scientiarum Fennicæ*, Vol. 50, No. 5 (1922), 46 pages.

PALEY, R. E. A. C., and WIENER, N. "Notes on the Theory and Application of Fourier Transforms. I, II," *Transactions of the American Mathematical Society*, Vol. 35 (1933), pp. 348–355.

PHRAGMÉN, E., and LINDELÖF, E. "Sur une Extension d'un Principe Classique de l'Analyse et sur quelques Propriétés des Fonctions Monogènes dans le Voisinage d'un Point Singulier," *Acta Mathematica*, Vol. 31 (1908), pp. 381–406.

PÓLYA, G. "Untersuchungen über Lücken und Singularitäten von Potenzreihen," *Mathematische Zeitschrift*, Vol. 29 (1929), pp. 549–640; Part 2, *Annals of Mathematics*, Series 2, Vol. 34 (1933), pp. 731–777.

RIESZ, F. and M. "Über die Randwerte einer analytischen Funktion," in *Compte Rendu du Quatrième Congrès des Mathématiciens Scandinaves tenu à Stockholm 1916*, pp. 27–44. Uppsala, 1920.

RIESZ, M. "Sur le problème des moments et le théorème de Parseval correspondant," *Acta Litterarum ac Scientiarum, Szeged*, Vol. 1 (1922–23), pp. 209–225.

———. "Sur les fonctions conjuguées," *Mathematische Zeitschrift*, Vol. 27 (1927), pp. 218–244.

For the theory of Laplace integrals see

DOETSCH, G. *Handbuch der Laplace-Transformation*, Vols. I, II, III, Verlag Birkhäuser, Basel, 1950, 1955, 1957.

WIDDER, D. V. *The Laplace Transform.* Princeton University Press, Princeton, New Jersey, 1941.

Section 19.4 is largely based on the following papers by the author, of which the first was written in collaboration with J. D. Tamarkin and the last with W. B. Caton:

"On the Theory of Laplace Integrals," I and II, *Proceedings of the National Academy of Sciences*, Vol. 19 (1933), pp. 908–912, and Vol. 20 (1934), pp. 140–144.

"On Laplace Integrals," in *Comptes Rendus du Huitième Congrès des Mathématiciens Scandinaves tenu à Stockholm 1934*, pp. 216–227. Lund, 1935.

"Bilinear Formulas in the Theory of the Transformation of Laplace," *Compositio Mathematica*, Vol. 6 (1938), pp. 94–102.

"Laguerre Polynomials and Laplace Integrals," *Duke Mathematical Journal*, Vol. 12 (1946), pp. 217–242.

Bibliography

AHLFORS, L. V. *Complex Analysis. An Introduction to the Theory of Analytic Functions of One Complex Variable.* McGraw-Hill Book Company, Inc., New York, 1953.

————, and SARIO, L. *Riemann Surfaces.* Princeton University Press, Princeton, New Jersey, 1960.

APPELL, P., and GOURSAT, É. *Théorie des Fonctions Algébriques et de leurs Intégrales.* Gauthier-Villars, Paris, 1895. Revised Edition by P. FATOU, 1929.

BEHNKE, H., and SOMMER, F. *Theorie der analytischen Funktionen einer komplexen Veränderlichen.* Springer-Verlag, Berlin, 1955.

BERGMAN, S. *The Kernel Function and Conformal Mapping.* Mathematical Surveys, Vol. V. American Mathematical Society, New York, 1950.

BIEBERBACH, L. *Analytische Fortsetzung.* Ergebnisse der Mathematik, New Series, No. 3. Springer-Verlag, Berlin, 1955.

BLISS, G. A. *Algebraic Functions.* Colloquium Publications, Vol. 16. American Mathematical Society, New York, 1933.

BLOCH, A. *Les Fonctions Holomorphes et Méromorphes dans le Cercle-unité.* Mémorial des Sciences Mathématiques, No. 20. Gauthier-Villars, Paris, 1926.

BOAS, R. P., JR. *Entire Functions.* Academic Press, New York, 1954.

————., and BUCK, R. C. *Polynomial Expansions of Analytic Functions.* Ergebnisse der Mathematik, New Series, No. 19. Springer-Verlag, Berlin, 1958.

BOREL, ÉMILE. *Leçons sur les Fonctions Monogènes Uniformes d'une Variable Complexe.* Gauthier-Villars, Paris, 1917.

CARATHÉODORY, C. *Conformal Representation.* Cambridge Tracts in Mathematics and Mathematical Physics, No. 28. Cambridge University Press, London, 1932.

COPSON, E. T. *An Introduction to the Theory of Functions of a Complex Variable.* Clarendon Press, Oxford, 1935.

DIENES, P. *The Taylor Series.* Clarendon Press, Oxford, 1931.

DOETSCH, G. *Handbuch der Laplace-Transformation,* Vols. I, II, III. Verlag Birkhäuser, Basel, 1950, 1955, 1957.

FORD, L. R. *Automorphic Functions,* Second Edition. Chelsea Publishing Company, Inc., New York, 1951.

GOLUSIN, G. M. *Geometrische Funktionentheorie.* VEB Deutscher Verlag der Wissenschaften, Berlin, 1957.

HURWITZ, A., and COURANT, R. *Vorlesungen über allgemeine Funktionentheorie und elliptische Funktionen,* Second Edition. Verlag von Julius Springer, Berlin, 1925.

JENKINS, J. A. *Univalent Functions and Conformal Mapping.* Ergebnisse der Mathematik, New Series, No 18. Springer-Verlag, Berlin, 1958.

481

JULIA, GASTON *Leçons sur les Fonctions Uniformes à Point Singulier Essentiel Isolé.* Gauthier-Villars, Paris, 1924.

――――. *Principes Géométriques d'Analyse.* Gauthier-Villars, Paris, 1930.

KLEIN, F., and FRICKE, R. *Vorlesungen über die elliptischen Modulfunktionen,* Vols. I, II. B. G. Teubner, Leipzig, 1890 and 1892.

KOBER, H. *Dictionary of Conformal Representations.* Dover Publications, Inc., New York, 1957.

LANDAU, E. *Darstellung und Begründung einiger neuerer Ergebnisse der Funktionentheorie,* Second Edition. Springer-Verlag, Berlin, 1929. Reprinted in: *Das Kontinuum.* Chelsea Publishing Company, New York, 1960, 1973.

LEAU, L. *Les Suites de Fonctions en Général (Domaine Complexe).* Mémorial des Sciences Mathématiques, No. 59. Gauthier-Villars, Paris, 1932.

LITTLEWOOD, J. E. *Lectures on the Théory of Functions.* Oxford University Press, London, 1944.

MONTEL, P. *Leçons sur les Familles Normales de Fonctions Analytiques et leurs Applications.* Gauthier-Villars, Paris, 1927. Chelsea Publishing Company, New York, 1973.

――――. *Leçons sur les Séries de Polynomes d'une Variable Complexe.* Gauthier-Villars, Paris, 1910.

NEHARI, Z. *Conformal Mapping.* McGraw-Hill Book Company, Inc., New York, 1952.

NEVANLINNA, R. *Le Théorème de Picard-Borel et la Théorie des Fonctions Méromorphes.* Gauthier-Villars, Paris, 1929. Chelsea Publishing Company, New York, 1973.

――――. *Eindeutige analytische Funktionen,* Second Edition. Grundlehren der Mathematischen Wissenschaften, Vol. XLVI. Springer-Verlag, Berlin, 1953.

PFLUGER, A. *Theorie der Riemannschen Flächen.* Springer-Verlag, Berlin, 1957.

PÓLYA, G., and SZEGÖ, G. *Aufgaben und Lehrsätze aus der Analysis,* Vols. I, II. Verlag von Julius Springer, Berlin, 1925.

PRIWALOW, I. I. *Randeigenschaften analytischer Funktionen.* VEB Deutscher Verlag der Wissenschaften, Berlin, 1956.

SAKS, S., and ZYGMUND, A. *Analytic Functions.* Monografie Matematyczne, Vol. 28, Warsaw and Wrocław, 1952.

SPRINGER, G. *Introduction to Riemann Surfaces.* Addison-Wesley Publishing Company, Inc., Reading, Massachusetts, 1957.

SZEGÖ, G. *Orthogonal Polynomials.* Colloquium Publications, Vol. 23. American Mathematical Society, New York, 1939.

TANNERY, J., and MOLK, J. *Éléments de la Théorie des Fonctions Elliptiques,* Vols. I–IV. Gauthier-Villars, Paris, 1893, 1896, 1898, 1902. Second Edition. Chelsea Publishing Company, New York, 1972.

TSUJI, M. *Potential Theory in Modern Function Theory.* Maruzen, Tokyo, 1959.

VALIRON, G. *Familles Normales et Quasi-Normales de Fonctions Méromorphes.* Mémorial des Sciences Mathématiques, No. 38. Gauthier-Villars, Paris, 1932.

————. *Lectures on the General Theory of Integral Functions.* Édouard Privat, Toulouse, 1923. Chelsea Publishing Company, New York, 1949.

VIVANTI, G., and GUTZMER, A. *Theorie der eindeutigen analytischen Funktionen.* B. G. Teubner, Leipzig, 1906.

WALKER, R. J. *Algebraic Curves.* Princeton University Press, Princeton, New Jersey, 1950.

WALSH, J. L. *Interpolation and Approximation by Rational Functions in the Complex Domain,* Revised Edition. Colloquium Publications, Vol. 20. American Mathematical Society, Providence, Rhode Island, 1956.

WEYL, H. *Die Idee der Riemannschen Fläche,* Third Edition. B. G. Teubner, Stuttgart, 1955.

WHITTAKER, E. T., and WATSON, G. N. *A Course of Modern Analysis,* Fourth Edition. Cambridge University Press, London, 1952.

WIDDER, D. V. *The Laplace Transform.* Princeton University Press, Princeton, New Jersey, 1941.

WITTICH, HANS. *Neuere Untersuchungen über eindeutige analytische Funktionen.* Ergebnisse der Mathematik, New Series, No. 8. Springer-Verlag, Berlin, 1955.

Index

FUNCTION THEORY

(See also: Analysis)

APPELL, Paul and Edouard GOURSAT: Fonctions Algebriques et leurs Integrales, 3rd ed. Vol. I (French) xxv + 526 pp. 5⅜ x 8. CIP. ISBN -0285-X, ∞G   Vol. II (French) xii + 521 pp. 5⅜ x 8. ISBN -0299-X, ∞G

CARATHEODORY, Constantin: Theory of Functions, 2nd ed. Vol. I, 310 pp. ISBN -0097-0, ∞G   'ol. II, 220 pp. 6 x 9. ISBN -0106-3, ∞G

FORD, Lester R.: Automorphic Functions. 343 pp. 5⅜ x 8. ISBN -0085-7, ∞G

HENSEL, Kurt & LANDSBERG, G.: Algebraische Funktionen. (Germ.) xvi + 707 pp. 6 x 9. ISBN -0179-9, ∞G

HILLE, Einar: Analytic Function Theory, 2nd ed. Vol. I, 319 pp. 6 x 9. CIP. ISBN -0269-8, ∞G

HOBSON, Ernest W.: Spherical and Ellipsoidal Harmonics. xi + 500 pp. 5⅜ x 8. ISBN -0104-7, ∞G

KRAZER, Adolf: Lehrbuch der Thetafunktionen. (Germ.) xxiv + 509 pp. 5⅜ x 8. ISBN -0244-2, ∞G

MARKUSHEVICH, Aleksei I.: Theory of Functions of a Complex Variable, 2nd ed., 3 vols. in 1. xxiv + 1,128 pp. CIP. ISBN -0296-5, ∞

MONTEL, Paul: Lecons sur les Familles Normales de Fonctions Analytiques. (French) xiii + 301 pp. 5⅜ x 8. CIP. ISBN -0271-X, ∞G

NEVANLINNA, Rolf: Le Theoreme de Picard-Borel. (French) x + 171 pp. 5⅜ x 8. CIP. ISBN -0272-8, ∞G

— — —and PAATERO, Viekko: Introduction to Complex Analysis, 2nd ed. ix + 350 pp. 6 x 9. ISBN -0310-4, ∞G

NIELSEN, Niels: Die Gammafunktion, 2 v. in 1. (Germ.) 448 pp. 6 x 9. ISBN -0188-8, ∞G

OSGOOD, William F.: Funktionentheorie. 5th ed. Vol. I (Germ.) 818 pp. 5 x 8. ISBN -0193-4, ∞G   2nd ed. Vol. II (Germ.) 685 pp. 5 x 8. ISBN -0182-9, ∞G

PICARD, Emile & SIMART, G.: Fonctions Algebriques. 2 v. in 1. (French) 784 pp. 5⅜ x 8. ISBN -0248-5, ∞G

SCHWARZ, Hermann Amandus: Gesammelte Mathematische Abhandlungen, 2 vols. in 1. (Germ.) 726 pp. 6 x 9. ISBN -0260-4, ∞G

SPRINGER, George: Introduction to Riemann Surfaces, 2nd ed. Viii + 307 pp. 6 x 9. ISBN -313-9, ∞G

TANNERY, Jules and Jules MOLK: Fonctions Elliptiques, 4 vols. in 2 (French) 2nd ed., 1,146 pp. 5⅜ x 8. ISBN -0257-4, ∞G

TSUJI, MASATSUGU: Potential Theory in Modern Function Theory, 2nd ed. x + 590 pp. 5⅜ x 8. CIP. ISBN -0281-7, ∞G

WEBER. Heinrich: Lehrbuch der Algebra, 3 vols. (Germ.) 3rd ed. 2,345 pp. 5⅜ x 8. ISBN -0144-6, ∞G   (Vol. III: Elliptische Funktionen.)

DIFFERENTIAL EQUATIONS

BATEMAN, Harry: Differential Equations. xi + 306 pp. 5⅜ x 8 ISBN -0190-X, ∞G

BLISS, Gilbert A.: Fundamental Existence Theorems. See: EVANS, G. C.

CARATHEODORY, Constantin: Calculus of Variations and Partial Differential Equations, 2nd (revised) ed. xvi + 401 pp. 6 x 9. ISBN -0318-X, ∞G

EVANS, Griffith C.: Logarithmic Potential, 2nd ed.; & BLISS, G. A. Fundamental Existence Theorems, & KASNER, E. Differential-Geometric Aspects of Dynamics. 3 vols in 1. 399 pp. 5⅜ x 8. ISBN -0305-8, ∞G

GARABEDIAN, Paul R.: Partial Differential Equations, xii + 672 pp. 6 x 9. ISBN -0325-2, ∞G

KAMKE, Erich: Differentialgleichungen: Loesungsmethoden und Loesungen. Vol. II. Partial D. E. (Germ.) xv + 243 pp. 6 x 9. CIP. ISBN -0277-9, ∞G

LAPPO-DANILEVSKII, J. A.: Systemes des Equations Differentielles, 3 vols. in 1 (French) 689 pp. 5⅜ x 8. ISBN -0094-6, ∞'G

GROUP THEORY
(Including: Continuous Groups)

CAMPBELL, John R.: Introductory Treatise on Lie's Theory. xx + 416 pp. 5⅜ x 8. ISBN -0183-7, T.O.P.

GORENSTEIN, Daniel: Finite Groups, 2nd (corr.) ed. xvii + 517 pp. 6 x 9. CIP., ISBN -0301-5, ∞G

HALL, Marshall: The Theory of Groups. xiii + 434 pp. 5⅜ x 8. CIP. ISBN -0288-4, ∞G

KOWALEWSKI, Gerhard: Kontinuierliche Gruppen. (Germ.) viii + 396 pp. 5⅜ x 8. ISBN -0070-9, G

KUROSH, Alexander G.: The Theory of Groups, Vol. I, 271 pp. 6 x 9. ISBN -0107-1, ∞G $12.50. Vol. II, 308 pp. 6 x 9. ISBN -109-8, ∞G

LIE, Sophus: Transformationsgruppen, 2nd ed., 3 vols. (Germ.) 2,090 pp. 6 x 9. ISBN -0232-9, ∞G

— — —Differentialgleichungen. (Germ.) xiv + 568 pp. 6 x 9. ISBN -0206-X, ∞G

— — —Continuierliche Gruppen (Germ.) xii + 810 pp. 6 x 9. ISBN -0199-3, ∞G

— — —Geometrie der Beruehrungstransformationen. (Germ.) 705 pp. 6 x 9. CIP. ISBN -0291-4, ∞G

ZASSENHAUS, Hans J.: The Theory of Groups, 2nd ed. viii + 265 pp. 6 x 9. ISBN -0053-9, ∞G